Image Processing and GIS for Remote Sensing

Image Processing and GIS for Remote Sensing

Techniques and Applications

Jian Guo Liu and Philippa J. Mason

Department of Earth Science and Engineering,
Imperial College London

SECOND EDITION

WILEY Blackwell

This edition first published 2016 © 2016 by John Wiley & Sons, Ltd.

Registered Office
John Wiley & Sons, Ltd., The Atrium, Southern Gate, Chichester, West Sussex, PO19 8SQ, UK

Editorial Offices
9600 Garsington Road, Oxford, OX4 2DQ, UK
The Atrium, Southern Gate, Chichester, West Sussex, PO19 8SQ, UK
111 River Street, Hoboken, NJ 07030-5774, USA

For details of our global editorial offices, for customer services and for information about how
to apply for permission to reuse the copyright material in this book please see our website at
www.wiley.com/wiley-blackwell.

Library of Congress Cataloging-in-Publication data applied for

ISBN: 9781118724200

A catalogue record for this book is available from the British Library.

Wiley also publishes its books in a variety of electronic formats. Some content that appears in print may not
be available in electronic books.

Cover image: The cover image is a 3D perspective view of a Landsat 8 OLI (Operational Land Imager)
image, draped over a Digital Elevation Model (SRTM3 1-arc-second 30 m), displaying rocks of Proterozoic
age outcropping over an area of the Atlas Mountains of Morocco, which also host many mineral deposits.
The colour composite is composed of OLI bands 6-4-2 in RGB, with a DDS (Direct Decorrelation Stretch)
enhancement; bands which are in the Short-Wave InfraRed (6), Red (4) and Blue (2) spectral regions.
The bright green features in this image would therefore actually appear distinctly red to the naked eye.
Geologically, the beautiful and complex colour patterns are caused by folds and faults systems which
have deformed a series of different rock types exposed in the semi-arid environment of Morocco.
(Courtesy to NASA/JPL and USGS).

Set in 8.5/12pt Meridien by SPi Global, Pondicherry, India
Printed and bound in Singapore by Markono Print Media Pte Ltd

1   2016

Contents

v

Overview of the book

Remote sensing is a mechanism for collecting raster data or images, and remotely sensed images represent an objective record of the spectrum relating to the physical properties and chemical composition of the earth surface materials. Information extraction from those images is, on the other hand, an entirely subjective process. People with differing application foci will derive very different thematic information from the same source image. Image processing thus becomes a vital tool for the extraction of thematic and/or quantitative information from raw image data. For more in-depth analysis, the images need to be analysed in conjunction with other complementary data, such as existing thematic maps of topography, geomorphology, geology and landuse, or with geochemical and geophysical survey data, or 'ground truth' data, logistical and infrastructure information; and so here comes GIS, a highly sophisticated tool for the management, display and analysis of all kinds of spatially referenced information.

Remote sensing, image processing and GIS are all extremely broad subjects in their own right; far too broad to be covered in one book. As illustrated in Fig. 0.1, our book is aimed at the overlap between the three disciplines, providing an overview of essential techniques and a selection of case studies in a variety of application areas, emphasising the close relationship between them. The application cases are biased toward the earth sciences but the image processing and GIS techniques are generic and independent of any software, and therefore transferable skills suited to all applications.

The book has been written with university students and lecturers in mind as a principal textbook. For students' needs in particular, we have tried to convey knowledge in simple words, with clear explanations and with conceptual illustrations. For image processing and GIS, mathematics is unavoidable, but we understand that this may be off-putting for some. To minimize such effects, we try to emphasize the concepts, explaining in common sense terms rather than in too much mathematical detail. The result is intended to be a comprehensive yet 'easy learning' solution to a fairly challenging topic.

There are sections providing extended coverage of some necessary mathematics and advanced materials for use by course tutors and lecturers; these sections will be marked as such. Hence the book is written for both students and teachers. With many author-developed techniques and recent research case studies, it is also an excellent reference book for higher level readers including researchers and professionals in remote sensing application sectors.

In this book, we have presented a unique combination of tools, techniques and applications that we hope will be of use to the full breadth of geoscientific and remote sensing communities. The book begins with the fundamentals of the core image processing tools used in remote sensing and GIS with adequate mathematical details in Part I, then it becomes slightly more applied and less mathematical in Part II to cover the wide scope of GIS where many of those core image processing tools are used in different contexts. Part III contains the entirely applied part of the book, where we describe a selection of cases where image processing and GIS have been used, by the authors, in teaching, research and industrial projects in which there is a dominant remote sensing component.

Since the publication of the first edition in 2009, we have been pleasantly delighted and encouraged by the interest in this book, and comments from students and colleagues alike have stimulated us to produce this second edition. In making the explanations of the book generic and not tied to any particular software, we attempted to 'future-proof' its content in the first edition. Inevitably, however, there are many small things that have needed to be updated in this second edition. We have implemented minor updates to all the main chapters in Parts I and II. A completely new chapter on sub-pixel technology and image phase correlation has been added in Part I (Chapter 11), which is based on new and ongoing research in this area. Part III has been considerably enhanced with more

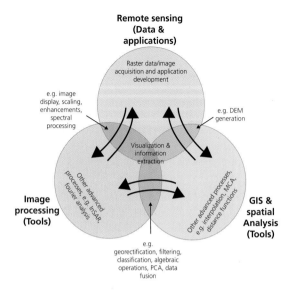

Fig. 0.1 Schematic illustration of the scope of this book.

recent case studies to include new data and/or other new complementary research where relevant.

We also take this opportunity to acknowledge the many data sources that we have made use of in this book. These include NASA's 40-year Landsat archive (without which much of our daily work would be impossible), ESA radar image archive (ERS and Envisat), Aster GDEM and SRTM global elevation data, ESRI's online knowledge base, and the NERC (Natural Environment Research Council, UK) airborne campaign (Airborne Thematic Mapper).

Images are both the principal input and product of remote sensing and GIS, and they are meant to be visualised! The e-version of the book therefore now contains hyperlinks to high resolution digital versions of the images and illustrations, allowing the reader to examine the details at pixel level and to appreciate the effects of particular processing techniques such as filtering, image fusion, and so forth.

PART I

Image processing

This part covers the most essential image processing techniques for image visualisation, quantitative analysis and thematic information extraction for remote sensing applications. A series of chapters introduce topics with increasing complexity from basic visualisation algorithms, which can be easily used to improve your digital camera pictures, to more complicated multi-dimensional transform-based techniques.

Digital image processing can improve image visual quality, selectively enhance and highlight particular image features and classify, identify and extract spectral and spatial patterns representing different thematic information from images. It can also arbitrarily change image geometry and illumination conditions to give different views of the same image. Importantly, *image processing cannot increase information from the original image data,* although it can indeed optimise the visualisation for us to see more from the enhanced images than from the original.

For real applications our considered opinion, based on years of experience, is that *simplicity is beautiful.* Image processing does not follow the well-established physical law of *energy conservation.* As shown in Fig. P.1, often the results produced using very simple processing techniques in the first 10 minutes of your project may actually represent 90% of the job done! This should not encourage you to abandon this book after the first three chapters since it is the remaining 10% that you achieve during the 90% of your time that will serve the highest level objectives of your project. The key point is that thematic image processing should be application driven, whereas our learning is usually technique driven.

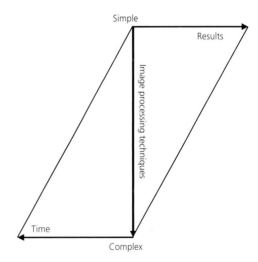

Fig. P.1 This simple diagram is to illustrate that the image processing result is not necessarily proportional to the time/effort spent. On the contrary, you may spend little time in achieving the most useful results and with simple techniques; on the other hand, you may spend a lot of time to achieve very little using complicated techniques.

Image Processing and GIS for Remote Sensing: Techniques and Applications, Second Edition. Jian Guo Liu and Philippa J. Mason.
© 2016 John Wiley & Sons, Ltd. Published 2016 by John Wiley & Sons, Ltd.

CHAPTER 1

Digital image and display

1.1 What is a digital image?

An image is a picture, photograph or any form of a two-dimensional (2D) representation of objects or a scene. The information in an image is presented in tones or colours. *A digital image is a two-dimensional array of numbers*. Each cell of a digital image is called a pixel, and the number representing the brightness of the pixel is called a digital number (DN) (Fig. 1.1). As a 2D array, a digital image is composed of data in lines and columns. The position of a pixel is allocated with the line and column of its DN. Such regularly arranged data, without x and y coordinates, are usually called *raster data*. As digital images are nothing more than data arrays, mathematical operations can be readily performed on the digital numbers of images. Mathematical operations on digital images are called *digital image processing*.

Digital image data can also have a third dimension: *layers* (Fig. 1.1). Layers are the images of the same scene but containing different information. In multi-spectral images, layers are the images of different spectral ranges called *bands* or *channels*. For instance, a colour picture taken by a digital camera is composed of three bands containing red, green and blue spectral information individually. The term *band* is more often used than *layer* to refer to multi-spectral images. Generally speaking, geometrically registered multi-dimensional data sets of the same scene can be considered as layers of an image. For example, we can digitise a geological map and then co-register the digital map with a Landsat TM image. Then the digital map becomes an extra layer of the scene beside the seven TM spectral bands. Similarly, if we have a dataset of digital elevation model (DEM) to which a SPOT image is rectified, then the DEM can be considered as a layer of the SPOT image beside its four spectral bands. In this sense, we can consider a set of co-registered digital images as a three-dimensional (3D) dataset and with the 'third' dimension providing the link between image processing and GIS.

A digital image can be stored as a file in a computer data store on a variety of media, such as a hard disk, memory stick, CD, etc. It can be displayed in black and white or in colour on a computer monitor as well as in hard copy output such as film or print. It may also be output as a simple array of numbers for numerical analysis. As a digital image, its advantages include:

- The images do not change with environmental factors as hard copy pictures and photographs do;
- the images can be identically duplicated without any change or loss of information;
- the images can be mathematically processed to generate new images without altering the original images;
- the images can be electronically transmitted from or to remote locations without loss of information.

Remotely sensed images are acquired by sensor systems onboard aircraft or spacecraft, such as earth observation satellites. The sensor systems can be categorised into two major branches: *passive sensors* and *active sensors*. Multi-spectral optical imaging systems are passive

Image Processing and GIS for Remote Sensing: Techniques and Applications, Second Edition. Jian Guo Liu and Philippa J. Mason.
© 2016 John Wiley & Sons, Ltd. Published 2016 by John Wiley & Sons, Ltd.

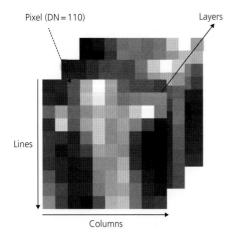

Pixel (DN = 110)   Layers

Lines

Columns

Fig. 1.1 A digital image and its elements.

sensors that use solar radiation as the principal source of illumination for imaging. Typical examples include across-track and push-broom multi-spectral scanners, and digital cameras. An active sensor system provides its own means of illumination for imaging, such as synthetic aperture radar (SAR). Details of major remote sensing satellites and their sensor systems are beyond the scope of this book, but we provide a summary in Appendix A for your reference.

1.2 Digital image display

We live in a world of colour. The colours of objects are the result of selective absorption and reflection of electromagnetic radiation from illumination sources. Perception by the human eye is limited to the spectral range of 0.38–0.75 μm, that is, a very small part of the solar spectral range. The world is actually far more colourful than we can see. Remote sensing technology can record over a much wider spectral range than human visual ability, and the resultant digital images can be displayed as either black and white or colour images using an electronic device such as a computer monitor. In digital image display, the tones or colours are visual representations of the image information recorded as digital image DNs, but they do not necessarily convey the physical meanings of these DNs. We will explain this further in our discussion on false colour composites later.

The wavelengths of major spectral regions used for remote sensing are listed below:

Visible light (VIS):	0.4–0.7 μm
Blue (B)	0.4–0.5 μm
Green (G)	0.5–0.6 μm
Red (R)	0.6–0.7 μm
Visible-photographic infrared:	0.5–0.9 μm
Reflective infrared (IR):	0.7–3.0 μm
Nearer infrared (NIR):	0.7–1.3 μm
Short-wave infrared (SWIR):	1.3–3.0 μm
Thermal infrared (TIR):	3–5 μm, 8–14 μm
Microwave:	0.1–100 cm

Commonly used abbreviations of the spectral ranges are denoted by the letters in the brackets in the list above. The spectral range covering visible light and nearer infrared is the most popular for broadband multi-spectral sensor systems and it is usually denoted as VNIR.

1.2.1 Monochromatic display

Any image, either a panchromatic image or a spectral band of a multi-spectral image, can be displayed as a black and white (B/W) image by a monochromatic display. The display is implemented by converting DNs to electronic signals in a series of energy levels that generate different grey tones (brightness) from black to white, and thus to formulate a B/W image display. Most image processing systems support an 8 bits graphical display, which corresponds to 256 grey levels and displays DNs from 0 (black) to 255 (white). This display range is wide enough for human visual capability. It is also sufficient for some of the more commonly used remotely sensed images, such as Landsat TM/ETM+, SPOT HRV and Terra-1 ASTER VIR-SWIR (see Appendix A); the DN ranges of these images are not wider than 0–255. On the other hand, many remotely sensed images have much wider DN ranges than 8 bits, such as Ikonos and QuickBird, whose images have an 11 bits DN range (0–2047), and Landsat 8 Operational Land Imager (OLI), of 12 bits. In this case, the images can still be visualised in an 8-bit display device in various ways, such as by compressing the DN range into 8 bits or displaying the image in scenes of several 8-bit intervals of the whole DN range. Many sensor systems offer wide dynamic ranges to ensure that the sensors can record across all levels of radiation energy without localised sensor adjustment. Since the received solar radiation does not normally vary significantly within an image scene of limited size, the actual DN range of the scene is usually much narrower than the full dynamic range of the sensor and thus can be well adapted into an 8-bit DN range for display.

In a monochromatic display of a spectral band image, the brightness (grey level) of a pixel is proportional to the reflected energy in this band from the corresponding ground area. For instance, in a B/W display of a red band image, light red appears brighter than dark red. This is also true for invisible bands (e.g. infrared bands), though the 'colours' cannot be seen. After all, any digital image is composed of DNs; the physical meaning of DNs depends on the source of the image. A monochromatic display visualises DNs in grey tones from black to white, while ignoring the physical relevance.

1.2.2 Tristimulus colour theory and RGB (red, green, blue) colour display

If you understand the structure and principle of a colour TV tube, you must know that the tube is composed of three colour guns of red, green and blue. These three colours are known as *primary colours*. The mixture of the lights of these three primary colours can produce any colour on a TV. This property of the human perception of colour can be explained by the *tristimulus colour theory*. The human retina has three types of cones and the response by each type of cone is a function of the wavelength of the incident light; they peak at 440 *nm* (blue), 545 *nm* (green) and 680 *nm* (red). In other words, each type of cone is primarily sensitive to one of the primary colours: blue, green or red. A colour perceived by a person depends on the proportion of each of these three types of cones being stimulated and thus can be expressed as a triplet of numbers (r, g, b) even though visible light is electromagnetic radiation in a continuous spectrum of 380–750 *nm*. A light of non-primary colour C will stimulate different portions of each cone type to form the perception of this colour:

$$C = rR + gG + bB \qquad (1.1)$$

Equal mixtures of the three primary colours ($r = g = b$) give white or grey, whilst equal mixtures of any two primary colours generate a complementary colour. As shown in Fig. 1.2, the complementary colours of red, green and blue are cyan, magenta and yellow. The three complementary colours can also be used as primaries to generate various colours, as in colour printing. If you have experience with colour painting, you know that

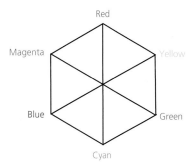

Fig. 1.2 The relation of the primary colours to their complementary colours.

any colours can be generated by mixing three colours: red, yellow and blue; this is based on the same principle.

Digital image colour display is based entirely on the tristimulus colour theory. A colour monitor, like a colour TV, is composed of three precisely registered colour guns, red, green and blue. In the red gun, pixels of an image are displayed in reds of different intensity (i.e. dark red, light red, etc.) depending on their DNs. The same is true of the green and blue guns. Thus if the red, green and blue bands of a multi-spectral image are displayed in red, green and blue simultaneously, a colour image is generated (Fig. 1.3) in which the colour of a pixel is decided by the DNs of red, green and blue bands (r, g, b). For instance, if a pixel has red and green DNs of 255 and blue DN of 0, it will appear in pure yellow on display. This kind of colour display system is called an *Additive RGB Colour Composite System*. In this system, different colours are generated by additive combinations of red, green and blue components.

As shown in Fig. 1.4, consider the components of an RGB display as the orthogonal axes of a 3D colour space; the maximum possible DN level in each component of the display defines the *RGB colour cube*. Any image pixel in this system may be represented by a vector from the origin to somewhere within the colour cube. Most standard RGB display systems can display 8 bits/pixel/channel, up to 24 bits = 256^3 different colours. This capacity is enough to generate a so-called 'true colour' image. The line from the origin of the colour cube to the opposite convex corner is known as the *grey line* because pixel vectors that lie on this line have equal components in red, green and blue (i.e. $r = g = b$). If the same band is used as red, green and blue components, all the pixels will lie on the grey line. In this case, a B/W image will be produced even though a colour display system is used.

RED

Green

Blue

RGB color composite

Fig. 1.3 Illustration of RGB additive colour image display.

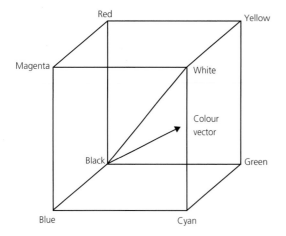

Fig. 1.4 The RGB colour cube.

As mentioned before, although colours lie in the visible spectral range of 380–750 *nm*, they are used as a tool for information visualisation in colour display of all digital images. Thus, for digital image display, the assignment of each primary colour for a spectral band or layer can arbitrarily depend on the requirements of the application, which may not necessarily correspond to the actual colour of the spectral range of the band. If we display three image bands in the red, green and blue spectral ranges in RGB, then a *true colour composite* (TCC) image is generated (Fig. 1.5, bottom left). Otherwise, if the image bands displayed in red, green and blue do not match the spectra of these three primary colours, a *false colour composite* (FCC) image is produced. A typical example is the so-called *standard false colour composite*

(SFCC) in which the NIR band is displayed in red, red band in green and green band in blue (Fig. 1.5, bottom right). The SFCC effectively highlights any vegetation distinctively in red. Obviously, we could display various image layers, which are without any spectral relevance, as a false colour composite. The false colour composite is the general case of RGB colour display whilst the true colour composite is only a special case of it.

1.2.3 Pseudo-colour display

The human eye can recognise far more colours than it can grey levels, so colour may be used very effectively to enhance small grey-level differences in a B/W image. The technique to display a monochrome image as a colour image is called *pseudo-colour display*. A pseudo-colour image is generated by assigning each grey level to a unique colour (Fig. 1.6). This can be done by interactive colour editing or by automatic transformation based on certain logic. A common approach is to assign a sequence of grey levels to colours of increasing spectral wavelength and intensity.

The advantage of pseudo-colour display is also its disadvantage. When a digital image is displayed in grey scale, using its DNs in a monochromic display, the sequential numerical relationship between different DNs is effectively presented. This crucial information is lost in a pseudo-colour display because the colours assigned to various grey levels are not quantitatively related in a numeric sequence. Indeed, the image in a pseudo-colour display is an image of symbols; it is no longer a digital image! We can regard the grey-scale

Band 1: Blue   Band 2: Green   Band 3: Red   Band 4: Near infrared

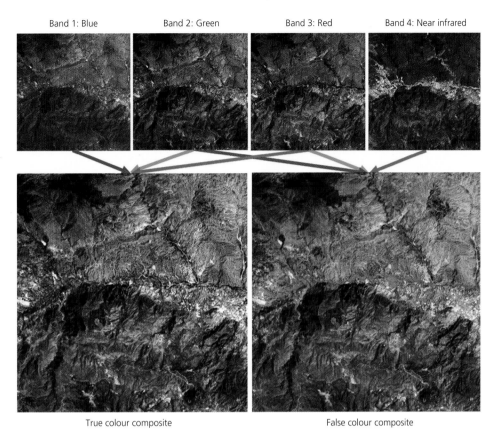

True colour composite   False colour composite

Fig. 1.5 True colour and false colour composites of blue, green, red and NIR bands of a Landsat-7 ETM+ image. If we display the blue band in blue, green band in green and red band in red, then a true colour composite is produced as shown in the bottom left. If we display the green band in blue, red band in green and NIR band in red, then a so-called standard false colour composite is produced as shown in the bottom right.

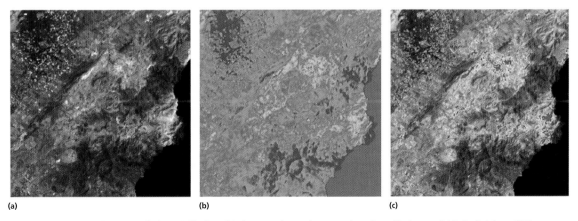

(a)   (b)   (c)

Fig. 1.6 (a) An image in grey-scale (B/W) display; (b) the same image in a pseudo-colour display; and (c) the brightest DNs are highlighted in red on a grey-scale background.

B/W display as a special case of pseudo-colour display in which a sequential grey scale based on DN levels is used instead of a colour scheme. Often, we can use a combination of B/W and pseudo-colour display to highlight important information in particular DN ranges in colours over a grey-scale background as shown in Fig. 1.6(c).

1.3 Some key points

In this chapter, we learnt what a digital image is and the elements that make up a digital image, and we also learnt about B/W and colour displays of digital images. It is important to remember these key points:

- A digital image is a raster dataset or a 2D array of numbers;
- Our perception of colours is based on the tristimulus theory of human vision. Any colour is composed of three primary colours: red, green and blue.
- Using an RGB colour cube, a colour can be expressed as a vector of the weighted summation of red, green and blue components.
- In image processing, colours are used as a tool for image information visualisation. From this viewpoint,

the true colour display is a special case of the general false colour display.

- Pseudo-colour display results in the loss of the numerical sequential relationship of the image DNs. It is therefore no longer a digital image; it is an image of symbols.

1.4 Questions

1 What is a digital image and how is it composed?
2 Describe the tristimulus colour theory and principle of RGB additive colour composition.
3 Explain the relationship between primary colours and complementary colours using a diagram.
4 Illustrate the colour cube in a diagram. How is a colour composed of RGB components? Describe the definition of the grey line in the colour cube.
5 What is a false colour composite? Explain the principle of using colours as a tool to visualise spectral information of multi-spectral images.
6 Describe how to generate a pseudo-colour display. What are the merits and disadvantages of pseudo-colour displays?

CHAPTER 2

Point operations (contrast enhancement)

Contrast enhancement, sometimes called radiometric enhancement or histogram modification, is the most basic but also the most effective technique to optimise the image contrast and brightness for visualisation or to highlight information in particular DN ranges.

Let X represent a digital image; x_{ij} is the digital number (DN) of any a pixel in the image at line i and column j. Let Y represent the image derived from X by a function f; y_{ij} is the output value corresponding to x_{ij}. Then a contrast enhancement can be expressed in a general form:

$$y_{ij} = f\left(x_{ij}\right) \qquad (2.1)$$

This processing transforms a single input image X to a single output image Y, through a function f, in such a way that the DN of an output pixel y_{ij} depends on and only on the DN of the corresponding input pixel x_{ij}. This type of processing is called a *point operation*. Contrast enhancement is a point operation that modifies the image brightness and contrast but does not alter the image size.

2.1 Histogram modification and lookup table

Let x represent a DN level of an image X; the number of pixels of each DN level $h(x)$ is called the *histogram* of the image X. The $h(x)$ can also be expressed as a percentage of the pixel number of a DN level x against the total number of pixels in the image X. In this case, in statistical terms, $h(x)$ is a *probability density function*.

A histogram is a good presentation of contrast, brightness and data distribution of an image. Every image has a unique histogram, but the reverse is not necessarily true because a histogram does not contain any spatial information. As a simple example, imagine how many different patterns you can form on a 10×10 grid chessboard using 50 white pieces and 50 black pieces. All these patterns have the same histogram!

It is reasonable to call point operation *histogram modification* because the operation only alters the histogram of an image but not the spatial relationship of image pixels. In eqn. 2.1, point operation is supposed to be performed pixel by pixel. For the pixels with the same input DN but different locations ($x_{ij} = x_{kl}$), the function f will produce the same output DN ($y_{ij} = y_{kl}$). Thus the point operation is independent of pixel position. The point operation on individual pixels is the same as that on DN levels:

$$y = f\left(x\right) \qquad (2.2)$$

As shown in Fig. 2.1, suppose $h_i(x)$, the histogram of an input image X, is a continuous function; as a point operation does not change image size, the number of pixels in the DN range δx in the input image X should be equal to the number of pixels in the DN range δy in the output image Y, thus we have:

$$h_i\left(x\right)\delta x = h_o\left(y\right)\delta y \qquad (2.3)$$

Let $\delta x \to 0$ then $\delta y \to 0$,

$$h_i\left(x\right)dx = h_o\left(y\right)dy \qquad (2.4)$$

Therefore,

$$h_o\left(y\right) = h_i\left(x\right)\frac{dx}{dy} = h_i\left(x\right)\frac{dx}{f'\left(x\right)dx} = \frac{h_i\left(x\right)}{f'\left(x\right)} \qquad (2.5)$$

Image Processing and GIS for Remote Sensing: Techniques and Applications, Second Edition. Jian Guo Liu and Philippa J. Mason.
© 2016 John Wiley & Sons, Ltd. Published 2016 by John Wiley & Sons, Ltd.

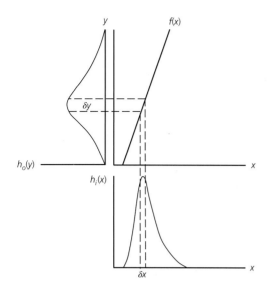

Fig. 2.1 The principles of the point operation by histogram modification.

We can also write eqn. 2.5 as

$$h_o(y) = \frac{h_i(x)}{y'}$$

Equation 2.5 describes that the histogram of output image can be derived from the histogram of input image divided by the first derivative of point operation function.

For instance, given a linear function: $y = 2x - 6$ then $y' = 2$, from eqn. 2.5 we have

$$h_o(y) = \frac{1}{2}h_i(x)$$

This linear function will produce an output image with a flattened histogram twice as wide and half as high as that of the input image and with all the DNs shifted to the left by three DN levels. This linear function stretches image DN range to increase its contrast.

As $f'(x)$ is the gradient of the point operation function $f(x)$, eqn. 2.5 indicates

- when the gradient of a point operation function is greater than 1, it is a stretching function that increases the image contrast;
- when the gradient of a point operation function is less than 1 but positive, it is a compression function that decreases the image contrast;
- if the gradient of a point operation function is negative, then the image becomes negative with black and white inversed.

For a nonlinear point operation function, this stretches and compresses different sections of DN levels, depending on its gradient at different DN levels. This is shown later in the discussion on logarithmic and exponential point operation functions.

In the real case of an integer digital image, both $h_i(x)$ and $h_o(y)$ are discrete functions. Given a point operation $y = f(x)$, the DN level x in the image X is converted to a DN level y in output image Y and the number of pixels with DN value x in X is equal to that of pixels with DN value y in Y. Thus,

$$h_i(x) = h_o(y) \qquad (2.6)$$

Equation 2.6 seems contradictory to eqn. 2.3: $h_i(x)\delta x = h_o(y)\delta y$ for the case of continuous function. In fact, eqn. 2.6 is a special case of eqn. 2.3 for $\delta x = \delta y = 1$, where 1 is the minimal DN interval for an integer digital image. Actually the point operation modifies the histogram of a digital image by moving the 'histogram bar' of each DN level x to a new DN level y according to the function f. The length of each histogram bar is not changed by the processing and thus no information is lost, but the distances between histogram bars are changed. For the given example above, the distance between histogram bars is doubled and thus the equivalent histogram averaged by the gap is flatter than the histogram of the input image (Fig. 2.2). In this sense, eqn. 2.3 always holds while eqn. 2.6 is true only for individual histogram bars but not for the equivalent histogram. A point operation may merge several DN levels of an input image into one DN level of the output image. Equation 2.6 is then no longer true for some histogram bars and the operation results in information loss.

As point operation is in fact a histogram modification, it can be performed more efficiently using a *lookup table* (LUT). A LUT is composed of DN levels of an input image X and their corresponding DN levels in the output image Y; an example is shown in Table 2.1. When applying a point operation function to enhance an image, first the LUT is generated by applying function $y = f(x)$ to every DN level x of the input image X to generate the corresponding DN level y in the output image Y. Then, the output image Y is produced by just replacing x with its corresponding y for each pixel. In this case for an 8-bit image, $y = f(x)$ needs to be calculated for no more than 256 times. If a point operation is performed without using a LUT, $y = f(x)$

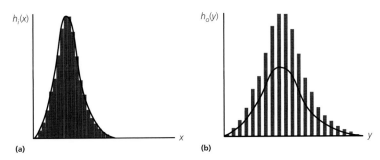

Fig. 2.2 Histograms (a) before and (b) after linear stretch for integer image data. Though the histogram bars in the histogram of the stretched image on the right are the same height as those in the original histogram on the left, the equivalent histogram drawn in the curve is wider and flatter because of the wider interval of these histogram bars.

Table 2.1 An example LUT for a linear point operation function $y = 2x - 6$.

x	y
3	0
4	2
5	4
6	6
7	8
8	10
...	...
130	254

needs to be calculated as many times as the total number of pixels in the image. For a large image, the LUT approach speeds up processing dramatically especially when the point operation function $y = f(x)$ is a complicated one.

As most display systems can only display 8-bit integers in 0–255 grey levels, for image visualization, it is important to configure a point operation function in such a way that the value range of an output image Y is within 0–255.

2.2 Linear contrast enhancement (LCE)

The point operation function for linear contrast enhancement is defined as

$$y = ax + b \qquad (2.7)$$

It is the simplest and one of the most effective contrast enhancement techniques. In this function, coefficient a controls the contrast of output images and b modifies the overall brightness by shifting the zero position of the histogram of y to $-\dfrac{b}{a}$ (to the left if negative and to the right if positive). LCE improves image contrast without distorting the image information if the output DN range is wider than the input DN range. In this case, the LCE does nothing but widen the increment of DN levels and shift histogram position along the image DN axis. For instance, LCE function $y = 2x - 6$ shifts the histogram $h_i(x)$ to the left by three DN levels and doubles the DN increment of x to produce an output image Y with a histogram $h_o(y) = h_i(x)/2$ that is two times wider than but half the height of the original.

There are several popular LCE algorithms available in most image processing software packages:

1 Interactive linear stretch: This changes a and b of eqn. 2,7 interactively to optimise the contrast and brightness of the output image based on the user's visual judgement.

2 Piecewise linear stretch (PLS): This uses several different linear functions to stretch different DN ranges of an input image [Fig. 2.3(a)–(c)]. PLS is a very versatile point operation function: it can be used to simulate a non-linear function that cannot be easily defined by a mathematical function. Most image processing software packages have interactive PLS functionality allowing users to configure PLS for optimised visualisation. Thresholding can be regarded as a special case of PLS as shown in Fig. 2.3(d) and (e), though in concept it is a conditional logic operation.

3 Linear scale: This automatically scales the DN range of an image to the full dynamic range of the display system (8 bits) based on the maximum and minimum of the input image X.

$$y = 255 \left[x - Min(x) \right] / \left[Max(x) - Min(x) \right] \qquad (2.8)$$

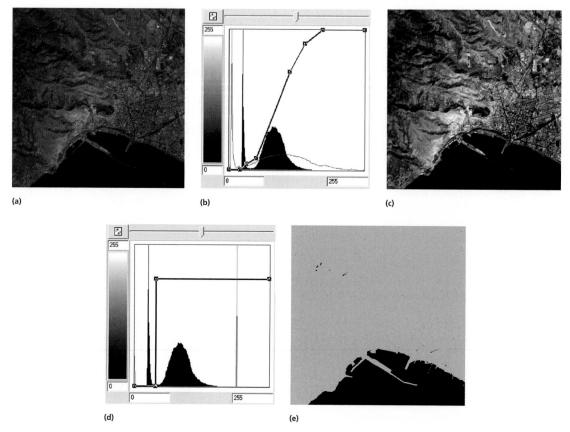

Fig. 2.3 Interactive PLS function for contrast enhancement and thresholding: (a) the original image; (b) the PLS function for contrast enhancement; (c) the enhanced image; (d) the PLS function for thresholding; and (e) the binary image produced by thresholding.

In many modern image processing software packages, this function is largely redundant as the operation specified in eqn. 2.8 can be easily done using an interactive PLS. However, eqn. 2.8 helps us to understand the principle.

4 Mean/standard deviation adjustment: This linearly stretches an image to make it satisfy a given mean (E_o) and standard deviation (SD_o).

$$y = E_o + SD_o \frac{x - E_i}{SD_i} \quad \text{or} \quad y = \frac{SD_o}{SD_i} x + E_o - \frac{SD_o}{SD_i} E_i \quad (2.9)$$

where E_i and SD_i are the mean and standard deviation of the input image X.

These last two linear stretch functions are often used for automatic processing, while for interactive processing PLS is the obvious choice.

2.2.1 Derivation of a linear function from two points

As shown in Fig. 2.4, a linear function $y = ax + b$ can be uniquely defined by two points (x_1, y_1) and (x_2, y_2) based on the formula

$$\frac{y - y_1}{x - x_1} = \frac{y_2 - y_1}{x_2 - x_1}$$

Given $x_1 = Min(x), x_2 = Max(x)$ and $y_1 = 0, y_2 = 255$, we can then derive the linear scale function as below

$$\frac{y}{x - Min(x)} = \frac{255}{Max(x) - Min(x)}$$

Thus

$$y = \frac{255[x - Min(x)]}{Max(x) - Min(x)}$$

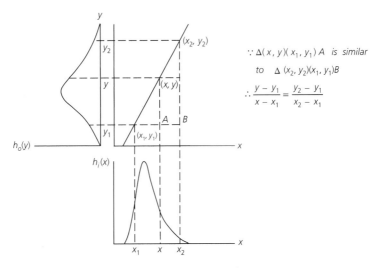

Fig. 2.4 Derivation of a linear function from two points of input image X and output image Y.

Similarly, linear functions for mean and standard deviation adjustment defined in (2.9) can be derived from

Either $x_1 = E_i, x_2 = E_i + SD_i, y_1 = E_o, y_2 = E_o + SD_o$

Or $x_1 = E_i, x_2 = E_i + SD_i, y_1 = E_o, y_2 = E_o + SD_o$

2.3 Logarithmic and exponential contrast enhancement

Logarithmic and exponential functions are inverse operations of one another. For contrast enhancement, the two functions modify the image histograms in opposite ways. Both logarithmic and exponential functions change shapes of image histograms and distort the information in original images.

2.3.1 Logarithmic contrast enhancement

The general form of the logarithmic function used for image processing is defined as below:

$$y = b\ln(ax + 1) \qquad (2.10)$$

Here a (>0) controls the curvature of the logarithmic function, while b is a scaling factor to make the output DNs fall within a given value range, and the shift 1 is to avoid the zero value at which the logarithmic function loses its meaning. As shown in Fig. 2.5, the gradient of the function is greater than 1 in the low DN range and thus it spreads out low DN values, whilst in the high DN range the gradient of the function is less than 1 and so it compresses high DN values. As a result, logarithmic contrast enhancement shifts the peak of the image

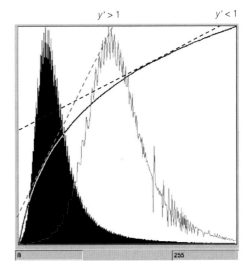

Fig. 2.5 Logarithmic contrast enhancement function.

histogram to the right and highlights the details in dark areas in an input image. Many images have histograms similar in form to logarithmic normal distributions. In such cases, a logarithmic function will effectively modify the histogram to the shape of a normal distribution.

We can slightly modify eqn. 2.10 to introduce a shift constant c:

$$y = b\ln(ax + 1) + c \qquad (2.11)$$

This function allows the histogram of the output image to shift by c.

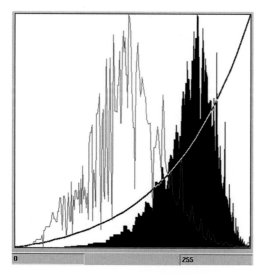

Fig. 2.6 Exponential contrast enhancement function.

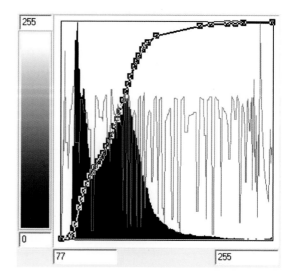

Fig. 2.7 Histogram of histogram equalization.

2.3.2 Exponential contrast enhancement

The general form of the exponential function used for image processing is defined as below:

$$y = be^{ax+1} \tag{2.12}$$

Here again, a (>0) controls the curvature of the exponential function, while b is a scaling factor to make the output DNs fall within a given value range, and the exponential shift 1 is to avoid the zero value because $e^0 \equiv 1$. As the inverse of the logarithmic function, exponential contrast enhancement shifts the image histogram peak to the left by spreading out high DN values and compressing low DN values to enhance detail in light areas at the cost of suppressing the tone variation in the dark areas (Fig. 2.6). Again, we can introduce a shift parameter c to modify the exponential contrast enhancement function as below:

$$y = be^{ax+1} + c \tag{2.13}$$

2.4 Histogram equalisation (HE)

Histogram equalisation is a very useful contrast enhancement technique. It transforms an input image to an output image with a uniform (equalised) histogram. The key point of HE is to find the function that converts $h_i(x)$ to $h_o(y) = A$, where A is a constant. Suppose image X has N pixels and the desired output DN range is L (the number of DN levels), then

$$h_o(y) = A = \frac{N}{L} \tag{2.14}$$

According to eqn. 2.4,

$$dy = h_i(x)dx / h_o(y) = \frac{L}{N}h_i(x)dx \tag{2.15}$$

Thus, the histogram equalisation function is as below:

$$y = \frac{L}{N}\int h_i(x)dx = \frac{L}{N}H_i(x) \tag{2.16}$$

As the histogram $h_i(x)$ is essentially the probability density function of X, the $H_i(x)$ is the *cumulative distribution function* of X. The calculation of $H_i(x)$ is simple for a discrete function in the case of digital images. For a given DN level x, $H_i(x)$ is equal to the total number of those pixels with DN values no greater than x.

$$H_i(x) = \sum_{k=0}^{x} h_i(k) \tag{2.17}$$

Theoretically, histogram equalisation can be achieved if $H_i(x)$ is a continuous function. However, as $H_i(x)$ is a discrete function for an integer digital image, HE can only produce a relatively flat histogram mathematically equivalent to an equalised histogram, in which the distance between histogram bars is proportional to their heights (Fig. 2.7).

The idea behind HE contrast enhancement is that the data presentation of an image should be evenly distributed across the whole value range. In reality, however, HE often produces images with contrast too high. This is because natural scenes are more likely to follow normal (Gaussian) distributions, and consequently the human eye is adapted to be more sensitive for discriminating subtle grey-level changes of intermediate brightness than of very high and very low brightness.

2.5 Histogram matching (HM) and Gaussian stretch

Histogram matching (HM) is a point operation that transforms an input image to make its histogram match a given shape defined by either a mathematical function or the histogram of another image. It is particularly useful for image comparison and differencing. If the two images in question are modified to have similar histograms, the comparison will be on a fair basis.

HM can be implemented by applying HE twice. Equation 2.14 implies that an equalised histogram is only decided by image size N and the output DN range L. Images of the same size always have the same equalized histogram for a fixed output DN range, and thus HE can act as a bridge to link images of the same size but different histograms (Fig. 2.8). Consider $h_i(x)$ is the histogram of an input image and $h_o(y)$ the reference histogram to be matched. Suppose $z = f(x)$ is the HE function to transform $h_i(x)$ to an equalised histogram $h_e(z)$, and $z = g(y)$ the HE function to transform the reference histogram $h_o(y)$ to the same equalised histogram $h_e(z)$, then

$$z = g(y) = f(x)$$

Thus

$$y = g^{-1}(z) = g^{-1}\{f(x)\} \tag{2.18}$$

Recall eqn. 2.16; $f(x)$ and $g(y)$ are the cumulative distribution functions of $h_i(x)$ and $h_o(y)$ individually. Thus the HM can be easily implemented by a three-column LUT containing corresponding DN levels of x, z and y. An input DN level x will be transformed to an output DN level y sharing the same z value. As shown in Table 2.2, for $x = 5$, $z = 3$, while for $y = 0$, $z = 3$. Thus for an input $x = 5$, the LUT coverts to an output $y = 0$ and so on. The output image Y will have a histogram that matches the reference histogram $h_o(y)$.

If the reference histogram $h_o(y)$ is defined by a Gaussian distribution function:

$$h_o(y) = \frac{1}{\sigma\sqrt{2\pi}} \exp\left(\frac{-(x - \bar{x})^2}{2\sigma^2}\right) \tag{2.19}$$

where σ is the standard deviation and $\bar{x}$ the mean of image X, the HM transformation is then called **Gaussian stretch** since the resultant image has a histogram in the shape of a Gaussian distribution.

Table 2.2 An example LUT for histogram matching.

x	z	y
5	3	0
6	4	2
7	5	4
8	6	5
...	...	...

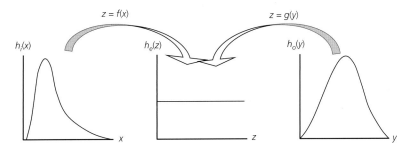

Fig. 2.8 Histogram equalization acts as a bridge for histogram matching.

2.6 Balance contrast enhancement technique (BCET)

Colour bias is one of the main causes of poor colour composite images. For RGB colour composition, if the average brightness of one image band is significantly higher or lower than the other two, the composite image will show obvious colour bias. To eliminate this, the three bands used for colour composition must have an equal value range and mean. The balance contrast enhancement technique (BCET) is a simple solution to this problem. Using a parabolic function derived from an input image, BCET can stretch (or compress) the image to a given value range and mean without changing the basic shape of the image histogram. Thus three image bands for colour composition can be adjusted to the same value range and mean to achieve a balanced colour composite.

The BCET based on a parabolic function is as below:

$$y = a(x - b)^2 + c \qquad (2.20)$$

This general form of parabolic function is defined by three coefficients, a, b and c. It is therefore capable of adjusting three image parameters: minimum, maximum and mean. The coefficients a, b and c can be derived based on the minimum, maximum and mean (l, h and e) of the input image X and the given minimum, maximum and mean (L, H and E) for the output image Y as below:

$$b = \frac{h^2(E-L) - s(H-L) + l^2(H-E)}{2[h(E-L) - e(H-L) + l(H-E)]}$$

$$a = \frac{H-L}{(h-l)(h+l-2b)} \qquad (2.21)$$

$$c = L - a(l-b)^2$$

where s is the mean square sum of input image X,

$$s = \frac{1}{N}\sum_{i=1}^{N} x_i^2$$

Figure 2.9 illustrates a comparison between RGB colour composites using the original band 5, 4 and 1 of an ETM+ sub-scene and the same bands after BCET stretch. The colour composite of the original bands [Fig. 2.9(a)] shows strong colour bias to magenta as the result of much lower brightness in band 4, displayed in green. This colour bias is completely removed by BCET, which stretches all the bands to the same value range 0–255 and mean 110 [Fig. 2.9(b)]. The BCET colour composite in Fig. 2.9(b) presents various terrain materials (rock types, vegetation etc.) in much more distinctive colours than those in the colour composite of the original image bands in Fig. 2.9(a). An interactive PLS may achieve similar results but without quantitative control.

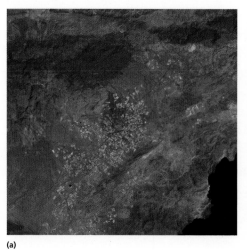
(a)

(b)

Fig. 2.9 Colour composites of ETM+ bands 5, 4 and 1 in red, green and blue: (a) colour composite of the original bands showing magenta cast as the result of colour bias; and (b) BCET colour composite stretching all the bands to an equal value range of 0–255 and mean of 110.

2.6.1 Derivation of coefficients *a*, *b* and *c* for a BCET parabolic function (Liu 1991)*

Let x_i represent any pixel of an input image X, with N pixels, then the minimum, maximum and mean of X are:

$$l = \min(x_i), \quad h = \max(x_i), \quad e = \frac{1}{N}\sum_{i=1}^{N} x_i, \quad i = 1,2,\dots,N$$

Suppose L, H and E are the desired minimum, maximum and mean for the output image Y, then we can establish the following equations:

$$L = a(l-b)^2 + c$$
$$H = a(h-b)^2 + c \qquad (2.22)$$
$$E = \frac{1}{N}\sum_{i=1}^{N}\left[a(x_i - b)^2 + c\right]$$

Solving *b* from eqn. 2.22:

$$b = \frac{h^2(E-L) - s(H-L) + l^2(H-E)}{2\left[h(E-L) - e(H-L) + l(H-E)\right]} \qquad (2.23)$$

where $s = \frac{1}{N}\sum_{i=1}^{N} x_i^2$.

With *b* known, *a* and *c* can then be resolved from eqn. 2.22 as

$$a = \frac{H-L}{(h-l)(h+l-2b)} \qquad (2.24)$$

$$c = L - a(l-b)^2 \qquad (2.25)$$

The parabolic function is an even function [Fig. 2.10(a)]. Coefficients *b* and *c* are the coordinates of the turning point of the parabola that determine the section of the parabola to be utilised by the BCET function. In order to perform BCET, the turning point and its nearby section of the parabola should be avoided, so that only the section of the monotonically increasing branch of the curve is used. This is possible for most cases of image contrast enhancement.

From the solutions of *a*, *b* and *c* in eqns. 2.23–2.25, we have the following observations:

1 If $b < l$ then $a > 0$, the parabola is open upward and a section of the right (monotonically increasing) branch of the parabola is used in BCET.

2 If $b > h$ then $a < 0$, the parabola is open downward and a section of left (monotonically increasing) branch of the parabola is used in BCET.

3 If $l < b < h$, then BCET fails to avoid the turning point of the parabola and becomes malfunctional.

For example, Table 2.3 shows the minimum (*l*) maximum (*h*) and mean (*e*) of seven band images of a Landsat TM sub-scene and the corresponding coefficients of the BCET parabolic functions. Using these parabolic functions, images of bands 1–5 and 7 are all

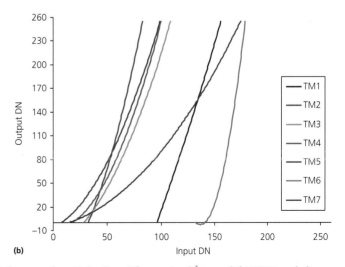

Standard parabolic functions   **BCET parabolic functions of 7 Bands of a TM Sub-scene**

Fig. 2.10 (a) Standard parabolas $y = x^2$ and $y = -x^2$, the cases of $a = \pm1$, $b = 0$, $c = 0$ for $y = a(x-b)^2 + c$; and (b) BCET parabolic functions for 7 band images of a TM sub-scene. The parabola for the band 6 image in red involves the turning point and both branches and is therefore not usable.

Table 2.3 Derivation of BCET parabolic functions for the 7 band images of a Landsat TM sub-scene to stretch each band to $L = 0$, $H = 255$ and $E = 100$.

Image bands	Input image			BCET coefficients			BCET output image		
	Min (l)	Max(h)	Mean (e)	a	b	c	Min (L)	Max(H)	Mean (E)
TM1	96	156	120.78	0.00857	−121.91	−407.03	0	255	100.00
TM2	33	83	55.47	0.03594	−12.95	−75.88	0	255	99.90
TM3	29	109	66.47	0.01934	−13.4	−34.78	0	255	99.99
TM4	18	100	59	0.02776	2.97	−6.26	0	255	100.05
TM5	15	175	94.33	0.00718	−15.97	−6.89	0	255	99.96
TM6	132	179	157.88	0.14302	*136.53*	−2.94	Malfunction, not used		
TM7	8	99	52.63	0.02072	−14.11	−10.13	0	255	100.02

successfully stretched to the given value range and mean: $L = 0$, $H = 255$ and $E = 100$ as shown in the right part of the table. The only exception is the band 6 image because $l < b < h$ and the BCET malfunctions. As illustrated in Fig. 2.10(b), the BCET parabolic function for band 6 involves the turning point and both branches of the parabola fall in the value range of this image, unlike all the other bands where only one monotonic branch is used.

2.7 Clipping in contrast enhancement

In digital images, a few pixels (often representing noise) may occupy a wide value range at the low and high ends of histograms. In such cases, setting a proper cut-off to clip both ends of the histogram in contrast enhancement is necessary to make effective use of the dynamic range of a display device. Clipping is often given as a percentage of the total number of pixels in an image. For instance, if 1% and 99% are set as the cut-off limits for the low and high ends of the histogram of an image, the image is then stretched to set the DN levels $< x_l$, where $H_i(x_l) = 1\%$, to 0 and DN levels $> x_h$, and where $H_i(x_h) = 99\%$, to 255 for an 8-bit/pixel/channel display in the output image.

This simple treatment often improves image display quality significantly, especially when the image looks hazy because of atmospheric scattering. When using BCET, the input minimum (l) and maximum (h) should be determined based on appropriate cut-off levels of x_l and x_h.

2.8 Tips for interactive contrast enhancement

The general purpose of contrast enhancement is to optimise visualisation. Often after quite complicated image processing, you need to apply interactive contrast enhancement to view the results properly. After all, you need to be able to see the image! Visual observation is always the most effective way to judge image quality. This doesn't sound technical enough for digital image processing but this golden rule is quite true! On the other hand, the histogram gives you a quantitative description of image data distribution and so can also effectively guide you to improve the image visual quality. As said before, the business of contrast enhancement is histogram modification, and so you should find the following guidelines useful.

1 Make full use of the dynamic range of the display system. This can be done by specifying the actual limits of the input image to be displayed in 0 and 255 for an 8-bit display. Here percentage clipping is useful to avoid large gaps in either end of the histogram.

2 Adjust the histogram to make its peak near the centre of the display range. For many images, the peak may be slightly skewed toward the left to achieve the best visualisation unless the image is dominated by bright features, in which case the peak could skew to the right.

3 As implied by eqn. 2.5, a point operation function modifies an image histogram according to the function's gradient or slope $f'(x)$.

If gradient = 1 (slope = 45°), the function does nothing and the image is not changed.

If gradient > 1 (slope > 45°), the function stretches the histogram to increase image contrast.

If gradient < 1 (slope < 45°) and non negative, the function compresses the histogram to decrease image contrast.

A common approach in the PLS is therefore to use functions with slope > 45° to spread the peak section and those with slope < 45° to compress the tails at the both ends of the histogram.

2.9 Questions

1 What is a point operation in image processing? Give the mathematical definition.

2 Using a diagram explain why a point operation is also called histogram modification.

3 Given the following point operation functions, derive the output histograms $h_o(y)$ from the input histogram $h_i(x)$:

$$y = 3x - 8; \quad y = 2.5x^2 - 3x + 2; \quad y = \sin(x)$$

4 Try to derive the *linear scale* functions and the *mean and standard deviation adjustment* functions defined by eqns. 2.8 and 2.9. (See the answer at the end in Fig. 2.11.)

5 Given Fig. 2.6 of exponential contrast enhancement, roughly mark the section of the exponential function that stretches the input image and the section that compresses the input image and explain why (refer to Fig. 2.5).

6 How is HE is achieved? How is HE used to achieve histogram matching?

7 What type of function does a BCET use and how is balanced contrast enhancement achieved?

8 Try to derive the coefficients a, b and c in the BCET function $y = a(x - b)^2 + c$.

9 What is clipping and why is it often essential for image display?

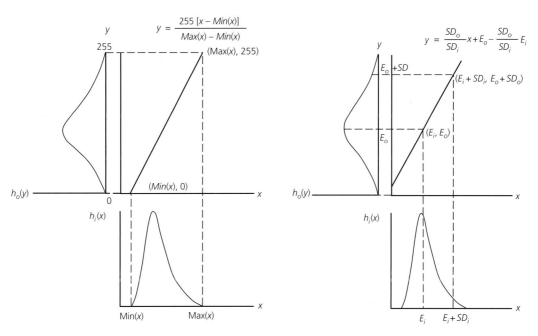

Fig. 2.11 Derivation of the linear stretch function and mean/standard deviation adjustment function.

CHAPTER 3

Algebraic operations (multi-image point operations)

For multi-spectral or, more generally, multi-layer images, algebraic operations such as the four basic arithmetic operations $(+, -, \times, \div)$, logarithmic, exponential, sin and tan can be applied to the digital numbers (DNs) of different bands for each pixel to produce a new image. Such processing is called image algebraic operation. Algebraic operations are performed pixel by pixel among DNs of spectral bands (or layers) for each pixel without involving neighbourhood pixels. They can therefore be considered as *multi-image point operations* defined as below:

$$y = f\left(x_1, x_2, \ldots x_n\right) \qquad (3.1)$$

where n is the number of bands or layers.

Obviously, all the images involving algebraic operations should be precisely co-registered.

To start with, let us consider the four basic arithmetic operations: addition, subtraction, multiplication and division. In multi-image point operations, arithmetic processing is sometimes the same as matrix operations, such as addition and subtraction, but sometimes totally different from and much simpler than matrix operations, such as image multiplication and division. As the image algebraic operation is position relevant, that is, pixel-to-pixel based, we can generalise the description. Let $X_i, i = 1, 2, \ldots, n$ represent both the ith band image and any a pixel in the ith band image of an n bands imagery dataset $\mathbf{X}$ $(X_i \in \mathbf{X})$, and Y the output image as well as any a pixel in the output image.

3.1 Image addition

This operation produces a weighted summation of two or more images:

$$Y = \frac{1}{k} \sum_{i=1}^{n} w_i X_i \qquad (3.2)$$

where w_i is the weight of image X_i and k is a scaling factor. If $w_i = 1$ for $i = 1, \ldots, n$ and $k = n$, eqn. 3.2 defines an average image.

An important application of image addition is to reduce noise and increase the signal-to-noise ratio (SNR). Suppose each image band of an n band multi-spectral image is contaminated by an additive noise source $N_i (i = 1, 2, \ldots, n)$. The noise pixels are not likely to occur at the same position in different bands, and thus a noise pixel DN in band i will be averaged with the non-noise DNs in the other n-1 bands. As a result the noise will be largely suppressed. It is proved from signal processing theory that of n duplications of an image, each contaminated by the same level of random noise, the SNR of the sum image of these n duplications equals the square root n times the SNR of any individual duplication:

$$SNR_y = \sqrt{n} \cdot SNR_i \qquad (3.3)$$

Equation 3.3 implies that for an n bands multi-spectral image, the summation of all the bands can increase SNR by about $\sqrt{n}$ times. For instance, if we average bands 1–4 of a Landsat TM image, the SNR of this average image is about two times ($\sqrt{4} = 2$) of that of each individual band.

You may notice in our later chapters on topics of RGB-IHS (red, green and blue to intensity, hue and saturation) transformation and principal component analysis (PCA) that an intensity component derived from RGB-IHS transformation is an average image of the R, G and B component images, and in most cases, the first principal component is a weighted sum image of all the images involving PCA operations.

Image Processing and GIS for Remote Sensing: Techniques and Applications, Second Edition. Jian Guo Liu and Philippa J. Mason.
© 2016 John Wiley & Sons, Ltd. Published 2016 by John Wiley & Sons, Ltd.

3.2 Image subtraction (differencing)

Image subtraction produces a difference image from two input images:

$$Y = \frac{1}{k}\left(w_i X_i - w_j X_j\right) \qquad (3.4)$$

The weights w_i and w_j are important to ensure that balanced differencing is performed. If the brightness of X_i is significantly higher than that of X_j, for instance, the difference image $X_i - X_j$ will be dominated by X_i, and as a result, the true difference between the two images will not be effectively revealed. To produce a 'fair' difference image, balance contrast enhancement technique (BCET) or histogram matching (matching the histogram of X_i to that of X_j) may be applied as a pre-processing step. Whichever method is chosen, the differencing that follows should then be performed with equal weighting ($w_i = w_j = 1$).

Subtraction is one of the simplest and most effective techniques for selective spectral enhancement, and it is also useful for change detection and removal of background illumination bias. However, in general, subtraction reduces the image information and decreases image SNR because it removes the common features while retaining the random noise in both images.

Band differences of multi-spectral images are successfully used for studies of vegetation, landuse and geology. As shown in Fig. 3.1, TM band difference of TM3-TM1 (R-B) highlights iron oxides; TM4-TM3 (NIR-Red) enhances vegetation; and TM5-TM7 is effective for detecting hydrated (clay) minerals (i.e. those containing the OH- ion; refer to Table A.1 in Appendix A for the spectral wavelengths of Landsat TM). These three difference images can be combined to form an RGB colour composite image to highlight iron oxides, vegetation and clay minerals in red, green and blue, as well as other ground objects in various colours. In many cases, subtraction can achieve similar results to division (ratio) and the operation is simpler and faster.

The image subtraction technique is also widely used for background noise removal in microscopic image analysis. An image of the background illumination field (as a reference) is captured before the target object is placed in the field. The second image is then taken with the target object in the field. The difference image between the two will retain the target while the effects of the illumination bias and background noise are cancelled out.

3.3 Image multiplication

Image multiplication is defined as below:

$$Y = X_i \cdot X_j \qquad (3.5)$$

Here the image multiplication is performed pixel by pixel; at each image pixel, its band i DN is multiplied with band j DN. This is fundamentally different from matrix multiplication. A digital image is a two-dimensional (2D) array, but it is **not** a matrix.

A product image of multiplication often has much greater DN range than the dynamic range of the display devices and thus needs to be re-scaled before display. Most image processing software packages can display any image based on its actual value range, which is then fit into a 0–255 display range.

One application of multiplication is *masking*. For instance, if X_i is a mask image composed of DN values 0 and 1, the pixels in image X_j corresponding to 0 in X_i will become 0 (masked off) and others will remain unchanged in the product image Y. This operation could be achieved more efficiently using a logical operation of a given condition. Another application is image modulation. For instance, topographic features can be added back to a colour coded classification image by using a panchromatic image (as an intensity component) to modulate the three colour components (red, green and blue) of the classification image as below:

1 Produce red (R), green (G) and blue (B) component images from the colour coded classification image.
2 Use the relevant panchromatic image (I) to modulate R, G and B components: $R \times I$, $G \times I$ and $B \times I$.
3 Colour composition using $R \times I$, $G \times I$ and $B \times I$.

This process is, in some image processing software packages, automated by draping a colour coded classification image on an intensity image layer (Fig. 3.2).

3.4 Image division (ratio)

Image division is a very popular technique, also known as an image *ratio*. The operation is defined as

$$Y = \frac{X_i}{X_j} \qquad (3.6)$$

In order to carry out image division, certain protection is needed to avoid overflow, in case a number is divided by zero. A commonly used trick in this context is to change 0 to 1 whenever a 0 becomes a divisor. A better

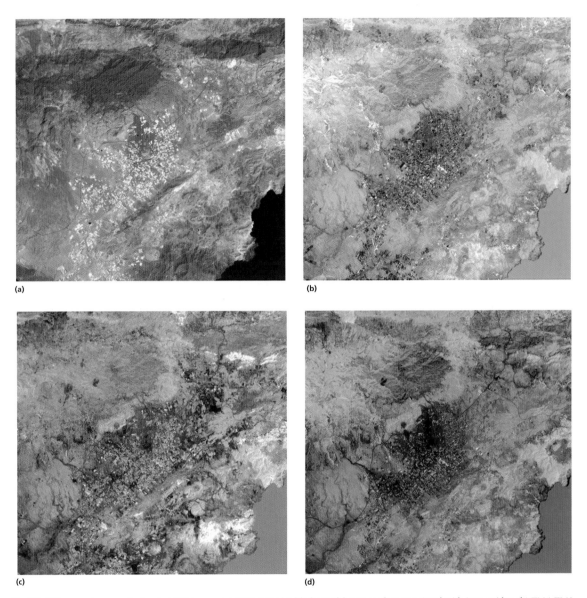

Fig. 3.1 Difference images of a Landsat TM image: (a) TM3-TM1 highlights red features often associated with iron oxides; (b) TM4-TM3 detects the diagnostic 'red edge' features of vegetation; (c) TM5-TM7 enhances the clay and hydrate mineral absorption features in the short-wave infrared (SWIR) spectral range; and (d) the colour composite of TM3-TM1 in red, TM4-TM3 in green and TM5-TM7 in blue that highlights iron oxide, vegetation and clay minerals in red, green and blue colours.

approach is to shift the value range of the denominator image upwards, by 1, to avoid zero. For an 8-bit image, this shift changes the image DN range from 0–255 to 1–256 that just exceeds 8 bits. This was a problem for the older generation of image processing systems before the 1990s but is no longer a problem for most modern image processing software packages where the image

processing is performed based on the double precision floating point data type.

A ratio image Y is an image of real numbers instead of integers. If both X_i and X_j are 8-bit images, the possible maximum value range of Y is 0, [1/255, 1], (1, 255]. Instead of a much simpler notation, [0, 255], we deliberately write the value range in such a way to emphasise

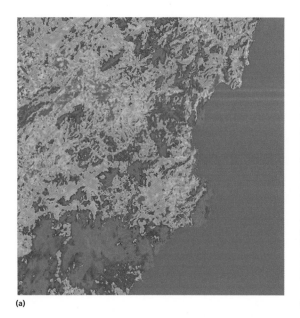
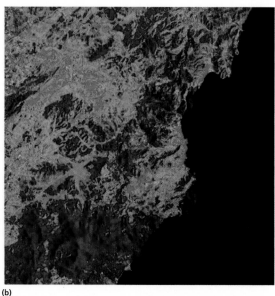

(a)   (b)

Fig. 3.2 Multiplication for image modulation: (a) a colour coded classification image; and (b) the intensity modulated classification image.

that the value range [1/255, 1] may contain just as much information as that in the much wider value range (1, 255]! A popular approach for displaying a ratio image on an 8-bit/pixel/channel display system is to scale the image into a 0–255 DN range, and many image processing software packages may perform the operation automatically. This may result in up to 50% information loss because the information recorded in value range [1/255, 1] could just be in a few DN levels.

If we consider an image ratio as a coordinate transformation from a Cartesian coordinate system to a polar coordinate system (Fig, 3.3) rather than a division operation, then

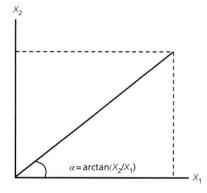

Fig. 3.3 Ratio as a coordinate transformation from a Cartesian coordinates system to a polar coordinates system.

$$Y = \frac{X_i}{X_j} = \tan(\alpha)$$

$$\alpha = \arctan\left(\frac{X_i}{X_j}\right)$$
(3.7)

Ratio image Y is actually a tangent image of the angle α. The information of a ratio image is evenly presented by angle α in value range $[0, \pi/2]$ instead of by $Y = \tan(\alpha)$ in value range [0,255]. Therefore, to achieve a 'fair' linear scale stretch of a ratio image, it is necessary to convert Y to α by eqn. 3.7. A linear scale can then be performed as

$$\beta = 255 \frac{\alpha - \mathrm{Min}(\alpha)}{\mathrm{Max}(\alpha) - \mathrm{Min}(\alpha)}$$
(3.8)

After all, the above transform may not be always necessary. Ratios are usually designed to highlight the target features as high ratio DNs. In this case, the direct stretch of ratio image Y may enhance the target features well but at the expense of the information represented by low ratio DNs. From this sense and as an example, it is important to notice that although ratios TM1/TM3 and TM3/TM1 are reciprocals of one another mathematically and so contain the same information, they are different in terms of digital image display after linear scale! Remember, when you design a ratio, make

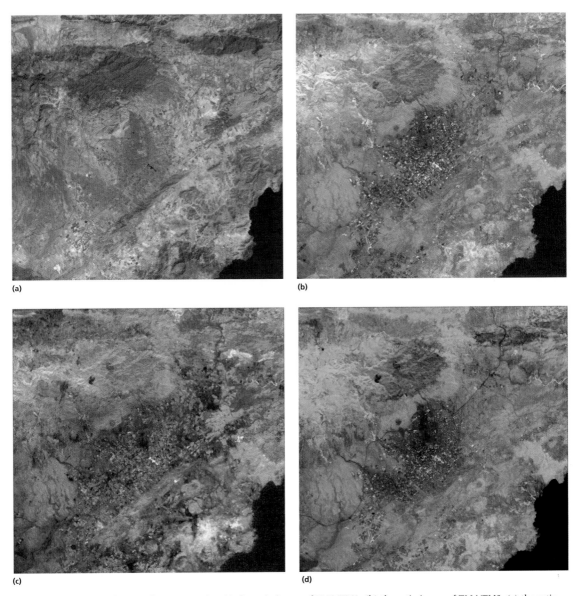

(a) (b) (c) (d)

Fig. 3.4 Ratio images and ratio colour composite: (a) the ratio image of TM3/TM1; (b) the ratio image of TM4/TM3; (c) the ratio image of TM5/TM7; and (d) the ratio colour composite of TM5/TM7 in blue, TM4/TM3 in green and TM3/TM1 in red.

sure that the target information is highlighted by high values in the ratio image.

Ratio is an effective technique to selectively enhance spectral features. Ratio images derived from different band pairs are often used to generate ratio colour composites in an RGB display. For instance, a colour composite of TM5/TM7 (blue), TM4/TM3 (green) and TM3/TM1 (red) may highlight clay mineral in blue, vegetation in green and iron oxide in red (Fig. 3.4). It is interesting to compare

Fig. 3.1(d) with Fig. 3.4(d) to notice the similarity between differencing and ratio techniques for selective enhancement. Many indices, such as the Normalised Difference Vegetation Index (NDVI), have been developed based on both differencing and ratio operations.

Ratio is also well known as an effective technique to suppress topographic shadows. For a given incident angle of solar radiation, the radiation energy received by a land surface depends on the angle between the land surface and the incident radiation. Therefore, solar illumination on a land surface varies with terrain slope and aspect, which results in topographic shadows. In a remotely sensed image, the spectral information is often occluded by sharp variations of topographic shadowing. The DNs in different spectral bands of a multi-spectral image are proportional to the solar radiation received by land surface and its spectral reflectance. Let $DN(\lambda)$ represent the digital number of a pixel in an image of spectral band λ, then:

$$DN(\lambda) = \rho(\lambda)E(\lambda) \qquad (3.9)$$

where $\rho(\lambda)$ and $E(\lambda)$ are the spectral reflectance and solar radiation of spectral band λ received at the land surface corresponding to the pixel.

As shown in Fig. 3.5, suppose a pixel representing a land surface facing the sun receives n times the radiation energy of that received by another pixel of land surface facing away from the sun, then the DNs of the two pixels in spectral bands i and j are as below:

Pixel in shadow:
$$DN1(i) = \rho(i)E(i) \text{ and } DN1(j) = \rho(j)E(j)$$

Pixel facing illumination :
$$DN2(i) = n\rho(i)E(i) \text{ and } DN2(j) = n\rho(j)E(j)$$

Thus the ratio between band i and j for both pixels will be:

$$R1_{i,j} = \frac{DN1(i)}{DN1(j)} = \frac{\rho(i)E(i)}{\rho(j)E(j)}$$

$$R2_{i,j} = \frac{DN2(i)}{DN2(j)} = \frac{n\rho(i)E(i)}{n\rho(j)E(j)} = \frac{\rho(i)E(i)}{\rho(j)E(j)} \qquad (3.10)$$

Therefore $R1_{i,j} = R2_{i,j}$

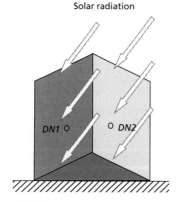

Solar radiation

DN1 ○   ○ DN2

Fig. 3.5 Principle of shadow suppression function of ratio images.

The 3.10 equations indicate that band ratios are independent of the variation of solar illumination caused by topographic shadowing and are decided only by the spectral reflectance of the image pixels. The pixels of the objects with the same spectral signature will result in the same band ratio values no matter whether they are under direct illumination or in shadow. Unfortunately, the real situation is more complicated than this simplified model because of atmospheric effects that often add different constants to different spectral bands. This is why the ratio technique can suppress topographic shadows but may not be able to eliminate their effects completely. Shadow suppression means losing topography that often accounts for more than 90% information of a multi-spectral image; ratio images therefore reduce SNRs significantly.

3.5 Index derivation and supervised enhancement

Infinite combinations of algebraic operations can be derived from basic arithmetic operations and algebraic functions. Aimless combinations of algebraic operations may mean an endless and potentially fruitless game, that is, you may spend a very long time without achieving any satisfactory result. Alternatively, you may happen upon a visually impressive image without being able to explain or interpret it. To design a meaningful and effectively combined operation, the knowledge of the spectral properties of targets is essential. The formulae should be composed on the basis of spectral or physical principles and designed for the enhancement of particular targets; these are then referred to as *spectral indices*, such as the NDVI. An index can be considered as supervised enhancement. Here we briefly introduce a few commonly used examples of indices based on Landsat TM/ETM+ image data. You may design your own indices for a given image processing task based on spectral analysis. In Part III of this book, you may find several examples of this kind of supervised enhancement in the teaching and research case studies.

3.5.1 Vegetation indices
As shown in Fig. 3.6, healthy vegetation has a high reflection peak in the near infrared (NIR) and an absorption trough in the red. If we could see NIR, vegetation would be NIR rather than green. This significant difference between red and NIR bands is known as the *red edge*; it is a unique

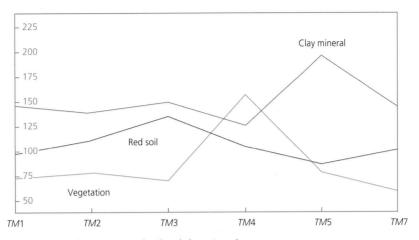

Fig. 3.6 Image spectral signatures of vegetation, red soil and clay minerals.

spectral property that makes vegetation different from all other ground objects. Obviously, this diagnostic spectral feature of vegetation can be very effectively enhanced by differencing and ratio operations. Nearly all the vegetation indices are designed to highlight the red edge in one way or another.

The NDVI is one of the most popular vegetation indices:

$$NDVI = \frac{NIR - Red}{NIR + Red} \qquad (3.11)$$

This index is essentially a difference between NIR and Red spectral band images. The summation of NIR and Red in the denominator is a factor to normalize the NDVI to a value range $[-1,1]$ by ratio.

$$\text{The NDVI for TM imagery}: \quad Y = \frac{TM4 - TM3}{TM4 + TM3} \qquad (3.12)$$

Vegetation can also be enhanced using a ratio index:

$$Y = \frac{NIR - \text{Min}(NIR)}{Red - \text{Min}(Red) + 1} \qquad (3.13)$$

And so for TM images: $\quad Y = \frac{TM4 - \text{Min}(TM4)}{TM3 - \text{Min}(TM3) + 1} \qquad (3.14)$

The effect of the subtraction of the band minimum is to roughly remove the added constants of atmospheric scattering effects (refer to the recommended remote sensing textbooks) so as to improve topography suppression by ratio. The value of 1 added to the denominator is to avoid a zero value. Figure 3.7 illustrates these vegetation indices derived from a TM image.

3.5.2 Iron oxide ratio index

Iron oxides and hydroxides are some of the most commonly occurring minerals in the natural environment. They appear red or reddish brown to the naked eye because of high reflectance in red and absorption in blue (Fig. 3.6). Typical red features on land surfaces, such as red soils, are closely associated with the presence of iron bearing minerals. We can enhance iron oxides using the ratio between red and blue spectral band images [Fig. 3.8(a)]:

$$Y = \frac{Red - \text{Min}(Red)}{Blue - \text{Min}(Blue) + 1} \qquad (3.15)$$

$$\text{For TM imagery}: \quad Y = \frac{TM3 - \text{Min}(TM3)}{TM1 - \text{Min}(TM1) + 1} \qquad (3.16)$$

3.5.3 TM clay (hydrated) mineral ratio index

Clay minerals are characteristic of hydrothermal alteration in rocks and are therefore very useful indicators for mineral exploration using remote sensing. The diagnostic spectral feature of clay minerals, which differentiates them from unaltered rocks, is that they all have strong absorption in the spectral range around 2.2 μm (corresponding to TM band 7) in contrast to high reflectance in the spectral range around 1.65 μm (corresponding to TM band 5), as shown in Fig. 3.6. Thus clay minerals can be generally enhanced by the ratio between these two SWIR bands [Fig. 3.8(b)]:

$$Y = \frac{TM5 - \text{Min}(TM5)}{TM7 - \text{Min}(TM7) + 1} \qquad (3.17)$$

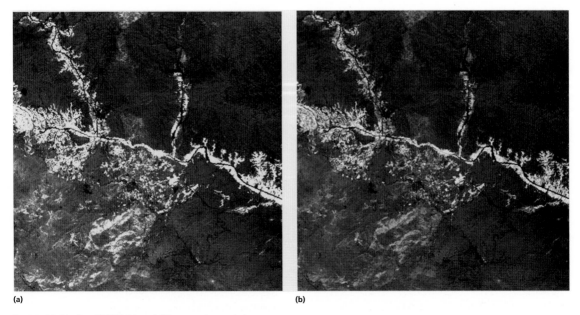

(a)   (b)

Fig. 3.7 (a) Landsat TM NDVI; and (b) vegetation ratio images.

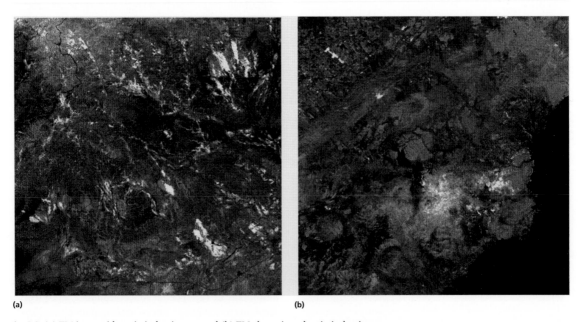

(a)   (b)

Fig. 3.8 (a) TM iron oxide ratio index image; and (b) TM clay mineral ratio index image.

This index can only achieve a general enhancement of all clay minerals using TM, ETM+ or Landsat Operational Land Imager (OLI) images. ASTER imagery, on the other hand, offers 5 SWIR bands and enables more specific discrimination of different clay minerals (though still not specific identification). You may try to design ASTER indices to target various clay minerals yourself.

3.6 Standardization and logarithmic residual

A typical example of a combined algebraic operation is the so-called *standardisation*:

$$Y_i = \frac{X_i}{\frac{1}{k}\sum_{\lambda=1}^{k}X_\lambda} \quad (3.18)$$

where X_i represents band i image, Y_i the standardised band i image and k the total number of spectral bands.

This ratio type operation can suppress topographic shadows based on the same principle as explained in § 3.4. The denominator in the formula is the average image of all the bands of a multi-spectral image; this allows the ratio for every band to be produced using the same divisor. The standardization enables the spectral variation among different bands, at each pixel, to be better enhanced using the ratio to the same denominator.

If we consider eqn. 3.18 as an arithmetic mean-based standardization, then another technique called *logarithmic residual* (Green & Craig 1985) can be considered as a geometric mean-based standardization:

$$\ln\left(R_{i\lambda}\right) = \ln\left(x_{i\lambda}\right) - \ln\left(x_{i.}\right) - \ln\left(x_{.\lambda}\right) + \ln\left(x_{..}\right) \quad (3.19)$$

where $x_{i\lambda}$ is the DN of pixel i in band λ.

$x_{i.} = \left(\prod_{\lambda=1}^{k}x_{i\lambda}\right)^{\frac{1}{k}}$ Geometric mean of pixel i over all the k bands.

$x_{.\lambda} = \left(\prod_{i=1}^{n}x_{i\lambda}\right)^{\frac{1}{n}}$ Geometric mean of band λ.

$x_{..} = \left(\prod_{i=1}^{n}\prod_{\lambda=1}^{k}x_{i\lambda}\right)^{\frac{1}{kn}}$ Global geometric mean of all the pixels in all the bands.

Then,

$$y_{i\lambda} = e^{\ln(R_{i\lambda})} \quad (3.20)$$

where $y_{i\lambda}$ is the logarithmic residual of $x_{i\lambda}$.

We can re-write eqn. 3.19 in the following form:

$$\ln\left(R_{i\lambda}\right) = \ln\frac{x_{i\lambda}}{x_{i.}} + \ln\frac{x_{..}}{x_{.\lambda}} \quad (3.21)$$

The first item in eqn. 3.21 has a similar form to 3.18 but the denominator is a geometric mean instead of an arithmetic mean for pixel i over k bands. The second

item in 3.21 is equivalent to a band spectral offset; it is a constant for all the pixels in one spectral band but varies with different spectral bands.

The logarithmic residual technique suppresses topographic shadows more effectively than other techniques but the resulting images are not often visually impressive, even after a proper stretch, because of their rather low SNR.

3.7 Simulated reflectance

Many image processing techniques have been developed on the basis of fairly sophisticated physical or mathematical models, but the actual constituent operations are simple arithmetic operations. Simulated reflectance technique (Liu *et al.* 1997b) is an example of this type.

3.7.1 Analysis of solar radiation balance and simulated irradiance

Suppose the solar radiation incident upon a solid land surface, of unit area equivalent to an image pixel, is irradiance E. This energy is partially reflected and absorbed by the terrain material depending on the reflectance (or albedo) ρ and absorptance α:

$$M_r = \rho E \quad (3.22)$$

$$M_a = \alpha E \quad (3.23)$$

where M_r is the reflected solar radiation and M_a the absorbed.

Considering the land surface as the surface of a solid medium of considerable thickness (which is generally true) and to satisfy conservation of energy, we have

$$\rho + \alpha = 1 \quad (3.24)$$

Based on the concept of the radiation balance (Robinson 1966) the solar radiation balance, B, on the earth is described by

$$\begin{aligned} B &= E(1-\rho) - M_e \\ &= \alpha E - M_e \quad (3.25) \\ &= M_a - M_e \end{aligned}$$

where M_e is the radiation emitted (thermal emission) from the land surface.

Then,

$$E = \rho E + \alpha E = M_r + M_a = M_r + M_e + B$$

or,

$$E - B = M_r + M_e \quad (3.26)$$

A dark (low albedo) ground object absorbs more solar radiation energy (mostly in the visible to NIR spectral region) than a bright (high albedo) object and becomes warmer because of the complicated thermodynamic processes within the terrain material; this means that the dark object eventually re-emits more thermal radiation in the thermal spectral region 8–14 μm. This general complementary relationship between reflected radiation M_r and emitted radiation M_e from the land surface can be easily observed in TM or ETM+ images in which dark subjects in visible band images are bright in thermal band images and vice versa. The phenomenon implies that the sum of M_r and M_e, the right side of eqn. 3.26, can be treated roughly as a constant for a given irradiance E, and therefore is independent of the spectral properties (albedo) of land surface.

Irradiance E varies with topography only. Suppose that the sun is a 'parallel' radiation source to the earth with constant incident radiant flux density M_s, then the solar irradiance upon the land surface, E, varies with the angle between the land surface and the incident solar radiation, γ. When the land surface is perpendicular to the incident solar radiation M_s, E is at its maximum and equal to M_s. If the solar radiation has a zenith angle θ_1 and azimuth angle ϕ_1, then the irradiance upon a land surface with slope angle θ_2 and aspect direction ϕ_2 can be calculated as

$$E = M_s \sin\gamma$$
$$= M_s\left[\sin\theta_1\sin\theta_2\cos(\phi_1 - \phi_2) + \cos\theta_1\cos\theta_2\right] \quad (3.27)$$

As stated in eqn. 3.25, B is dependent on solar irradiance E and is therefore affected by topography in the same way as eqn. 3.27. Thus $E - B$ varies mainly with topography whilst invariant to land surface albedo ρ. We refer to $E - B$ defined by 3.26 as the *simulated irradiance*, as it behaves like irradiance but with modified energy by B.

3.7.2 Simulated spectral reflectance image
For the reflected spectral radiation of a particular wavelength λ, eqn. 3.22 becomes

$$M_r(\lambda) = \rho(\lambda)E(\lambda) \quad (3.28)$$

where λ is the spectral wavelength and $\rho(\lambda)$ spectral reflectance and

$$E = \int_0^\infty E(\lambda)d\lambda \quad (3.29)$$

The albedo ρ, as the total reflectance, is the integral of reflected spectral radiation, over the entire spectral range, divided by irradiance:

$$\rho = \frac{\int_0^\infty \rho(\lambda)E(\lambda)d\lambda}{E} \quad (3.30)$$

We define the *simulated spectral reflectance* of band λ as

$$\rho_{sim}(\lambda) = \frac{M_r(\lambda)}{M_r + M_e} = \rho(\lambda)\frac{E(\lambda)}{E - B} \quad (3.31)$$

The right side of this equation comprises two components, the reflectance $\rho(\lambda)$ and the ratio of the spectral irradiance of band λ to the simulated irradiance: $E(\lambda)/(E - B)$. This irradiance ratio is approximately a constant for all pixels in the image band λ because both $E(\lambda)$ and $E - B$ vary with topography in a similar way to that defined by eqn. 3.27. As a result, topographic features are suppressed and the image defined by 3.31 is directly proportional to the spectral reflectance image by a constant factor.

Similarly, a *simulated thermal emittance* is defined as

$$\varepsilon_{sim}(\lambda) = \frac{M_e(\lambda)}{M_r + M_e} = \frac{M_e(\lambda)}{E - B} \quad (3.32)$$

Many airborne sensor systems have both multi-spectral bands and thermal bands with the same spatial resolution, such as Airborne Thematic Mapper (ATM) images. In this case the simulated reflectance can be derived from these bands without degrading the spatial resolution. For ATM images, M_e is recorded in a broad thermal band *ATM11* (8–14 μm) and M_r is split and recorded in 10 reflective spectral bands *ATM1–ATM10*. A simulated panchromatic band image M_r can therefore be generated as the weighted sum of the 10 reflective spectral bands:

$$M_r = \sum_{i=1}^{10} w_i ATMi$$

The weights w_i can be calculated either from the sensor gain factors and offsets or using the solar radiation curve and the spectral bandwidths, as described later.

Thus, based on eqn. 3.31, we have an ATM simulated reflectance image:

$$\rho_{sim}(\lambda) = \frac{M_r(\lambda)}{E - B} = \frac{ATM\lambda}{ATM11 + \sum_{i=1}^{10} w_i ATMi} \quad (3.33)$$

where $\lambda = 1 \sim 10$ is the band number.

And a broadband thermal emittance image (ε_{sim}) for ATM is given by

$$\varepsilon_{sim} = \frac{M_e}{E - B} = \frac{ATM11}{ATM11 + \sum\limits_{i=1}^{10} w_i ATMi} \qquad (3.34)$$

Similarly, we can derive the simulated reflectance/emittance images for Landsat TM, ETM+, OLI and ASTER datasets but with degraded spatial resolution because the thermal band resolution of these sensor systems is significantly lower than that of reflective multi-spectral bands.

TM simulated reflectance/emittance:

$$TM_{sim\_\rho\_\varepsilon}(\lambda) = \frac{TM\lambda}{TM6 + \sum\limits_{i=1}^{1\sim5,7} w_i TMi} \qquad (3.35)$$

where $TM_{sim\_\rho\_\varepsilon}$ is the simulated reflectance $\rho_{sim}(\lambda)$ for bands $\lambda = 1 \sim 5,7$ and simulated emittance $\varepsilon_{sim}(\lambda)$ for band $\lambda = 6$.

For ETM+ images we can use the same formula as above. We can also use the panchromatic band (ETM+ Pan) image to replace bands 2, 3 and 4 in eqn. 3.35 because the spectral range of the ETM+ Pan covers the same range of these three bands.

ASTER simulated reflectance/emittance:

$$ASTER_{sim\_\rho\_\varepsilon}(\lambda) = \frac{ASTER\lambda}{\sum\limits_{i=1}^{9} w_i ASTERi + \sum\limits_{j=10}^{14} w_j ASTERj} \qquad (3.36)$$

where $ASTER_{sim\_\rho\_\varepsilon}$ is the simulated reflectance $\rho_{sim}(\lambda)$ for bands $\lambda = 1 \sim 9$ and simulated emittance $\varepsilon_{sim}(\lambda)$ for bands $\lambda = 10 \sim 14$.

3.7.3 Calculation of weights

As described above, the simulated panchromatic ATM image is generated from a weighted sum of all the spectral bands of ATM. In practice, this involves image pre-processing and calculation of weights and it can be carried out in several ways:

1 The standard de-calibration procedure to convert the image DN in each spectral band to radiance using sensor gain and offset:

$$Radiance = \frac{DN}{Gain} - Offset$$

The same conversion should also be performed on the thermal band. Careful atmospheric correction is needed before the summation is implemented.

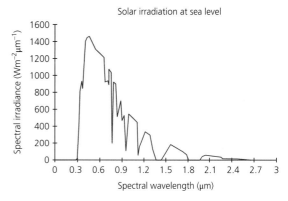

Fig. 3.9 Solar spectral irradiation at sea level (after Fraster 1975).

2 Use of the solar radiation curve (Fig. 3.9) to calculate the weights. The average height of a spectral band in the solar radiation curve is measured and then multiplied by the bandwidth. The product, after re-scaling to a percentage based on the summation of the products for all spectral bands, is the weight for the band. Each image band should be linearly stretched with a proper cut-off at both high and low ends of the histogram before the weight is applied. This stretch roughly removes the effects of atmospheric scattering and makes effective use of the whole DN range of the image (0–255 for 8-bit data). With all the image bands having the same DN range after the stretch, the weights calculated from the solar radiation curve can then be applied to all bands on an equal basis. The resultant simulated panchromatic image should, in principle, be further re-scaled to optimise the albedo cancellation in the summation with the thermal band image so as to generate the simulated irradiance image (the denominator of eqn. 3.34). A real irradiance image should represent variation in topography only, without the influence of albedo. Effective cancellation of albedo in the simulated irradiance image is the key factor for retaining albedo information in the subsequent simulated reflectance image. In practice, the weights can be modified arbitrarily to achieve enhancement of particular spectral features. Table 3.1 and 3.2 give the weights of ATM and ETM+ bands.

3.7.4 Example: ATM simulated reflectance colour composite

Figure 3.10 shows a colour composite of ATM bands 9, 4 and 2 in red, green and blue (a) and the simulated reflectance colour composite of the same bands (b). The

Table 3.1 Weights for ATM bands in the calculation of simulated reflectance.

ATM Bands	1	2	3	4	5	6	7	8	9	10	11
Wavelength (μm)	0.42–0.45	0.45–0.52	0.52–0.60	0.60–0.62	0.63–0.69	0.69–0.75	0.76–0.90	0.91–1.05	1.55–1.75	2.08–2.35	8.50–14.0
$E\left(Wm^{-2}\mu m^{-1}\right)$	1440	1420	1307	1266	1210	909	710	505	148	43	
Weights	0.07	0.16	0.17	0.04	0.12	0.09	0.16	0.12	0.05	0.02	1

Table 3.2 Weights for TM/ETM+ bands in calculation of simulated reflectance.

TM/ETM+ Bands	1	2	3	4	5	6	7
Wavelength (μm)	0.45–0.53	0.52–0.60	0.63–0.69	0.76–0.90	1.55–1.75	10.4–12.5	2.08–2.35
Weights	0.2	0.3	0.2	0.1	0.1	1	0.1

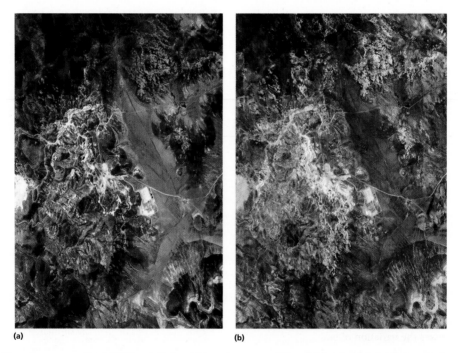

(a)　　　　　　　　　　　　　　　　(b)

Fig. 3.10 (a) The colour composite of ATM bands 9, 4 and 2 in red, green and blue; (b) the simulated reflectance colour composite of ATM bands 9, 4 and 2.

normal colour composite was prepared using the BCET for optimum colour presentation. For the simulated reflectance colour composite, all image bands were linearly stretched with clipping at both ends of the image histograms to make full use of the 8-bit (0–255) value range, and then the weights in Table 3.1 were applied. The resultant simulated reflectance images were linearly re-scaled to 0–255 for colour display.

The colour composite of ATM bands 9, 4 and 2 is generally good for visual interpretation of rock types, alteration minerals and soil/regolith. Figure 3.10 gives the overall impression that the simulated reflectance colour composite (b) has a very similar colour appearance to the normal colour composite (a) but with topography suppressed. This general similarity makes the visual interpretation easy since the image colours relate directly to spectral signatures. Further examination indicates that the simulated reflectance image has more spectral (colour) variations than the normal colour composite. In the simulated reflectance colour composite, the main contribution to image contrast is given by spectral variation rather than topography. This enables the spectral features within very light or very dark (low albedo or topographic shadow) areas to be enhanced effectively.

3.7.5 Comparison with ratio and logarithmic residual techniques

As mentioned before, many techniques, such as ratio, standardization and logarithmic residual, have been developed to enhance information relating to the spectral reflectance of ground objects by suppressing topographic shadowing. All these techniques suffer the limitation that albedo variation is suppressed together with the topographic shadow. These techniques cannot, therefore, effectively differentiate objects with similar spectral profiles but different albedo, such as separation between black and grey. This is because they involve ratio type operations either between the spectral reflectance of different bands or between band spectral reflectance and albedo.

$$Ratio:\quad R_{i,j} = \frac{Band(i)}{Band(j)} = \frac{\rho(i)E(i)}{\rho(j)E(j)}$$

The spectral irradiance ratio $E(i)/E(j)$ is a constant for any pixel in the image so that topographic shading is removed. The reflectance ratio $\rho(i)/\rho(j)$ varies with the spectral signatures of the pixels but cancels out albedo variation because a high albedo object will have a similar reflectance ratio to that of a low albedo object if the two have similar spectral profiles. In other words, the band ratio technique cannot separate albedo from irradiance on the land surface and consequently, the method suppresses the variation of both.

For logarithmic residual, eqn. 3.19 of logarithmic residual operation can be rewritten as

$$R_{i\lambda} = \frac{x_{i\lambda}/x_{i.}}{x_{.\lambda}/x_{..}} = \frac{\rho_i(\lambda)E_i(\lambda)/\rho_i E_i}{\rho_.(\lambda)E(\lambda)/\rho_. E} \qquad (3.37)$$

where
$\rho_i(\lambda)$ is the spectral reflectance of band λ of pixel i.
ρ_i is the albedo of the pixel i.
$\rho_.(\lambda)$ is the average reflectance of the spectral band λ.
$\rho_{..}$ is the average albedo of the whole image.
Other variables are defined as in eqn. 3.19.

The irradiance ratio $E_i(\lambda)/E_i$ is a constant for all pixels and thus independent of position i,

$$\frac{E_i(\lambda)}{E_i} = \frac{E(\lambda)}{E}$$

Thus eqn. 3.37 can be simplified as

$$R_{i\lambda} = \frac{\rho_{..}}{\rho_.(\lambda)} \times \frac{\rho_i(\lambda)}{\rho_i} \qquad (3.38)$$

Equation 3.38 is irrelevant to irradiance, and topographic shadows are therefore eliminated by logarithmic residual processing. This formula is actually a product of two ratios: the ratio of the average albedo of the whole image against the average reflectance of spectral band λ, $\rho_{..}/\rho_.(\lambda)$; and the ratio of pixel spectral reflectance and pixel albedo, $\rho_i(\lambda)/\rho_i$. The ratio $\rho_{..}/\rho_.(\lambda)$ is independent of position and constant for all the pixels in the band λ logarithmic residual image. The variation in a logarithmic residual image is therefore decided only by the ratio $\rho_i(\lambda)/\rho_i$. In a similar way to ratio images, this ratio cancels the variation of albedo.

The advantage of the simulated reflectance technique is that it does not involve ratio operations between spectral reflectance of different bands or between spectral reflectance and albedo. By using the thermal band image to generate a simulated irradiance component, the simulated reflectance technique suppresses topographic shadows but retains albedo information. Thus the spectral information can be better enhanced.

3.8 Summary

In this chapter, we learned about simple arithmetic operations between images and discussed their main applications in image spectral enhancement. The key point to understand is that all image algebraic operations are point-based and performed among the corresponding pixels in different images without involvement of neighbouring pixels. We can therefore regard algebraic operations as *multi-image point operations*.

A major application of image algebraic operations is the selective enhancement of the spectral signatures of intended targets in a multi-spectral image. For this purpose, investigating the spectral properties of these targets is essential to the composition of effective algebraic operations; random attempts are unlikely to be fruitful. This procedure, from spectral analysis to composing an algebraic formula, is generally referred to as *supervised enhancement*. If such a formula is not image dependent and so can be widely used, it is called an *index* image; for instance, the NDVI is a very well-known vegetation index image.

One important issue for spectral enhancement is the suppression of topographic shadowing effects. The ratio-based techniques, standardization and logarithmic residual are based on numerical cancellation of variations in solar illumination. The results are related to ratios of spectral reflectance between different bands whilst variation in albedo will be suppressed together with topography. The simulated reflectance, in contrast, which is based on a simplified physical model of solar radiation on the land surface, presents the properties of true spectral reflectance, including albedo, with only a constant difference.

3.9 Questions

1 Why is an image algebraic operation also known as a multi-image point operation? Write down the mathematical definition of the multi-image point operation.

2 Why can image addition improve the image SNR? If nine photographs of the same scene are taken, using a stationary camera under identical illumination conditions, and then summed to generate an average image, by how many times is the SNR improved in comparison with an individual picture?

3 Describe image difference (subtraction) and ratio (division) operations and then compare the two techniques in terms of change detection, selective enhancement and processing efficiency.

4 How important are the weights in image subtraction? Suggest a suitable pre-processing technique for image differencing.

5 Why does image differencing decrease the SNR?

6 Describe image multiplication and its main application.

7 Explain the characteristics of the value range of a ratio image. Do you think that two reciprocal ratio images contain the same information when displayed after linear scale? Explain why.

8 Using a diagram, describe a ratio image in terms of a coordinate transformation between Cartesian and polar coordinates.

9 Explain the principles of topographic suppression using the image ratio technique.

10 What is NDVI and how it is designed? Is it a ratio index or difference index and why?

11 Describe the design and functionality of Landsat TM or ETM+ iron oxide and clay indices.

12 Try the normalised differencing approach, similar to NDVI, to enhance iron oxide and clay minerals. Compare the results with the relevant ratio indices and explain why the ratio-based indices are more effective for these two types of minerals. (Key: the red edge signal for vegetation is much stronger than the difference between red and blue for iron oxide between the SWIR bands for clay minerals).

13 Describe and compare the standardization and logarithmic residual techniques and their functionalities.

14 What is simulated reflectance? What is the essential condition for the derivation of a simulated reflectance image?

15 Referring to the physical model for the derivation of simulated reflectance, explain why it can be regarded as a true simulation of reflectance.

16 What are the major advantages of simulated reflectance over the ratio, standardization and logarithmic residual techniques?

CHAPTER 4

Filtering and neighbourhood processing

Filtering is a very important research field of digital image processing. All filtering algorithms involve so-called 'neighbourhood processing' because they are based on the relationship between neighbouring pixels rather than a single pixel in point operations.

Digital filtering is useful for enhancing lineaments that may represent significant geological structures such as faults, veins or dykes. It can also enhance image texture for discrimination of lithologies and drainage patterns. For landuse studies, filtering may highlight the textures of urbanisation, road systems and agricultural areas and, for general visualisation, filtering is widely used to sharpen images. However, care should be taken because filtering is not so 'honest' in retaining the information of the original image. It is advisable to use filtered images in close reference to colour composite images or black and white single band images for interpretation.

Digital filtering can be implemented either by 'box filters' based on the concept of convolution in the spatial domain or using Fourier transform (FT) in the frequency domain. In the practical applications of remote sensing, convolution-based box filters are the most useful for their computing efficiency and reliable results. In giving a clear explanation of the physical and mathematical meanings of filtering, FT is essential for understanding the principle of convolution. FT is less computationally efficient for raster data, in terms of speed and computing resources, but it is more versatile than convolution in accommodating various filtering functions.

For point operations, we generally regard an image as a raster data stream and denote $x_{ij} \in X$ as a pixel at line i and column j in image X. As the pixel coordinates in an image are irrelevant for point operations, the subscripts ij

are not involved in the processing and can be ignored. For filtering and neighbourhood processing, however, the pixel coordinates are very relevant and, in this sense, we regard an image as a two-dimensional (2D) function. We therefore follow the convention of denoting an image with a pixel at image column x and line y as a 2D function $f(x, y)$, when introducing essential mathematical concepts of filtering in this chapter. On the other hand, for the simplicity, the expression of $x_{ij} \in X$ is still used in describing some filters and algorithms.

4.1 FT: Understanding filtering in image frequency

The information in an image can be considered as the spatial variations at various frequencies or the assembly of spatial information of various frequencies, as illustrated in Fig. 4.1. Smooth gradual tonal variations represent low frequency information whilst sharp boundaries represent high frequency information. FT is a powerful tool to convert image information from the spatial domain into the frequency domain. Filtering can thus be performed on selected frequencies, and this is particularly useful for removing periodical noise, such as that induced by scanning lines.

First, let us try to understand the physical meaning of Fourier transform based on a simple optical filtering system before we start the slightly boring mathematics. As shown in Fig. 4.2, it is a so-called 4f optical system. The f is the focus length of the lens performing FT and inverse Fourier transform (IFT). As shown in Fig. 4.2(a), given an image $f(x, y)$, where x and y are the spatial coordinates of pixel position, the first lens performs an FT to

Image Processing and GIS for Remote Sensing: Techniques and Applications, Second Edition. Jian Guo Liu and Philippa J. Mason.
© 2016 John Wiley & Sons, Ltd. Published 2016 by John Wiley & Sons, Ltd.

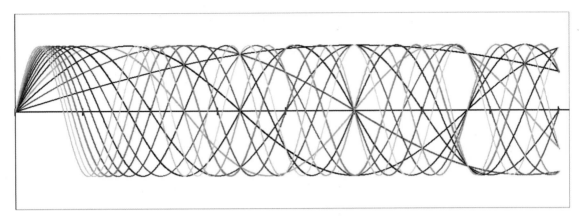

Fig. 4.1 An image can be considered as an assembly of spatial information at various frequencies.

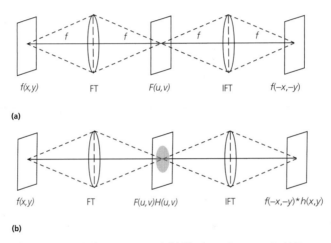

Fig. 4.2 Optical FT system for filtering. (a) Optical FT system; and (b) filtering using an optical FT system.

transform $f(x, y)$, at its front focal plane, to a FT $F(u, v)$, a frequency spectrum with frequencies u in horizontal and v in vertical direction, at its rear focal plane. The second lens then performs IFT to transform $F(u, v)$ at its front focal plane back to the image at its rear focal plane but with a 180° rotation, $f(-x, -y)$. Figure 4.3 shows an image (a) and its FT spectrum plane (b); the frequency increases from the zero at the centre of the spectrum plane to higher and higher frequencies toward the edge of the plane. If a filter (mask) $H(u, v)$ is placed at the rear focal plane of the FT lens, to mask off the signals of particular frequencies in $F(u, v)$, which is equivalent to the operation $F(u, v)H(u, v)$, then a filtered image (with 180° rotation), $g = f(-x, -y) * h(x, y)$, is produced, as shown in Fig. 4.2(b). Such an optical FT system can perform filtering very efficiently and with great flexibility

since the filter can be designed in various arbitrary shapes that are very difficult or impossible to be defined by mathematical functions. Unfortunately, such an analogue approach is limited by the requirements of delicate and expensive optical instruments and very strict laboratory conditions. On the other hand, with rapid progress in computing technology and the development of more efficient and accurate fast Fourier transform (FFT) algorithms, frequency domain filtering has become a common function for most image processing software packages.

From the illustration in Fig. 4.2, we know that the FT-based filtering has three steps:

- FT to transfer an image into the frequency domain;
- remove or alter the data of particular frequencies using a filter;

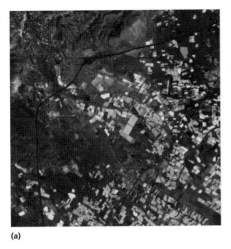 

(a)   (b)

Fig. 4.3 (a) An image $f(x,y)$; and (b) its FT frequency spectrum $F(u,v)$.

- IFT to transfer the filtered frequency spectrum back to the spatial domain to produce a filtered image.

Suppose $f(x,y)$ is an input image and $F(u,v)$ is the 2D FT of $f(x,y)$; the 2D FT and IFT in continuous (C) form are

$$\text{C2D-FT:}\quad F(u,v) = \int_{-\infty}^{+\infty}\int_{-\infty}^{+\infty} f(x,y)e^{-i2\pi(ux+vy)}dxdy \quad (4.1)$$

$$\text{C2D-IFT:}\quad F(x,y) = \int_{-\infty}^{+\infty}\int_{-\infty}^{+\infty} F(u,v)e^{i2\pi(ux+vy)}dudv \quad (4.2)$$

where $i = \sqrt{-1}$.

The 2D FT and IFT in discrete (D) form are:

$$\text{D2D-FT:}\quad F(u,v) = \frac{1}{n}\sum_{x=0}^{n-1}\sum_{y=0}^{n-1} f(x,y)e^{-i2\pi(ux+vy)/n} \quad (4.3)$$

$$\text{D2D-IFT:}\quad f(x,y) = \frac{1}{n}\sum_{u=0}^{n-1}\sum_{v=0}^{n-1} F(u,v)e^{i2\pi(ux+vy)/n} \quad (4.4)$$

where n is the calculation window size.

The operations of FT and IFT are essentially the same, but one is from the image domain to the frequency domain and the other from the frequency domain to the image domain. An important property of FT is known as *convolution theorem*. It states that if F and H are the FT of function f and h, then the FT of the convolution $f*h$ is equal to the product of F and H:

$$\text{FT}(f*h) = \text{FT}(f)\text{FT}(h) = FH \quad (4.5)$$

The inverse form:

$$f*h = \text{IFT}\big[\text{FT}(f)\text{FT}(h)\big] = \text{IFT}(FH) \quad (4.6)$$

or

$$G = FH \quad \text{for} \quad g = f*h \quad \text{and} \quad G = \text{FT}(g) \quad (4.7)$$

This is the key concept of filtering based on FT. F is the frequency representation of the spatial information of image f. If H is a filtering function to change the power of particular frequencies or make off to zero, eqn. 4.6 performs the filtering that changes or removes these frequencies and produces a filtered image $g = f*h$.

4.2 Concepts of convolution for image filtering

From the convolution theorem, we know that the image filtering using FT is equivalent to a convolution between an image $f(x,y)$ and a function $h(x,y)$ that is usually called the *point spread function* (PSF). A 2D convolution is defined as below:

$$g(x,y) = f(x,y)*h(x,y) = \iint f(u,v)h(x-u,y-v)dudv \quad (4.8)$$

Comparing eqn. 4.8 with the convolution theorem (4.6), it is clear that filtering in the image domain by a PSF defined as $h(x,y)$ is equivalent to that in frequency domain by a frequency filtering function $H(u,v)$. The $h(x,y)$ is actually the FT or the 'image' of the frequency filtering function $H(u,v)$. Filtering can therefore be performed directly in the image domain by convolution without involving time-consuming FT and IFT, if the image presentation of a frequency filter can be found. For many standard frequency filtering functions, such as high pass and low pass filters, their images can be derived easily by IFT, as illustrated in Fig. 4.4. It is clear then that convolution is a shortcut for filtering operations.

In the case of discrete integer digital images, the integral form of eqn. 4.8 becomes a summation:

$$g(x,y) = \sum_{u=-\infty}^{+\infty} \sum_{v=-\infty}^{+\infty} f(u,v)h(x-u,y-v) \qquad (4.9)$$

If the range over which the PSF $h(x,y)$ is non-zero is $(-w, +w)$ in one dimension and $(-t, +t)$ in the other, then eqn. 4.9 can be written as

$$g(x,y) = \sum_{u=x-w}^{x+w} \sum_{v=y-t}^{y+t} f(u,v)h(x-u,y-v) \qquad (4.10)$$

In digital image filtering, w and t are the half size of a filter kernel in the horizontal and vertical directions. The pixel of a filtered image, $g(x,y)$, is created by a summation over the neighbourhood pixels $f(u,v)$ surrounding the input image pixel $f(x,y)$ weighted by $h(x-u,y-v)$.

As explained before, the filter kernel, or the PSF $h(x,y)$, is a 2D array or the image of a frequency filter $H(u,v)$ according to the convolution theorem (eqn. 4.6). The numbers in a filter kernel $h(x,y)$ are the weights for summation over the neighbourhood of $f(x,y)$. For the whole image, the convolution filtering is performed by shifting the filter kernel, pixel by pixel, to apply eqn. 4.10 to every pixel of the image being filtered. Although the kernel size may be either an odd number or even number,

an odd number is preferred to ensure the symmetry of the filtering process. A kernel of even number size results in half pixel shift in the filtering result. Commonly used filter kernel sizes are 3 × 3, 5 × 5 or 7 × 7. Rectangular kernels are also used according to particular needs.

Convolution is the theoretical foundation of image domain spatial filtering, but many spatial filters are not necessarily based on the mathematical definition of convolution but on neighbourhood relationships.

For neighbourhood processing in image domain, the image margins of half the size of the processing window can either be excluded from the output or processed together with a mirror copy of the half window size block on the inside of the margins.

4.3 Low pass filters (smoothing)

Smoothing filters are designed to remove high frequency information and retain low frequency information thus reducing the noise but at the cost of degrading image detail. Figure 4.4(a) illustrates a typical low pass filter $H(u,v)$ and the corresponding PSF $h(x,y)$. Most kernel filters for smoothing involve weighted averages among the pixels within the kernel. The larger the kernel, the lower the frequency of information retained. Smoothing based on averaging is effective for eliminating noise pixels, which are often distinguished by very different digital numbers (DNs) from their neighbours but, on the other hand, the process blurs the image as the result of removing the high frequency

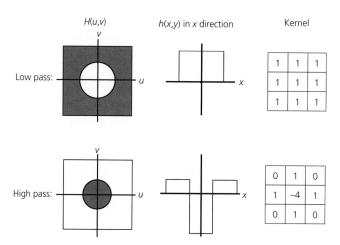

Fig. 4.4 Illustration of low pass and high pass frequency filters and their PSFs.

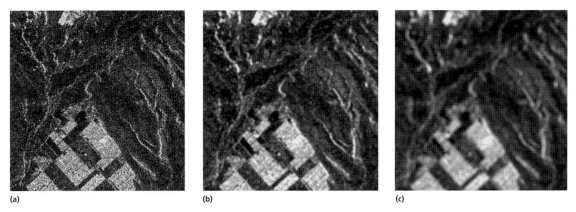

(a)   (b)   (c)

Fig. 4.5 (a) Original image; (b) 5 × 5 mean filter result; and (c) 9 × 9 mean filter result.

information. As illustrated in Fig. 4.5, a synthetic aperture radar (SAR) multi-look image appears noisy because of radar speckle [Fig. 4.5(a)]; the speckle is effectively removed using a 5 × 5 mean filter, so better revealing the ground features [Fig. 4.5(b)]. These features are blurred when a 9 × 9 mean filter is applied [Fig. 4.5(c)]. For removing noise without blurring images, *edge-preserve smoothing* becomes an important research topic of filtering.

The following are examples of 3×3 low pass filter kernels. The size and shape of the kernels can be varied.

Mean filters:

$$\frac{1}{9}\begin{pmatrix} 1 & 1 & 1 \\ 1 & 1 & 1 \\ 1 & 1 & 1 \end{pmatrix} \quad \frac{1}{5}\begin{pmatrix} 0 & 1 & 0 \\ 1 & 1 & 1 \\ 0 & 1 & 0 \end{pmatrix}$$

Weighted mean filters:

$$\frac{1}{16}\begin{pmatrix} 1 & 2 & 1 \\ 2 & 4 & 2 \\ 1 & 2 & 1 \end{pmatrix} \quad \frac{1}{6}\begin{pmatrix} 0 & 1 & 0 \\ 1 & 2 & 1 \\ 0 & 1 & 0 \end{pmatrix}$$

4.3.1 Gaussian filter

A Gaussian filter is a smoothing filter with a 2D Gaussian function as its PSF:

$$G(x,y) = \frac{1}{\sqrt{2\pi}\sigma} e^{\left(\frac{x^2+y^2}{2\sigma^2}\right)} \quad (4.11)$$

where σ is the standard deviation of the Gaussian function.

Figure 4.6 presents the PSF of Gaussian filters with $\sigma = 0.5$, $\sigma = 1.0$ and $\sigma = 2.0$. The Gaussian function is a continuous function. For discrete raster images, the Gaussian filter kernel is a discrete approximation for a given σ. For $\sigma = 0.5$, a Gaussian filter is approximated by a 3×3 kernel, as below:

$$G_{\sigma=0.5} = \begin{pmatrix} 0.0113 & 0.0838 & 0.0113 \\ 0.0838 & 0.6193 & 0.0838 \\ 0.0113 & 0.0838 & 0.0113 \end{pmatrix}$$

Obviously, it is essentially a weighted mean filter.

4.3.2 *K* nearest mean filter

This involves the re-assignment of the value of a pixel x_{ij} of image X to the average of the k neighbouring pixels in the kernel window whose DNs are closest to that of x_{ij}. A typical value of k is 5 for a 3×3 square window. This approach avoids extreme DNs, which are likely to be noise, and ensures their removal. On the other hand, if the pixel in the kernel window is an edge pixel, taking the average of k nearest DNs favours preserving the edge. The k nearest mean is therefore an edge-preserving smoothing filter. As shown in Fig. 4.7(a), the central pixel 0 is very likely to represent noise. The five DNs nearest to 0 are 0, 54, 55, 57 and 58 and the mean is 44.8. The suspected noise DN, 0, at the central pixel is then replaced with the k nearest mean 44.8 that is nearer to 0 than the average 53.4 produced by a mean filter. For the case in Fig. 4.7(b), the central pixel DN 156 is replaced with 158.6, the k nearest mean of 156, 155, 159, 161 and 162. As 158.6 is much nearer to 156 than the average 113 produced by a mean filter, the edge between the pixels in the DN range of 54–58 and those in 155–162 is better preserved and the image is less blurred.

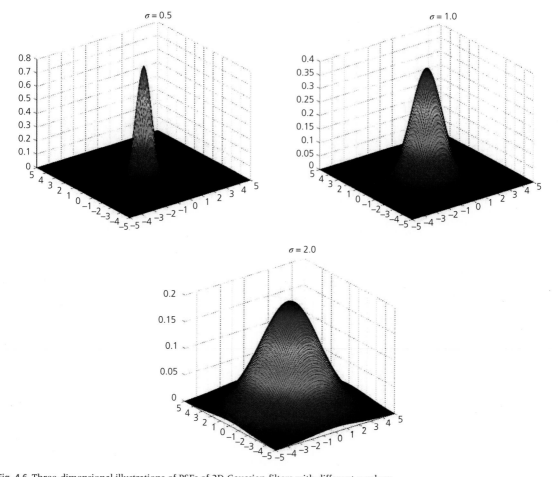

Fig. 4.6 Three-dimensional illustrations of PSFs of 2D Gaussian filters with different σ values.

<table>
<tr><td>55</td><td>58</td><td>65</td></tr>
<tr><td>57</td><td>0</td><td>63</td></tr>
<tr><td>54</td><td>61</td><td>68</td></tr>
</table>

(a)

<table>
<tr><td>55</td><td>58</td><td>155</td></tr>
<tr><td>57</td><td>156</td><td>159</td></tr>
<tr><td>54</td><td>161</td><td>162</td></tr>
</table>

(b)

Fig. 4.7 Two image templates in a 3×3 kernel window to illustrate the effects of edge-preserving filters: (a) a template with a noise pixel 0; and (b) a template with an edge between DNs of 54–58 and 155–162.

4.3.3 Median filter

Here the value of a pixel x_{ij} of image X is re-assigned to the median DN of its neighbouring pixels in a kernel window (e.g. 3×3). We use the same examples in Fig. 4.7 to explain. For image template (a), the DNs are ranked: 0, 54, 55, 57, **58**, 61, 63, 65, 68; the median in this neighbourhood

is 58. The central pixel 0, a suspected isolated noise pixel, is replaced by the median 58, which is a reasonable estimation based on the neighbouring pixels.

For image template (b), the DNs are ranked: 54, 55, 57, 58, **155**, 156, 159, 161, 162; the central DN 156 is replaced with the median value 155, which is very close in value to the original DN. Thus the sharp edge between pixels with DNs in the range of 54–58 and those with DNs ranging from 155 to 162 is preserved. If a mean filter were used, the central pixel DN would be replaced by a value of 113, which is significantly lower than the original DN 156, and as a result, the edge would be blurred. The median filter is therefore an edge-preserving smoothing filter.

In summary, the median filter can remove isolated noise without blurring the image too much. Median filtering can be performed in the vertical or horizontal direction only, if the filter window is one line or one column width instead of a square box.

4.3.4 Adaptive median filter

The adaptive median filter is designed from the basic principle of median filter as follows:

| Median1 | Median2 | Median3 |

It involves the re-assignment of the value of a pixel x_{ij} of image X to the median of the above medians in its three different 3 × 3 neighbourhood patterns. This filter is unlikely to change the DN of a pixel if it is not noise and thus is very effective for edge preservation. Using the same examples in Fig. 4.7, for image template (a), we have:

Median1

55		65
	0	
54		68

Median2

	58	
57	0	63
	61	

Median3

	0	

Thus, Median1 = 55, Median2 = 58 and Median3 = 0, and the median of these three medians is 55; thus the central DN 0 is replaced with 55 and the isolated noise is removed.

For image template (b), we have:

Median1

55		155
	156	
54		162

Median2

	58	
57	156	159
	161	

Median3

	156	

Here, Median1 = 155, Median2 = 156 and Median3 = 156, and the median of these three medians is 156; thus the central DN 156 remains unchanged and the edge is preserved. Obviously, this is a strong edge-preserving smoothing filter. Larger window sizes can also be used with median and adaptive median filters.

4.3.5 *K* nearest median filter

The design of this filter combines the principles of the k nearest mean filter and the median filter. It involves the re-assignment of the value of a pixel x_{ij} of image X to the median of the k neighbour pixels in the kernel window

whose DNs are closest to that of x_{ij}. A typical value of k is 5 for a 3 × 3 square window. Taking the same example in Fig. 4.7, for template (a), the five nearest DNs to the central pixel 0 are: 0, 54, **55**, 57 and 58. The suspected noise DN, 0, at the central pixel is then replaced with the k nearest median 55. This is a more reasonable replacement value and is closer to the neighbourhood of x_{ij} than 44.8 generated by the K nearest mean filter. For template (b), the five nearest DNs to the central pixel 156 are: 155, 156, **159**, 161 and 162. Thus the central pixel DN 156 is replaced by 159. The K nearest median filter is more effective for removing noise and preserving edges than the K nearest mean filter.

4.3.6 Mode (majority) filter

This is a rather 'democratic' filter based on election. A pixel value is re-assigned to the most popular DN among its neighbourhood pixels. This filter performs smoothing based on the counting of pixels in the kernel rather than numerical calculations. Thus it is suitable for smoothing images of non-sequential data (symbols) such as classification images or other discrete raster data. For a 3 × 3 kernel, the recommended majority number is 5. If there is no majority found within a kernel window, then the central pixel in the window remains unchanged.

For example:

6	6	6
6	2	6
5	5	6

There are six pixels with DN = 6, therefore the central DN 2 is replaced by 6. For a classification image, then the numbers in this window are the class numbers and their meaning is no different from class symbols A, B and C. If we use a mean filter, the average of the DNs in the window will be 5.3, but class 5.3 has no meaning in a classification image!

4.3.7 Conditional smoothing filters

Filters of this type have the following general form:

If (some condition)
    Apply filter 1
Else
    Apply filter 2
Endif

Typical examples of this type of filters are *noise cleaning filters*. These filters are all designed based on the assumption that a bad pixel, a bad line or a bad column (data missing or CCD malfunction) in an image will have significantly different DNs from its neighbourhood pixels, lines or columns.

For a 3×3 window neighbourhood of pixel x_{ij}:

$$
\begin{array}{ccc}
x_{i-1,j-1} & x_{i-1,j} & x_{i-1,j+1} \\
x_{i,j-1} & x_{i,j} & x_{i,j+1} \\
x_{i+1,j-1} & x_{i+1,j} & x_{i+1,j+1}
\end{array}
$$

Clean pixels filter: The difference between a bad pixel and the mean of either of its two alternative neighbourhood pixels will be greater than the difference between the two neighbourhood means and thus can be identified and replaced by the mean of its four nearest pixels.

$$
AVE1 = \frac{1}{4}\left(x_{i-1,j-1} + x_{i+1,j+1} + x_{i+1,j-1} + x_{i-1,j+1}\right)
$$

$$
AVE2 = \frac{1}{4}\left(x_{i-1,j} + x_{i+1,j} + x_{i,j+1} + x_{i,j+1}\right)
$$

$$
DIF = |AVE1 - AVE2|
$$

If $:|x_{i,j} - AVE1| > DIF$ and $|x_{i,j} - AVE2| > DIF$,

then $: y_{i,j} = AVE2$

otherwise $: y_{i,j} = x_{i,j}$

Clean lines filter: If an image has a bad line, the difference between a pixel in this line and the mean of either the line above or beneath within the processing window will be greater than the difference between the means of these two lines. Then the bad line pixel can be replaced by the average of the pixels above and beneath it. As such, a bad line is replaced by the average of the pixels above and beneath it. Figure 4.8 illustrates the effects of the filter.

$$
AVE1 = \frac{1}{3}\left(x_{i-1,j-1} + x_{i-1,j} + x_{i-1,j+1}\right)
$$

$$
AVE2 = \frac{1}{3}\left(x_{i+1,j-1} + x_{i+1,j} + x_{i+1,j+1}\right)
$$

$$
DIF = |AVE1 - AVE2|
$$

If $:|x_{i,j} - AVE1| > DIF$ and $|x_{i,j} - AVE2| > DIF$,

then $: y_{i,j} = \left(x_{i-1,j} + x_{i+1,j}\right)/2$

otherwise $: y_{i,j} = x_{i,j}$

Clean columns filter: This filter is designed with the same logic as the above but in the vertical direction.

$$
AVE1 = \frac{1}{3}\left(x_{i-1,j-1} + x_{i,j-1} + x_{i+1,j-1}\right)
$$

$$
AVE2 = \frac{1}{3}\left(x_{i-1,j+1} + x_{i,j+1} + x_{i+1,j+1}\right)
$$

$$
DIF = |AVE1 - AVE2|
$$

If $:|x_{i,j} - AVE1| > DIF$ and $|x_{i,j} - AVE2| > DIF$,

then $: y_{i,j} = \left(x_{i,j-1} + x_{i,j+1}\right)/2$

otherwise $: y_{i,j} = x_{i,j}$

4.4 High pass filters (edge enhancement)

Edges and textures in an image are typical examples of high frequency information. High pass filters remove low frequency image information and therefore enhance high frequency information such as edges. Most commonly used edge enhancement filters are based on first and second derivatives or Gradient and Laplacian. Given an image $f(x, y)$,

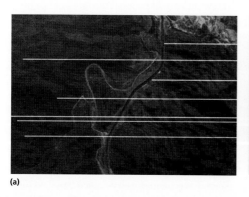

(a)

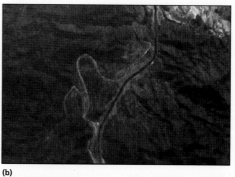

(b)

Fig. 4.8 (a) An Airborne Thematic Mapper (ATM) image is degraded by several bad lines as the result of occasional sensor failure; and (b) the bad lines are successfully removed by the clean lines filter.

Gradient : $\quad \nabla f = \dfrac{\partial f(x,y)}{\partial x}\,\overline{i} + \dfrac{\partial f(x,y)}{\partial y}\,\overline{j}$ (4.12)

where $\overline{i}$ and $\overline{j}$ are unit vectors in the x and y directions.

Laplacian : $\quad \nabla^2 f = \dfrac{\partial^2 f(x,y)}{\partial x^2} + \dfrac{\partial^2 f(x,y)}{\partial y^2}$ (4.13)

It is important to notice that the two types of high pass filters work in different ways. Gradient is the first derivative at pixel $f(x,y)$ and as a measurement of DN change rate, it is a vector characterising the maximum magnitude and direction of the DN slope around the pixel $f(x,y)$. Laplacian, as the second derivative at pixel $f(x,y)$, is a scalar that measures the rate of change in gradient. In plain words, Laplacian describes the curvature of a slope but not its magnitude and direction (discussed again in § 17.4.2). As shown in Fig. 4.9, a flat DN slope has a constant gradient but zero Laplacian because the change rate of a flat slope

is zero. For a slope with a constant curvature (an arc of a circle), the gradient is a variable while the Laplacian is a constant. Only for a slope with varying curvature, both gradient and Laplacian are variables. This is why Laplacian suppresses all the image features except sharp boundaries where DN gradient changes dramatically, whilst gradient filtering retains boundary as well as slope information.

An output image of high pass filtering is usually no longer in the 8-bit positive integer range and so must be rescaled based on the actual limits of the image to 0–255 for display.

4.4.1 Gradient filters

The numerical calculation of gradient based on eqn. 4.12 is a simple differencing between a pixel under filtering and its neighbour pixels divided by the distance in between. Gradient filters are always in pairs to produce x component (g_x) and y component (g_y) or components in diagonal directions.

$$g_x = \frac{f(x,y)-f(x+\delta x,y)}{\delta x} \quad g_y = \frac{f(x,y)-f(x,y+\delta y)}{\delta y}$$

(4.14)

For instance, the gradient between a pixel $f(x,y)$ and its next neighbour on the right $f(x+1,y)$ is $g_x = f(x,y)-f(x+1,y)$; here the increment for a raster data is $\delta x = 1$ [Fig. 4.10(a)].

We present several of the most commonly used gradient filter kernels that are based on the principles of the equations in 4.14, assuming an odd number kernel size.

Gradient filters:

$$g_x = \begin{pmatrix} 0 & -1 & 1 \end{pmatrix} \quad g_y = \begin{pmatrix} 0 \\ -1 \\ 1 \end{pmatrix}$$

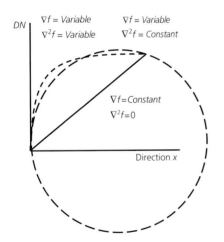

Fig. 4.9 Geometric meaning of first and second derivatives.

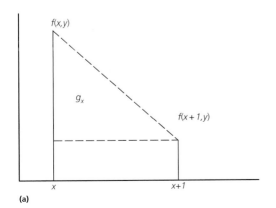

(a)

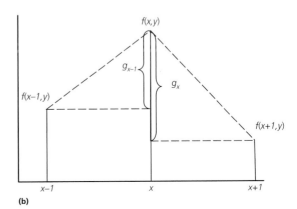

(b)

Fig. 4.10 (a) Calculation of gradient in x direction g_x; and (b) the calculation of the x component of Laplacian, $\nabla^2 f_x$.

Prewitt filters:

$$\begin{pmatrix} -1 & 0 & 1 \\ -1 & 0 & 1 \\ -1 & 0 & 1 \end{pmatrix} \begin{pmatrix} -1 & -1 & -1 \\ 0 & 0 & 0 \\ 1 & 1 & 1 \end{pmatrix}$$

$$\text{Or} \quad \begin{pmatrix} -1 & -1 & 0 \\ -1 & 0 & 1 \\ 0 & 1 & 1 \end{pmatrix} \begin{pmatrix} 0 & 1 & 1 \\ -1 & 0 & 1 \\ -1 & -1 & 0 \end{pmatrix}$$

Sobel filters:

$$\begin{pmatrix} -1 & 0 & 1 \\ -2 & 0 & 2 \\ -1 & 0 & 1 \end{pmatrix} \begin{pmatrix} -1 & -2 & -1 \\ 0 & 0 & 0 \\ 1 & 2 & 1 \end{pmatrix}$$

$$\text{Or} \quad \begin{pmatrix} -1 & -2 & 0 \\ -2 & 0 & 2 \\ 0 & 2 & 1 \end{pmatrix} \begin{pmatrix} 0 & 2 & 1 \\ -2 & 0 & 2 \\ -1 & -2 & 0 \end{pmatrix}$$

The magnitude g_m and orientation g_a of a gradient can be computed from g_x and g_y:

$$g_m = \sqrt{g_x^2 + g_y^2} \quad g_a = \arctan\left(g_y / g_x\right) \quad (4.15)$$

If we apply eqn. 4.15 to a digital elevation model (DEM), the g_m produces a slope map while g_a an aspect map of topography (see also § 17.4.1).

Figure 4.11 illustrates the results of Sobel filters. The g_x image in (b) enhances the vertical edges while the g_y image in (c) enhances horizontal edges. A filtered image is no longer a positive integer image; it is composed of both positive and negative real numbers, as gradient can be either positive or negative depending on if DN is changing from dark to bright or vice versa.

4.4.2 Laplacian filters

As the second derivative, we can consider Laplacian as the difference in gradient. Equation 4.13 is composed of two parts: secondary partial derivative in x direction and y direction. We can then rewrite 4.13 as

$$\nabla^2 f = \nabla^2 f_x + \nabla^2 f_y \quad (4.16)$$

Let's consider the x direction as shown in Fig. 4.10(b).
• The gradient at position $x-1$: $g_{x-1} = f(x-1,y) - f(x,y)$
• The gradient at position x: $g_x = f(x,y) - f(x+1,y)$

$$\nabla^2 f_x = g_{x-1} - g_x$$

• Thus we have: $\qquad = f(x-1,y) + f(x+1,y)$
$\qquad\qquad\qquad -2f(x,y)$

• Similarly in the y direction, we have: $\nabla^2 f_x = g_{y-1} - g_y$
$= f(x,y-1) + f(x,y+1) - 2f(x,y)$

• Thus: $\nabla^2 f = f(x-1,y) + f(x+1,y) + f(x,y-1) +$
$f(x,y+1) - 4f(x,y)$

The above calculation equation can be translated into a standard Laplacian filter kernel:

$$\begin{pmatrix} 0 & 1 & 0 \\ 1 & -4 & 1 \\ 0 & 1 & 0 \end{pmatrix}$$

A more commonly used equivalent form of Laplacian filter is:

$$\begin{pmatrix} 0 & -1 & 0 \\ -1 & 4 & -1 \\ 0 & -1 & 0 \end{pmatrix}$$

(a)

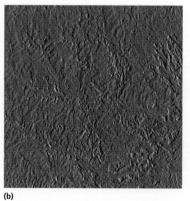
(b)

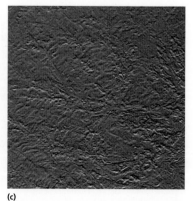
(c)

Fig. 4.11 Illustration of an image (a) and its g_x and g_y of Sobel filters in (b) and (c).

If we also consider the diagonal directions, then the Laplacian filter is modified as

$$\begin{pmatrix} -1 & -1 & -1 \\ -1 & 8 & -1 \\ -1 & -1 & -1 \end{pmatrix}$$

In general, we can consider Laplacian filtering for raster image data as a summation of all the differences between a pixel $f(x,y)$ and its neighbouring pixels $f(x+\delta x, y+\delta y)$.

$$\nabla^2 f = \sum_{\delta x=-1}^{1} \sum_{\delta y=-1}^{1} \left[f(x,y) - f(x+\delta x, y+\delta y) \right] \quad (4.17)$$

The Laplacian filter produces an image of edges (Fig. 4.12). The histogram of such an image is typically symmetrical to a high peak at zero with both positive and negative values [Fig. 4.12(c)]. It is important to remember that both very negative and very positive values are edges. As implied by the Laplacian kernels, if the DN of the central pixel in the Laplacian kernel is higher than those of its neighbouring pixels, the Laplacian is positive, indicating a convex edge; otherwise, if the central pixel in the Laplacian kernel is lower than those of its neighbour pixels, the Laplacian is negative, indicating a concave edge.

4.4.3 Edge-sharpening filters

Increasing the central weight of the Laplacian filter by k is equivalent to adding k times of original image back to the Laplacian filtered image. The resultant image is similar to the original image but with sharpened edges. The commonly used add-back Laplacian filters (also called edge-sharpening filters) are:

$$\begin{pmatrix} 0 & -1 & 0 \\ -1 & 10 & -1 \\ 0 & -1 & 0 \end{pmatrix} \begin{pmatrix} -1 & -1 & -1 \\ -1 & 14 & -1 \\ -1 & -1 & -1 \end{pmatrix}$$

The central weight can be changed arbitrarily to control the proportion between the original image and the edge image. This simple technique is popular not only for remote sensing imagery but also to commercial digital picture enhancement for photographic products.

4.5 Local contrast enhancement

The PSF $h(x,y)$ can be dynamically defined by the local statistics. In this case, $h(x,y)$ is no longer a pre-defined fixed function as the cases of smoothing, Gradient, and Laplacian filters. It varies with $f(x,y)$ according to image local statistics. This branch of filtering techniques is called adaptive filtering. Adaptive filtering can be used for contrast enhancement, edge enhancement and edge-preserving smoothing as well as noise removal.

One typical adaptive filter is the local contrast enhancement. The purpose of local contrast enhancement is to produce the same contrast in every local region throughout an image. An adaptive algorithm is employed to adjust parameters of a point operation function pixel by pixel, based on local statistics, so as to achieve contrast enhancement. This technique represents the combination of a point operation and neighbourhood processing. We therefore introduce it in this chapter rather than in Chapter 2 on point operations.

The simplest local contrast enhancement is the local mean adjustment technique. Here we use the same notations as in Chapter 2 to better illustrate the

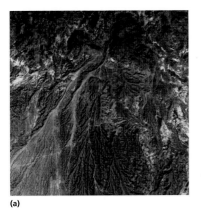

(a)

(b)

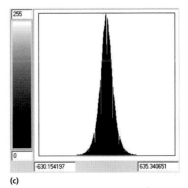

(c)

Fig. 4.12 Illustration of an image (a) and its Laplacian filtering result (b), together with the histogram of the Laplacian image (c).

processing as a contrast enhancement. Let $\bar{x}_{ij}$ be the local mean in some neighbourhood of pixel x_{ij}, say a 31×31 square window centred at (i, j), then

$$y_{ij} = x_{ij} + m_o - \bar{x}_{ij} \qquad (4.18)$$

This technique adjusts local brightness to the global mean m_o of the image while leaving the local contrast unchanged. For the whole image, the processing may reduce the image global contrast (standard deviation) but will maintain the average brightness (global mean).

Let $\bar{x}_{ij}$ and σ_{ij} be the local mean and local standard deviation in some neighbourhood of pixel x_{ij}, then a local contrast enhancement algorithm using a linear function is defined as follows:

$$y_{ij} = \bar{x}_{ij} + \left(x_{ij} - \bar{x}_{ij} \right) \frac{\sigma_o}{\sigma_{ij} + 1} \qquad (4.19)$$

where 1 in the denominator is to prevent overflow when σ_{ij} is almost 0.

This local enhancement function stretches $x(i, j)$ to achieve a pre-defined local standard deviation σ_o. In addition, the local mean can also be adjusted by modifying eqn. 4.19 as below:

$$y_{ij} = am_o + \left(1 - a\right)\bar{x}_{ij} + \left(x_{ij} - \bar{x}_{ij} \right) \frac{\sigma_o}{\sigma_{ij} + 1} \qquad (4.20)$$

where m_o is the mean to be enforced locally and $0 \leq \alpha \leq 1$ is a parameter to control the degree to which it is enforced.

The function defined by eqn. 4.20 will produce an image with a local mean m_o and local standard deviation σ_o everywhere in the image. The actual local mean m_o and local standard deviation σ_o vary from pixel to pixel in a certain range depending on the strength of parameter α.

It is important to keep in mind that local contrast enhancement is not a point operation. It is essentially a neighbourhood processing. This technique may well enhance localised subtle details in an image but it will not preserve the original image information. As illustrated in Fig. 4.13, the global contrast and the obvious regional tonal variation in image (a) is suppressed with some subtle local textures enhanced in image (b).

4.6 FFT selective and adaptive filtering*

This section presents our research (Liu & Morgan 2006) illustrating how filters are designed based on the spatial pattern of targeted periodic noise in order to remove it.

Remotely sensed images or products derived from these images can be contaminated by systematic noise of particular frequency or frequencies that vary according to some function relating to the sensor or imaging

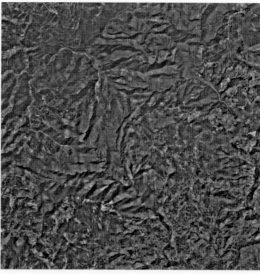

(a)   (b)

Fig. 4.13 Effects of local enhancement: (a) original image; and (b) local enhancement image.

configuration. To remove this type of noise patterns, FFT filtering in frequency domain is the most effective approach.

As shown in Fig. 4.14(a), the image is a feature matching result of a pair of Landsat-7 ETM+ panchromatic images across a major earthquake event, aiming to reveal the co-seismic shift at sub-pixel accuracy (Liu *et al.* 2006). Severe horizontal and vertical striping noise patterns plague the image and seriously obscure the desired surface shift information.

4.6.1 FFT selective filtering

In-depth investigation indicates that the horizontal striping is caused by Landsat ETM+'s two-way scanning in conjunction with the orbital drift between two image acquisitions. Figure 4.15 (right) shows a small sub-scene of an area in the image that is relatively homogenous. In general, the clear striping pattern is representative of the entire image, displaying equally spaced lighter and darker bands. This regular noise pattern was significantly enhanced by applying a 1 line × 101 column smoothing filter to remove much of the scene content (Fig. 4.15 left), revealing the striping noise of fixed frequency. This type of noise in a certain frequency can be removed by frequency domain filtering. After converting the image $f(x,y)$ into a frequency spectrum $F(u,v)$ via FFT, the key point for a successful filtering is to locate the representative frequency spectrum 'spots' corresponding to the periodic noise pattern and mask them off with a function $H(u,v)$ before making the inverse FFT back to an image. Such FFT selective filtering comprises the following steps (Fig. 4.16).

Procedure for FFT selective filtering:

1 A single sub-scene (e.g. 1024 × 1024 pixels) of the image is selected from an area with relatively homogenous scene content and clear striping.

2 The stripes are enhanced by a 1 × 101 one-dimensional horizontal smoothing filter.

3 The stripe-enhanced sub-scene is then thresholded to create a black and white binary image.

4 The binary stripe image is transformed to the frequency domain, using the 2D FFT, and the power spectrum is calculated to give the absolute magnitude of the frequency components.

5 The spectrum magnitude is normalized to 0–1 after masking off the dominant zero and very near zero frequencies that always form the highest magnitudes

in the frequency spectrum but carry no periodic noise information. The frequencies of the highest magnitude peaks that relate to specific noise components can then be located and retained by thresholding.

6 Inverting this result (i.e. interchanging 0s for 1s and 1s for 0s) creates a binary mask for selective frequency filtering. The zero and very near zero frequencies are set to 1 in the mask to retain the natural scene content in filtering. Additionally, this initial binary filter function is convolved with an appropriately sized Gaussian pulse to eliminate the possibility of ringing artefacts.

7 The mask generated from the selected sub-scene can then be used on the entire noisy image, because the identified striping frequencies are independent of position in the image. The filtering operation is done simply by multiplying the mask with the 2D FFT of the entire image.

8 Finally, the frequency filtering result is transformed back to the image domain, via the 2D inverse FFT, to produce a de-striped image.

4.6.2 FFT adaptive filtering

As shown in Figs. 4.17 and 4.14(b), after the successful removal of the horizontal stripes, the vertical noise pattern in the horizontally filtered image is more clearly revealed than before. The noise is not simple stripe-like but instead forms a series of parallel, vertically aligned wavy lines with progressively increasing frequency from the middle to both edges of the image [Fig. 4.14(b)]. Analysis and simulation of the changing frequency of the noise indicate that it is caused by the transition from the fixed angular speed (2.21095 rads/sec) of ETM+ scanner to the line speed on the curved earth surface, as depicted in Fig. 4.18. The surface scanning speed, with a fixed angular speed of the rotating scanner mirror, varies symmetrically across the swath width about the nadir point, accelerating out towards the edges as defined by eqn. 4.21 derived based on the diagram in Fig. 4.18.

The scanning speed on the earth's surface dl/dt varies as a function of θ or the scanning angular speed $d\theta/dt$.

By the sine theorem: $\dfrac{R}{\sin\theta} = \dfrac{R+r}{\sin\alpha}$

Then, $\alpha = \arcsin\left[\dfrac{R+r}{R}\sin\theta\right]$

As $\varphi = \pi - \alpha - \theta$, thus $\varphi = \pi - \theta - \arcsin\left[\dfrac{R+r}{R}\sin\theta\right]$

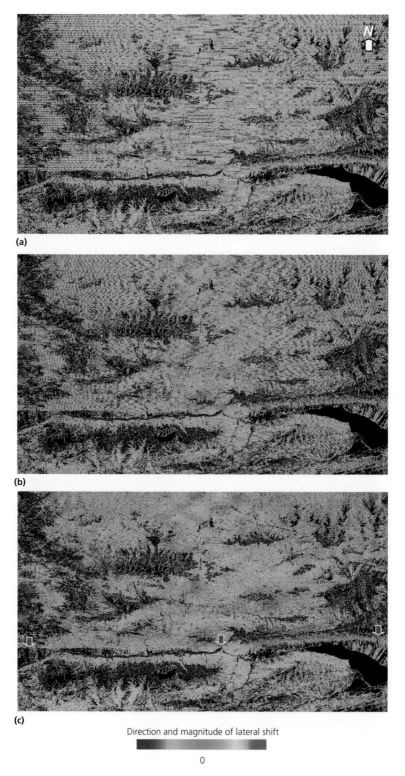

(a)

(b)

(c)

Direction and magnitude of lateral shift

0

Fig. 4.14 (a) The original co-seismic shift image; (b) the FFT selective filtering result – the horizontal stripes have been successfully removed; however, the vertical noise is now very noticeable; and (c) the FFT adaptive filtering result – the multiple frequency wavy patterns of vertical stripes have been successfully removed and thus clearly revealed the phenomena of regional co-seismic shift along the Kunlun fault line as indicated by three arrows. Yellow-red indicates movement to the right and cyan-blue indicates movement to the left.

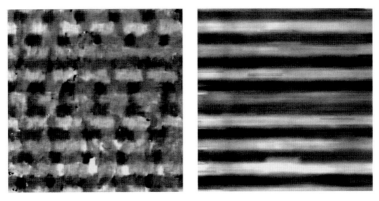

Fig. 4.15 A 500 × 500 pixel sub-scene of the image (left); and (right) the image has been filtered with a one-dimensional, horizontal smoothing filter (kernel size 1 × 101) to isolate the noise and aid analysis.

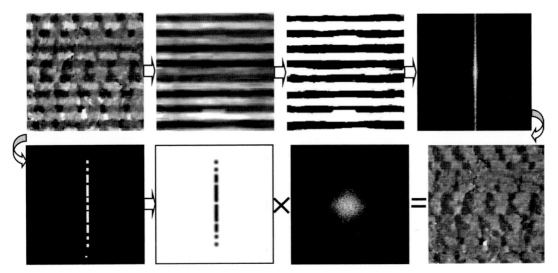

Fig. 4.16 A graphical illustration, step-by-step, of the selective frequency-filtering algorithm for the removal of horizontal stripes in the image.

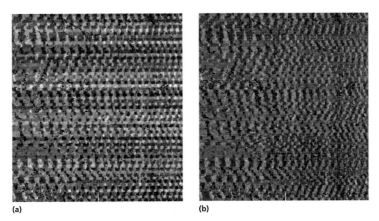

(a)   (b)

Fig. 4.17 After the selective horizontal filtering applied to the original image (left) to remove horizontal stripes successfully, the vertical wavy stripe noise becomes more obvious in the filtered image (right). Image size: 2000 × 2000.

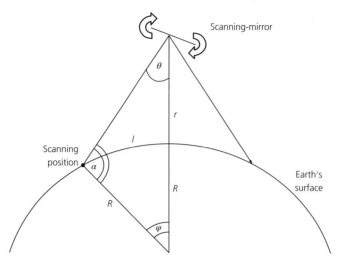

Fig. 4.18 The geometric relationship between a constant angular speed of cross-track scanning and the corresponding line speed on the curved surface of the earth.

Given the arc length $l = R\varphi$, then $\dfrac{dl}{dt} = R\dfrac{d\varphi}{dt} = R\dfrac{d\varphi}{d\theta}\cdot\dfrac{d\theta}{dt}$, thus

$$\frac{dl}{dt} = -R\frac{d\theta}{dt}\left(1 + \frac{(R+r)\cos\theta}{R\sqrt{1 - \left(\dfrac{R+r}{R}\sin\theta\right)^2}}\right) \quad (4.21)$$

where

θ: Scanning angle (variable).

φ: Angle subtended by two earth radii, corresponding to θ.

α: Angle between the scanning line of sight and the earth's radius at the scanning position.

r: Height of sensor above the earth's surface.

R: Radius of earth.

l: Scanning length on earth's surface for angle θ from nadir to scanning position, equivalent to the arc length between earth's radii.

Within the scanning range of the ETM+ sensor system, the function defined by eqn. 4.21 can be precisely simulated by a second order polynomial least square fit, resulting in a parabolic curve (Fig. 4.19):

$$f(x) = 4.586\times10^{-10}x^2 - 1.368\times10^{-7}x + 0.0019 \quad (4.22)$$

where x is the image column position.

For processing efficiency, this simple parabolic function is used to adapt the filter aperture in the frequency domain to remove vertical noise patterns of differing frequencies at different image columns. The idea is that each image column should be filtered according to a noise frequency calculated from eqn. 4.22 at the corresponding scanning angle θ, so the image will be adaptively filtered, column by column, with decreasing frequency from left to middle, and then increasing frequency from middle to right.

In practice, the wavy form of the stripes makes this a truly 2D filtering problem, in which frequency components will be located not only on the horizontal frequency axis but also at locations diagonal to the spectrum centre. The adaptive filter design must therefore be based on the concept of a circularly symmetric band-reject filter of the Butterworth type. The transfer function for a Butterworth band-reject filter of order n is defined as below:

$$H(u,v) = \frac{1}{1 + \left[\dfrac{D(u,v)W}{D^2(u,v) - D_0^2}\right]^{2n}} \quad (4.23)$$

where $D(u,v)$ is the distance from the point (u,v) to the centre origin of the frequency spectrum, D_0 is the radius of the filter defined as the principal frequency to be filtered out and n is the order of the filter roll-off either side of the central ring defined by D_0 forming the bandwidth W.

For an $M \times N$ image section and its subsequent FT, $D(u,v)$ is given by

$$D(u,v) = \left[(u - M/2)^2 + (v - N/2)^2\right]^{1/2} \quad (4.24)$$

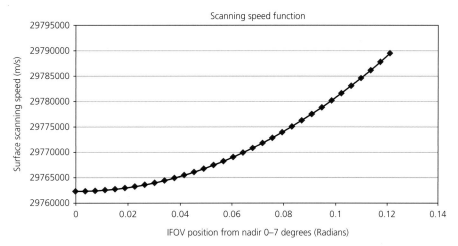

Fig. 4.19 The plot of the earth's surface scanning speed function as defined by eqn. 4.21 in the scanning angular range 0°–7° (in diamond markers) fitted with the parabolic curve defined in eqn. 4.22. IFOV, instant field of view.

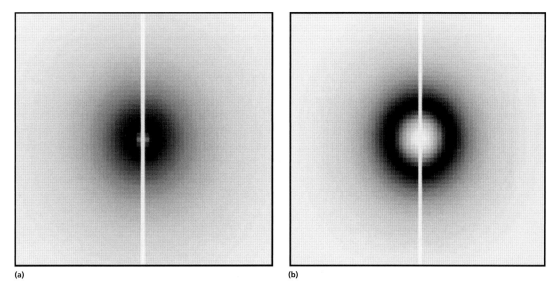

(a)   (b)

Fig. 4.20 Two 512×512 adaptive Butterworth band-reject filters defined by eqn. 4.23. Left: $D_0 = 3, W = 3, n = 3$; and right: $D_0 = 8, W = 3, n = 3$. The filters have been edited to allow the central vertical frequencies through without suppression because these components have undergone previous horizontal filtering. The radius of the filter D_0 varies with the image column position according to eqn. 4.22.

The filter function, $H(u, v)$, thus removes a band of frequencies at a radial distance, D_0, from the frequency spectrum centre whilst smoothly attenuating the frequencies either side of D_0 to reduce the possibility of ringing. The key aspect of this adaptive filter design is that D_0 is decided by the scanning function eqn. 4.21 or 4.22. Thus D_0 varies column by column given the

scanning angle θ corresponding to the image column position (Fig. 4.20).

The FFT frequency adaptive filtering procedure is as below:

1 Starting from the left edge of the image, FFT for a 512×512 neighbourhood is applied to produce $F(u, v)$.

2 The noise frequency at the central column of the 512 × 512 neighbourhood is calculated using formula (4.22), given the column position.

3 In the Fourier domain, the noise frequency in $F(u,v)$ is masked off by multiplication with the filter transfer function $H(u,v)$ to produce $F(u,v)H(u,v)$.

4 Inverse FFT to transform the filtered frequency spectrum $F(u,v)H(u,v)$ back to an image $f(x,y)*h(x,y)$, but only retaining the central column.

5 Move to the next column and repeat steps 1–4 till the last column is reached.

6 Move down to the next 512 block below and repeat steps 1-5 till the end of the image is reached.

After the FFT selective and adaptive filtering the horizontal and vertical noise patterns are effectively removed, as shown in Fig. 4.14(c), and the left lateral co-seismic shift along the Kunlun fault is clearly revealed.

4.7 Summary

Image filtering is a process to remove image information of particular frequencies. Within this context, it is typical of signal processing in the frequency domain via FT. On the basis of the convolution theorem, FT-based filtering can be performed in the image domain, using convolution and realised by neighbourhood processing in an image. The most commonly used filters for digital image filtering rely on the concept of convolution and operate in the image domain for reasons of simplicity and processing efficiency.

Low pass filters are mainly used to smooth image features and to remove noise but often at the cost of degrading image spatial resolution (blurring). To remove random noise with the minimum degradation of resolution, various edge-preserving filters have been developed, such as the adaptive median filter. A classification image is a 'symbol' image rather than a digital image and therefore should not be subject to any numerical operations. The mode (majority) filter is suitable for smoothing a classification image, as the filtering process is based on the election of local majority within the processing window without numerical operations.

There are two different types of high pass filters: gradient filters and Laplacian. As the first derivative of DN change in a direction, the gradient gives a measurement of DN slope. Gradient is a vector and so gradient filters are directional; they are commonly used as orthogonal pairs for directional enhancement. Images representing the magnitude and orientation of gradient can be calculated from the pair of images derived from

the orthogonal filters. Laplacian, as the second derivative, is a scalar that measures the change rate of DN slope. Image edge features are characterised as significant DN slope changes, and Laplacian is therefore very effective for enhancing and extracting them. One of the most common applications of Laplacian for not only remote sensing image processing but also general graphic enhancement is the so-called *edge-sharpening filter*.

Combining neighbourhood processing with point operations for contrast enhancement formulates a new method of image processing, local contrast enhancement. It adjusts the image contrast based on local statistics calculated in a neighbourhood. This is based on the concept of a more general branch of neighbourhood processing: adaptive filters. As examples, image characteristics-based derivation of FFT selective and adaptive filters is introduced at the end of this chapter for advanced readers.

Our general advice on all filtering and neighbourhood processing is not to trust them blindly! Artefacts can be introduced so use of the original images as a reference is always recommended.

4.8 Questions

1 Use a diagram to illustrate the $4f$ optical image filtering system and explain the principle of image filtering based on FT.

2 What is the convolution theorem and what is its importance in digital image filtering?

3 Explain the relationship between the filtering function $H(u,v)$ in the frequency domain and the PSF function $h(x,y)$ in spatial (image) domain.

4 If the range over which the PSF $h(x,y)$ is non-zero is $(-w,+w)$ in one dimension and $(-t,+t)$ in the other, write down the discrete form of convolution $f(x,y)*h(x,y)$.

5 What is a low pass filter for digital image filtering, and what are its effects? Give some examples of low pass filter kernels.

6 Discuss the major drawback of mean filters and the importance of edge-preserving smoothing filters.

7 To smooth a classification image, what filter is appropriate and why? Describe this filter with an example.

8 Give a general definition of conditional filters.

9 Describe the clean pixels filter and explain how it works.

10 Describe K nearest mean filter, median filter and adaptive median filter and discuss their merits based on the filtering results of the sample image below:

173	140	124	113	100
167	145	136	18	83
138	252	122	96	117
144	134	83	87	116
137	115	95	119	142

Median filter			Adaptive median filter			K nearest mean filter		

11 What is it meant by high pass filtering?

12 Describe the mathematical definitions of image gradient and Laplacian together with examples of gradient filters and Laplacian filters.

13 Use a diagram to illustrate and explain the different functionalities of gradient and Laplacian-based high pass filters.

14 Given a DEM, how does one calculate the slope and aspect of topography using gradient filters?

15 Why is the histogram of a Laplacian filtered image symmetrical to a high peak at zero with both positive and negative values?

16 What is an edge-sharpening filter? What are the major applications of edge-sharpening filters?

17 Describe local contrast enhancement technique as a neighbourhood processing procedure and explain why it is not a point operation.

CHAPTER 5

RGB-IHS transformation

In this chapter, we first describe the principles of the RGB-IHS and IHS-RGB transformations. Two de-correlation stretch techniques, both based on saturation stretch, are then discussed. Finally, a hue RGB (HRGB) colour composition technique is introduced. The RGB-IHS transformation is also a powerful tool for data fusion, but we leave this part to Chapter 6, along with the discussion of several other data fusion techniques.

5.1 Colour co-ordinate transformation

A colour is expressed as a composite of three primaries: **R**ed, **G**reen and **B**lue, according to the tristimulus theory. For colour perception, on the other hand, a colour is quantitatively described in terms of three variables: **I**ntensity, **H**ue and **S**aturation, which are measurements of the brightness, spectral range and purity of a colour. There are several variants of RGB-IHS transformation based on different models. For the RGB additive colour display of digital images, a simple RGB colour cube is the most appropriate model. The RGB-IHS colour co-ordinate transformation in a colour cube is similar to a three-dimensional (3D) Cartesian-conical co-ordinate transformation.

As shown in Fig. 5.1, any colour in a three-band colour composite is a vector $\mathbf{P}(r, g, b)$ within a colour cube of 0–255 in three dimensions (for 24 bits RGB colour display). The major diagonal line connecting the origin and the farthest vertex is called the *grey line* because the pixels lying on this line have equal components in red, green and blue ($r = g = b$). The intensity of a colour vector $\mathbf{P}$ is defined as the length of its projection on the grey line, OD, the hue the azimuth angle around the grey line, α, and the saturation the angle between the colour vector $\mathbf{P}$ and the grey line, φ. Let the hue angle of pure blue colour be zero; we then have the following RGB-IHS transformation:

$$I(r, g, b) = \frac{1}{\sqrt{3}}(r + g + b) \tag{5.1}$$

$$H(r, g, b) = \arccos\frac{2b - g - r}{2V} \tag{5.2}$$

$$where \quad V = \sqrt{\left(r^2 + g^2 + b^2\right) - \left(rg + rb + gb\right)}$$

$$S(r, g, b) = \arccos\frac{r + g + b}{\sqrt{3\left(r^2 + g^2 + b^2\right)}} \tag{5.3}$$

Saturation as defined in eqn. 5.3 can then be re-written as a function of intensity:

$$S(r, g, b) = \arccos\frac{I(r, g, b)}{\sqrt{r^2 + g^2 + b^2}} \tag{5.4}$$

For the same intensity, if $r = g = b$, saturation reaches its minimum, $S(r, g, b) = \arccos 1 = 0$; if two of r, g, b are equal to zero, the colour is a pure primary and saturation reaches its maximum, $S(r, g, b) = \arccos\frac{1}{\sqrt{3}} \approx 54.7356°$. Actually, saturation of a colour is the ratio between its achromatic and chromatic components; so saturation increases with the increase of difference between r, g and b. It can therefore be defined by the maximum and minimum of r, g and b in a value range from no saturation (0) to full saturation (1), as in the formula below (Smith 1978):

$$S(r, g, b) = \frac{Max(r, g, b) - Min(r, g, b)}{Max(r, g, b)} \tag{5.5}$$

Image Processing and GIS for Remote Sensing: Techniques and Applications, Second Edition. Jian Guo Liu and Philippa J. Mason.
© 2016 John Wiley & Sons, Ltd. Published 2016 by John Wiley & Sons, Ltd.

Equation 5.5 implies that a colour vector reaches full saturation if at least one of its r, g and b components is equal to 0 while not all of them are 0. For instance, colour $\mathbf{P}(r, g, b) = (255,0,0)$ is pure red with full saturation and $\mathbf{P}(r, g, b) = (0,200,150)$ is a greenish cyan with full saturation.

The value range of hue is $0\sim 2\pi$ or $0°\sim 360°$, while the value range of the arccosine function of hue in eqn. 5.2 is $0\sim\pi$, but the 2π range of hue can be determined based on the relationship between r, g, and b. For instance, if $b > r > g$, the actual hue angle is $hue(r, g, b) = 2\pi - H(r, g, b)$.

Given intensity I, hue angle α and saturation angle φ, we can also derive the IHS-RGB transformation based on the same 3D geometry depicted in Fig. 5.1:

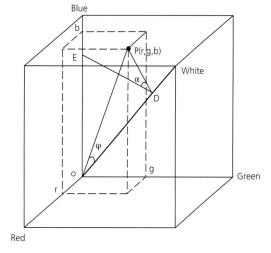

Blue, b, E, P(r,g,b), White, α, D, φ, o, g, Green, r, Red

Fig. 5.1 The colour cube model for RGB-IHS transformation.

$$B(I, \alpha, \varphi) = \frac{I}{\sqrt{3}}\left(1 + \sqrt{2}\tan\varphi\cos\alpha\right) \tag{5.6}$$

$$G(I, \alpha, \varphi) = \frac{I}{\sqrt{3}}\left[1 - \sqrt{2}\tan\varphi\cos\left(\frac{\pi}{3} + \alpha\right)\right] \tag{5.7}$$

$$R(I, \alpha, \varphi) = \frac{I}{\sqrt{3}}\left[1 + \sqrt{2}\tan\varphi\cos\left(\frac{2\pi}{3} + \alpha\right)\right] \tag{5.8}$$

Equivalently, but with a slight difference, the RGB-IHS transformation can also be derived from matrix operations by a co-ordinate rotation of the colour cube, and aided by sub-co-ordinates of v_1 and v_2. As shown in Fig. 5.2, the sub-axis v_1 is perpendicular to the grey line starting from the intensity I; it is in the plane decided by the blue axis and the grey line. The sub-axis v_2 is perpendicular to both the grey line and v_1. Thus v_1 and v_2 formulate a plane perpendicular to the grey line, and the end point of the colour vector $\mathbf{P}(r, g, b)$ is in this plane. Thus, considering the Cartesian-polar co-ordinate transformation of the sub-co-ordinate system of v_1 and v_2, the following matrix operation between I, v_1, v_2 and R, G, B can be established:

$$\begin{pmatrix} I \\ v_1 \\ v_2 \end{pmatrix} = \begin{pmatrix} 1/3 & 1/3 & 1/3 \\ -1/\sqrt{6} & -1/\sqrt{6} & 2/\sqrt{6} \\ 1/\sqrt{6} & -2/\sqrt{6} & 0 \end{pmatrix} \begin{pmatrix} R \\ G \\ B \end{pmatrix} \tag{5.9}$$

The hue and saturation can then be derived based on their relationships with v_1 and v_2 (Fig. 5.2):

$$H = \arctan\left(v_2/v_1\right) \tag{5.10}$$

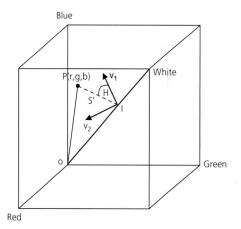

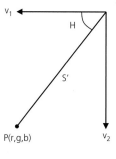

Blue, P(r,g,b), v₁, White, S', H, I, v₂, o, Green, Red; v₁, H, S', P(r,g,b), v₂

Fig. 5.2 The model of the matrix RGB-IHS transformation (adapted from Niblack 1986).

$$S' = \sqrt{v_1^2 + v_2^2} \qquad (5.11)$$

$$S = \arctan \frac{S'}{\sqrt{3}I} \qquad (5.12)$$

Here S' is the saturation for a given intensity I, while S is the intensity scaled angular saturation as depicted in Fig. 5.1. Depending on the RGB-IHS model used, there are several different definitions for saturation. Many publications define saturation in the form of eqn. 5.11; however, this definition is correct only for a fixed intensity. For digital image RGB additive colour composite display based on the RGB colour cube model, the definitions of saturation in angle φ given in eqns. 5.4 and 5.5 are the most appropriate. Equation 5.12 is essentially the same as 5.4.

An IHS-RGB transformation can then be derived from the inversion of eqn. 5.9:

$$\begin{pmatrix} R \\ G \\ B \end{pmatrix} = \begin{pmatrix} 1 & -1/2\sqrt{6} & 3/2\sqrt{6} \\ 1 & -1/2\sqrt{6} & -3/2\sqrt{6} \\ 1 & 1/\sqrt{6} & 0 \end{pmatrix} \begin{pmatrix} I \\ v_1 \\ v_2 \end{pmatrix} \qquad (5.13)$$

$$\begin{aligned} v_1 &= S' \cos 2\pi H \\ v_2 &= S' \sin 2\pi H \end{aligned} \qquad (5.14)$$

RGB-IHS and IHS-RGB transformations allow us to manipulate intensity, hue and saturation components separately and thus enable some innovative processing for de-correlation stretch and image data fusion.

5.2 IHS de-correlation stretch

High correlation generally exists among spectral bands of multi-spectral images. As a result, the original image bands displayed in RGB formulate a slim cluster along the grey line occupying only a very small part of the space of the colour cube [Fig. 5.3(a)]. Contrast enhancement on individual image bands can elongate the cluster in the colour cube, but it is not effective for increasing the volume of the cluster since it is equivalent to stretching the intensity only [Fig. 5.3(b)]. To increase the volume, the data cluster should expand in both directions along and perpendicular to the grey line. This is equivalent to stretching both intensity and saturation components [Fig. 5.3(c)]. The processing is called *IHS de-correlation stretch* (*IHSDS*) because the correlation among the three bands is reduced to generate a spherical data cluster in the RGB cube, as indicated in Table 5.1. In comparison with an ordinary contrast stretch, the IHSDS is essentially a saturation stretch. As proposed by Gillespie *et al.* (1986), the IHS de-correlation stretch technique involves the following steps:

- RGB-IHS transformation;
- stretch **I**ntensity and **S**aturation components;
- IHS-RGB transformation.

In the second step, the hue component can also be stretched. However, when transforming back to RGB display, the resultant colours may not be comparable to those of the original image and this makes image interpretation potentially difficult.

The limited hue range of a colour composite image is mainly caused by colour bias. If the average brightness of

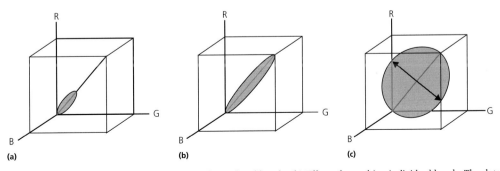

Fig. 5.3 (a) Distribution of pixels in RGB cube for typical correlated bands. (b) Effect of stretching individual bands. The data cluster is elongated along the grey line but not stretched to fill the RGB cube by this operation. (c) De-correlation stretch expands the data cluster in the direction perpendicular to the grey line to fill the 3D space of the RGB cube.

Table 5.1 The correlation coefficients before and after de-correlation stretch of the Landsat 7 ETM+ bands 5, 3 and 1 RGB colour composite shown in Fig. 5.4.

Correlation Matrix before DS				Correlation Matrix after DS			
Correlation	Band 1	Band 3	Band 5	Correlation	Band 1	Band 3	Band 5
Band 1	1.00	0.945	0.760	Band 1	1.00	0.842	0.390
Band 3	0.945	1.00	0.881	Band 3	0.842	1.00	0.695
Band 5	0.760	0.881	1.00	Band 5	0.390	0.695	1.00

one band is significantly higher than those of the other two bands, the colour composite will have an obvious colour 'cast' of the primary colour assigned to the band of highest intensity. As we discussed in Chapter 2, the balance contrast enhancement technique (BCET) was developed to solve just this problem. BCET removes inter-band colour bias and therefore increases the hue variation of a colour composite. As a result, the hue component derived from a BCET stretched colour composite will have a much wider value range than that derived from the original colour biased composite. The wider hue value range achieved by BCET means more spectral information is presented as hue rather than as intensity and saturation, which are fundamentally different from the wide hue range that would be achieved by stretching the hue component. In many cases, a simple linear stretch with automatic clipping or an interactive piece-wise linear stretch can also effectively eliminate the colour bias. An optimised IHSDS can be achieved by performing BCET or linear stretch as a pre-processing step, as summarised below:

- BCET stretch (or linear stretch with appropriate clipping);
- RGB-IHS transformation;
- saturation component stretching;
- IHS-RGB transformation.

The DN ranges of the images converted from stretched IHS components back to RGB co-ordinates may exceed the maximum range of the display device (usually 8 bits or 0–255 per channel) and so may need to be adjusted to fit the maximum 8-bit DN range. This can be done easily in any image processing system, such as ER Mapper; the image will be automatically displayed within 8 bits per channel using the actual limits of input image DNs.

The effect of the IHSDS-based saturation stretch is similar to that of the de-correlation stretch based on principal component (PC) analysis (to be introduced in

Chapter 7). The difference between them is that PC de-correlation stretch is based on scene statistics whilst the IHS de-correlation stretch is interactive, flexible and based on user observation of the saturation image and its histogram.

De-correlation stretch enhances the colour saturation of a colour composite image and thus effectively improves the visual quality of the image spectral information, without significant distortion of its spectral characteristics, as illustrated in Fig. 5.4. De-correlation stretch-enhanced colour images are easy to understand and interpret and have been successfully used for many applications of remote sensing.

5.3 Direct de-correlation stretch technique

This technique performs a *direct de-correlation stretch (DDS)* without using RGB-IHS and IHS-RGB transformations (Liu & Moore 1996). The DDS achieves the same effect as the IHS de-correlation stretch. As DDS involves only simple arithmetic operations and can be controlled quantitatively, it is much faster, more flexible and more effective than the IHS de-correlation stretch technique.

As shown in Fig. 5.1, a colour vector, **P**, and the grey line together define a plane or a slice of the RGB cube. If we take this slice out as shown in Fig. 5.5, the grey line, the full saturation line and the maximum intensity line formulate a triangle that includes all the colours with the same hue but various intensity and saturation. The colour vector **P** is between the grey (achromatic) line and the maximum saturation (chromatic) line and it can be considered as the sum of two vectors: a vector **a** representing the achromatic (zero saturation) component, the white light in the colour, and a vector **c** representing the chromatic (full

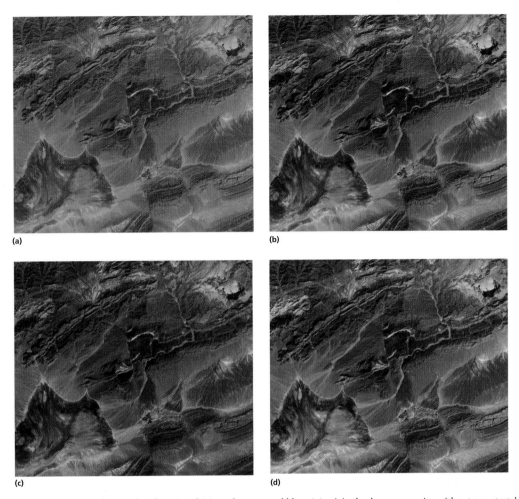

Fig. 5.4 Colour composites of ETM+ bands 5, 3 and 1 in red, green and blue: (a) original colour composite without any stretch; (b) BCET stretched colour composite; (c) IHS de-correlation stretched colour composite after BCET stretch; and (d) DDS ($k = 0.5$) colour composite after BCET stretch.

saturation) component that is relevant to the pure colour of the hue.

Given $\mathbf{P} = (r, g, b)$, let $a = Min(r, g, b)$, then

$$\begin{aligned} \mathbf{a} &= (a, a, a) \\ \mathbf{c} &= (r-a, g-a, b-a) \\ &= \mathbf{P} - \mathbf{a} \end{aligned} \qquad (5.15)$$

or

$$\mathbf{P} = \mathbf{a} + \mathbf{c} \qquad (5.16)$$

A direct de-correlation stretch is achieved by reducing the achromatic component $\mathbf{a}$ of the colour vector $\mathbf{P}$, as defined below:

$$\mathbf{P}_k = \mathbf{P} - k\mathbf{a} \qquad (5.17)$$

where k is an achromatic factor and $0 < k < 1$.

As shown in Fig. 5.5, the operation shifts the colour vector $\mathbf{P}$ away from the achromatic line to form a new colour vector $\mathbf{P}_k$ with increased saturation ($\varphi_k > \varphi$) and decreased intensity ($OD_k < OD$). To restore the intensity to a desired level, linear stretch can then be applied to each image in red, green and blue layers. This will elongate $\mathbf{P}_k$ to $\mathbf{P}_{ks}$, which has the same hue and saturation as $\mathbf{P}_k$ but increased intensity ($OD_{ks} > OD_k$). The operation does not affect the hue since it only reduces the achromatism of the colour and leaves the hue information, $\mathbf{c}$, unchanged.

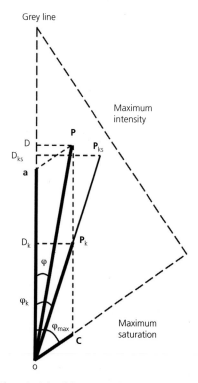

Grey line

Maximum
intensity

Maximum
saturation

Fig. 5.5 The principle of the DDS technique.

This is easy to understand if we re-write eqn. 5.17 as below:

$$\mathbf{P}_k = \mathbf{P} - k\mathbf{a} = \mathbf{c} + \mathbf{a} - k\mathbf{a} = \mathbf{c} + (1-k)\mathbf{a}$$

The algebraic operations for vector eqn. 5.17 are

$$r_k = r - ka = r - k\min(r, g, b)$$
$$g_k = g - ka = g - k\min(r, g, b) \qquad (5.18)$$
$$b_k = b - ka = b - k\min(r, g, b)$$

Again, as already indicated in the IHS de-correlation stretch, the three bands for colour composition must be well stretched (e.g. BCET or linear stretch with appropriate clipping) before the DDS is applied.

The DDS performs a de-correlation stretch essentially the same as that based on IHS transformation as illustrated in Fig. 5.4. We can prove the following properties of the DDS (refer to § 5.6 for details):

1 DDS is controlled by the achromatic factor k.
2 For a given k, the amount of saturation stretch is dependent on the initial colour saturation; a lower saturation image is subject to stronger saturation stretch than a higher saturation image for a given k.

3 DDS does not alter the relationship between those colours with the same saturation but different intensities.
4 For colours with the same intensity but different saturation, DDS results in higher intensity for more saturated (purer) colours.

The value k is specified by users. It should be set based on the saturation level of the original colour composite. The lower the saturation of an image is, the greater the k value should be given (within the range of 0–1), and $k = 0.5$ is generally good for most cases. Figure 5.6 illustrates the initial BCET colour composite and the DDS colour composites with $k = 0.3, k = 0.5, k = 0.7$; these DDS composites all show increased saturation without distortion of hues, in comparison with the original BCET colour composite, and their saturation increases with the increase of k. The merits of simplicity and quantitative control of DDS are obvious.

5.4 Hue RGB colour composites

As shown in Fig. 5.7, with the RGB-IHS transformation, three hue images can be derived from three different band triplets of a multi-spectral image. In each hue image, the brightness (the pixel DN) changes with hues that are determined by the spectral profiles of the source bands of the triplet. If three hue images are displayed in red, green and blue using an RGB additive colour display system, an HRGB false colour composite image is produced (Liu & Moore 1990). Colours in an HRGB image are controlled by the hue DNs of the three component hue images. An HRGB image can therefore incorporate spectral information of up to nine image bands. Pixel colours in an HRGB image are unique presentations of the spectral profiles of all the original image bands. The merits of an HRGB image are two fold:

- It suppresses topographic shadows more effectively than ratio;
- it condenses and displays spectral information of up to nine image bands in a colour composite of three hue images.

From the definition of hue, it is easy to prove that the H component is independent of illumination and therefore is free of topographic shadows. Suppose the irradiance upon a sunlit terrain slope, E_t, is n times of that upon a terrain slope in shade, E_b, then,

$$E_t = nE_b, \quad r_t = nr_b, \quad g_t = ng_b, \quad b_t = nb_b$$

where r, g and b represent the radiance of the three bands used for RGB colour composition.

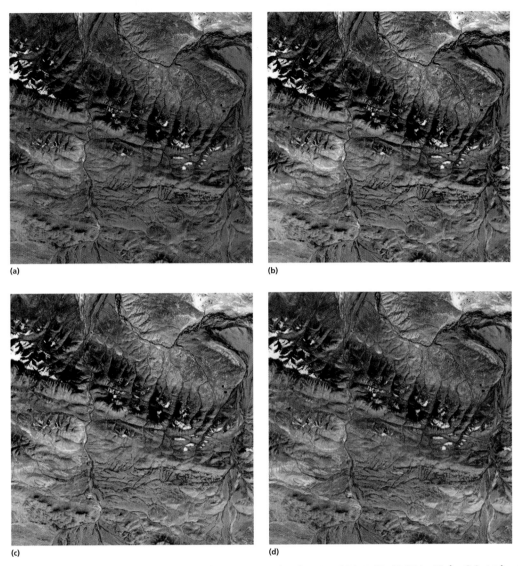

Fig. 5.6 (a) A BCET standard false colour composite of Terra-1 ASTER bands 3, 2 and 1 in RGB; (b) DDS with $k = 0.3$; (c) $k = 0.5$; and (d) $k = 0.7$.

From eqn. 5.2 we have:

$$H\left(r_t, g_t, b_t\right) = \arccos \frac{2b_t - g_t - r_t}{2V_t}$$

$$= \arccos \frac{2nb_b - ng_b - nr_b}{2nV_b}$$

$$\qquad (5.19)$$

$$= \arccos \frac{2b_b - g_b - r_b}{2V_b}$$

$$= H\left(r_b, g_b, b_b\right)$$

where

$$V_t = \sqrt{\left(r_t^2 + g_t^2 + b_t^2\right) - \left(r_t g_t + r_t b_t + g_t b_t\right)}$$

$$= \sqrt{\left[\left(nr_b\right)^2 + \left(ng_b\right)^2 + \left(nb_b\right)^2\right] - \left(n^2 r_b g_b + n^2 r_b b_b + n^2 g_b b_b\right)}$$

$$= nV_b$$

Thus hue is independent of illumination and not affected by topographic shadows. More generally, we can prove that:

$$H(r_i, g_i, b_i) = H(r_j, g_j, b_j)$$
$$if \quad (5.20)$$
$$E_j = nE_i + a$$

This equation implies that if a hue image is derived from three spectrally adjacent bands (thus the atmospheric effects on each band are not significantly different and

can be represented by a), it is little affected by shadows as well as atmospheric scattering.

With topography completely removed, an HRGB image has low signal-to-noise ratio (SNR) and it is actually difficult for visual interpretation of ground objects. As shown in Fig. 5.8, an HRGB image is very like a classification image without topographic features. It can

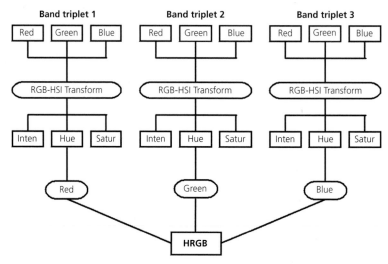

Fig. 5.7 Schematic illustration of the production of an HRGB image.

(a)

(b)

Fig. 5.8 (a) An HRGB colour composite of an ATM (Airborne Thematic Mapper) image: red – hue derived from bands 10, 9, 8; green – hue derived from bands 7, 6, 5; and blue – hue derived from bands 4, 3, 2. (b) An ordinary colour composite of bands 8, 5 and 2 in RGB for comparison.

therefore be used as pre-processing in preparation for classification. For visual interpretation of an HRGB image, it is advisable to use ordinary colour composites as reference images.

5.5 Derivation of RGB-IHS and IHS-RGB transformation based on 3D geometry of the RGB colour cube*

5.5.1 Derivation of RGB-IHS transformation

As shown in Fig. 5.9, the *Intensity* OD is the projection of colour vector $\mathbf{P}(r, g, b)$ onto the grey line OW or vector $\mathbf{W}(a, a, a)$, where a can be any a value within the colour cube. Then, according to the vector projection rule:

$$I(r, g, b) = \frac{\mathbf{P} \cdot \mathbf{W}}{|\mathbf{W}|} = \frac{ra + ga + ba}{\sqrt{3a^2}} = \frac{1}{\sqrt{3}}(r + g + b) \quad (5.21)$$

The *Hue* angle α (or $\angle PDE$) is the angle between two planes defined by triangles OBW and OPW that intercept along the grey line OW. Both planes can be defined by the three corner points of the two triangles.

$$\text{OBW}: \begin{vmatrix} B & G & R \\ a & a & a \\ a & 0 & 0 \end{vmatrix} = 0, \quad G - R = 0 \quad (5.22)$$

$$\text{OPW}: \begin{vmatrix} B & G & R \\ a & a & a \\ b & g & r \end{vmatrix} = 0, \quad B(r - g) + G(b - r) + R(g - b) = 0 \quad (5.23)$$

Thus the angle between planes OBW and OPW, the hue angle α, can be decided:

$$\cos\alpha = \frac{2b - g - r}{\sqrt{2}\sqrt{(r - g)^2 + (b - r)^2 + (g - b)^2}}$$

$$= \frac{2b - g - r}{2\sqrt{(r^2 + g^2 + b^2) - (rg + rb + gb)}} \quad (5.24)$$

$$H(r, g, b) = \arccos \frac{2b - g - r}{2\sqrt{(r^2 + g^2 + b^2) - (rg + rb + gb)}}$$

The *Saturation* is the angle φ between colour vector $\mathbf{P}(r, g, b)$ and grey line vector $\mathbf{W}(a, a, a)$. Thus according to vector dot product, we have,

$$\cos\varphi = \frac{\mathbf{P} \cdot \mathbf{W}}{|\mathbf{P}| \, |\mathbf{W}|} = \frac{a(r + g + b)}{\sqrt{r^2 + g^2 + b^2} \, \sqrt{3a^2}} = \frac{r + g + b}{\sqrt{3(r^2 + g^2 + b^2)}} \quad (5.25)$$

$$S(r, g, b) = \arccos \frac{r + g + b}{\sqrt{3(r^2 + g^2 + b^2)}}$$

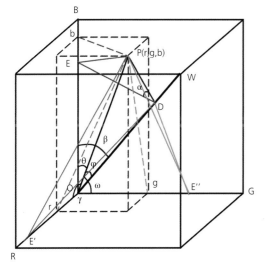 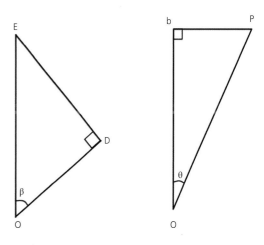

Fig. 5.9 The relationship between RGB and IHS in the colour cube 3D geometry.

5.5.2 Derivation of IHS-RGB transformation

Given Intensity I, Hue α and Saturation φ, we can derive $R(I,\alpha,\varphi)$, $G(I,\alpha,\varphi)$ and $B(I,\alpha,\varphi)$ as depicted in Fig. 5.9.

To find $B(I,\alpha,\varphi)$, the key is to find the angle θ between colour vector **P** and B axis.

The angle between the grey line and any of the RGB axes is identical. For instance, the angle β between the grey line and the B axis is:

$$\cos\beta = \frac{\mathbf{B}\cdot\mathbf{W}}{|\mathbf{B}|\,|\mathbf{W}|} = \frac{ab}{\sqrt{b^2}\sqrt{3a^2}} = \frac{1}{\sqrt{3}} \qquad (5.26)$$

As $\cos\beta = \dfrac{OD}{OE} = \dfrac{1}{\sqrt{3}}$, $OD = I$ thus $OE = \sqrt{3}I$ (see the triangle OED in Fig. 5.9)

$$ED = \sqrt{OE^2 - I^2} = \sqrt{2}I$$

The length of the colour vector **P** is $OP = \dfrac{I}{\cos\varphi}$, while the distance between **P** and the grey line is $PD = I\tan\varphi$.

From the triangle EPD, we then have:

$$
\begin{aligned}
EP^2 &= ED^2 + PD^2 - 2ED\cdot PD\cos\alpha \\
&= 2I^2 + I^2\tan^2\phi - 2\sqrt{2}I^2\tan\phi\cos\alpha
\end{aligned} \qquad (5.27)
$$

From the triangle OEP, we can also find EP as

$$
\begin{aligned}
EP^2 &= OE^2 + OP^2 - 2OE\cdot OP\cos\theta \\
&= 3I^2 + \frac{I^2}{\cos^2\varphi} - 2\sqrt{3}\frac{I^2\cos\theta}{\cos\varphi}
\end{aligned} \qquad (5.28)
$$

Take the right sides of eqns. 5.27 and 5.28; we can then solve $\cos\theta$:

$$\cos\theta = \frac{1}{\sqrt{3}}\cos\varphi + \sqrt{\frac{2}{3}}\sin\varphi\cos\alpha \qquad (5.29)$$

As shown in the triangle ObP in Fig. 5.9:

$$b = OP\cos\theta = \frac{\cos\theta}{\cos\varphi}I = \frac{I}{\cos\varphi}\left(\frac{1}{\sqrt{3}}\cos\varphi + \sqrt{\frac{2}{3}}\sin\varphi\cos\alpha\right)$$

Thus,

$$B(I,\alpha,\varphi) = b = \frac{I}{\sqrt{3}}\left(1 + \sqrt{2}\tan\varphi\cos\alpha\right) \qquad (5.30)$$

If we look into the RGB colour cube along the grey line, the projection of this cube on the plane perpendicular to the grey line is a hexagon, as shown in Fig. 5.10. Given a hue angle α of colour vector **P** starting from the B axis,

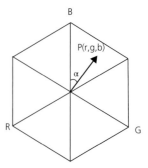

Fig. 5.10 Given that hue is the angle α between the B axis and colour vector **P**, the angle between **P** and R is $\dfrac{2\pi}{3} + a$ and that between P and G is $\dfrac{4\pi}{3} + a$.

the angle between **P** and the R axis is $\dfrac{2}{3}\pi + \alpha$ and that between **P** and the G axis is $\dfrac{4}{3}\pi + \alpha$. On the other hand, the intensity I and saturation angle φ are both independent of the orientation of the RGB co-ordinate system. Thus, we can solve $R(I,\alpha,\varphi)$ and $G(I,\alpha,\varphi)$ in a similar way to solving $B(I,\alpha,\varphi)$, as above, based on the central symmetry of the grey line to the RGB axes. As shown in Fig. 5.9, if we consider starting the hue angle from R axis, then

$$OE' = OE = \sqrt{3}I \quad \text{and} \quad E'D = ED = \sqrt{2}I$$

From triangles $E'PD$ and $OE'P$, again we can establish two equations of $E'P$:

$$
\begin{aligned}
E'P &= E'D^2 + PD^2 - 2E'D\cdot PD\cos\alpha \\
&= 2I^2 + I^2\tan^2\varphi - 2\sqrt{2}I^2\tan\varphi\cos\left(\frac{2}{3}\pi + \alpha\right)
\end{aligned} \qquad (5.31)
$$

$$
\begin{aligned}
E'P &= OE'^2 + OP^2 - 2OE'\cdot OP\cos\gamma \\
&= 3I^2 + \frac{I^2}{\cos^2\varphi} - 2\sqrt{3}\frac{I^2\cos\gamma}{\cos\varphi}
\end{aligned} \qquad (5.32)
$$

where γ is the angle between **P** and R axis.

Solve eqn. 5.31 and 5.32 for $\cos\gamma$:

$$\cos\gamma = \frac{1}{\sqrt{3}}\cos\varphi + \sqrt{\frac{2}{3}}\sin\varphi\cos\left(\frac{2}{3}\pi + \alpha\right) \qquad (5.33)$$

As shown in the triangle OrP in Fig. 5.9,

$$
\begin{aligned}
r = OP\cos\gamma &= \frac{\cos\gamma}{\cos\varphi}I \\
&= \frac{I}{\cos\varphi}\left[\frac{1}{\sqrt{3}}\cos\varphi + \sqrt{\frac{2}{3}}\sin\varphi\cos\left(\frac{2}{3}\pi + \alpha\right)\right]
\end{aligned}
$$

Finally,

$$R(I, \alpha, \varphi) = r = \frac{1}{\sqrt{3}} \left[1 + \sqrt{2} \tan \varphi \cos \left(\frac{2}{3} \pi + \alpha \right) \right] \quad (5.34)$$

In the same way and considering $\cos \left(\frac{4}{3} \pi + \alpha \right) = -\cos \left(\frac{\pi}{3} + \alpha \right)$, we have,

$$G(I, \alpha, \varphi) = g = \frac{1}{\sqrt{3}} \left[1 - \sqrt{2} \tan \varphi \cos \left(\frac{\pi}{3} + \alpha \right) \right] \quad (5.35)$$

5.6 Mathematical proof of DDS and its properties*

5.6.1 Mathematical proof of DDS

The geometrically obvious fact of the saturation stretch of the DDS in Fig. 5.5 can be easily proven in simple algebra.

Let $v = \max(r, g, b)$. From eqn. 5.5, the saturation components for $\mathbf{P}$ and $\mathbf{P}_k$ are

$$S = \frac{v - a}{v} = 1 - \frac{a}{v}$$

$$S_k = \frac{(v - ka) - (a - ka)}{v - ka} = 1 - \frac{a - ka}{v - ka}$$

The difference between them is

$$\delta S = S_k - S = \frac{a}{v} - \frac{a - ka}{v - ka} = \frac{ka(v - a)}{v(v - ka)} \geq 0 \quad (5.36)$$

Therefore $S_k \geq S$.

There are three cases:

1 If $a = 0$, then $S_k = S = 1$, and colours with full saturation (pure colours) are not affected.

2 If $a = v$, then $S_k = S = 0$, and the saturation of grey tones (achromatic vectors) remains zero though the intensity is scaled down.

3 Otherwise, $S_k > S$, when the colour vectors between the achromatic line and the maximum saturation line are shifted (stretched) away from the grey line depending on k, $\mathbf{a}$ and $\mathbf{c}$, as further discussed in the next subsection.

5.6.2 The properties of DDS

DDS is independent of hue component. It enhances saturation with intensity preserved. This can be further verified by an investigation of the properties of DDS as below.

1 DDS is controlled by the achromatic factor k.

The saturation increment of DDS, defined by eqn. 5.36, is a monotonically increasing function of k viz.:

$$\frac{d\delta S}{dk} = \frac{a(v - a)}{(v - ka)^2} > 0 \quad (5.37)$$

where $v > a$ and $0 < k < 1$.

Let $k = 1$, δS reaches its maximum: $\delta S_{max} = \frac{a}{v}$, and

$$S_k = S + \delta S_{max} = \frac{v - a}{v} + \frac{a}{v} = 1$$

This is the case for an image of chromatic component $\mathbf{c}$.

As shown in Fig. 5.11(a), a large value of k (near 1) results in a great increase in saturation (δS) for a non-saturated colour vector. Such an overstretch of saturation compresses the colour vectors into a narrow range near the maximum saturation line. Conversely, a small k of value (approaching 0) has little effect on saturation. In general, $k = 0.5$ gives an even stretch of saturation between 0 and 1. The value of k can be adjusted according to requirements. In general, a large k value is desirable for an image with very low saturation and vice versa.

2 For a given k, the saturation stretch is dependent on colour saturation.

Consider δS as a function of achromatic element a, then the first derivative of δS for a is:

$$\frac{d\delta S}{da} = \frac{1}{v} - \frac{v - kv}{(v - ka)^2} \quad (5.38)$$

The second derivative is:

$$\frac{d^2 \delta S}{da^2} = -\frac{2k(1 - k)v}{(v - ka)^3} < 0 \quad (5.39)$$

where $0 < k < 1$ and $v \geq a$.

Therefore, as a function of a, δS has a maximum when

$$\frac{d\delta S}{da} = \frac{1}{v} - \frac{v - kv}{(v - ka)^2} = 0$$

From the above equation, we have

$$a = \frac{1 - \sqrt{1 - k}}{k} v \quad (5.40)$$

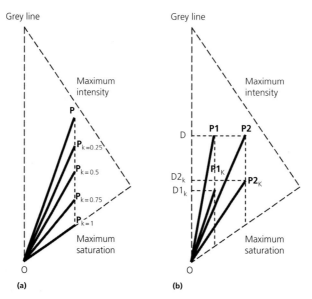

Fig. 5.11 The properties of the DDS technique: (a) variation of DDS with the achromatic factor k; and (b) for the colours with the same intensity, DDS results in higher intensity for a colour with higher saturation.

where $\dfrac{1-\sqrt{1-k}}{k} < 1$ for $0 < k < 1$.

The saturation for the case of eqn. 5.40 is then

$$S = \frac{v-a}{v} = 1 - \frac{1-\sqrt{1-k}}{k} \qquad (5.41)$$

Therefore, the saturation increment δS reaches its maximum when S satisfies eqn. 5.41.

For $k = 0.5$, the saturation stretch reaches the maximum: $\delta S_{max} \approx 0.172$, when $S \approx 0.414$. The saturation stretch of DDS becomes less when saturation is either greater or less than 0.414.

The criteria for the maximum stretch of saturation can be easily controlled by modifying the value of k. This property of DDS has a self-balancing effect that optimises the stretch of the most condensed saturation range.

3 DDS does not alter the relationship between the colours with the same saturation but different intensities.

Any colour with the same hue and saturation as a colour vector $\mathbf{P}$ but different intensity can be defined as a colour vector $n\mathbf{P}$,

$$n\mathbf{P} = (nr, ng, nb)$$

where n is a positive real number.

Then,

$$\min(nr, ng, nb) = na$$
$$n\mathbf{a} = (na, na, na)$$

In the same way of the derivation of $\mathbf{P}_k$, we have,

$$(n\mathbf{P})_k = n\mathbf{P} - kn\mathbf{a} = n(r - ka, g - ka, b - ka) = n\mathbf{P}_k$$

$\mathbf{P}_k$ and $(n\mathbf{P})_k$ have the same orientation (saturation and hue). The magnitude (intensity) relationship between them is the same as that between $\mathbf{P}$ and $n\mathbf{P}$. This means that the DDS transforms one saturation value to another uniquely. DDS reduces colour intensity but does not alter the intensity relationship among colours with the same hue and saturation. For a given hue and saturation, the relative brightness of colours remains unchanged after DDS.

4 For colours with the same intensity but different saturation, DDS results in higher intensity for more saturated (purer) colours.

According to the definition of colour intensity of a colour $\mathbf{P}$ in eqn. 5.1, DDS shifts the colour $\mathbf{P}$ to a colour $\mathbf{P}_k$ with intensity

$$I_k = \frac{1}{\sqrt{3}}\left(r - ka + g - ka + b - ka\right)$$
$$= \frac{1}{\sqrt{3}}\left(r + g + b - 3ka\right) \qquad (5.42)$$
$$= I - \sqrt{3}ka$$

Colours with the same intensity but different saturation values have the same sum of r, g and b but different values for a (the minimum of r, g and b). The higher the saturation, the smaller is a and the greater is the I_k produced by the DDS according to eqn. 5.42. This effect is illustrated in Fig. 5.11(b) where colours P1 and P2 have the same intensity (OD). After reducing the same proportion of their achromatic components by DDS, the intensity of the less saturated colour $\mathbf{P1}_k$ is lower than that of the more saturated colour $\mathbf{P2}_k$ ($OD1_k < OD2_k$). For information enhancement, this property has the positive effect of increasing the variation of colours with the same intensity but different saturation in terms of differences in both intensity and saturation values.

5.7 Summary

The composition of three primaries, red, green and blue, produces any colours according to the tristimulus theory while the colour quality is described as intensity, hue and saturation. The RGB-IHS transformation and the inverse transformation IHS-RGB are similar to a 3D Cartesian-conical co-ordinate transformation and can be derived from either 3D geometry or matrix operations for co-ordinate rotations of the RGB colour cube.

The RGB-IHS and IHS-RGB transformations allow us to manipulate colour intensity, hue and saturation components separately and with great flexibility. One major application is the saturation stretch-based de-correlation stretch technique that enhances image colour saturation without altering the hues of the colours. The effects are the same as reducing the inter-band correlation between the three bands for the RGB colour

composition. For the same purpose, a shortcut algorithm of processing is found, the DDS (direct de-correlation stretch). Based on colour vector de-composition in to achromatic and chromatic components, the DDS performs the saturation stretch directly in the RGB domain without involving the RGB-IHS and IHS-RGB transformations.

Both intensity and saturation are either defined by or affected by the illumination impinged on the objects that are imaged. The hue is, however, by definition entirely independent of illumination condition (irradiance) and therefore topographic shading. The hue of a colour is actually the spectral property coding. An HRGB colour composite technique is thus introduced that can code the spectral property of up to nine spectral bands into various colours to generate an information-rich colour image without effects of topographic shadows though the image is actually difficult for visual perception and interpretation.

5.8 Questions

1 Use a diagram of the RGB colour cube to explain the mathematical definition and physical meaning of intensity, hue and saturation.
2 What are the value ranges of intensity, hue and saturation according to the RGB colour cube model of RGB-IHS transformation?
3 Why is RGB-IHS a useful image processing technique?
4 Describe the principle of IHS de-correlation stretch with the aid of diagrams.
5 Describe the major steps of IHS de-correlation stretch.
6 What is the drawback of stretching the hue component in the IHS de-correlation stretch? Can the value range of the hue component be increased without stretching the hue component directly, and how can it be achieved?
7 Use a diagram to explain the principle of DDS. In what senses are the DDS and the IHSDS similar as well as different?

CHAPTER 6
Image fusion techniques

The term 'image fusion' has become very widely used in recent years, but often to mean quite different things. Some people regard all image enhancement techniques as image fusion but in general, image fusion refers specifically to techniques for integrating images or raster datasets of different spatial resolutions, or with different properties, to formulate new images. In this book, we take the latter, narrower definition and, in this chapter, following directly from the topics discussed in the last chapters, we introduce several commonly used simple image fusion techniques for multi-resolution and multi-source image integration.

6.1 RGB-IHS transformation as a tool for data fusion

The RGB-IHS transformation can be used as a tool for data fusion as well as enhancement. A typical application is to fuse a low resolution colour composite with a high resolution panchromatic image to improve spatial resolution. With regard to optical sensor systems, image spatial resolution and spectral resolution are contradictory quantities. For a given signal noise ratio, a higher spectral resolution (narrower spectral band) is usually achieved at the cost of spatial resolution. Image fusion techniques are therefore useful for integrating a high spectral resolution image with a high spatial resolution image, such as Landsat TM (six spectral bands with 30 m resolution) and SPOT pan (panchromatic band with 10 m resolution), to produce a fused image with high spectral and spatial resolutions. Using the RGB-IHS transformation this can be done easily by replacing the intensity component of a colour composite with a high resolution image as follows:

- Rectify the low resolution colour composite image (e.g. a TM image) to the high resolution image of the same scene (e.g. a SPOT panchromatic image). The rectification can be done the other way around, that is, from high resolution to low resolution but, in some image processing software packages, the lower resolution image will need to be interpolated to the same pixel size as the high resolution image.
- Perform the RGB-IHS transformation on the low resolution colour composite image.
- Replace the intensity component, I, by the high resolution image.
- Perform the reverse IHS-RGB transformation.

The resultant fused image is a mixture of spectral information from the low resolution colour composite and high spatial resolution of image textures and patterns. Figure 6.1 shows a colour composite of TM bands 541 in RGB with a balance contrast enhancement technique (BCET) stretch in Fig. 6.1(a), SPOT pan in Fig. 6.1(b) and the TM-SPOT pan fusion image in Fig. 6.1(c). The fused image presents more detailed topographic/textural information, introduced by the SPOT pan image together with the spectral information from the three TM bands; unfortunately this may occur with considerable spectral (colour) distortion. The colour distortion can be significant if the spectral range of the three TM bands for colour composition is very different from that of the panchromatic band. In this case the intensity component, calculated as the summation of the three TM bands according to eqn. 5.1,

Image Processing and GIS for Remote Sensing: Techniques and Applications, Second Edition. Jian Guo Liu and Philippa J. Mason.
© 2016 John Wiley & Sons, Ltd. Published 2016 by John Wiley & Sons, Ltd.

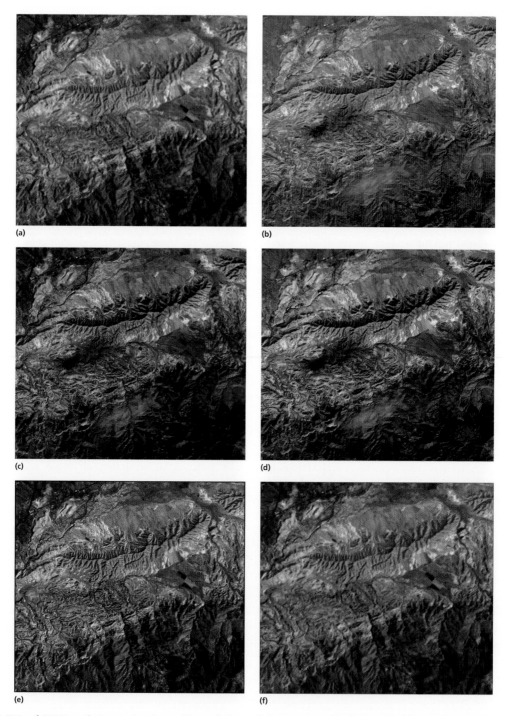

Fig. 6.1 TM and SPOT pan fusion results of several image fusion techniques: (a) a colour composite of TM bands 541RGB with BCET; (b) SPOT pan image of the same area; (c) the IHS fusion image of TM 541 and SPOT pan; (d) the Brovey transform fusion image; (e) the SFIM fusion image with 5 × 5 smoothing filter; and (f) the SFIM fusion image with 3 × 3 smoothing filter.

will be different from the SPOT pan replacing it and, as a consequence, colour distortion is introduced.

In the same way, RGB-IHS can also be used for multi-source data integration such as the fusion of multi-spectral image data with raster geophysical or geochemical data, as outlined below:

- Co-register the datasets to be fused;
- RGB-IHS transformation;
- replacement of I component by a geophysical or geochemical dataset;
- IHS-RGB transformation.

The resultant image contains both spectral information of the original image bands and geophysical or geochemical information as intensity variation. The interpretation of such fused images demands thorough understanding of the input datasets. A more productive method is to use the so called 'colour-drape' technique in which the geophysical or geochemical dataset is used as if it were a raster surface, such as a digital elevation model (DEM), with a colour composite image draped over it in a three-dimensional (3D) perspective view. This concept will be discussed further in Part II.

6.2 Brovey transform (intensity modulation)

The Brovey transform is a shortcut to image fusion, compared with the IHS image fusion technique, and is based on direct intensity modulation. Let R, G and B represent three image bands displayed in red, green and blue, and let P represent the image to be fused as the intensity component of the colour composite. The Brovey transform is then defined by the following:

$$R_b = \frac{3RP}{R+G+B}$$

$$G_b = \frac{3GP}{R+G+B} \tag{6.1}$$

$$B_b = \frac{3BP}{R+G+B}$$

It is obvious that the sum of the three bands in the denominator is equivalent to the intensity component of the colour composite, and the Brovey transform can then be re-written as

$$R_b = R \times P/I$$
$$G_b = G \times P/I \tag{6.2}$$
$$B_b = B \times P/I$$

The operations of the Brovey transform are therefore simply each band multiplied by the ratio of the replacement image over the intensity of the corresponding colour composite. If image P is a higher resolution image, then eqn. 6.2 performs image fusion to improve spatial resolution and, if P is a raster dataset of different source, then 6.2 performs multi-source data integration. The Brovey transform achieves a similar result to that of the IHS fusion technique without carrying out the whole process of RGB-IHS and IHS-RGB transformations and is thus far simpler and faster. It does, however, also introduce colour distortion, as shown in Fig. 6.1(d).

6.3 Smoothing filter-based intensity modulation

Both IHS and Brovey transform image fusion techniques can cause colour distortion if the spectral range of the intensity replacement (or modulation) image is different from that of the three bands in the colour composite. This problem is inevitable in colour composites that do not use consecutive spectral bands, and it may become serious in vegetated and agricultural scenes if the images to be fused were acquired in different growing seasons. Preserving the original spectral properties is very important in remote sensing applications that rely on spectral signatures, such as lithology, soil and vegetation. The spectral distortion introduced by these fusion techniques is uncontrolled and not quantified because the images for fusion are often acquired by different sensor systems, on different dates and/or in different seasons. Fusion in this context cannot, therefore, in any way be regarded as spectral enhancement and should be avoided to prevent unreliable interpretations. In seeking a spectral preservation image fusion technique that also improves spatial resolution, a smoothing filter-based intensity modulation (SFIM) image fusion technique has been developed (Liu 2000).

6.3.1 The principle of SFIM

The DN value of a daytime optical image of reflective spectral band λ is mainly determined by two factors: the solar radiation impinging on the land surface, irradiance $E(\lambda)$, and the spectral reflectance of the land surface $\rho(\lambda)$: $DN(\lambda) = \rho(\lambda)E(\lambda)$.

Let $DN(\lambda)_{low}$ represent a DN value in a lower resolution image of spectral band λ and $DN(\gamma)_{high}$ the DN value of the corresponding pixel in a higher resolution image of spectral band γ, and the two images are taken in similar solar illumination conditions (such as the case of TM and SPOT), then,

$$DN(\lambda)_{low} = \rho(\lambda)_{low} E(\lambda)_{low} \text{ and } DN(\gamma)_{high} = \rho(\gamma)_{high} E(\gamma)_{high}$$

After co-registering the lower resolution image to the higher resolution image precisely and meanwhile interpolating the lower resolution image to the same pixel size of the higher resolution image, the SFIM technique is defined as

$$
\begin{aligned}
DN(\lambda)_{sim} &= \frac{DN(\lambda)_{low} \times DN(\gamma)_{high}}{DN(\gamma)_{mean}} \\
&= \frac{\rho(\lambda)_{low} E(\lambda)_{low} \times \rho(\gamma)_{high} E(\gamma)_{high}}{\rho(\gamma)_{low} E(\gamma)_{low}} \\
&\approx \rho(\lambda)_{low} E(\lambda)_{high}
\end{aligned}
\tag{6.3}
$$

where $DN(\lambda)_{sim}$ is the simulated higher resolution pixel corresponding to $DN(\lambda)_{low}$ and $DN(\gamma)_{mean}$ the local mean of $DN(\gamma)_{high}$ over a neighbourhood equivalent to the resolution of $DN(\lambda)_{low}$.

For a given solar radiation, irradiance upon a land surface is controlled by topography. If the two images are quantified to the same DN range, we can presume that $E(\lambda) \approx E(\gamma)$ for any given resolution because both vary with topography in the same way as denoted in eqn. 3.27. We can also presume that $\rho(\lambda)_{low} \approx \rho(\gamma)_{high}$ if there is no significant spectral variation within the neighbourhood for calculating $DN(\gamma)_{mean}$. Thus in

$$\frac{\rho(\lambda)_{low} E(\lambda)_{low} \times \rho(\gamma)_{high} E(\gamma)_{high}}{\rho(\gamma)_{low} E(\gamma)_{low}}$$

$E(\lambda)_{low}$ and $E(\gamma)_{low}$ cancel each other; $\rho(\gamma)_{low}$ and $\rho(\gamma)_{high}$ also cancel each other; and $E(\gamma)_{high}$ can be replaced by $E(\lambda)_{high}$. We then have the final simple solution of eqn. 6.3.

The local mean $DN(\gamma)_{mean}$ is calculated for every pixel of the higher resolution image using a convolution smoothing filter. The filter kernel size is decided by the resolution ratio between the higher and lower resolution images. For instance, to fuse a 30 m resolution TM band image with a 10 m resolution SPOT pan image, the minimum smoothing filter kernel size for calculating

the local mean of the SPOT pan image pixels is 3×3 defined as

$$\frac{1}{9}\begin{pmatrix} 1 & 1 & 1 \\ 1 & 1 & 1 \\ 1 & 1 & 1 \end{pmatrix}.$$

The image of $DN(\gamma)_{mean}$ is equivalent to the image of $DN(\lambda)_{low}$ in topography and texture because they both have a pixel size of the higher resolution image and a spatial resolution of the lower resolution image. The crucial approximation, $E(\lambda)_{low} \approx E(\gamma)_{low}$, for simplifying eqn. 6.3, therefore stands and the approach of SFIM is valid. The final result, the image of $DN(\lambda)_{sim}$, is a product of the higher resolution topography and texture, $E(\lambda)_{high}$, introduced from the higher resolution image, and the lower resolution spectral reflectance of the original lower resolution image, $\rho(\lambda)_{low}$. It is therefore independent of the spectral property of the higher resolution image used for intensity modulation. In other words, the SFIM is a spectral preservation fusion technique. This is the major advantage of SFIM over the IHS and Brovey transform fusion techniques.

Since the spectral difference between the lower and the higher resolution images is not fundamental to the operations, eqn. 6.3 can be more concisely presented as a general SFIM processing algorithm:

$$IMAGE_{SFIM} = \frac{IMAGE_{low} \times IMAGE_{high}}{IMAGE_{mean}} \tag{6.4}$$

where $IMAGE_{low}$ is a pixel of a lower resolution image co-registered to a higher resolution image of $IMAGE_{high}$, and $IMAGE_{mean}$ a pixel of smoothed $IMAGE_{high}$ using averaging filter over a neighbourhood equivalent to the actual resolution of $IMAGE_{low}$.

The ratio between $IMAGE_{high}$ and $IMAGE_{mean}$ in eqn. 6.4 cancels the spectral and topographical contrast of the higher resolution image and retains the higher resolution edges only, as illustrated by a SPOT pan image in Fig. 6.2. The SFIM can thus be understood as a lower resolution image directly modulated by higher resolution edges, and the result is independent of the contrast and spectral variation of the higher resolution image. The SFIM is therefore reliable to the spectral properties as well as contrast of the original lower resolution image.

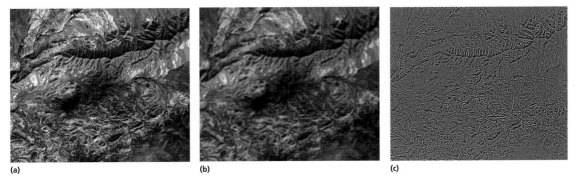

Fig. 6.2 (a) Original SPOT pan image; (b) smoothed SPOT pan image with a 5 × 5 smoothing filter; and (c) the ratio image between (a) and (b).

Unlike the RGB-IHS transform-based data fusion technique that can only operate on three multi-spectral bands each time, the SFIM defined by eqn. 6.4 can be applied directly to each individual band to generate a pan-sharpened multi-spectral imagery dataset.

6.3.2 Merits and limitations of SFIM

Figure 6.1 illustrates the TM/SPOT pan fusion results produced by IHS, Brovey transform and SFIM fusion techniques. It is clear that the SFIM result [Fig. 6.1(e)] demonstrates the highest spectral fidelity to the original TM band 541 colour composite [Fig. 6.1(a)], showing no noticeable colour differences, whilst the fusion results of both IHS and Brovey transforms [Fig. 6.1(c) and (d), respectively] present considerable colour distortion as well as contrast changes. In particular, a patch of thin cloud and shadow in the bottom right and central part of the SPOT pan image are fused into the IHS and Brovey transform fusion images, which are not shown in the SFIM fusion images because these are cancelled out by the ratio between the original SPOT pan image and its smoothed version as shown in Fig. 6.2(c).

The SFIM is sensitive to the accuracy of image co-registration. Edges with imperfect co-registration will become slightly blurred because the cancellation between $E(\lambda)_{low}$ and $E(\gamma)_{low}$ in eqn. 6.3 is no longer perfect in such a case. This problem can be eased by using a smoothing filter with a larger kernel than the resolution ratio. In such a case, $E(\gamma)_{low}$ represents lower frequency information than $E(\lambda)_{low}$ in eqn. 6.3. The division between the two does not lead to a complete cancellation and the residual is the high frequency information of the lower resolution image (relating to edges). Thus, in the fused image, the main edges appearing in both images will be sharpened whilst the subtle textural patterns, which are recognisable only in the higher resolution image, will be retained. Figure 6.1(e) is processed using the SFIM with a 5 × 5 smoothing filter; the blurring effects are effectively suppressed while the spatial details are significantly improved. In comparison, the SFIM with 3 × 3 smoothing filter in Fig. 6.1(f) is rather blurred, which subdued the improvement of spatial resolution. Thus a filter kernel at least one step larger than the resolution ratio between the low and high resolution images is recommended. Another way to improve the SFIM fusion quality is to achieve precise pixel-wise image co-registration, and this will be introduced in Chapter 11.

Please pay attention to the following issues:

1 As already mentioned, it is important that for the SFIM operations defined by eqn. 6.4, the lower resolution image must be interpolated to the same pixel size as the higher resolution image by the co-registration process; that is, the $IMAGE_{low}$ must have the same pixel size as $IMAGE_{high}$ though it is in a lower resolution. For the lower resolution image interpolation, simple pixel duplication must be avoided and instead, bilinear, biquadratic or bicubic re-sampling should be applied.

2 As SFIM is based on a solar radiation model, the technique is not applicable to the fusion of images with different illumination and imaging geometry, such as TM and ERS-1 synthetic aperture radar (SAR), or to integrate multi-source raster datasets.

3 If the spectral range of the higher resolution image is the same as that of the lower resolution colour composite and they are taken in similar solar radiation conditions,

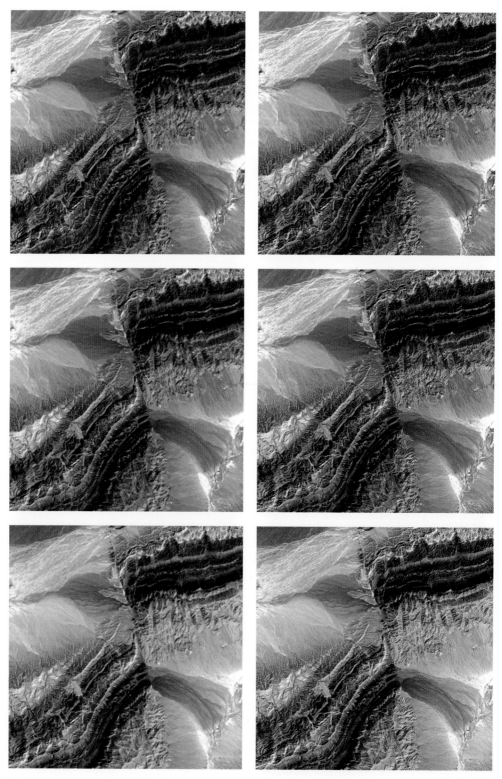

Fig. 6.3 Landsat 8 OLI colour composites before (left column) and after (right column) SFIM pan-sharpen. Row 1: bands 4-3-2RGB, true colour composite. Row 2: bands 5-4-3RGB, standard false colour composite. Row 3: bands 5-6-7RGB false colour composite.

none of the three fusion techniques will introduce significant spectral (colour) distortion. In this case, the IHS and Brovey transform fusion techniques are preferable to the SFIM for producing sharper images.

6.3.3 An example of SFIM pan-sharpen of Landsat 8 OLI image

It is a common design of many multi-spectral optical sensor systems onboard earth observation satellites to provide both multi-spectral images and panchromatic images, but the panchromatic images have higher spatial resolution than multi-spectral images that have higher spectral resolution. In this case, the images of panchromatic band and multi-spectral bands supplied to users are often precisely co-registered at pixel accuracy, thus making SFIM pan-sharpen straightforward without the need of image co-registration. Many earth observation satellite imagery data suppliers use the SFIM to produce their standard pan-sharpened imagery data product in mess production such as the Airbus-ASTRUM (Pleiades Imagery User Guide 2012, § 2.2.4 pan-sharpened, p. 17).

Here we present an example of Landsat 8 OLI (Operational Land Imager) pan-sharpen. As shown in Table A.1 in Appendix A, the panchromatic band (band 8) of OLI has 15 m resolution; multi-spectral bands (bands 1–7) have 30 m resolution. Figure 6.3 presents three pairs of colour composites of OLI multi-spectral bands before and after the pan-sharpen using the SFIM. These colour composites are bands 4-3-2 RGB true colour composites in row 1, bands 5-4-3 RGB standard false colour composites in row 2 and bands 5-6-7 RGB false colour composites in row 3. Comparing with the original images in the left column, the SFIM pan-sharpened images in the right column have improved spatial resolution without spectral distortion. The false colour composite of bands 5-6-7 RGB is chosen to be presented because the spectral ranges of these three bands are not overlapped with that of the panchromatic band, thus the spectral fidelity of the SFIM is fully verified.

6.4 Summary

Image fusion is an active research field. In this chapter, we introduced the three simplest and most popular image fusion techniques that can be performed by commonly used image processing software packages.

For image fusion aiming to improve the spatial resolution of multi-spectral images using panchromatic images, the minimisation of spectral distortion is an important issue, whilst achieving the sharpest image textures is another, not to mention economising on the processing speed. The SFIM provides a spectral preservation image fusion solution, but its high requirements in terms of image co-registration accuracy often result in slightly blurred edge textures. This weakness can be easily amended by using a larger smoothing filter. A thorough solution to overcome the weakness is a new method developed for pixel-wise image co-registration at sub-pixel accuracy, as introduced in Chapter 11. The image co-registration is only required for image fusion from multiple sensors/platforms such as Landsat and SPOT. For optical sensor systems with both multi-spectral bands and a higher resolution panchromatic band, the multi-spectral and pan images are already precisely co-registered at sensor level. Nevertheless, the RGB-IHS and Brovy transform-based techniques remain popular for their simplicity and robustness, though spectral distortion is often unavoidable.

In seeking robust spectral preservation image fusion techniques, some techniques have been developed based on wavelet transforms. Considering the current general availability, robustness, mathematic complexity and processing efficiency of the wavelet transform-based image fusion techniques, we decided not to cover this branch so as to keep the contents concise and essential. On the other hand, the SFIM is widely regarded as one of the best spectral preservation image fusion techniques so far; it is used to produce standard pan-sharpened imagery products by several large EO satellite imagery data suppliers such as Airbus-ASTRUM (Pleiades Imagery User Guide 2012, § 2.2.4 pan-sharpened, p. 17).

6.5 Questions

1 How could you improve the spatial resolution of a 30 m resolution TM colour composite with a 10 resolution SPOT panchromatic image, using RGB-IHS and Brovey transformations?

2 Explain the major problem of image fusion using RGB-IHS and Brovey transformations.

3 Describe the principle and derivation of the SFIM method, explaining why the SFIM is a spectral preservation image fusion method.

4 What is the main problem for the SFIM and how should it be dealt with?

CHAPTER 7

Principal component analysis

The principal component analysis (PCA) is a general method of analysis for correlated multi-variable datasets. Remotely sensed multi-spectral imagery is typically of such datasets for which PCA is an effective technique for spectral enhancement and information manipulation. The PCA is based on linear algebraic matrix operations and multi-variable statistics. Here we focus on the principles of the PCA technique and its applications and avoid going into the mathematical details since these comprise fairly standard linear algebraic algorithms that are implemented in most image processing software packages.

Relying on the concept of PCA as a coordinate rotation, we expand our discussion to the general concept of physical property orientated image coordinate transformation. This discussion also leads to the widely used tasselled cap transformation in the derivation of multispectral indices of brightness, greenness and wetness.

PCA can effectively concentrate the maximum information of many correlated image spectral bands into a few uncorrelated principal components (PCs) and therefore can reduce the size of a dataset and enable effective image RGB display of its information. This links to the statistical methods for band selection that aim to select optimum band triplets with minimal interband correlation and maximum information content.

7.1 Principle of the PCA

As shown in Table 7.1, the six reflective spectral bands of a TM image are highly correlated. For instance, the correlation between band 5 and band 7 is 0.993. This means that there is 99.3% information redundancy between these two bands and only 0.7% of unique information! This is the general case for multi-spectral earth observation imagery data because topography represents the image features common to all bands. The narrower the spectral ranges of the image bands, the higher the correlation between the adjacent bands. As such, multi-spectral imagery is not efficient for information storage.

Consider an m band multi-spectral image as an m-dimensional raster dataset in an m dimension orthogonal coordinate system, forming an m-dimensional ellipsoid cluster. Then the coordinate system is oblique to the axes of the ellipsoid data cluster if the image bands are correlated. The axes of the data ellipsoid cluster formulate an orthogonal coordinate system in which the same image data are represented by $n\,(n \le m)$ independent *principal components*. In other words, the PCs are the image data representation in the coordinate system formulated by the axes of the ellipsoid data cluster. Thus, *principal component analysis is a coordinate rotation operation to rotate the coordinate system of the original image bands to match the axes of the ellipsoid of the image data cluster.* As shown by a two-dimensional (2D) illustration in Fig. 7.1(a), suppose the image data points form an elliptic cluster; the aim of PCA is to rotate the orthogonal co-ordinate system of band 1 and band 2 to match the two axes of the ellipsoid, the PC1 and PC2. The coordinates of each data point in the PC co-ordinate system will be the DNs of the corresponding pixels in the PC images. The first PC is represented by the longest axis of the data cluster and the second PC the second longest, and so on. The axes representing high order PCs may be too short to represent any substantial information, and then the apparent m-dimensional

Image Processing and GIS for Remote Sensing: Techniques and Applications, Second Edition. Jian Guo Liu and Philippa J. Mason.
© 2016 John Wiley & Sons, Ltd. Published 2016 by John Wiley & Sons, Ltd.

ellipsoid is effectively degraded to n ($n < m$) independent dimension. For instance, as shown in Fig. 7.1(b), the three-dimensional (3D) data cluster is effectively 2D, as the PC3 axis is very short, representing little independent information. The same data can then be effectively represented by PC1 and PC2 in a 2D coordinate system with little information loss. In this way, PCA reduces image dimensionality and represents nearly the same image information with fewer independent dimensions in a smaller dataset without redundancy. In summary, the PCA is a linear transformation converting m correlated dimensions to n ($n \le m$) independent (uncorrelated) dimensions. This is equivalent to a co-ordinate rotation transform to rotate the original m axes oblique to the ellipsoid data cluster to match the orientation of the axes of the ellipsoid in n independent dimensions, and thus the image data represented by each dimension is orthogonal (independent) to all the other dimensions.

For image processing, PCA generates uncorrelated PC images from the originally correlated image bands.

Let **X** represent an m band multi-spectral image; its *covariance matrix* Σ_x is a full representation of the m-dimensional ellipsoid cluster of the image data. The covariance matrix is a non-negative definite matrix, and it is symmetrical along its major diagonal. Such a matrix can be converted into a diagonal matrix via basic matrix operations. The elements on the major diagonal of the covariance matrix are the variance of each image band, whilst the symmetrical elements off the major diagonal are the covariance between two different bands. For instance, a covariance matrix of a four bands image is as below:

$$\begin{pmatrix} \sigma_{11} & \sigma_{12} & \sigma_{13} & \sigma_{14} \\ \sigma_{21} & \sigma_{22} & \sigma_{23} & \sigma_{24} \\ \sigma_{31} & \sigma_{32} & \sigma_{33} & \sigma_{34} \\ \sigma_{41} & \sigma_{42} & \sigma_{43} & \sigma_{44} \end{pmatrix}$$

For the elements not on the major diagonal in this matrix, $\sigma_{ij} = \sigma_{ji}$. The elements σ_{12} and σ_{21} both are the covariance between band 1 and band 2 and so on. If band 1 and band 2 are independent, then their covariance $\sigma_{12} = \sigma_{21} = 0$. This means that independent variables in a multi-dimensional space should have a diagonal covariance matrix. Thus an image dataset of n independent PCs should have a diagonal covariance matrix.

In mathematics, the PCA is simply to find a transformation **G** that diagonalizes the covariance matrix Σ_x of

Table 7.1 Correlation matrix of bands 1–5, 7 of a TM sub-scene.

Correlation	TM1	TM2	TM3	TM4	TM5	TM7
TM1	1.000	0.962	0.936	0.881	0.839	0.850
TM2	0.962	1.000	0.991	0.965	0.933	0.941
TM3	0.936	0.991	1.000	0.979	0.955	0.964
TM4	0.881	0.965	0.979	1.000	0.980	0.979
TM5	0.839	0.933	0.955	0.980	1.000	0.993
TM7	0.850	0.941	0.964	0.979	0.993	1.000

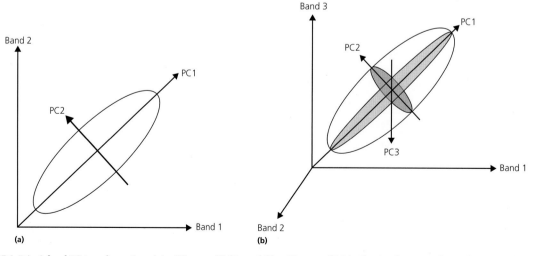

Fig. 7.1 Principle of PC transformation: (a) a 2D case of PCA; and (b) a 3D case of PCA. The 3D cluster is effectively 2D, as the value range of the PC3 is very narrow and the data distribution is mainly in an elliptic plate.

the m bands image $\mathbf{X}$ to produce an n PCs image $\mathbf{Y}$ with a diagonal covariance matrix Σ_y. The rank of Σ_y is n and $n = m$ if Σ_x is a full rank matrix; otherwise, $n < m$ with reduced dimensionality.

The covariance matrix of an m bands image $\mathbf{X}$ is defined as below:

$$\sum_x \varepsilon\left\{(\mathbf{x} - \mathbf{m}_x)(\mathbf{x} - \mathbf{m}_x)^t\right\} \approx \frac{1}{N-1}\sum_{j=1}^{N}(\mathbf{x}_j - \mathbf{m}_x)(\mathbf{x}_j - \mathbf{m}_x)^t \tag{7.1}$$

where $\mathbf{x}_j = (x_{j1}, x_{j2}, \dots x_{jm})^t$, $(\mathbf{x}_j \in \mathbf{x}, j = 1, 2, \dots, N)$ is any an m-dimensional pixel vector of an m bands image $\mathbf{X}$, N the total number of pixels in the image $\mathbf{X}$ and $\mathbf{m}_x$ the mean vector of the image $\mathbf{X}$. The operation ε is a mathematical expectation.

$$\mathbf{m}_x = \varepsilon\{\mathbf{x}\} = \frac{1}{N-1}\sum_{j=1}^{N}\mathbf{x}_j \tag{7.2}$$

Since the covariance matrix Σ_x is a symmetrical, non-negative definite matrix, there exists a linear transformation $\mathbf{G}$ that diagonalizes Σ_x. Let

$$\mathbf{y} = \mathbf{Gx} \tag{7.3}$$

subject to the constraint that the covariance matrix of $\mathbf{y}(\mathbf{y}_j \in \mathbf{y}, j = 1, 2, \dots, N)$ is diagonal. In $\mathbf{Y}$ space the covariance matrix is, by definition,

$$\sum_y = \varepsilon\left\{(\mathbf{y} - \mathbf{m}_y)(\mathbf{y} - \mathbf{m}_y)^t\right\} \tag{7.4}$$

where $\mathbf{m}_y$ is the mean vector of the transformed image $\mathbf{Y}$. Thus we have

$$\mathbf{m}_y = \varepsilon(\mathbf{y}) = \varepsilon\{\mathbf{Gx}\} = \mathbf{G}\varepsilon\{\mathbf{x}\} = \mathbf{Gm}_x \tag{7.5}$$

$$\sum_y = \varepsilon\left\{(\mathbf{Gx} - \mathbf{Gm}_x)(\mathbf{Gx} - \mathbf{Gm}_x)^t\right\}$$
$$= \mathbf{G}\varepsilon\left\{(\mathbf{x} - \mathbf{m}_x)(\mathbf{x} - \mathbf{m}_x)^t\right\}\mathbf{G}^t \tag{7.6}$$
$$= \mathbf{G}\Sigma_x\mathbf{G}^t$$

As Σ_y is the diagonal matrix derived from Σ_x, according to the rules of matrix operations we can prove that the transformation $\mathbf{G}$ is the $n \times m$ transposed matrix of the eigenvectors of Σ_x.

$$\mathbf{G} = \begin{pmatrix} g_{11} & g_{12} & \cdots & g_{1m} \\ g_{21} & g_{22} & \cdots & g_{2m} \\ \cdots & \cdots & \cdots & \cdots \\ g_{n1} & g_{n2} & \cdots & g_{nm} \end{pmatrix} = \begin{pmatrix} \mathbf{g}_1^t \\ \mathbf{g}_2^t \\ \cdots \\ \mathbf{g}_n^t \end{pmatrix} \tag{7.7}$$

The Σ_y is a diagonal matrix with the eigenvalues of Σ_x as non-zero elements along the diagonal:

$$\Sigma_y = \begin{pmatrix} \lambda_1 & & & 0 \\ & \lambda_2 & & \\ & & \ddots & \\ 0 & & & \lambda_n \end{pmatrix} \tag{7.8}$$
$$\lambda_1 > \lambda_2 > \cdots > \lambda_n$$

The eigenvalue λ_i is the variance of PC_i image and is proportional to the information contained in PC_i. As indicated in eqn. 7.8, the information content decreases with the increment of the PC rank.

In computing, the key operation of PCA is to find eigenvalues of Σ_x from which the eigenvector matrix $\mathbf{G}$ is derived. The eigenvalues of Σ_x can be calculated from its *characteristic equation*:

$$|\Sigma_x - \lambda\mathbf{I}| = 0 \tag{7.9}$$

where $\mathbf{I}$ is an m dimension identity matrix.

Any eigenvector of matrix Σ_x is defined as a vector $\mathbf{g}(\mathbf{g} \in \mathbf{G})$ that satisfies

$$\Sigma_x\mathbf{g} = \lambda\mathbf{g} \quad \text{or} \quad (\Sigma_x - \lambda\mathbf{I})\mathbf{g} = 0 \tag{7.10}$$

This formula is called the *characteristic polynomial* of Σ_x. Thus once the ith eigenvalue λ_i is known then the ith eigenvector $\mathbf{g}_i$ is determined. There are several standard computing algorithms for the numerical solutions of the characteristic equation (7.9) but the mathematics is beyond the scope of this book.

Eigenvector $\mathbf{G}$ determines how each PC is composed from the original image bands. In fact, each component image is a linear combination (a weighted summation) of the original image bands:

$$PC_i = \mathbf{g}_i^t\mathbf{X} = \sum_{k=1}^{m}g_{ik}Band_k \tag{7.11}$$

where g_{ik} is the element of $\mathbf{G}$ at the ith row and kth column, or the kth element of the ith eigenvector $\mathbf{g}_i^t = (g_{i1}, g_{i2}, \dots g_{ik} \dots, g_{im})$.

7.2 PC images and PC colour composition

PC images are useful for reducing data dimensionality, condensing topographic and spectral information, improving image colour presentation and enhancing specific spectral features. Here we discuss some

Table 7.2 The covariance matrix of bands 1–5, 7 of a TM sub-scene.

Covariance	TM1	TM2	TM3	TM4	TM5	TM7
TM1	232.202	196.203	305.763	348.550	677.117	345.508
TM2	196.203	178.980	284.415	335.185	660.570	335.997
TM3	305.763	284.415	460.022	545.336	1083.993	551.367
TM4	348.550	335.185	545.336	674.455	1347.927	678.275
TM5	677.117	660.570	1083.993	1347.927	2802.914	1402.409
TM7	345.508	335.997	551.367	678.275	1402.409	711.647

Table 7.3 The eigenvector matrix and eigenvalues of the covariance matrix of bands 1–5, 7 of a TM sub-scene.

Eigenvectors	PC1	PC2	PC3	PC4	PC5	PC6
TM1	0.190	−0.688	−0.515	−0.260	−0.320	−0.233
TM2	0.183	−0.362	0.032	0.050	0.136	0.902
TM3	0.298	−0.418	0.237	0.385	0.638	−0.354
TM4	0.366	−0.136	0.762	−0.330	−0.389	−0.079
TM5	0.751	0.433	−0.296	−0.318	0.242	0.013
TM7	0.378	0.122	−0.093	0.756	−0.511	0.011
Eigenvalues	4928.731	102.312	15.581	9.011	3.573	1.012
Information	97.4%	2.02%	0.31%	0.18%	0.07%	0.02%

characteristics of PC images using an example. The covariance matrix and eigenvector matrix of six reflective spectral bands of a small sub-scene of a Landsat TM image are presented in Table 7.2 and Table 7.3 and the inter-band correlation matrix is shown in Table 7.1. The six PC images are shown in Fig. 7.2. We make the following observations:

1 The elements of $\mathbf{g}_1$ are all positive and therefore *PC1* [Fig. 7.2(a)] is a weighted summation of all the original image bands. In this sense, it resembles a broad spectral range panchromatic image. It has a very large eigenvalue of 4928.731 (*PC1* variance) and accounts for 97.4% of the information from all six bands. For a fixed DN range, more information means a higher signal-to-noise ratio (SNR). This conforms to the conclusion that image summation increases SNR stated in § 3.1.

2 *PC1* concentrates features common to all six bands. For earth observation satellite images, this common information is usually topography.

3 The elements of $\mathbf{g}_i (i > 1)$ are usually a mixture of positive and negative values and thus PC images of higher rank (>1) are linear combinations of positive and negative images of the original bands.

4 The higher ranked PCs lack topographic features and show more spectral variation. They all have significantly

smaller eigenvalues (PC variances) than the *PC1*. The eigenvalues decrease rapidly with the increment of PC rank and so have progressively lower SNR, which is illustrated by their increasingly noisy appearance. The *PC6* image is almost entirely noise and contains little information, as indicated by very small variance of 1.012. In this sense, *PC6* can be disregarded from the dataset and thus the effective dimensionality is reduced to 5 from the original 6 with negligible information loss of 0.02%.

We can look at individual PC images or display three PCs as a colour composite. As *PC1* is mainly topography, colour composites excluding *PC1* may better present spectral information with topography subdued. PCs represent condensed and independent image information and therefore produce more colourful (i.e. informative) colour composites. However, here we have a problem in that a PC is a linear combination of the original spectral bands, so its relationship to the original spectral signatures of targets representing ground objects is no longer apparent.

To solve this problem, a feature-oriented PC selection (*FPCS*) method for colour composition was proposed by Crosta and Moore (1989). The technique provides a simple way to select PCs from the spectral signatures of

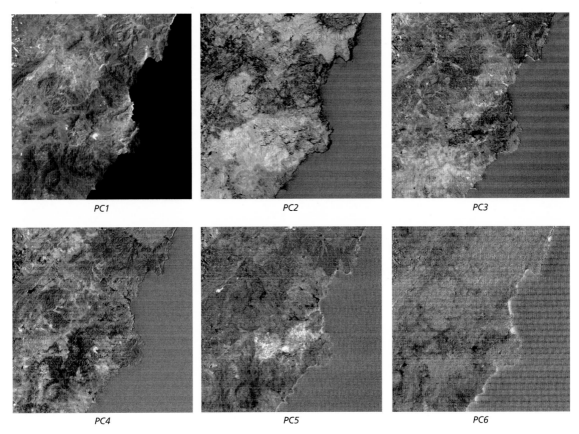

Fig. 7.2 PC images derived from six reflective spectral bands of a sub-scene of a TM image. The *PC1–PC6* images are arranged from the top left to the bottom right.

significant spectral targets (e.g. minerals) to enhance the spectral information of these minerals in the PC colour composite. The technique involves examination of the eigenvectors to identify the contributions from original bands (either negative or positive) to each PC. Specific PCs can then be selected on the basis of the major contributors that are likely to display the desired targets (spectral features).

Let us look at the eigenvectors in Table 7.3. The *PC3* is dominated by large positive loading (0.762) from TM4 caused by high reflectance of vegetation in near infrared (NIR) and large negative loading (-0.515) of the blue band TM1. The red band TM3 and the short-wave infrared (SWIR) band TM5 give the second largest positive and negative contribution, respectively. The PC3 therefore highlights vegetation and particularly vegetation on red soils. The largest positive contribution to *PC4* is from the clay absorption SWIR band TM7 (0.756), and the other elements of the *PC4* include a

positive contribution from TM3 and negative contributions of similar amounts from TM1, TM4 and TM5. The *PC4* therefore enhances ferric iron and lacks the signature of clay minerals. The *PC5* can be considered as a combination of the difference between TM5 and TM7, highlighting clay minerals, and between TM3 and TM1, highlighting iron oxides, together with a negative contribution from vegetation. *PC5* is therefore effective for indicating hydrothermal alteration minerals by its strong co-occurrence of clay minerals and iron oxide. Figure 7.3(a) is a colour composite displaying the *PC4* in red, *PC3* in green and *PC5* in blue. Apart from showing vegetation in green and red soils/regoliths in red, the image effectively highlights hydrothermal alteration zones of a known epithermal gold deposit distinctively in blue. Geologically the alteration zone is characterised by high concentrations of alteration-induced clay minerals and gossaniferous iron oxides on the surface of outcrops.

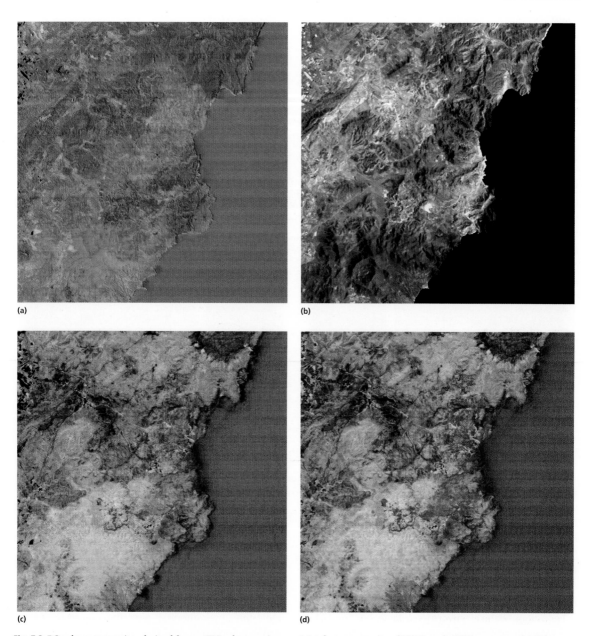

Fig. 7.3 PC colour composites derived from a TM sub-scene image: (a) colour composite of *PC4* in red, *PC3* in green and *PC5* in blue: (b) SPCA colour composite using *PC1*s derived from bands groups listed in Table 7.4: (c) SPCA (spectral contrast mapping) colour composite using *PC2*s derived from band groups listed in Table 7.5: (d) FPCS spectral contrast mapping using *PC2*s and *PC3* derived from band groups listed in Table 7.6.

7.3 Selective PCA for PC colour composition

The eigenvalues representing the variances of PCs in Table 7.3 indicate that for commonly used remotely sensed multi-spectral image data, a very large portion of information (data variance) is concentrated in *PC1*, and this relates to the irradiance variation on topography. Higher rank PCs contain significantly less information but it is more relevant to the spectral signatures of specific ground objects. Colour composites of PCs are often very effective for highlighting such ground objects and minerals that may not be distinguishable in colour composites of the original bands. In PC colour composites noise may be exaggerated because the high rank PCs contain significantly less information than lower rank PCs and they have very low SNRs. When PC images are stretched and displayed in the same value range, the noise in higher rank PCs is improperly enhanced.

We would like to use three PCs with comparable information levels for colour composite generation. Chavez (1989) introduced a general approach, referred to as selective principal component analysis (SPCA), to produce PC colour composites in which the maximum information of either topographic or spectral features is condensed and in which the information content from each PC displayed is better balanced. There are two types of SPCA: *dimensionality and colour confusion reduction* and *spectral contrast mapping*.

7.3.1 Dimensionality and colour confusion reduction

The spectral bands of a multi-spectral image are arranged into three groups and each group is composed of highly correlated bands. PCA is performed on each group and then the three *PC1*s derived from these three groups are used to generate an RGB colour composite. As bands in each group are highly correlated, the *PC1* concentrates the maximum information of each group. For six reflective spectral bands of a TM or ETM+ image, the technique may condense more than 98% variance (information) in the derived *PC1* colour composite. The recommended groups for six reflective spectral bands of TM or ETM+ images are shown in Table 7.4.

The approach sounds clever but it is actually equivalent to generating a colour composite using broader spectral bands, given that a *PC1* is a positively weighted summation of the bands involved in the PCA. As illustrated in Fig. 7.3(b), it is essentially a colour composite of a broad visible band in blue, an NIR band in green and broad SWIR band in red.

7.3.2 Spectral contrast mapping

A more interesting and useful approach is spectral contrast mapping, where the primary objective is to map the contrast between different parts of the spectrum, and so to identify information unique to each band rather than information in common. For this purpose, *PC2*s derived from band pairs are used instead of *PC1*s. By using only two bands as inputs, the information that is common to both bands is mapped to *PC1* and the unique information is mapped to *PC2*. In general, low or medium correlation between the bands in each pair is preferred for this approach. The recommended grouping for the six reflective spectral bands of TM/ETM+ is listed in Table 7.5 as an example.

Based on the above principle, groups of three or more bands may also be used for spectral contrast mapping. The technique can generate spectrally informative colour composites with significantly reduced topographic shadow effects, as illustrated in Fig. 7.3(c). In this colour composite, the *PC2* from bands 1 and 3 shows red spectral contrast relating to red soils and iron oxides, and the *PC2* from bands 2 and 4 relates to vegetation while the *PC2* from bands 5 and 7 effectively highlights the spectral contrast of clay alteration minerals. The striped pattern in the sea in this image is less obvious than that in the *PC4-3-5* RGB colour composite in Fig. 7.3(a), implying better SNR.

7.3.3 FPCS spectral contrast mapping

The outcome of the spectral contrast mapping largely depends on the spectral band groupings. Knowing the spectral signatures of intended targets, we can use the FPCS method to decide the grouping of bands and then the selection of PCs for the final RGB display. We demonstrate the principle of this approach with the same example as above.

From the eigenvector matrix in Table 7.3, we see that none of the PCs picks up the 'red edge' feature diagnostic to vegetation nor the absorption feature of clay minerals characterised by the difference between TM5 and TM7. We therefore consider the grouping listed in

Table 7.4 Dimensionality and colour confusion reduction for TM or ETM+.

Groups	PCA	Colour
TM1,2,3	$PC1_{1,2,3}$	Blue
TM4		Green
TM5,7	$PC1_{5,7}$	Red

Table 7.5 TM or ETM+ spectral bands grouping for spectral contrast mapping.

Groups	PCA	Colour
TM1,3	$PC2_{1,3}$	Red
TM2,4	$PC2_{2,4}$	Green
TM5,7	$PC2_{5,7}$	Blue

Table 7.6 FPCS spectral contrast mapping for TM/ETM+.

Groups	Intended Targets	FPCS	Colour
TM 1,2,3	Red soils and iron oxide	$PC2_{1,2,3}$	Red
TM 2,3,4	Vegetation	$PC2_{2,3,4}$	Green
TM 3,5,7	Clay minerals	$PC3_{3,5,7}$	Blue

Table 7.7 Eigenvector matrixes of the three band groups for FPCS spectral contrast mapping.

Eigenvector	PC1	PC2	PC3
TM1	0.507	*−0.834*	−0.218
TM2	0.458	*0.046*	0.888
TM3	0.731	*0.549*	−0.405
TM2	0.366	*−0.510*	−0.778
TM3	0.593	*−0.517*	0.617
TM4	0.717	*0.687*	−0.113
TM3	0.331	0.920	*0.210*
TM5	0.843	−0.388	*0.372*
TM7	0.424	0.054	*−0.904*

Table 7.6. From the eigenvector matrixes in Table 7.7, we make the following observations:

1 The *PC2* derived from TM bands 1, 2 and 3 is essentially the difference between red (TM3) and blue (TM1) and it therefore enhances red features like iron oxides. This *PC2* is chosen to display in red.

2 The *PC2* derived from TM bands 2, 3 and 4 is dominated by the positive contribution of NIR (TM4) and balanced by a negative contribution from red (TM3) and green (TM2). It therefore highlights healthy vegetation. This *PC2* is chosen to display in green.

3 The *PC3* derived from TM bands 3, 5 and 7 is a summation of TM3 and TM5 subtracting TM7 and therefore enhances clay alteration minerals and iron oxides. This *PC3* is chosen to display in blue.

The resultant FPCS spectral contrast mapping colour composite in Fig. 7.3(d) resembles the simple SPCA spectral contrast mapping colour composite in Fig. 7.3(c), but the signatures of red soils/regoliths, vegetation and clay minerals are more distinctively displayed in red, green and blue.

After all the effort of SPCA, Fig. 7.3 indicates that both the spectral contrast mapping and the FPCS spectral contrast mapping images are less colourful than the simple colour composite of PCs. One of the reasons for this is that

the selected PCs from the three different bands groups are not independent. They may be well correlated even though the PCs within each group are independent from each other. For instance, if we group TM bands 1 and 5 as one group and bands 2 and 7 as another, then the *PC2*s derived from the two groups will be highly correlated because both bands 1 and 2 and bands 5 and 7 are highly correlated. The way the image bands are grouped will control the effectiveness of spectral contrast mapping.

7.4 De-correlation stretch

A very important application of PCA is the *de-correlation stretch* (DS). We have already learnt two saturation stretch-based DS techniques in Chapter 5, but the initial concept of the de-correlation stretch, as proposed by Taylor (1973), was based on PCA and further developed by Soha and Schwartz (1978).

The interpretation of PC colour composites is not straightforward, and ordinary colour composites of the original image bands are often needed for reference. The PCA-based DS (PCADS) generates a colour composite from three image bands with reduced inter-band correlation and thus presents image spectral information in more distinctive and saturated colours without distorting hues. The idea of PCADS is to stretch multi-dimensional image data along their PC axes (the axes of the data ellipsoid cluster) rather than original axes representing image bands. In this way, the volume of data cluster can be effectively increased and the inter-band correlation is reduced, as illustrated in a 2D case in Fig. 7.4. The PCADS is achieved in three steps:

- PCA to transform the data from the original image bands to PCs;
- contrast enhancement on each of the PCs (stretching the data cluster along PC axes);
- inverse PCA to convert the enhanced PCs back to the corresponding image bands.

According to the PCA defined in the eqn. 7.6, the inversed PCA is defined as below:

$$\Sigma_x = \mathbf{G}^{-1}\Sigma_y\left(\mathbf{G}^t\right)^{-1} \tag{7.12}$$

The PCADS technique effectively improves colour saturation of a colour composite image without changing its hue characteristics. It is similar to the IHSDS in its result but is based on quite different principles. PCADS is statistically scene dependent as the whole operation

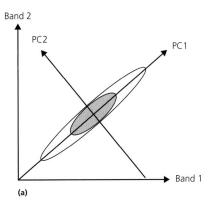 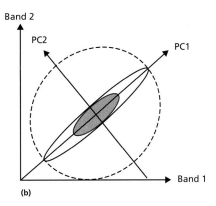

Fig. 7.4 Illustration of the DS in a 2D case: (a) stretching of the original bands is equivalent to stretching along the *PC1* axis to elongate the data cluster; and (b) stretching along both the *PC1* and *PC2* axes expands the elliptic data cluster in 2D as denoted by the dashed line ellipse.

starts from the image covariance matrix, and it can be operated on all image bands simultaneously. The IHSDS, in contrast, is not statistically scene dependent and only operates on three image bands. Both techniques involve complicated forward and inverse coordinate transformations. In particular, the PCADS requires quite complicated inverse operations on the eigenvector matrix and is therefore not computationally efficient. The direct de-correlation stretch (DDS) is the most efficient technique and it can be quantitatively controlled based on the saturation level of the image.

7.5 Physical property orientated coordinate transformation and tasselled cap transformation

The image PCA is a rotational operation of an m dimension orthogonal coordinate system for an m band multi-spectral image **X**. The rotation is scene dependent and determined by the eigenvector matrix **G** of the covariance matrix Σ_x. Consider the rotation transform defined in eqn. 7.3; in a general sense, we can arbitrarily rotate the m-dimensional orthogonal coordinate system in any direction as defined by a transformation **R**.

$$\mathbf{y} = \mathbf{R}\mathbf{x} \tag{7.13}$$

Here **y** is a linear combination of **x** specified by the coefficients (weights) in **R**.

For example, a 3-D rotation from $\mathbf{x}(x_1, x_2, x_3)$ to $\mathbf{y}(y_1, y_2, y_3)$ is defined as

$$\begin{pmatrix} y_1 \\ y_2 \\ y_3 \end{pmatrix} = \begin{pmatrix} \cos\alpha_1 & \cos\beta_1 & \cos\gamma_1 \\ \cos\alpha_2 & \cos\beta_2 & \cos\gamma_2 \\ \cos\alpha_3 & \cos\beta_3 & \cos\gamma_3 \end{pmatrix} \begin{pmatrix} x_1 \\ x_2 \\ x_3 \end{pmatrix} \tag{7.14}$$

where the subscript 1 denotes the rotation angles between y_1 axis and the x_1, x_2, x_3 axes in the positive direction; the subscript 2 the rotation angles between y_2 axis and the x_1, x_2, x_3 axes in the positive direction, and so on for the subscript 3.

In addition to the rotation, we can also consider a coordinate shift, as defined by a shift vector **C**, and thus eqn. 7.13 is modified as

$$\mathbf{y} = \mathbf{R}\mathbf{x} + \mathbf{C} \tag{7.15}$$

Based on image data analysis of spectral signatures of particular targets, we can learn the data distribution of specific physical properties of ground objects, such as vegetation greenness, soil brightness and land surface wetness. We can then rotate the original image coordinate system to orientate to the directions of the maximum variation of these physical properties if these properties are orthogonal. This operation is quite similar to the PCA, but the rotation is decided by the transformation **R** as derived from sample image data representative of the intended physical properties rather than the eigenvector matrix **G**, which is derived from the image covariance matrix. As such, the rotational transformation **R** is invariant to the images taken by the same multi-spectral sensor system, on the one hand, but is constrained and biased by its empirical nature, on the other hand.

One of the most successful examples of the physical property orientated coordinated transformation is the *tasselled cap transformation* initially derived by Kauth and Thomas (1976) for Landsat MSS and then further developed by Crist and Cicone (1984) for Landsat TM. As shown in Fig. 7.5, the goal of the tasselled cap transformation is to transform the six reflective spectral bands (1–5, 7) of TM or ETM+ in visible light and nearer infrared (VNIR) and SWIR spectral ranges, into three orthogonal components orientated as three key properties of the land surface: *brightness, greenness* (vigour of green vegetation) and *wetness*. The axes of brightness and greenness define the plane of vegetation presenting the 2D scattering of vegetation of varying greenness and grown on soils of different brightness, while the axes of wetness and brightness define the plane of soil presenting the 2D scattering of soil brightness in relation to soil moisture. Mainly based on a TM image of North Carolina taken on 24 September 1982 together with several other TM scenes and simulated TM images, Crist and Cicone derived the TM tasselled cap transformation as below:

and wetness using the FPCS approach. However, for a multi-spectral image of a barren region where rock and regolith are dominant, the data variance representing vegetation will be very limited and not recognised as an axis of the data ellipsoid cluster. As a result, none of the PCs can represent greenness, but using the tasselled cap transformation, the physical properties of greenness will be shown with a low and narrow value range. The tasselled cap transformation 'pinpoints' the pre-defined physical property and is therefore scene independent and not affected by variation in land cover.

The tasselled cap transformation is widely used for its high relevance to the crop growth cycle, soil property and surface moisture (both soil and vegetation). Since it is derived from sample TM image data of particular areas, although quite representative in general terms, it is not a universal model nor is it correct for all the regions of the earth; caution and critical assessment must be applied when using it. Considering variations in imaging conditions and environments, several variants of the tasselled cap transformation

$$\begin{pmatrix} Brightness \\ Greenness \\ Wetness \end{pmatrix} = \begin{pmatrix} 0.3037 & 0.2793 & 0.4343 & 0.5585 & 0.5082 & 0.1863 \\ -0.2848 & -0.2435 & -0.5436 & 0.7243 & 0.0840 & -0.1800 \\ 0.1509 & 0.1793 & 0.3299 & 0.3406 & -0.7112 & -0.4572 \end{pmatrix} \begin{pmatrix} TM1 \\ TM2 \\ TM3 \\ TM4 \\ TM5 \\ TM7 \end{pmatrix} \qquad (7.16)$$

The transformation in eqn. 7.16 can also be expressed in Table 7.8. Obviously, as a positive weighted summation of all the bands, the brightness is equivalent to a *PC1* image. From an image with vegetation and soils in various wetness levels as dominant land cover, we may locate higher rank PCs equivalent to greenness

have been proposed. One example is the tasselled cap transformation for Landsat 7 ETM+ at-satellite reflectance, as shown in Table 7.9 (Huang *et al.* 2002). It was derived in consideration that effective atmospheric correction is often not feasible for regional applications.

Table 7.8 Crist and Cicone TM tasselled cap transformation coefficients.

TM Band	1	2	3	4	5	7
Brightness	0.3037	0.2793	0.4343	0.5585	0.5082	0.1863
Greenness	−0.2828	−0.2435	−0.5436	0.7243	0.0840	−0.1800
Wetness	0.1509	0.1793	0.3299	0.3406	−0.7112	−0.4572

Table 7.9 Landsat 7 ETM+ at-satellite reflectance tasselled cap transformation coefficients.

ETM+ Band	1	2	3	4	5	7
Brightness	0.3561	0.3972	0.3904	0.6966	0.2286	0.1596
Greenness	−0.3344	−0.3544	−0.4556	0.6966	−0.0242	−0.2630
Wetness	0.2626	0.2141	0.09926	0.0656	−0.7629	−0.5388

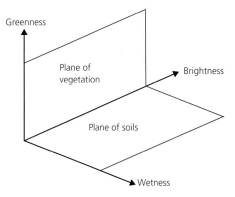

Fig. 7.5 The tasselled cap transformation coordinate system (Mather, 2004. Reproduced with permission of John Wiley & Sons.)

7.6 Statistical methods for band selection

With the increase in spectral resolution of remotely sensed data, more bands are available for colour composition, but only three bands can be used. For any particular application, it is not practical (or necessary) to exhaust all the possible three-band combinations for colour composition; band selection therefore becomes a quite important issue.

Band selection techniques can be divided in two major types: *statistical band selection* and *target-oriented band selection* techniques. Statistical techniques are normally used for selecting a few bands that may produce relatively high quality colour composites of optimised visualisation for general purposes. Target-oriented techniques are applied to highlight image features of particular interest. For instance, for vegetation studies image bands in the VNIR spectrum are essential, while for mineral mapping, SWIR bands are the most effective.

For 7 bands TM images, 11 bands ATM images and 14 bands ASTER images, statistical techniques are very effective tools to help users produce the few most informative colour composites from a large number of potential

3 bands combinations. For image data with much higher spectral resolution, such as AVIRIS (224 bands), the statistical approach becomes unfeasible. The band selection, in such cases, should be based on the spectral signatures of the particular targets. A combined approach could also be considered; that is, locating the relevant spectral range(s) from target spectral signatures first, and then applying statistical techniques to select the most informative 3- band groups for colour composition.

This section introduces a statistical band selection technique from Crippen (1989) that is a more sound technique than the two widely used techniques as briefly reviewed here.

7.6.1 Review of Chavez's and Sheffield's methods

Generally, bands with low inter-band correlation contain more information than highly correlated bands and therefore tend to produce colourful colour composites. Based on this principle, two band selection techniques were developed by Chavez *et al.* (1982) and Sheffield (1985).

Chavez's technique, the Optimal Index Factor (OIF), was initially designed for selecting Landsat MSS ratios for ratio colour composites and was applied later to more advanced multi-spectral imagery (e.g. Landsat TM and ATM) for band selection. The OIF is defined as

$$OIF = \frac{SD_i + SD_j + SD_k}{|r_{ij}| + |r_{ik}| + |r_{jk}|} \qquad (7.17)$$

where SD_i is the standard deviation for band i and r_{ij} is the correlation coefficient between band i and band j.

The largest OIF is considered to indicate the best band triplet for colour composition.

Sheffield's method is more general, and for selecting n bands from an m ($>n$) bands image. The selection is based on the volume of the n-dimensional ellipsoids, as defined by $n \times n$ principal sub-matrixes of the covariance matrix, and a larger volume indicates higher information content. For colour composition, the data distributions of band triplets ($n = 3$) are represented by 3×3 principal sub-matrixes in 3D ellipsoids. The band triplet having an ellipsoid of the maximum volume is considered to contain the maximum information. Since the volume of the ellipsoid representing the data of a band triplet is decided by the value of the determinant of the correspondent 3D principal sub-matrix, selection is performed by computing and ranking of the determinants of each 3×3 principal sub-matrix of the m band covariance matrix.

In both methods, the variances of image bands are considered as indicators of information content and are used as the basis for band selection. In fact, the variance of an image can be easily changed by contrast enhancement. A linear stretch does not change the information of an image and it is a common case for sensor calibration using gain factors, but it can affect band selection using these two techniques because a linear stretch increases the standard deviation in Chavez's method and the volume of the ellipsoid defined by the 3×3 principal sub-matrixes of the covariance matrix in Sheffield's method. A good band selection technique should be unaffected by linear stretch at the very least.

7.6.2 Index of three-dimensionality

If we leave aside the issue of image information content, we should realise that the main reason for a poor colour composite is not the lack of information in each band but the information redundancy caused by high correlation between the three bands, as illustrated in Table 7.1. An effective statistical band selection technique should therefore aim to choose band triplets with minimum inter-band correlation and therefore minimum information redundancy.

We can prove that the correlation coefficients among different bands are independent of linear operations. Let r_{ij} represent the correlation coefficient between two image bands, X_i and X_j, then,

$$r_{ij} = \frac{\sigma_{ij}}{\sqrt{\sigma_{ii}\sigma_{jj}}} \qquad (7.18)$$

where σ_{ii} and σ_{jj} are the variances of band i and j and σ_{ij} is the covariance between the two bands.

If band X_i is enhanced by a linear stretch, $Y_i = a_i X_i + b_i$ and band X_j, $Y_j = a_j X_j + b_j$

Then the variance and covariance of Y_i and Y_j are

$$a_i^2 \sigma_{ii} \quad a_j^2 \sigma_{jj} \quad a_i a_j \sigma_{ij}$$

Thus the correlation coefficient between Y_i and Y_j is

$$R_{ij} = \frac{a_i a_j \sigma_{ij}}{\sqrt{a_i^2 \sigma_{ii} a_j^2 \sigma_{jj}}} = \frac{\sigma_{ij}}{\sqrt{\sigma_{ii}\sigma_{jj}}} = r_{ij} \qquad (7.19)$$

Therefore, a band selection technique based on band correlation coefficients is not affected by linear contrast enhancement.

Based on this principle, Crippen (1989) proposed using the square root of the determinant of the three-band correlation matrix as a measurement of the three-dimensionality of the three-band data distribution. A high three-dimensionality indicates a spherical data distribution and so a low inter-band correlation. The determinant of the correlation matrix of band i, j and k is

$$\begin{vmatrix} r_{ii} & r_{ij} & r_{ik} \\ r_{ji} & r_{jj} & r_{jk} \\ r_{ki} & r_{kj} & r_{kk} \end{vmatrix} = 1 + 2r_{ij}r_{ik}r_{jk} - r_{ij}^2 - r_{ik}^2 - r_{jk}^2$$

The *three-dimensionality index* for band selection is thus defined as below:

$$3DIndex = \sqrt{1 + 2r_{ij}r_{ik}r_{jk} - r_{ij}^2 - r_{ik}^2 - r_{jk}^2} \qquad (7.20)$$

The value range of the *3DIndex* is [0, 1], where 1 indicates perfectly 3D and 0 not 3D at all. The higher the index value, the better statistical choice of a band triplet for colour composition.

7.7 Remarks

It is interesting to note that, although PCA is based on quite complex covariance matrix operations, in the end a PC image is simply a linear combination of the original image bands. In analysing the eigenvector of a PC image, with the FPCS technique, we are essentially selecting the PC through image differencing. High rank PCs are nothing more than compound difference images, but these are so composed as to be independent from each other. The PCA ensures orthogonal (independent) PCs on the basis of the data distribution, while differencing allows the targeting of specific spectral signatures of interest, although the resulting difference images are not themselves orthogonal. The FPCS combines the merits of both but may not always reveal the diagnostic spectral features.

Also noteworthy is that the average image, IHS intensity image and *PC1* image share a great deal in common. The three images all represent the sum of the spectral bands and all increase the image SNR. A band average is an equal weight summation of any number of image bands; the IHS intensity image is an average of three bands used for RGB-IHS transformation, while PC1 is a weighted summation of all image bands involving PCA.

The concept of the DS is rooted in PCA but the PCADS is less efficient and less widely used than the saturation DS techniques because it involves complicated matrix operations in its inverse PC transformation. Although the two types of DS techniques are based on different principles, their effects on the bands of an RGB colour composite triplet are entirely equivalent: they both increase the three-dimensionality of the data cluster.

7.8 Questions

1 Using a diagram explain the principle of PCA.
2 Discuss the data characteristics of PC images and their applications.
3 Compare and contrast the images of band average, IHS intensity and *PC1*, and discuss their relative merits.
4 What are the major advantages and disadvantages of PC colour composition?
5 Describe the FPCS method and discuss its application to PC colour composition.

6 Discuss the two SPCA techniques and their applications.
7 Describe and comment on the combined approach of the FPCS and SPCA for spectral enhancement.
8 What is a DS? Describe the major steps of PCADS.
9 Compare the three DS techniques, PCADS, IHSDS and DDS, in principle, results and processing efficiency.
10 What do PCA and the physical property orientated coordinate transformation have in common? How are they different in methodology and applications?
11 In what sense are the PCA and the tasselled cap transformation similar? What is the major difference between the two techniques? Comment on the merits and drawbacks of the two methods for earth observation.
12 Describe the two main approaches for band selection. Why is band selection necessary for visualisation of multi-spectral images?
13 Describe the principles of the index of three-dimensionality. What is the major consideration behind the design of this technique?
14 From the correlation matrix in Table 7.1, calculate the index of three-dimensionality using eqn. 7.20 for band triplets 321, 432, 531 and 541.

CHAPTER 8

Image classification

Image classification belongs to a very active field in computing research, *pattern recognition*. Image pixels can either be classified by their multi-variable statistical properties, such as the case of multi-spectral classification, or by segmentation based on both statistics and spatial relationships with neighbouring pixels. In this chapter, we will look at multi-variable statistical classification techniques for imagery data.

8.1 Approaches of statistical classification

Generally, statistical classification can be catalogued into two major branches: *unsupervised* and *supervised* classifications.

8.1.1 Unsupervised classification

This is entirely based on the statistics of the image data distribution and is often called *clustering*. The process is automatically optimised according to cluster statistics without the use of any knowledge-based control (i.e. ground truth). The method is therefore objective and entirely data driven. It is particularly suited to images of targets or areas where there is no ground truth knowledge or where such information is not available, such as in the case of planetary images. Even for a well-mapped area, unsupervised classification may reveal some spectral features that were not apparent beforehand. The result of an unsupervised classification is an image of statistical clusters, where the thematic contents of the clusters are not known. Ultimately, such a classification image still needs interpretation based on some knowledge of ground truth.

8.1.2 Supervised classification

This is based on the statistics of *training areas* representing different ground objects selected subjectively by users on the basis of their own knowledge or experience. The classification is controlled by users' knowledge but, on the other hand, is constrained and may even be biased by their subjective view. The classification can therefore be misguided by inappropriate or inaccurate training area information and/or incomplete user knowledge.

Realising the limitations of both major classification methods, a *hybrid classification* approach has been introduced. In the hybrid classification of a multi-spectral image, first an unsupervised classification is performed, then the result is interpreted using ground truth knowledge and finally, the original image is re-classified using a supervised classification with the aid of the statistics of the unsupervised classification as training knowledge. This method utilises unsupervised classification in combination with ground truth knowledge as a comprehensive training procedure and therefore provides more objective and reliable results.

8.1.3 Classification processing and implementation

A classification may be completed in one step, as a *single pass classification* or in an iterative optimisation procedure referred to as an *iterative classification*. The single pass method is the normal case for supervised classification, while the iterative classification represents the typical approach to unsupervised classification (clustering). The iterative method can also be incorporated into a supervised classification algorithm.

Image Processing and GIS for Remote Sensing: Techniques and Applications, Second Edition. Jian Guo Liu and Philippa J. Mason.
© 2016 John Wiley & Sons, Ltd. Published 2016 by John Wiley & Sons, Ltd.

Most image processing software packages perform image classification in the image domain by *image scanning classification*. This approach can classify very large image datasets with many spectral bands and high quantisation levels, with very low demands on computing resources (e.g. RAM), but it cannot accommodate sophisticated classifiers (decision rules). Image classification can also be performed in feature space by *feature space partition*. In this case, sophisticated classifiers incorporating data distribution statistics can be applied, but the approach demands a great deal of computer memory to cope with the high data dimensionality and quantisation. This problem is being overcome by increasingly powerful computing hardware and dynamic memory management in programming.

8.1.4 Summary of classification approaches
- Unsupervised classification;
- supervised classification;
- hybrid classification;
- single pass classification;
- iterative classification;
- image scanning classification;
- feature space partition.

8.2 Unsupervised classification (iterative clustering)

8.2.1 Iterative clustering algorithms
For convenience of description, let $\mathbf{X}$ be an n-dimensional feature space of n variables $(\mathbf{x}_1, \mathbf{x}_2, ..., \mathbf{x}_n)$, Y_i be an object of an object set Y (an image) defined by measurements of the n variables [e.g. digital numbers (DNs) of n spectral band], $Y_i = (y_{i1}, y_{i2}, ..., y_{in})$, $i = 1, 2, ..., N$. N is the total number of objects in Y or the total number of pixels in an image. As shown in Fig. 8.1, in the feature space $\mathbf{X}$, the object Y_i is represented by an observation vector, that is, a data point $\mathbf{X}_j \in \mathbf{X}$ at the coordinates $(x_{j1}, x_{j2}, ..., x_{jn})$, $j = 1, 2, ..., M$. M is the total number of data points representing N objects. If $\mathbf{X}$ is a Euclidean space, then $x_{jh} \sim y_{ih}$, $h = 1, 2, ..., n$. Obviously, a data point $\mathbf{X}_j$ in the feature space $\mathbf{X}$ can be shared by more than one image pixel Y_i and therefore $M \le N$.

The goal of the clustering process is to identify the objects of set Y in m classes. This is equivalent to the partition of the relevant data points in feature space $\mathbf{X}$ into m spatial clusters, $\omega_1, \omega_2, ..., \omega_m$. Generally, there are two

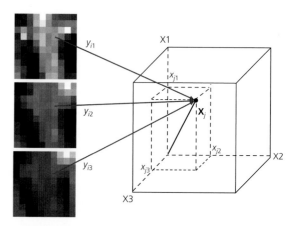

Fig. 8.1 A three-dimensional (3D) illustration of the relationship between a feature space point $\mathbf{X}_j$ and a multi-spectral image pixel $Y_i = (y_{i1}, y_{i2}, y_{i3})$.

principal iterative clustering algorithms labelled α and β (Diday & Simon 1976).

8.2.1.1 Algorithm α
1 *Initialisation*
   Let m elements $Y_q \in Y$, chosen at random or by a selection scheme, be the 'representation' of m clusters denoted as $\omega_1, \omega_2, ..., \omega_k, ..., \omega_m$.
2 *Clustering*
   For all i, assign any element $Y_i (Y_i \in Y)$ to a cluster ω_k, if the dissimilarity measurement $\delta(Y_i, \omega_k)$ is minimal.
3 *Update statistical representation*
   For all k, new statistics of cluster ω_k are computed as the renewed representation of the cluster ω_k.
4 *Stability*
   If no ω_k has changed above the given criteria then stop, else go to 2.

8.2.1.2 Algorithm β
1 As in step 1 of algorithm α.
2 One element $Y_i (Y_i \in Y)$ is assigned to cluster ω_k, if $\delta(Y_i, \omega_k)$ is minimal.
3 A new representation of ω_k is computed from all the elements of cluster ω_k, including the last element.
4 If all elements $Y_i (Y_i \in Y)$ have been assigned to a cluster then stop, else go to step 2.

The algorithm α may not necessarily converge if the criterion for terminating the iteration is too tight. The algorithm β ends when the last pixel is reached. The algorithm α is more commonly used for image classification because of its self-optimisation mechanism and processing efficiency. Cluster splitting and merging functions can be added in the algorithm α after step 4, which allows the algorithm to operate more closely with the true data distribution and to reach more optimised convergence. With the progress of the clustering iteration, the initial cluster centres are moving toward the true data cluster centres via the updating of their statistical representations at the end of each iteration. The only user control on clustering is the initial parameter setting, such as number and position of the starting centres of clusters, iteration times or termination criteria, maximum and minimum number and size of clusters, and so forth. The initial setting will affect the final result. In this sense, the clustering iteration mechanism can only ensure local optimisation, the optimal partition of clusters for the given initial parameter setting, but not the global optimisation, because the initial parameter setting cannot be optimal for the best possible clustering result.

For most image processing packages, image clustering using either of the two algorithms is executed on an object set Y, that is, the image. The processing is on a pixel-by-pixel basis by scanning the image but, with advances in computing power, the very demanding feature space partition clustering in the feature space $\mathbf{X}$ becomes feasible.

One of the most popular clustering algorithms for image classification, the ISODATA algorithm (Ball 1965; Ball & Hall 1967), is a particular case of the algorithm α in which the dissimilarity measure $\delta(Y_i, \omega_k)$ in step 2 is the square Euclidean distance. The assumption underlying this simple and efficient technique is that all the clusters have equal variance and population. This assumption is generally untrue in image classification, and as a result classification accuracy may be low. To improve ISODATA, more sophisticated measures of dissimilarity, such as maximum likelihood estimation and population weighted measurements, have been introduced. For all these different decision rules, within the ISODATA frame, the processing is performed by image scanning.

8.2.2 Feature space iterative clustering

As mentioned earlier, image classification can be performed by image scanning as well as by feature space partition. Most multi-variable statistical classification algorithms can be realised by either approach but, for more advanced decision rules, such as optimal multiple point re-assignment (OMPR), which will be introduced later, feature space partition is the only feasible method because all pixels sharing the same DN values in each image band must be considered simultaneously. Here we introduce a 3D feature space iterative clustering method (3D-FSIC), an algorithm that can be easily extended to more dimensions.

8.2.2.1 3D-FSIC

Step 1 Create a 3D scattergram of the input image   Read the input image, Y, pixel by pixel and record the pixel frequencies in a 3D scattergram, that is, a 3D array (Fig. 8.2a):

$$G\left(d1 \times d2 \times d3\right)$$

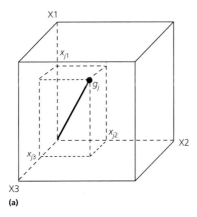 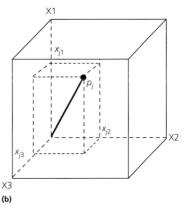

(a)   **(b)**

Fig. 8.2 (a) a 3D array G of a scattergram; and (b) a 3D array P of a feature space partition.

where *d1*, *d2* and *d3* are the array sizes in the three dimensions or the maximum DN values of the three bands of the image *Y*.

The value of any element g_j in G indicates how many pixels share the point $\mathbf{X}_j$ at the coordinates (x_{j1}, x_{j2}, x_{j3}) in the 3D feature space $\mathbf{X}$, or the number of pixels with the same DN values as pixel Y_i in the image *Y*, where $y_{ih} \sim x_{jh}, h = 1, 2, 3$.

Step 2 Initialisation Select *m* points in the 3D feature space $\mathbf{X}$ as the 'seeds' of *m* clusters and call them ω_k, $k = 1, 2, \ldots, m$. The choice could be made at random or via an automatic seed selection technique.

Step 3 Feature space clustering For all *j*, assign any point $\mathbf{X}_j (\mathbf{X}_j \in \mathbf{X}, j = 1, 2 \ldots N)$ to cluster ω_k if the dissimilarity $\delta(\mathbf{X}_j, \omega_k)$ is minimal. Thus all the pixels sharing the point $\mathbf{X}_j$ are assigned to cluster ω_k simultaneously. The size of cluster ω_k, N_k, increases by the value g_j while, if it is a re-assignment, the size of the cluster to which $\mathbf{X}_j$ was formerly assigned decreases by the same value. The cluster sequential number *k* of point $\mathbf{X}_j$ is recorded by a 3D feature space partition array $P(d1 \times d2 \times d3)$ in the element P_j at coordinates (x_{j1}, x_{j2}, x_{j3}) [Fig. 8.2 (b)].

Step 4 Update the statistical representation of each cluster For all *k* (*k* = 1, 2, …, *m*), statistical parameters, such as mean vector μ_k, covariance matrix Σ_k and so forth are calculated. These parameters make up the new representation of the cluster ω_k.

Step 5 Stability For all *k* (*k* = 1, 2, …, *m*), if the maximum spatial migration of the mean vector μ_k (the kernel of the cluster) is less than a user-controlled criterion, go to step 7, else go to step 6.

Step 6 Cluster splitting and merging Split the over-large and elongate clusters and merge clusters that are too small and/or too close to each other, according to user-controlled criteria; then update the statistical representations of the new clusters. Go to step 3.

Step 7 Transfer the clustering result from feature space to an image Read the input image *Y*, pixel by pixel. For all *i*, assign a pixel $Y_i(Y_i \in Y)$ to cluster ω_k if its corresponding data point $\mathbf{X}_j$ in feature space $\mathbf{X}$ is assigned to this cluster, according to the record in the feature space partition array *P*, that is

$$Y_i \to \omega_k \quad if \quad P_j = k$$

where P_j is at coordinates (x_{j1}, x_{j2}, x_{j3}) in *P* and $y_{ih} \sim x_{jh}, h = 1, 2, 3$.

Then assign the class number to the corresponding pixel in the output classification image Y_{class}.

8.2.3 Seed selection

The initial kernels (seeds) for unsupervised classification can be made randomly, evenly or by particular methods. Here we introduce an automatic seed selection technique (ASST) for 3D-FSIC.

In the 3D scattergram of the three images for classification, data will exhibit peaks (the points of high frequency) at the locations of spectral clusters. It is thus sensible to use these points as the initial kernels of clusters to start iterative clustering. Such a peak point has two properties:

* Higher frequency than all its neighbouring points in the feature space;
* relatively high frequency in the 3D scattergram.

These two properties are to be used to locate peak points. It is important to keep in mind that the multispectral image data and scattergram are discrete and that the DN value increment of an image may not necessarily be unity, especially after contrast enhancement. For instance, when an image of 7-bit DN range [0, 127] is linearly stretched to 8-bit DN range [0, 255], the increment of DN values becomes 2 instead of 1. In this case, any non-zero frequency DN level in the original image will have two adjacent zero frequency DN levels in the stretched image (Fig. 8.3); appearing as a pseudo-peak caused by data discontinuity. With these considerations in mind, ASST is composed of the following operations:

1 Locate and rank the first *N* points of highest frequency from the 3D scattergram to form a sequential set $\mathbf{X}_c$. *N* can be decided by setting a criterion frequency on the basis of experience and experimentation. For an 8-bit full scene TM image, 1024 is suggested. This operation will prevent the selection of isolated low frequency points (often representing noise) as seeds and thus satisfy the second property.

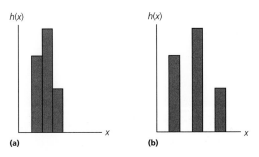

Fig. 8.3 Illustration of pseudo-peaks in an image histogram (a) caused by linear stretch (b).

2 The first element in the set $\mathbf{X}_c$ must be nominated as a seed because it cannot be surrounded by any elements of a higher frequency. Then, for the second element of $\mathbf{X}_c$, check if the first element is in its given neighbourhood range (the neighbourhood is used to avoid pseudo-peaks in the image with DN increment greater than 1) and if not, the second element is also selected as a seed. In general, for any element $\mathbf{X}_j$ in $\mathbf{X}_c$, check the coordinates of those elements ranked with higher frequency; $\mathbf{X}_j$ is selected as a seed if none of the higher frequency elements are within its neighbourhood. This operation makes the seed selection satisfy the first property.

8.2.4 Cluster splitting along *PC1*

In unsupervised classification (cluster partition), very large clusters may be generated. Such large clusters may contain several classes of ground objects. A function for cluster splitting is therefore necessary to achieve optimal convergence. In ISODATA, an over-large and elongate cluster ω is split according to the variable with greatest standard deviation. The objects (image pixels) in cluster ω are re-assigned to either of the two new clusters, $\omega 1$ and $\omega 2$, depending on whether their splitting variable values are above or below the mean of the splitting variable. As illustrated by the two-dimensional (2D) case in Fig. 8.4, splitting in this way may cause incorrect assignments of those objects in the grey area of the data ellipse. They are assigned to a new cluster that is farther from them rather than closer. This error can be avoided

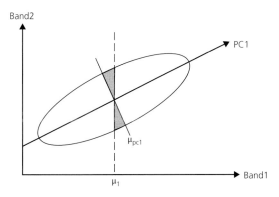

Fig. 8.4 A 2D example of cluster splitting based on band 1 (the variable with maximum standard deviation) and *PC1*. The grey areas indicate the misclassification resulted from the cluster splitting based on band 1. (Liu and Haigh 1994. Reproduced with permission of Taylor and Francis.)

if the cluster ω is split along its first principal component (*PC1*). Since *PC1* can be found without performing a PC transformation, not too many calculations are involved. The technique of cluster splitting based on *PC1* (Liu & Haigh 1994) includes two steps: finding the *PC1* followed by cluster splitting based on the *PC1*.

8.2.4.1 Find the first principal component *PC1*
The covariance matrix Σ of the cluster ω is a non-negative definite matrix. Thus the first eigenvalue and eigenvector of Σ, λ_1 and $\mathbf{a} = (a_1, a_2, ..., a_n)^t$, can be found by the iteration defined below.

$$\Sigma \mathbf{a}^{(s)} = \lambda_1^{(s+1)} \mathbf{a}^{(s+1)}$$
$$\mathbf{a}^{(0)} = \mathbf{I}$$

(8.1)

where s denotes the number of iterations and $\mathbf{I}$ is an identity vector.

The $\mathbf{a}$, as an eigenvector, is orthogonal, thus for each iteration s, we have

$$\left(\mathbf{a}^{(s)} \right)^t \mathbf{a}^{(s)} = 1$$

(8.2)

Then

$$\left(\Sigma \mathbf{a}^{(s)} \right)^t \Sigma \mathbf{a}^{(s)} = \lambda_1^{(s+1)} \left(\mathbf{a}^{(s+1)} \right)^t \lambda_1^{(s+1)} \mathbf{a}^{(s+1)} = \left(\lambda_1^{(s+1)} \right)^2$$

Thus,

$$\lambda_1^{(s+1)} = \left[\left(\Sigma \mathbf{a}^{(s)} \right)^t \Sigma \mathbf{a}^{(s)} \right]^{1/2}$$
$$\mathbf{a}^{(s+1)} = \frac{\Sigma \mathbf{a}^{(s)}}{\lambda^{(s+1)}}$$

(8.3)

After 5–6 iterations convergence with accuracy higher than 10^{-5} can be achieved and the first eigenvalue λ_1 and eigenvector $\mathbf{a}$ are found. Consequently, the first PC of cluster ω in the n-dimensional feature space $\mathbf{X}$ is derived as below:

$$PC1 = (\mathbf{a})^t \mathbf{X} = \sum_{h=1}^{n} a_h x_h$$

(8.4)

8.2.4.2 Cluster splitting
According to eqn. 8.4, the *PC1* coordinate of the mean vector $\mu = (\mu_1, \mu_2, ..., \mu_n)^t$ of cluster ω is

$$\mu_{pc1} = (\mathbf{a})^t \mu = \sum_{h=1}^{n} a_h \mu_h$$

(8.5)

For every data point $\mathbf{X}_j \in \omega$, calculate its *PC1* coordinate:

$$x_{j,pc1} = (\mathbf{a})^t \mathbf{X}_j = \sum_{h=1}^{n} a_h x_{jh} \qquad (8.6)$$

Assign $\mathbf{x}_j$ to $\omega 1$ if $x_{j,pc1} > \mu_{pc1}$ otherwise assign $\mathbf{X}_j$ to $\omega 2$.

Cluster splitting can also be performed on the objects (image pixels) instead of data points by replacing $\mathbf{X}_j$ by $Y_{i'}$ $(i = 1, 2, \ldots, N)$ in eqn. 8.6. After cluster splitting, the statistics of the two new clusters are calculated as the representations for the next clustering iteration.

8.3 Supervised classification

8.3.1 Generic algorithm of supervised classification

A supervised classification comprises three major steps, as follows:

Step 1 Training   Training areas representing different ground objects are manually and interactively defined on the image display. Statistics of the training areas are calculated to represent the relevant classes ω_k $(k = 1, 2, \ldots, m)$.

Step 2 Classification   For all i, assign any element $Y_i(Y_i \in Y)$ to a class ω_k, if the dissimilarity measurement $\delta(Y_{i'}, \omega_k)$ is minimal.

Step 3 Class statistics   Calculate the statistics of all resultant classes.

Iteration and class splitting/merging functions can also be accommodated into a supervised classification algorithm to provide an automated optimisation mechanism.

8.3.2 Spectral angle mapping classification

A pixel in an n bands multi-spectral image can be considered as a vector in the n-dimensional feature space $\mathbf{X}$. The magnitude (length) of the vector is decided by the pixel DNs of all the bands, while the orientation of the vector is determined by the shape of the spectral profile of this pixel. If two pixels have similar spectral properties but are under different solar illumination because of topography, the vectors representing the two pixels will have different length but very similar orientation. Therefore the classification of image pixels based on the spectral angles between them will be independent of topography (illumination) as well as any unknown linear translation factors (e.g. gain and offset). The spectral angle mapping (SAM) technique,

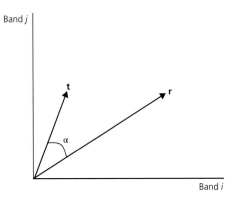

Band j

Band i

Fig. 8.5 A 2- illustration of two spectral vectors and the spectral angle (α) between them.

proposed by Kruse *et al.* (1993), is a supervised classification based on the angles between image pixels spectra and training data spectra or library spectra. The algorithm determines the similarity between two spectra by calculating the spectral angle between them as shown in a 2D diagram (Fig. 8.5). According to vector algebra, the angle between two vectors $\mathbf{r}$ and $\mathbf{t}$ is defined as

$$\alpha = \arccos\left(\frac{\mathbf{t} \cdot \mathbf{r}}{|\mathbf{t}| \cdot |\mathbf{r}|}\right) \qquad (8.7)$$

or

$$\alpha = \arccos\left(\frac{\sum_{i=1}^{m} t_i r_i}{\left(\sum_{i=1}^{m} t_i^2\right)^{1/2} \left(\sum_{i=1}^{m} r_i^2\right)^{1/2}}\right) \qquad (8.8)$$

where m is the number of spectral bands.

The value range of α is 0–π.

In general, for m reference spectral vectors $\mathbf{r}_k (k = 1, 2, \ldots, m)$, either from an existing spectral library or from training areas, the spectral vector $\mathbf{t}$ of an image pixel is identified as $\mathbf{r}_k$ if the angle α between them is minimal and is less than a given criterion.

The SAM classification is widely used in hyperspectral image data classification for mineral identification and mapping. It can also be used in broadband multi-spectral image classification. Within the framework of SAM, different dissimilarity functions can be implemented to assess the spectral angle, α.

8.4 Decision rules: Dissimilarity functions

Dissimilarity functions, based on image statistics, formulate decision rules at the core of both supervised and unsupervised classification algorithms, and these theoretically decide the accuracy of a classification algorithm. Here we introduce several commonly used decision rules of increasing complexity.

8.4.1 Box classifier

It is also called a parallel classifier. It is used for single-pass supervised classification. In principle, it is simply multi-dimensional thresholding [Fig. 8.6(a)].

For all i, assign any an element $Y_i(Y_i \in Y)$ to cluster ω_k, if

$$\min(\omega_k) \leq Y_i \leq \max(\omega_k) \qquad (8.9)$$

The 'boxes' representing the scopes of different classes may partially overlap one another, as in the shaded areas shown in Fig. 8.6(a). The pixels that fall in the overlap areas are treated as unclassified. This is a very crude but fast classifier.

8.4.2 Euclidean distance: Simplified maximum likelihood

The Euclidean distance is a special case of maximum likelihood that assumes equal standard deviation and population for all clusters. It is defined as

For all i, assign any element $Y_i(Y_i \in Y)$ to cluster ω_k, if

$$d(Y_i, \omega_k) = (Y_i - \mu_k)^t (Y_i - \mu_k) = \min\{d(Y_i, \omega_r)\} \quad (8.10)$$

for $r = 1, 2, \ldots, m$, where μ_k is the mean vector of cluster ω_k.

The Euclidian distance lies at the core of the ISODATA minimum distance classification.

8.4.3 Maximum likelihood

The maximum likelihood decision rule is based on Baye's theorem and assumes a normal distribution for all clusters. In this decision rule, the feature space distance between an image pixel Y_i and cluster ω_k is weighted by the covariance matrix Σ_k of ω_k with an offset relating to the ratio of N_k, the number of pixels in ω_k, to N, the total number of pixels of the image Y:

For all i, assign any element $Y_i(Y_i \in Y)$ to cluster ω_k, if

$$\delta(Y_i, \omega_k) = \ln|\Sigma_k| + (Y_i - \mu_k)^t$$
$$\sum_k^{-1}(Y_i - \mu_k) - \ln\frac{N_k}{N} = \min\{\delta(Y_i, \omega_r)\} \quad (8.11)$$

for $r = 1, 2, \ldots, m$.

As shown in Fig. 8.6(b), the minimum Euclidian distance classification will assign the pixel P(i,j) to the class centre ω_4, whereas the maximum likelihood minimum distance classification will be more likely to assign the pixel P(i,j) to the class centre ω_3 because this class is larger.

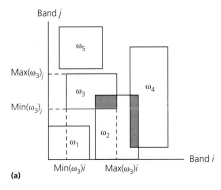

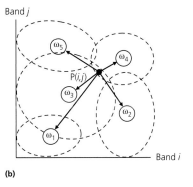

Fig. 8.6 Illustrations of 2D feature space partition of the box classifier and distance based classifications. (a) A box classifier is actually a simple multi-dimensional threshold – it cannot classify image pixels that fall in the value ranges of multiple classes as shown in the shaded areas. (b) The circles are the class centres and the ellipses represent the size of each class. The minimum Euclidean distance classification will assign the pixel P(i,j) to the class centre ω_4, whereas the maximum likelihood minimum distance classification will be more likely to assign the pixel P(i,j) to the class centre ω_3 because this class is larger.

8.4.4 Optimal multiple point re-assignment (OMPR)*

An advantage of 3D-FSIC is that the OMPR rule can be implemented if we let $\delta(\mathbf{X}_j, \omega_k)$ be an OMPR dissimilarity measurement at *step 3* of 3D-FSIC. The OMPR (Kittler & Pairman 1988) was developed based on the optimal point assignment rule (MacQueen 1967). By using OMPR, the cluster sizes and the number of the pixels sharing the same data point in feature space (point frequency) are taken into account when a re-assignment of these pixels is made. Thus the accuracy of the clustering partition can be reasonably improved.

Suppose a data point $\mathbf{X}_j$ currently allocated to cluster ω_l is shared by H pixels; the OMPR based on square Euclidean distance (Euclidean OMPR) for all these pixels to be re-assigned from cluster ω_l to cluster $\omega_{k'}$ shared by N_k pixels, will be achieved if ω_k satisfies:

$$\frac{N_k}{N_k + H} d\left(\mathbf{X}_j, \mu_k\right) = \min_{r \neq l} \frac{N_r}{N_r + H} d\left(\mathbf{X}_j, \mu_r\right)$$

$$< \frac{N_l}{N_l - H} d\left(\mathbf{X}_j, \mu_l\right) \qquad (8.12)$$

where N_r is the number of pixels in any a cluster ω_r and N_l that in cluster ω_l.

If the clusters are assumed to have a normal distribution (Gaussian model), the Gaussian OMPR is formed as follows:

For all j, assign any a data point $\mathbf{x}_j$ in cluster ω_l to cluster $\omega_{k'}$ if

$$\delta\left(\mathbf{X}_j, \omega_k\right) = \min_{r \neq l} \delta\left(\mathbf{X}_j, \omega_r\right)$$

$$< \ln|\Sigma_l| - \frac{N_l - H}{H} \ln\left[1 - \frac{H}{N_l - H} \Delta\left(\mathbf{X}_j, \omega_l\right)\right]$$

$$-2\ln\frac{N_l}{N} - (D+2)\frac{N_l - H}{H} \ln\frac{N_l}{N_l - H} \qquad (8.13)$$

where

$$\delta\left(\mathbf{X}_j, \omega_r\right) = \ln|\Sigma_r| + \frac{N_r + H}{H} \ln\left[1 + \frac{H}{N_r + H} \Delta\left(\mathbf{X}_j, \omega_r\right)\right]$$

$$- 2\ln\frac{N_r}{N} + (D+2)\frac{N_r + H}{H} \ln\frac{N_r}{N_r + H} \qquad (8.14)$$

and

$$\Delta\left(\mathbf{X}_j, \omega_r\right) = \left(\mathbf{X}_j - \mu_r\right)^t \sum_r^{-1} \left(\mathbf{X}_j - \mu_r\right)$$

D is the dimensionally of feature space $\mathbf{X}$.

In the OMPR method, data point inertia is considered. A data point shared by more pixels ('heavier') is more difficult to move from one cluster to another than a 'lighter' point.

8.5 Post-classification processing: Smoothing and accuracy assessment

8.5.1 Class smoothing process

A classification image appears to be a digital image in which the DNs are the class numbers, but we cannot perform numerical operations on class numbers. For instance the average of class 1 and class 2 cannot be a class 1.5! Indeed, the class numbers in a classification image do not have any sequential relationship; they are nominal values and can be treated as symbols such as *A*, *B* and *C* (see also § 13.3). A classification image is actually an image of symbols, not digital numbers; *it is therefore not a digital image* in the generally accepted sense. As such we cannot apply any numerical operation-based image processing to classification images.

A classification image often contains noise caused by the isolated pixels of some classes, within another dominant class that forms a sizable patch [Fig. 8.7(a)]. It is reasonable to presume that these isolated pixels are more likely to belong to this dominant class rather than to the classes that they are initially assigned to; these probably arise from classification errors. An appropriate smoothing process applied to a classification image will not only 'clean up' the image, making it visually less noisy, but will also improve the accuracy of classification.

Among the low pass filters that we have described so far, *the only filter that can be used to smooth a classification image is the mode (majority) filter*. The reason for this is simple since the mode filter smoothes an image without any numerical operations. For instance, if a pixel of class 5 is surrounded by pixels of class 2, the mode filter will re-assign this pixel to class 2 according to the majority class in the filtering kernel. Figure 8.7(b) illustrates the effect of mode filtering applied to an unsupervised classification image in Fig. 8.7(a).

8.5.2 Classification accuracy assessment

Ultimately there is no satisfactory method to assess the absolute accuracy of image classification for remote sensing earth observation applications (see also § 18.5.1).

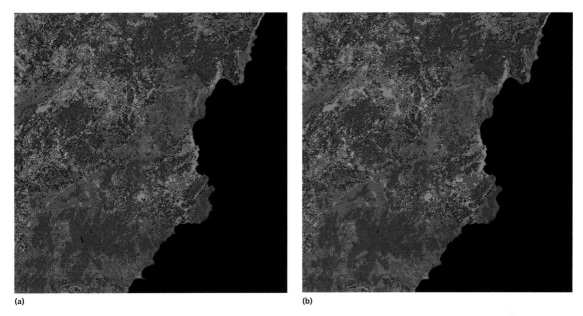

(a)   (b)

Fig. 8.7 Smoothing classification images: (a) ISODATA unsupervised classification with 24 classes; and (b) the classification image is smoothed using a mode filter with a majority of 5 in a 3 × 3 filtering kernel. Closely look at these images to observe the difference between them; image (b) is smoother than image (a).

The paradox is that we cannot conduct such an assessment without knowing the 100% of ground truth on one hand, while on the other hand, if we do have complete knowledge of ground truth, what is the point of the classification? However, an assessment or an estimation of relative accuracy of classification does provide valuable knowledge for us to accept or reject a classification result at a certain confidence level.

There are two generally accepted approaches to generate ground truth data.

1 Use field-collected data of typical classes as samples of ground truth. For rapidly temporally changing land cover classes, such as crops, field data should be collected simultaneously with the image acquisition. For temporally stable targets, such as rocks and soils, published maps as well as field data can be used. The classification accuracy of the sample areas with known classes gives an estimation of total classification accuracy. This seemingly straightforward approach is often impractical in reality because it is often constrained by errors in the recording of field observation, limited field accessibility and temporal irrelevance.

2 Another approach relies on image training. This uses typical spectral signatures and limited field experience, where a user can manually specify training areas of various classes using a multi-spectral image.

The pixels in these training areas are separated into two sets: one is used to generate class statistics for supervised classification and the other for subsequent classification accuracy assessment. For a given training area, we could take a selection of pixels sampled from a 2 × 2 grid as the training set, while the remaining pixels are used for the verification set (ground truth reference data), as shown in Fig. 8.8. The pixels in the verification set are assumed to belong to the same class as their corresponding training set. In another way, we can also select several training areas for the same class and use part of them for training and the rest for verification.

In practice the above two approaches are often used in combination.

Suppose that we have some kind ground truth reference data, a widely used method to describe the relative accuracy of classification is the *confusion matrix*.

$$\text{Confusion matrix:} \begin{pmatrix} C_{11} & C_{12} & \cdots & C_{1m} \\ C_{21} & C_{22} & \cdots & C_{2m} \\ \vdots & \vdots & \ddots & \vdots \\ C_{m1} & C_{m2} & \cdots & C_{m} \end{pmatrix} \quad (8.15)$$

Here, each of the elements, C_{ii}, in the major diagonal represents the number of pixels that are correctly

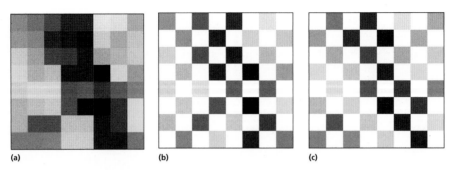

(a)   (b)   (c)

Fig. 8.8 A re-sampling scheme for classification accuracy assessment. An image (a) is re-sampled to formulate two images – (b) and (c) – and one is used as the training dataset while the other is used as the verification set.

classified for a class i. Any element off the major diagonal, C_{ij}, represents the number of pixels that should be in class i but are incorrectly classified as class j. Obviously, if all the image pixels are correctly classified, we should then have a diagonal confusion matrix where all non-diagonal elements become zero. The sum of all the elements in the confusion matrix is the total number of pixels in a classification image, N:

$$N = \sum_{i=1}^{m}\sum_{j=1}^{m} C_{ij}$$

The ratio between the summation of the major diagonal elements and the total number of pixels represents the percentage of the correct classification or *overall accuracy*:

$$Ratio_{correct}(\%) = \frac{1}{N}\sum_{i=1}^{m} C_{ii} \qquad (8.16)$$

The sum of any row i of the confusion matrix gives the total number of pixels that, according to ground truth reference, should be in class i, Nr_i:

$$Nr_i = \sum_{j=1}^{m} C_{ij}$$

Then the ratio C_{ii}/Nr_i is the percentage of correct classification of class i, according to the reference ground truth, and is often called *user's accuracy*.

The sum of any a column j of the confusion matrix gives the total number of pixels that have been classified as class j, Nc_j:

$$Nc_j = \sum_{i=1}^{m} C_{ij}$$

Then the ratio C_{jj}/Nc_j is the percentage of correct classification of class j, based on the classification result and is often called *producer's accuracy*.

Apart from the above accuracy measurements that are based on simple ratios, another commonly used statistical measure of classification accuracy and quality is the *kappa coefficient* (κ) that combines the above two class accuracy estimations, based on the rows and columns of the confusion matrix, to produce an estimation of total classification accuracy, as follows:

$$\kappa = \frac{N\sum_{i=1}^{m} C_{ii} - \sum_{i=1}^{m} Nr_i \cdot Nc_i}{N^2 - \sum_{i=1}^{m} Nr_i \cdot Nc_i} \qquad (8.17)$$

In the case of 100% agreement between the classification and the reference data, the confusion matrix is diagonal, that is, $\sum_{i=1}^{m} C_{ii} = N$. Thus,

$$\kappa = \frac{N^2 - \sum_{i=1}^{m} Nr_i \cdot Nc_i}{N^2 - \sum_{i=1}^{m} Nr_i \cdot Nc_i} = 1$$

While if there is no agreement at all, then all the elements on the diagonal of the confusion matrix are zero, that is, $\sum_{i=1}^{m} C_{ii} = 0$. In this case:

$$\kappa = \frac{-\sum_{i=1}^{m} Nr_i \cdot Nc_i}{N^2 - \sum_{i=1}^{m} Nr_i \cdot Nc_i} < 0$$

In summary, the maximum value of the kappa coefficient κ is 1, indicating perfect agreement between the classification and the reference data, while for no agreement, κ becomes negative. The minimum value

Table 8.1 An example confusion matrix.

Class reference	Class 1	Class 2	Class 3	Class 4	Class 5	Row sum Nr_i	C_{ii}/Nr_i (%)
Reference 1	*56*	*9*	*5*	*2*	*8*	80	70.0
Reference 2	*10*	*70*	*7*	*3*	*5*	95	73.7
Reference 3	*0*	*3*	*57*	*10*	*6*	76	75.0
Reference 4	*0*	*6*	*0*	*79*	*4*	89	88.8
Reference 5	*8*	*4*	*3*	*2*	*46*	63	73.0
Column sum Nc_j	74	92	72	96	69	403	
C_{jj}/Nr_j (%)	75.6	76.1	79.2	82.3	66.7		**76.4**

of κ is case dependent but as long as $\kappa \leq 0$, it indicates zero agreement between the classification and the reference data.

As illustrated in Table 8.1, the numbers in bold italic, in the central part, form the confusion matrix. Nr_i and C_{ii}/Nr_i are listed in the two right-hand columns, while Nc_j and C_{jj}/Nc_j are in the bottom two rows. The bold number in the bottom right corner is the total percentage of correct classification. The kappa coefficient (κ) can then be calculated from Table 8.1 by the following:

$$\kappa = \frac{403 \times 308 - 33023}{162409 - 33023} = \frac{91101}{129386} = 0.704$$

Despite the fact that the classification accuracy derived from the confusion matrix is very much a self-assessment and is by no means the true accuracy of classification, it does provide a useful measure of classification accuracy. The information in a confusion matrix is highly dependent on the quality of the training areas and field data. Accurately selected training areas can improve both the classification accuracy and the credibility of accuracy assessment, whereas poorly selected training areas will yield low classification accuracy and unreliable accuracy assessment. Strictly speaking, this method gives us only an estimation of the classification accuracy of the whole image.

8.6 Summary

In this chapter, we have introduced the most commonly used image classification approaches and algorithms. These methods are essentially multi-variable statistical classifications that achieve data partition in the multi-dimensional feature space of multi-layer image data, such as a multi-spectral remotely sensed image.

An iterative clustering method of unsupervised classification enables self-optimization toward a local optimal representative of the natural clusters in the data. How well the clustering converges to a local optimal depends on the dissimilarity function and clustering mechanism employed, whilst the quality of the local optimal is mainly affected by the initial cluster centres (the seeds) from where the iteration starts. Thus a seed selection technique, locating the peaks of data distribution, is introduced. A method for cluster splitting, based on the *PC1*, is also proposed to improve the clustering mechanism.

Though affected by the same factors, the accuracy of a supervised classification is largely controlled by user knowledge. High quality user knowledge could lead to correct classification of known targets, while poor user knowledge may mislead rather than help.

There are many methods of accuracy assessment, such as the well-known confusion matrix, but it is important to know the limitations of such methods that merely give a relative assessment rather than true accuracy of classification.

Finally, we must recognise that a classification image is not a true digital image but a symbol image presented in numbers. We could apply numerical operations to a classification image but the results do not really make any sense. We can, however, use logical operations to process classification images, such as to smooth a classification image using a mode (majority) filter because it does not involve any numerical operations.

8.7 Questions

1 What is multi-variable statistical classification? Describe the major approaches for image classification.

2 What are the advantages and disadvantages of unsupervised classification? Describe the algorithm α for iterative clustering.

3 Explain the self-optimisation mechanism of iterative clustering using a diagram.

4 Describe the main steps of the 3D-FSIC algorithm aided by diagrams. What are the main advantages and limitations of feature space iterative clustering?

5 What are the two properties for the design of the automatic seed selection technique?

6 What is the problem with cluster splitting along the axis of the variable with the maximum standard deviation? What is a better approach?

7 Describe the general steps of supervised classification.

8 Explain the principle of spectral angle classification and its merits.

9 In order to smooth a classified image, which filter do you think is appropriate and why? Explain the principle of this filter.

10 What is a confusion matrix? Based on the confusion matrix, give the definitions of *overall accuracy, user's accuracy* and *producer's accuracy.*

11 Comment on the issue of accuracy assessment for image classification.

CHAPTER 9

Image geometric operations

Geometric operations include the shift, rotation and warping of images to a given shape or framework. In remote sensing applications, geometric operations are mainly used for the co-registration of images of the same scene acquired by different sensor systems or at different times or from different positions, and for rectifying an image to fit a particular coordinate system (geocoding). Image mosaic is a geometric operation that was commonly used in the early days of remote sensing image processing when computer power was inadequate for the massive demands of the geocoding process, but this is no longer the case. Once a set of adjacent images are accurately rectified to a map projection system, such as a Universal Transverse Mercator (UTM) coordinate system (see Chapter 14 in Part II for details) the images, though separate, are effectively in a mosaic.

9.1 Image geometric deformation

An image taken from any sensor system is a distortion of the real scene. There are many sources of error since, for instance, any optical sensor system is a distorted filtered imaging system. Such source errors in sensor systems are usually corrected for in the sensor calibrations carried out by the ground segment of remote sensing platforms; they are beyond the scope of this chapter. Our main concerns lie on the user side of remote sensing applications, in the geometric distortions encountered during the imaging process, when a satellite or aircraft acquires images of the land surface.

9.1.1 Platform flight coordinates, sensor status and imaging position

As shown in Fig. 9.1, image geometry is fundamentally controlled by three sets of parameters:

- The platform flight coordinate system (x, y, z). x is in the flight direction, z is orthogonal to x, in the plane through the x axis and perpendicular to the earth's surface, and y is orthogonal to both x and z.
- The sensor 3D status is decided by orientation angles ω, ϕ, κ in relation to the platform flight coordinate system (x, y, z).
- The ground coordinates (X, Y, Z) of the imaging position usually conform to a standard coordinate system (defined by map projection and datum).

The preferred imaging geometry for the optical sensors of most earth observation satellites is that the satellite travels horizontally and parallel to the earth's curved surface with (x, y, z) matching (X, Y, Z) and sensor orientation angles (ω, ϕ, κ) all being equal to zero. This is the configuration of nadir (vertical) imaging, which introduces minimal geometric distortion.

For an optical imaging system, the focal length f of the lenses is another important parameter that decides the characteristics of the central projection distortion. For the same imaging area (field of view), a shorter focal length will result in greater topographic distortion (Fig. 9.2). Nadir view imaging is achieved when the principal of the optical system is vertical to the plane of the scene to be imaged. Otherwise, an oblique view imaging configuration is formed, depending on the sensor status; this can be side-looking, forward-looking, backward-looking or oblique-looking in any

Image Processing and GIS for Remote Sensing: Techniques and Applications, Second Edition. Jian Guo Liu and Philippa J. Mason.
© 2016 John Wiley & Sons, Ltd. Published 2016 by John Wiley & Sons, Ltd.

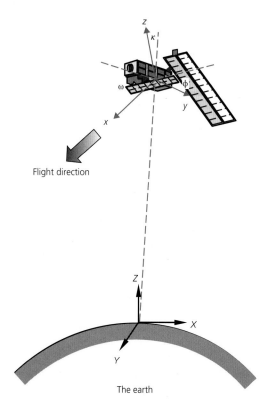

Flight direction

The earth

Fig. 9.1 The earth observation satellite imaging geometry. For the platform flight coordinate system (x, y, z), x is in the flight direction, z is orthogonal to x and in the plane through the x axis and perpendicular to the earth surface, and y is orthogonal to both x and z. The sensor status is decided by orientation angles (ω, ϕ, κ) in relation to the platform flight coordinate system (x, y, z). The coordinates (X, Y, Z) of the imaging position usually conform to a standard coordinate system (of defined map projection and datum).

direction. The geometric distortion for a nadir view image is central and symmetrical, increasing from the centre of the image to its edge, whereas that for an oblique view image increases from the near range to the far range of the image, in which all topographic features appear to fall away from the sensor look direction (Fig. 9.3). For satellite-borne vertical imaging using a scanner system, the usual sensor status is that the orientation of the sensor principal is vertical to the land surface and that the scanning direction is perpendicular to the flight direction. As illustrated in Fig. 9.4, when the scanning direction is skewed to the flight direction ($\omega \neq 0$), the swath of the image scan line will be narrower than it is perpendicular to the flight direction. With deliberate configuration in this way, the platform flight with a rotation angle from the flight direction can help to achieve higher spatial resolution at the cost of a narrower image scene.

The sensor status (ω, ϕ, κ) parameters are the most sensitive to the image geometry. A tiny displacement of sensor status can translate to significant distortion in image pixel position and scale. Great technical effort has been devoted to achieve very precise status control for modern earth observation satellites, to reduce the sensor status distortion to a minimum and thus to satisfy most applications. For airborne sensor systems, the stability of platform status is often problematic, as demonstrated in an Airborne Thematic Mapper (ATM) image in Fig. 9.5. This type of error can be corrected using onboard flight status parameters, to some extent, though the errors are often too severe to be corrected for.

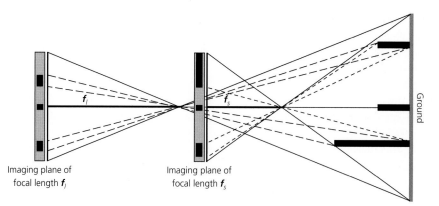

Imaging plane of focal length $\textbf{\textit{f}}_l$

Imaging plane of focal length $\textbf{\textit{f}}_s$

Ground

Fig. 9.2 The relationship between focal length and geometric distortion. A sensor with a shorter focal length (f_s) can cover the same field of view in much shorter distance than that with a longer focal length (f_l) but with more significant geometric distortion of tall ground objects and high terrain relief.

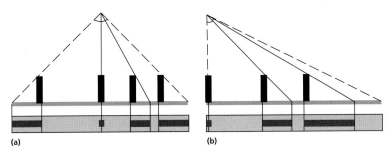

Fig. 9.3 Central projection distortion for (a) nadir view and (b) oblique view imaging.

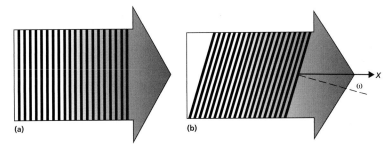

Fig. 9.4 The effects of sensor status in relation to image scanner orientation: (a) the scanning direction is perpendicular to the flight direction; and (b) the scanning direction is oblique to the flight direction with a sensor/platform rotation angle ω relating to the flight direction. Consequently, the swath of the image becomes narrower.

Fig. 9.5 Image scan line distortion in an ATM image caused by aircraft yaw.

9.1.2 Earth rotation and curvature

For spaceborne earth observation remote sensing, the sensor system onboard a satellite images a three-dimensional (3D) spherical land surface with topographic relief onto a two-dimensional (2D) flat image. Geometric distortion and position inaccuracy are therefore inevitable. Fortunately, for centuries, photogrammetric (or geomatic) engineering has developed many effective map projection models to achieve the optimised translation of the earth's spherical surface to a flat surface of maps and images. The most widely used map projection system is the UTM with either the WGS84 global datum or a local datum. We are going to revisit the topic of map projection in greater detail in Chapter 14 of Part II of the book. One of the major tasks of image geometric operation is to rectify an image to a given map projection system. The process is often called *geocoding* or *geo-referencing*.

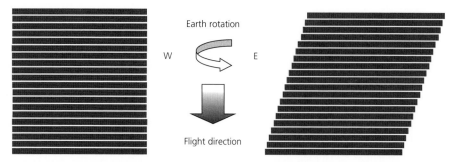

Fig. 9.6 The skew effect of earth rotation on a push-broom scanner onboard an earth observation satellite in a circular, near-polar, sun-synchronous orbit. The image is recorded as left but the actual area covered on the surface of the earth is as right.

Many earth observation satellites are configure to fly in *circular, near-polar, sun-synchronous orbits*, so to image nearly every part of the earth at about the same local time. On such an orbit, a satellite travels nearly perpendicular to the earth rotation direction. This is not a problem for an instant image taken by a camera, as all the pixels in the scene are taken simultaneously. This becomes a problem for a scanner, however, which is still (so far) the dominant design of spaceborne sensor systems; the image is built line by line in a time sequence with the satellite flying above the rotating earth. As shown in Fig. 9.6, for an along-track push-broom scanner (Appendix A), the earth surface is imaged in consecutive swathes in the time interval of a scanning cycle (or scanning frequency) to build a sequence of image lines in the flight direction of the satellite. The earth's rotation causes the longitude position to move westward in each scanning cycle and, as a result, the image built up is not a rectangular stripe but a parallelogram stripe that is skewed to the west in the imaging advancing direction. For the scenario of an across-track two way scanner (e.g. Thematic Mapper, Appendix A), the distortion pattern is more complicated because every pixel along a swath is imaged at a different time and the actual scanning speed at the earth surface changes not only from nadir to the edge of a swath but also between swathes for and against earth rotation (ref § 4.6). The processing to compensate for the earth rotation in relation to the scanning mechanism is usually done in bulk geometric correction by the ground assembly facilities – that is, the receiving station – using batch processing algorithms based on sensor/platform configuration parameters. The images after bulk processing to de-skew and crudely geocode are usually labelled as L-1B data.

9.2 Polynomial deformation model and image warping co-registration

Based on the above discussion, we can generally assume that a remotely sensed image has some geometric distortion as a result of image acquisition. Given a reference map projection or a reference image that is either geometrically correct or regarded as a geometric basis of a set of images, the main task of geometric operations is to establish a deformation model between the *input image* and the *reference* and then rectify or co-register the input image to the reference to generate an *output image*. In this way, either the geometric distortion of the input image is corrected or it is co-registered to a standard basis for further analysis.

The so-called 'rubber sheet warping', based on a polynomial deformation model, is the most important and commonly used geometric transformation for remotely sensed image data. There are several types of image warping:

- Image to map projection system (e.g. UTM);
- image to map (e.g. a topographic map);
- image to image.

The geometric transformation includes two major steps:

1 Establish the polynomial deformation model. This is usually done using *ground control points* (GCPs).
2 Image re-sampling based on the deformation model. This includes re-sampling image pixel positions (coordinates) and digital numbers (DNs).

9.2.1 Derivation of deformation model

A deformation model can be derived by fitting a polynomial function to GCP locations. This is done by selecting many GCPs representing the same ground positions in

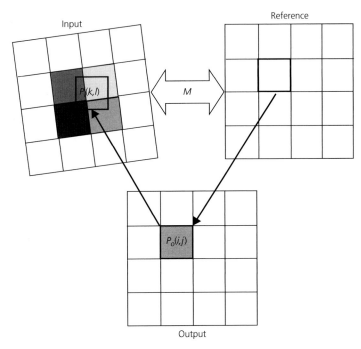

Fig. 9.7 Output-to-input mapping.

both the input image and the reference (an image or a map) to establish a deformation model, and then transforming the input image to the output image that is forced to fit the reference geometry.

In simple terms, this means transforming each pixel of the input image to the output image (input-to-output mapping) based on the deformation model, but a pixel position in an image is given in integers of line and column, while the transformation between the input and the output images may not always correspond exactly to integer positions for every pixel. Many pixels in the input image may take decimal positions, overlap or be apart from each other, resulting in 'black holes' in the output image. The input-to-output mapping cannot therefore generate an output image as a proper, regular raster image. To solve the problem, a commonly used approach is the *output-to-input mapping*.

Suppose transformation M is an output-to-input mapping that maps an output pixel at position (i,j) back to the input image at location (k,l), then the output image can be generated as shown in Fig. 9.7. For each output pixel position (i,j), compute from $M(i,j)$ to give the corresponding position (k,l) in the input image and then, at this position, pick up a pixel value $P_i(k,l)$ and

assign it to the output pixel $P_o(i,j)$. The output image is completed when all the pixels have been assigned.

The question now is how to derive a deformation model or the transformation M. Let (k,l) represent the position in the input image corresponding to the output position (i,j); the general form of the polynomial approximation for k and l is

$$M: \begin{cases} k = Q(i,j) = q_0 + q_1 i + q_2 j + q_3 i^2 + q_4 ij + q_5 j^2 + \cdots \\ l = R(i,j) = r_0 + r_1 i + r_2 j + r_3 i^2 + r_4 ij + r_5 j^2 + \cdots \end{cases} \quad (9.1)$$

Equation 9.1 defines the transformation M that calculates the approximation of the input position (k,l) from a given output position (i,j), if the coefficients $\mathbf{Q} = (q_0, q_1, q_2, \ldots)^t$ and $\mathbf{R} = (r_0, r_1, r_2, \ldots)^t$ are known. For n pairs of GCPs, we already know both (k,l) and (i,j) for every GCP, so the least squares solutions for $\mathbf{Q}$ and $\mathbf{R}$ can be derived. From n GCPs, we can establish n pairs of polynomials based on 9.1 written in matrix format as below:

$$\mathbf{K} = \mathbf{MQ}$$
$$\mathbf{L} = \mathbf{MR} \quad (9.2)$$

where

$$\mathbf{K} = \begin{pmatrix} k_1 \\ k_2 \\ \vdots \\ k_n \end{pmatrix}, \quad \mathbf{L} = \begin{pmatrix} l_1 \\ l_2 \\ \vdots \\ l_n \end{pmatrix}, \quad \mathbf{Q} = \begin{pmatrix} q_0 \\ q_1 \\ q_2 \\ \vdots \end{pmatrix}$$

$$\mathbf{R} = \begin{pmatrix} r_1 \\ r_2 \\ r_3 \\ \vdots \end{pmatrix}, \quad \mathbf{M} = \begin{pmatrix} 1 & i_1 & j_1 & i_1^2 & j_1^2 & \cdots \\ 1 & i_2 & j_2 & i_2^2 & j_2^2 & \cdots \\ \vdots & \vdots & \vdots & \vdots & \vdots & \vdots \\ 1 & i_n & j_n & i_n^2 & j_n^2 & \cdots \end{pmatrix}$$

The least squares solution for $\mathbf{Q}$ is

$$\mathbf{Q} = \left(\mathbf{M}^t \mathbf{M}\right)^{-1} \mathbf{M}^t \mathbf{K} \qquad (9.3)$$

Similarly, for $\mathbf{R}$

$$\mathbf{R} = \left(\mathbf{M}^t \mathbf{M}\right)^{-1} \mathbf{M}^t \mathbf{L} \qquad (9.4)$$

Once the coefficients $\mathbf{Q} = (q_0, q_1, q_2, \ldots)^t$ and $\mathbf{R} = (r_0, r_1, r_2, \ldots)^t$ are derived from the GCPs, the pixel position relationship between the input and output images is fully established through the transformation M. Given a location (i,j) in the output image, the corresponding position (k,l) in the input image can then be calculated from the transform M using eqn. 9.1. Theoretically, the higher the order of the polynomials, the higher the accuracy of warping that can be achieved, but the more control points are needed. A linear fit needs at least three GCPs, quadric fitting six and cubic fitting 10.

9.2.2 Pixel DN re-sampling

In the output-to-input mapping model, the output pixel at position (i,j) is mapped to its corresponding position (k,l) in the input image by the transform M. In most cases, (k,l) is not an integer position and there is no pixel DN value ready for this point. Re-sampling is an interpolation procedure used to find the DN for position (k,l) in the input image so as to assign it to the pixel $P_o(i,j)$ in the output image (Fig. 9.7). The simplest re-sampling function is the *nearest neighbour* method (Fig. 9.8), in which pixel $P_o(i,j)$ is assigned the DN of the input image pixel nearest to position (k,l).

A more accurate and widely used method is *bilinear interpolation*, as defined below (Fig. 9.9):

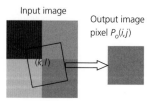

Fig. 9.8 Illustration of nearest neighbour DN re-sampling. The nearest neighbour of the output pixel $P_o(i,j)$ in the input image is the pixel of the bottom right, and therefore the DN of this pixel is assigned to the $P_o(i,j)$.

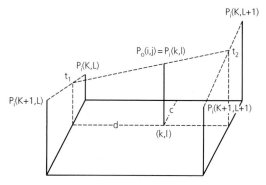

Fig. 9.9 Illustration of bilinear interpolation.

$$\begin{aligned} t_1 &= P_i(K,L)(1-c) + P_i(K+1,L)c \\ t_2 &= P_i(K,L+1)(1-c) + P_i(K+1,L+1)c \\ P_o(i,j) &= t_1(1-d) + t_2 d \end{aligned} \qquad (9.5)$$

where

$P_i(K,L)$ is an input pixel at an integer position (K,L) in the input image.

$P_o(i,j)$ is the output pixel at an integer position (i,j) in the output image, which corresponds to the $P_i(k,l)$ at a decimal position (k,l) in the input image and here,

K = the integer part of k

L = the integer part of l

$c = k{-}K$

$d = l{-}L$

Quadric and cubic polynomials are also popular interpolation functions for re-sampling with more complexity and improved accuracy.

Nowadays earth observation satellite images are often provided with standard map projection; this means that the images of the same scene are already co-registered and the demand for image co-registration becomes much less than before; however, the principle of image co-registration algorithms remains the same and the operation is often needed when using images and maps with unknown or incompatible map projections or without rectification.

9.3 GCP selection and automation of image co-registration

9.3.1 Manual and semi-automatic GCP selection

Ground control points can be selected manually, semi-automatically or automatically. These points are typically distinctive points with sharp contrast to their surroundings, such as corners of buildings, road junctions, sharp river bends and distinctive topographic features. For the manual method, the accuracy of the GCPs depends totally on your shaking hands. It is often not easy to accurately locate the corresponding points in the input and reference images. A simple approach can improve the efficiency and accuracy for manual GCP selection. First, select four GCPs spread in the four corners of the input and the reference images, and thus an initial geocoding frame (a linear fitting transformation M) is set up based on these four GCPs. Using four instead of the minimum three GCPs for the linear fitting can produce initial error estimation. After this initial setup, once a GCP is selected in the reference image, the corresponding position in the input image is roughly located based on the initial geocoding frame. A user only needs to fine-tune the position and confirm the selection. The transformation M will be continuously updated as GCPs are added, via the least squares solution.

The semi-automatic method allows users to identify GCPs in corresponding (input and reference) images and then automatically optimise point positions from one image to the other using local correlation. Suppose the GCP in the input image is at a position (k,l); the optimal coordinates of this point in the reference image are then decided when $r(k,l)$, the normalised cross-correlation (NCC) coefficient between the reference image and input image at position (k,l), reaches the maximum in an $l_w \times s_w$ calculation window used to roam in an $l_s \times s_s$ searching area in the reference image surrounding the roughly selected position of the GCP.

$$r(k,l) = \frac{\sum_{i=1}^{l_w}\sum_{j=1}^{s_w}\left(w_{i,j}-\overline{w}\right)\left(s_{k-1+i,l-1+j}-\overline{s}_{k,l}\right)}{\left[\sum_{i=1}^{l_w}\sum_{j=1}^{s_w}\left(w_{i,j}-\overline{w}\right)^2\right]^{1/2}\left[\sum_{i=1}^{l_w}\sum_{j=1}^{s_w}\left(s_{k-1+i,l-1+j}-\overline{s}_{k,l}\right)^2\right]^{1/2}} \Rightarrow \max$$

(9.6)

where the $(w_{i,j}-\overline{w})$ is calculated from the input image while $(s_{k-1+i,l-1+j}-\overline{s}_{k,l})$ the reference image.

Instead of NCC, a relatively new technique, phase correlation, can be used to locate the matching pixels in corresponding images at sub-pixel accuracy directly without roaming search, as detailed in Chapter 11.

With rapid development of computer vision technology, GCP selection and image co-registration has been largely automated between images of similar illumination condition and spectral/spatial property as briefly introduced in the following sub-section. However, the manual and semi-automatic GCP selection is still needed for co-registration between images from different sources and with very different spectral property and spatial characteristics (such as temporal changes in urban areas), and for co-registration between images and maps. Manual operation is slow and not very accurate, but it works when the automatic methods fail.

9.3.2 Automatic image co-registration

In a general term, GCP selection is to locate corresponding points or features between two images for matching. This belongs to a very active research field in computer vision: feature matching. Automatic selection of corresponding points (GCP) enables the automation of image co-registration. A workable automatic GCP selection method must able to accurately select adequate high quality GCPs that are evenly spread across the whole image to be co-registered. In general, it comprises two steps:

1 Automatic selection of feature points from one of the two corresponding images (either the input image or the reference image) as candidate GCPs.

2 Automatic localization of the corresponding points in the other image via local correlation. High quality GCPs are finally selected by their correlation level and spatial distribution.

In the last 15 years, quite a few effective techniques have been developed to locate corresponding points for automatic image co-registration. A very widely used algorithm is called SIFT (scale invariant feature transform) proposed by D. Lowe (2004). Since the method was first published, it has gained wide application; meanwhile, many variants of SIFT have been proposed, but they all use SIFT as a benchmark algorithm for performance assessment. SIFT is a *feature descriptor* with a multi-step processing procedure, including potential interest points identification, scale invariant stable keypoint selection, keypoint feature orientation and keypoint description. As such, SIFT is robust to considerable variation of feature scale orientation, shape distortion and global illumination (image contrast and brightness). It can therefore locate corresponding point pairs between images with certain affined deformation to achieve image co-registration. The details of the SIFT algorithm are beyond the scope of this textbook but can be found in the references given. A major further development of feature descriptors after SIFT is SURF (speed up robust features) proposed by Bay *et al.* (2008). With more or less the same robust performance, the major merit of SURF is its much higher computing efficiency, about three times faster than SIFT.

Once adequate and well-distributed corresponding points are selected automatically, the deformation model between the reference image and the input image can be derived, and the image co-registration can then be carried out using the output-to-input mapping described in § 9.2.1. The whole processing flow is automatic.

The image co-registration based on deformation model can be automated by automation of GCP selection, but it cannot assure precise matching between every pair of corresponding pixels in an image. If we can estimate the disparity between every pair of corresponding pixels, we can then force the image co-registration pixel-wise without a deformation model. In Chapter 11, an automatic method of pixel-wise image co-registration will be introduced as one of the applications of a relatively new technique: phase correlation.

9.4 Summary

After brief discussion of the major sources of geometric deformation in remotely sensed images acquired by earth observation satellites, we have introduced the details of image warping co-registration based on a polynomial deformation model derived from GCPs. The key points for this popular method are:

1 The accuracy of a polynomial deformation model largely depends on the quality, number and spatial distribution of GCPs, but also on the order of the polynomials. A higher order polynomial requires more GCPs and more computing time with improved accuracy. The most commonly used polynomial deformation models are linear, quadratic and cubic.

2 The co-registered image is generated based on an *output-to-input mapping* procedure to avoid pixel overlaps and holes in the output image. In this case, pixel DN re-sampling is necessary to draw DNs in non-integer positions in the input image for a DN in the corresponding integer position in the output image via nearest neighbour or bilinear interpolation.

Given that GCP selection is of vital importance for the accuracy and automation of image warping co-registration, we then addressed techniques for manual, semi-automatic and automatic GCP selections. The automatic GCP selection is an active research field of computer vision. We very briefly introduced the most widely used feature descriptors, SIFT and SURF, for automatic GCP selection.

The classical approach to image warping co-registration, based on a polynomial deformation model derived from GCPs, can be quite accurate if the GCPs are of high positional accuracy, but registration by this method is not achieved at pixel-to-pixel level. The registration error within a scene may vary from place to place depending on the local relative deformation between the images because GCP deformation model-based approaches are not adequate to correct the irregular deformation caused by sensor optic quality, platform status precision and so forth. In Chapter 11, we are going to introduce a pixel-wise image co-registration method as one of the applications of the phase correlation sub-pixel image matching technology. The method ensures the high accuracy of image co-registration at every image pixel regardless of the irregular deformations between the two images, and it is entirely automatic.

9.5 Questions

1 Describe the relationship between satellite flight direction, sensor status and imaging position relating to the ground, using a diagram.

2 Give an example of how the instability of sensor status produces geometric errors.

3 What is the best sensor status that introduces minimal geometric distortion for earth observation remote sensing?

4 Explain the relationship between focal length f and topographic distortion using a diagram.

5 Explain the relationship between imaging geometry and topographic distortion using a diagram.

6 Explain why de-skewing processing is essential for images acquired by a scanner onboard a satellite in a circular, near-polar, sun-synchronous orbit. If a camera instead of scanner is used for imaging, do you think that de-skewing processing is necessary and why?

7 What is output-to-input mapping (explain using a diagram), and why is it necessary for image warping co-registration?

8 How is the transformation, M, established in a polynomial deformation model?

9 How many GCPs are required to establish a linear, quadratic and cubic polynomial deformation model? Can you write down these polynomials?

10 With a diagram, derive the bilinear interpolation for pixel DN re-sampling. Calculate $P_i(5.3, 4.6)$ given $P_i(5,4) = 65$, $P_i(6,4) = 60$, $P_i(5,5) = 72$, and $P_i(6,5) = 68$.

CHAPTER 10

Introduction to interferometric synthetic aperture radar technique*

In this chapter, we introduce some interferometric synthetic aperture radar (InSAR) techniques for three-dimensional (3D) terrain representation, for quantitative measurements of terrain deformation and for the detection of random land surface changes. InSAR is not normally covered by the general scope of image processing, but it has become a widely used application of SAR data analysis in remote sensing. Many InSAR image processing software packages have been developed and some popular image processing systems now include InSAR functionality.

10.1 The principle of a radar interferometer

For many remote sensing applications, we use processed *multi-look* SAR images. These products represent images of the averaged intensity (or amplitude) of multiple radar looks to reduce radar speckles. The original SAR image representing all the information from the return radar signals is a *single look complex* (SLC) image. An SLC image is composed of complex pixel numbers that record not only the intensity (the energy of microwave signals returned from targets) but also the phase of the signal, which is determined by the distance between the target and the radar antenna.

Given a complex number of an SLC pixel: $c = a + ib$, $i = \sqrt{-1}$, the magnitude of c is $M_c = \sqrt{a^2 + b^2}$, which formulates SAR intensity image, while the phase angle of c is $\varphi = \arctan\left(\dfrac{b}{a}\right)$.

InSAR technology exploits the phase information in SAR SLC images for earth and planetary observations.

A SAR interferogram shows the phase differences between the corresponding pixels of the same object in two SAR images taken from near-repeat orbits. It represents topography as *fringes of interference*. Based on this principle, InSAR technology has been developed and used successfully for topographic mapping and measurement of terrain deformation caused by earthquakes, subsidence, volcano deflation and glacial flow motion (Zebker & Goldstein 1986; Gabriel *et al.* 1989; Goldstein *et al.* 1993; Massonnet & Adragna 1993; Massonnet *et al.* 1993, 1994, 1995; Zebker *et al.* 1994a, 1994b). The vertical accuracy is several 10s of metres for digital elevation model (DEM) generation and at centimetre level for measurement of terrain surface deformation.

A radar beam is nominally a mono-frequency electromagnetic wave. Its properties are similar to those of monochromatic coherent light. When two nearly parallel beams of coherent light illuminate the same surface, an interferogram can be generated showing the phase shift induced by the variation of position and topography of the surface, as a result of the interference between the two beams. The same principle applies to the return radar signals. A SAR interferometer acquires two SLC images of the same scene with the antenna separated by a distance B called the *baseline*. For a single-pass SAR interferometer, such as a SAR interferometer onboard an aircraft or the space shuttle (e.g. SRTM mission), two images are acquired simultaneously via two antennas separated by baseline B; one sends and receives the signals while the other receives only [Fig. 10.1(a)]. In contrast, a repeat-pass SAR interferometer acquires a single image of the same area twice

Image Processing and GIS for Remote Sensing: Techniques and Applications, Second Edition. Jian Guo Liu and Philippa J. Mason.
© 2016 John Wiley & Sons, Ltd. Published 2016 by John Wiley & Sons, Ltd.

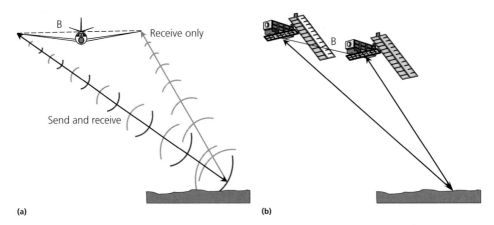

(a)   **(b)**

Fig. 10.1 (a) The single-pass SAR interferometer with both an active antenna, sending and receiving radar signals, and a passive antenna (separated by a distance B) to receive signals only; and (b) a repeat pass SAR interferometer to image the same area from two visits with a minor orbital drift B.

from two separate orbits with minor drift, which forms the baseline B [Fig. 10.1(b)]; this is the case for ERS1 & ERS2 SAR, ENVISAT ASAR, RADARSAT, ALOS PALSAR and Terra SAR X/Tandem X.

The purpose of InSAR is to derive a SAR interferogram, ϕ, which is the phase difference between the two coherent SLC images (often called *fringe pair*). First, the two SLC images are precisely co-registered pixel by pixel at sub-pixel accuracy based on local correlation in combination with embedded position data in SAR SLCs. The phase difference ϕ between the two corresponding pixels is then calculated from the phase angles φ_1 and φ_2 of these two pixels through their complex numbers:

$$\phi = \varphi_1 - \varphi_2 \tag{10.1}$$

To understand the relationship between phase difference and the InSAR imaging geometry, let us consider a SAR system observing the same ground swath from two positions, A1 and A2, as illustrated in Fig. 10.2. The ground point C is then observed twice from distance r (slant range) and $r + \delta$. The distance difference between the return radar signals for a round-trip is 2δ and the measured phase difference ϕ (interferogram) is

$$\phi = \frac{4\pi}{\lambda}\delta \tag{10.2}$$

or 2π times the round-trip difference, 2δ, in radar wavelength λ.

From the triangle A1-A2-C in Fig. 10.2, cosine theorem permits a solution for δ in terms of the imaging geometry as follows:

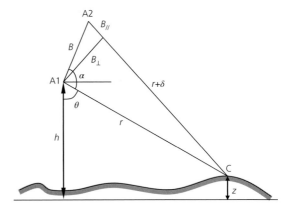

Fig. 10.2 The geometry of radar interferometry.

$$\left(r+\delta\right)^2 = r^2 + B^2 - 2rB\cos\left[\frac{\pi}{2}-\left(\theta-\alpha\right)\right]$$

or

$$\left(r+\delta\right)^2 = r^2 + B^2 - 2rB\sin\left(\theta-\alpha\right) \tag{10.3}$$

where B is the baseline length, r the radar slant range to a point on the ground, θ the SAR look angle and α the angle of the baseline with respect to horizontal at the sensor.

The baseline B can be de-composed to two components that are perpendicular $B_\perp$ and parallel $B_{//}$ to the look direction.

$$B_\perp = B\cos\left(\theta-\alpha\right) \tag{10.4}$$

$$B_{//} = B\sin\left(\theta-\alpha\right) \tag{10.5}$$

The InSAR data processing accurately calculates the phase difference ϕ between corresponding pixels between a fringe pair of SLC SAR images to produce an interferogram. The applications of InSAR are largely based on the relationships between the interferogram ϕ, topography and terrain deformation, for which the baseline B, especially the perpendicular baseline $B_\perp$, plays a key role.

10.2 Radar interferogram and DEM

One major application of InSAR is to generate a DEM. It is clear from Fig. 10.2 that the elevation of the measured point C can be defined as

$$z = h - r\cos\theta \qquad (10.6)$$

where h is the height of the sensor above the reference surface (datum).

It looks simple, but in this formula, the exact look angle θ is not directly known from the SLC images. We have to find these unknowns from the data that InSAR provides. From a SAR interferogram ϕ, we can express δ by re-arranging eqn. 10.2:

$$\delta = \frac{\lambda\phi}{4\pi} \qquad (10.7)$$

Modifying eqn. 10.3 as a sine function of $\theta - \alpha$,

$$\sin(\theta - \alpha) = \frac{r^2 + B^2 - (r + \delta)^2}{2rB} \qquad (10.8)$$

In this equation, the baseline B and slant range r are known and constants for both entire fringe pair images, whilst the only variable δ can be easily calculated from phase difference ϕ (SAR interferogram) using eqn. 10.7. Thus $\sin(\theta - \alpha)$ is resolved.

Expressing $\cos\theta$ in eqn. 10.6 as functions of α and $\sin(\theta - \alpha)$,

$$\begin{aligned} z &= h - r\cos\theta \\ &= h - r\cos(\alpha + \theta - \alpha) \\ &= h - r\cos\alpha\cos(\theta - \alpha) + r\sin\alpha\sin(\theta - \alpha) \qquad (10.9) \\ &= h - r\cos\alpha\sqrt{1 - \sin^2(\theta - \alpha)} + r\sin\alpha\sin(\theta - \alpha) \end{aligned}$$

In eqn. 10.9, the angle of the baseline with respect to the horizontal at the sensor, α, is a constant for the SAR fringe pair images and is determined by the imaging status, whereas $\sin(\theta - \alpha)$ can be derived from the interferogram ϕ using eqns. 10.7 and 10.8 and the elevation z can therefore be resolved.

In principle, we can measure the phase difference at each point in an image and apply the above three equations based on our knowledge of imaging geometry to produce the elevation data. There is, however, a problem in this: the InSAR measured phase difference is a variable in the 2π period or is 2π wrapped. Figure 10.3 shows an interferogram generated from a fringe pair of ERS-2 SAR images; the fringe patterns are like contour

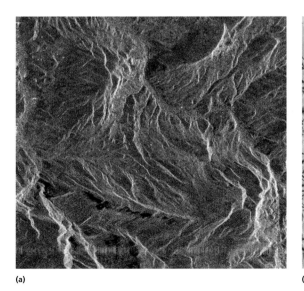

(a)

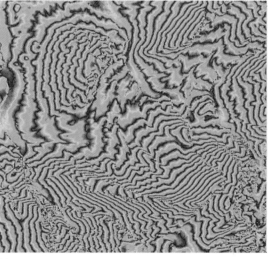

(b)

Fig. 10.3 (a) An ERS-2 SAR multi-look image; and (b) the SAR interferogram of the same scene in which the fringes, in 2π cycles, are like contours showing the topography.

lines representing the mountain terrain, but numerically, these fringes occur in repeating 2π cycles and do not give the actual phase differences, which could be n times 2π plus the InSAR measured phase difference. The phase information is recorded in the SAR data as complex numbers, and only the principal phase values (ϕ_p) within 2π can be derived. The actual phase difference should therefore be

$$\phi = \phi_p + 2n\pi \qquad (10.10)$$

Expressed in terms of the slant range difference:

$$\delta = \frac{\lambda\phi}{4\pi} = \frac{\lambda}{4\pi}\left(\phi_p + 2n\pi\right) = \frac{\lambda\phi_p}{4\pi} + \frac{n\lambda}{2} \qquad (10.11)$$

The interferometric phase therefore needs to be unwrapped to remove the modulo-2π ambiguity so as to generate DEM data. For a perfect interferogram in 2π-modulo, unwrapping can be achieved accurately via a spatial searching-based scheme, but various de-correlation factors between the fringe pair of images mean that SAR interferograms are often noisy. In such cases, unwrapping is an ill-portrayed problem. There are many well-established techniques for the unwrapping of noisy InSAR interferograms, each with its own merits and weaknesses, but the search for better techniques continues. The details of unwrapping are beyond the scope of this book. There are also other corrections necessary, such as the removal of the ramps caused by the earth's curvature and by the direction angle between the two paths, as they are usually not perfectly parallel. All the relevant functionalities are built in for commercially available InSAR software packages.

We now prove that the elevation resolution is proportional to the perpendicular baseline $B_\perp$. The partial derivative of elevation z to the slant range increment δ is

$$\frac{\partial z}{\partial \delta} = \frac{\partial z}{\partial \theta} \times \frac{\partial \theta}{\partial \delta} \qquad (10.12)$$

From eqn. (10.2),

$$\frac{\partial z}{\partial \theta} = r\sin\theta \qquad (10.13)$$

From eqn. (10.3),

$$\frac{\partial \theta}{\partial \delta} = -\frac{r+\delta}{Br\cos(\theta-\alpha)} \qquad (10.14)$$

Consider δ is very small in comparison with r, thus,

$$\frac{\partial z}{\partial \delta} = r\sin\theta\left[-\frac{r+\delta}{Br\cos(\theta-\alpha)}\right]$$
$$\approx -\frac{r\sin\theta}{B\cos(\theta-\alpha)} \qquad (10.15)$$
$$= -\frac{r\sin\theta}{B_\perp}$$

We can then derive the partial derivative of elevation z to the change in phase difference (interferogram) ϕ. From eqn. 10.7,

$$\frac{\partial \delta}{\partial \phi} = \frac{\lambda}{4\pi},$$
$$\frac{\partial z}{\partial \phi} = \frac{\partial z}{\partial \delta} \times \frac{\partial \delta}{\partial \phi} \approx \frac{\lambda r\sin\theta}{4\pi B_\perp} \qquad (10.16)$$

Therefore, for a phase increment $\Delta\phi$, we have

$$\Delta z = -\frac{\lambda r\sin\theta}{4\pi B_\perp}\Delta\phi \qquad (10.17)$$

For one fringe cycle, $\Delta\phi = 2\pi$, in this case,

$$\Delta z_{2\pi} = -\frac{\lambda r\sin\theta}{2B_\perp} \qquad (10.18)$$

The numerator in eqn. 10.18 is constant and thus the greater the $B_\perp$, the less the elevation increment with one cycle of 2π and the higher the DEM resolution. Given

$$r = \frac{h}{\cos\theta}$$

we can re-write eqn. 10.18 as

$$\Delta z_{2\pi} = -\frac{\lambda h\tan\theta}{2B_\perp} \qquad (10.19)$$

In the case of the ENVISAT SAR, where $\lambda = 0.056\,\text{m}$, $h = 800,000\,\text{m}$, look angle $\theta = 23°$ and $B_\perp = 300\,\text{m}$, we then have $|\Delta z_{2\pi}|=31.7\,\text{m}$. We find that the elevation resolution of InSAR is not very high even with a 300 m perpendicular baseline. This is because the translation from phase difference to elevation is indirect, according to the geometry shown in Fig. 10.2.

As hinted in eqn. 10.19, the $B_\perp$ enables the observation of the same ground position of height, from different view angles. This is the key factor for DEM generation using InSAR, and it is similar to DEM generation from a pair of stereo images, although based on quite different principles. As a simple demonstration, consider a special case, $B_\perp = 0$, where a point is observed from two SAR images in the same view angle, as shown in Fig. 10.4.

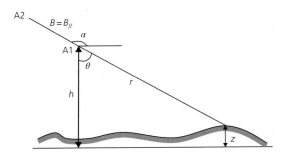

Fig. 10.4 A special case of InSAR for $B_\perp = 0$.

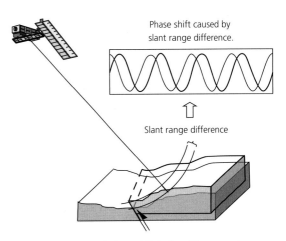

Fig. 10.5 Illustration of phase shift induced from terrain deformation, measured by differential InSAR.

Then $\delta = B_{//} = B$ is a constant that is independent of the view angle θ. In this unique case, the interferogram is a constant invariant with respect to position and it therefore contains no topographic information. A SAR fringe pair with very small $B_\perp$ is therefore insensitive to elevation. On the other hand, a very large $B_\perp$, though more sensitive to elevation, will significantly reduce the coherence level, and thus the signal-to-noise ratio (SNR) of the interferogram, because the two radar beams become less parallel. Most InSAR applications require the ratio $\dfrac{B_\perp}{r}$ to be less than 1/1000. Translating this to ERS-2 and ENVISAT with an orbital altitude of ca 800 km and a nominal view angle of 23°, the desirable $B_\perp$ for effective InSAR should be less than 1000 m.

10.3 Differential InSAR and deformation measurement

Satellite differential InSAR (DInSAR) is an effective tool for the measurement of terrain deformation as caused by earthquakes, subsidence, glacial motion and volcanic deflation. The DInSAR is based on a repeat-pass spaceborne radar interferometer configuration. As shown in Fig. 10.5, if the terrain is deformed by an earthquake on a fault, then the deformation is translated directly as the phase difference between two SAR observations, made before and after the event. If the satellite orbit is precisely controlled to make the two repeat observations from exactly the same position, or at least $B_\perp = 0$, as illustrated in Fig. 10.4, the phase difference measured from InSAR is entirely produced by the deformation in the slant range direction. This ideal zero baseline case is, however, unlikely to be the real situation in most earth observation satellites with SAR sensor systems. In general, the across-event SAR observations are made with a baseline $B_\perp \neq 0$,

and as a result, the phase difference caused by terrain deformation is jumbled with the phase difference caused by topography. Logically, the difference between the interferogram generated from two SAR observations before terrain deformation and that from two SAR observations across the deformation event should cancel out the topography and retain the phase difference, representing only the terrain deformation. Topographic cancellation can be achieved from the original or from the unwrapped interferograms. The results of DInSAR can then be presented either as differential interferograms or deformation maps (unwrapped differential interferograms). For simplicity in describing the concepts of DInSAR processing, the phase difference in the following discussion refers to the true phase difference, not the 2π wrapped principal phase value.

With two pre-event SAR images and one post-event image (within the feasible baseline range), a pre-event and a cross-event fringe pair can be formulated, and so a differential interferogram can be derived directly. The DInSAR formula, after the earth curvature flattening correction, is stated below (Zebker *et al.* 1994a):

$$\delta\phi = \phi_{2flat} - \frac{B2_\perp}{B1_\perp}\phi_{1flat} = \frac{4\pi}{\lambda}D \qquad (10.20)$$

$$\phi_{flat} = \frac{4\pi}{\lambda}(\theta - \theta_0) \cdot B_\perp \qquad (10.21)$$

where θ_0 is the look angle to each point in the image assuming zero local elevation on the ground (the reference ellipsoid of the earth); ϕ_{1flat} and ϕ_{2flat} are the unwrapped pre-event and cross-event interferograms after the earth curvature flattening correction; and D represents the displacement of the land surface in the radar line of sight (LOS).

The ratio between the perpendicular baselines of the two fringe pairs is necessary because the same 2π phase difference represents different elevations depending on $B_\perp$ according to eqn. 10.19.

The operation cancels out the stable topography and reveals the geometric deformation of the land surface. Equation 10.20 indicates that the deformation D is directly proportional to the differential phase difference $\delta\phi$, thus DInSAR can provide measurements of terrain deformation at better than half the wavelength of SAR at millimetre accuracy. For instance, the wavelength of the C-band SAR onboard ENVISAT is 56 mm, and thus 28 mm deformation along the slant range will be translated to 2π phase difference in a cross-event C-band SAR interferogram.

As an alternative approach, if a high quality DEM is available for an area under investigation, ϕ_1 can be generated from the DEM with an artificially given baseline equal to the baseline of the across-event fringe pair and simulated radar imaging geometry (Massonnet *et al.* 1993, 1994, 1995). In this case, the DInSAR is a simple difference between the cross-event SAR interferogram ϕ_2 and the simulated interferogram of topography ϕ_{1sim}.

$$\delta\phi = \phi_2 - \phi_{1sim} = \frac{4\pi}{\lambda} D \qquad (10.22)$$

The advantages of using a DEM to generate ϕ_{1sim} are that it is not restricted by the availability of suitable SAR fringe pairs and that the quality of ϕ_{1sim} is guaranteed without the need of unwrapping. As further discussed in the next section, the quality of a SAR interferogram is often significantly degraded by de-coherence factors, and as a result, unwrapping will unavoidably introduce errors.

The InSAR measured deformation D in slant range direction is often called LOS displacement in literatures. It is very important to notice that because of the side-looking imaging geometry of SAR, the same LOS displacement can be produced by deformation in either vertical or horizontal or both directions. As illustrated in Fig. 10.6, the ground deformation moving point A to position A' or to position A'' will produce the same LOS displacement in the same orientation. In fact any combination of horizontal and vertical motion moving

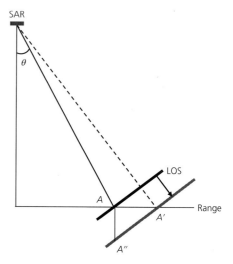

Fig. 10.6 The geometry of SAR imaging and the relationship between LOS displacement and ground deformation in horizontal and vertical directions.

point A to a position on the *line A''-A'* will produce the same LOS displacement.

Obviously, one crucial condition for DInSAR is that the satellite position for SAR imaging is controlled in high precision and is frequently recorded to ensure accurate baseline calculation. If the satellite can be controlled to repeat exactly, providing an identical orbit, then the so-called zero baseline InSAR is achieved; which, without topographic information, is in fact the same as a DInSAR measurement of deformation directly. In many applications, it is not always necessary to go through this rather complicated processing to generate differential SAR interferograms. Since the fringe patterns induced from topography and from terrain deformation are based on different principles, they are spatially very different and can often be visually separated for qualitative analysis. Also, since the terrain deformation-induced fringes are the direct translation from deformation to phase difference, they are often localised and show significantly higher density than the topographic fringes, especially in an interferogram with a short $B_\perp$. For a flat area without noticeable topographic relief, any obvious fringe patterns in a cross-event SAR interferogram should be the result of terrain deformation. In such cases, we can use InSAR for the purpose of DInSAR with great simplicity.

Figure 10.7 is an ENVISAT ASAR interferogram (draped over a Landsat 7 ETM+ true colour composite

Fig. 10.7 An ENVISAT interferogram draped over an ETM true colour composite showing the terrain deformation produced by an Mw 7.3 earthquake that occurred in the Siberian Altai on 27 September 2003.

image) showing the terrain deformation caused by an Mw 7.3 earthquake that occurred in the Siberian Altai on 27 September 2003. The high quality fringes are produced mainly in the basin, where elevation varies gently in a range of less than 250 m. The $B_\perp = 168$ *m* for this fringe pair, of wavelength $\lambda = 56$ mm, orbital attitude $h = 800$ km and look angle $\theta = 23°$, we can calculate from eqn. 10.19, so that each 2π phase shift represents 56.6 m elevation. Thus the 250 m elevation range translates to no more than five 2π fringes. The dense fringe patterns in this interferogram are dominantly produced by the earthquake deformation.

Certainly, when we say that the DInSAR can measure deformation on a millimetre scale, we mean the average deformation in the SAR slant range direction (i.e. LOS) in the image pixel, which is about 25 m spatial resolution for ERS SAR.

10.4 Multi-temporal coherence image and random change detection

So far the application of repeat-pass InSAR is considerably restricted from any areas subject to continuous random changes, such as in dense vegetation cover. To generate a meaningful interferogram, two SAR images must be highly coherent or have high coherence. The InSAR phase coherence is a measurement of local spatial correlation between two SAR images. Any random changes to the scatterers on the land surface between the two acquisitions will result in irregular variation in phase, which will reduce the coherence. In particular, the random variation exceeding half a wavelength of the radar beam in the slant range direction will cause a total loss of coherence.

As shown in Fig. 10.8, the terrain deformation caused by earthquakes and so forth typically comprises a block 3D shift that largely does not randomly alter the surface scatterers. Therefore, as long as there are no other factors causing random changes on the land surface, the phases of return SAR signal from ground scatterers will all shift in the same way as the block movement [Fig. 10.8(d)]; this collectively coherent phase shift can be recorded as a high quality SAR interferogram giving quantitative measurements of the deformation [Fig. 10.8(f)]. If, however, the land surface is subject to random changes, the phases of scatterers are altered randomly and will no longer be collectively coherent [Fig. 10.8(e)]; the coherence is consequently lost and the interferogram becomes chaotic [Fig. 10.8(f)].

Any land surface is subject to continuous random change caused by many de-correlation factors such as

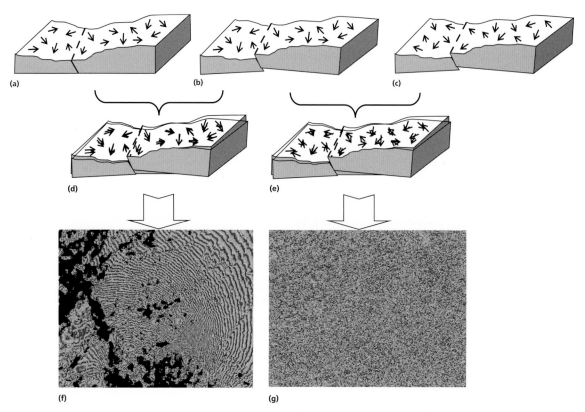

Fig. 10.8 Illustration of the effects of terrain deformation and random surface changes on interferograms. The small arrows in diagrams (a) to (e) represent the phase angles of return SAR signals from scatterers on land surface. The block movement along a fault between SAR acquisitions (a) and (b) results in coherent phase shift in the same direction as the motion between the two blocks (d), but there are no random phase shifts within each of the faulting blocks, and thus a good interferogram (f) is produced that records the terrain deformation. In the case that random changes are involved in addition to the faulting motion, as shown between (a) and (c), random phase shifts that are not related to the fault movement are introduced and the return signal phases between the two acquisitions are no longer coherent (e). As a result, the interferogram (g) is chaotic and does not show any meaningful interferometric patterns.

vegetation growth, wind-blown sands and erosion. Since random changes on a land surface are cumulative, a SAR fringe pair with a long temporal separation is likely to have low coherence and cannot therefore be used to produce a high quality interferogram. The low coherence means a lack of interferometric information, but this may still be useful in indicating random changes on the land surface, although it cannot give direct information of the nature of those changes. With the development of InSAR technique and applications, the value of multi-temporal coherence imagery as an information source for surface change detection has been widely recognised and many successful application cases have been reported (Corr & Whitehouse 1996; Liu *et al.* 1997a; Ichoku *et al.* 1998; Lee & Liu 2001; Liu *et al.* 2001, 2004).

Phase coherence of two SLC SAR images represents the local correlation of the radar reflection characteristics of the surface target between two observations. This can be estimated by ensemble averaging N neighbouring pixels of complex SAR data as

$$\rho = \frac{\left| \sum_{l=1}^{N} z_{1l} z_{2l}^{*} \right|}{\left[\sum_{l=1}^{N} z_{1l} z_{1l}^{*} \sum_{l=1}^{N} z_{2l} z_{2l}^{*} \right]^{1/2}} \qquad (10.23)$$

where ρ is the coherence magnitude estimation, z_1, z_2 the complex values of the two SAR images, and * the conjugate of a complex.

Obviously, the value range of ρ is [0, 1], ranging from totally incoherent to completely coherent.

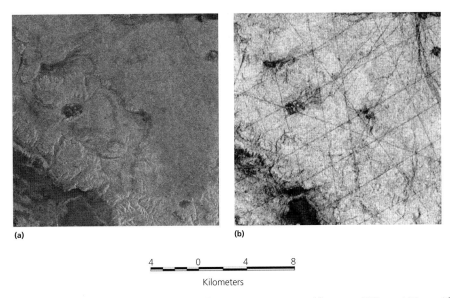

(a) (b)

4   0   4   8
Kilometers

Fig. 10.9 An ERS SAR multi-look image (a) and an InSAR coherence image generated from two SAR acquisitions with 350 days temporal separation (b) of an area in the Sahara desert. The coherence image reveals a mesh of seismic survey lines made during the period between the two acquisitions. These seismic lines are not shown in the multi-look image of the second acquisition of this InSAR pair.

Besides providing quality assessment for SAR interferograms, the interferometric coherence technique has been widely used for surface random change detection of phenomena such as rapid erosion, human activity-induced land disturbance and earthquake destruction assessment. Figure 10.9 presents an application example showing an ERS-2 SAR multi-look image (a) of an area in the Sahara desert and a coherence image (b) of the area derived from two SAR acquisitions with a temporal separation of 350 days. The coherence image reveals a mesh of dark straight lines that represent seismic survey lines made during the 350 days between the two SAR acquisitions, while the multi-look image (a) of the second acquisition of this InSAR pair shows nothing. For the seismic survey, trenches were dug to place the seismic sensors and, since then, the land surface has recovered to appear as before. Thus we can see no sign of these seismic survey lines in the multi-look image. On the other hand, the land surface materials along the survey lines are significantly disturbed with their scattering properties altered, randomly causing the loss of coherence as revealed in the coherence image.

10.5 Spatial de-correlation and ratio coherence technique

Apart from temporal random land surface changes during the period between two SAR acquisitions, there are other factors that can cause the loss of coherence in InSAR coherence images. Among these de-correlation factors, baseline and topographic factors are often called spatial de-correlation, as they are relevant to the geometric relations of sensor position and target distribution. In terrain of considerable relief, topographic de-correlation due to slope angle is often the dominant de-correlation factor. This type of de-correlation is an intrinsic property of a side-looking and ranging SAR system. The de-correlation is overwhelming, particularly on a foreshortened or layover slope where the coherence drops dramatically toward zero. The low coherence features of such slopes can easily be misinterpreted in the coherence imagery as an unstable land surface subject to rapid random change, even on a highly stable slope. It is important to separate topographic spatial de-correlation from the temporal de-correlation to achieve effective detection of random land surface changes.

Phase coherence decreases with the increase of $B_\perp$ as characterised in the formula below (Zebker & Villasenor 1992):

$$\rho_{spatial} = 1 - \frac{2R_y \cos\beta}{\lambda r} B_\perp \qquad (10.24)$$

where β is the incidence angle, R_y the ground range resolution, λ the radar wavelength and r the distance from the radar sensor to the centre of a resolution element (slant range).

In general, SAR fringe pairs with small $B_\perp$ are desirable for coherence image applications. From eqn. 10.24 we can further prove that the influence of spatial de-correlation varies with topography.

Let θ_0 represent the nominal look angle of the SAR (23° for ERS-2) and α the local terrain slope measured upward from horizontal, then the radar incident angle $\beta = \theta_0 - \alpha$ [Fig. 10.10(a)]. The ground range resolution is thus a function of the local terrain slope:

$$R_y = \frac{c}{2B_w \left| \sin(\theta_0 - \alpha) \right|} \qquad (10.25)$$

where c is the speed of light, and B_w the frequency bandwidth of the transmitted chirp signal.

R_y, the ground range resolution value, increases rapidly when the surface is nearly orthogonal to the radar beam and becomes infinite if the terrain slope is equal to the nominal look angle (i.e. $\alpha = \theta_0$) [Fig. 10.10(b)]. Note R_y is practically limited to a finite value because the terrain slope is not an infinite plane. The effect of a large value of R_y on the de-correlation is therefore significant in the case of the surface slope facing the radar. Substituting 10.25

into 10.24 results in a modified spatial de-correlation expression, as a function of perpendicular baseline and terrain slope (topography) (Lee & Liu 2001):

$$\rho_{spatial} = 1 - AB_\perp \left| \cot(\theta_0 - \alpha) \right| \qquad (10.26)$$

where $A = c/\lambda\pi B_w$, a constant for a SAR system.

This spatial de-correlation function describes the behaviour of topographic de-correlation as well as baseline de-correlation. For a given baseline, the correlation decreases as the local terrain slope approaches the nominal incidence angle, while the increase in baseline will speed up the deterioration of the correlation.

We now introduce a method of ratio coherence for analysing and separating spatial and temporal de-correlation phenomena (Lee & Liu 2001). The approximate total observed correlation of the returned radar signals can be generalised as a product of temporal and spatial correlation:

$$\rho = \rho_{temporal} \cdot \rho_{spatial}$$
$$= \rho_{temporal} \cdot \left(1 - AB_\perp \left| \cot(\theta_0 - \alpha) \right| \right) \qquad (10.27)$$

Consider three SAR observations named 1, 2, and 3 in time sequence. A ratio coherence image can then be established by dividing a coherence image with long temporal separation by the other with relatively short temporal separation as

$$\frac{\rho^{13}}{\rho^{12}} = \frac{\rho_{temporal}^{13}}{\rho_{temporal}^{12}} \cdot \frac{\rho_{spatial}^{13}}{\rho_{spatial}^{12}}$$
$$= \frac{\rho_{temporal}^{13}}{\rho_{temporal}^{12}} \cdot \frac{1 - AB_\perp^{13} \left| \cot(\theta_0 - \alpha) \right|}{1 - AB_\perp^{12} \left| \cot(\theta_0 - \alpha) \right|} \qquad (10.28)$$

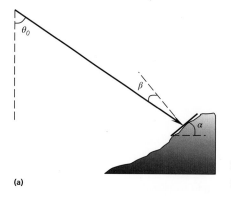
(a)

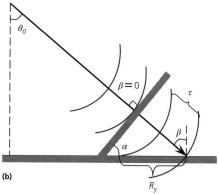
(b)

Fig. 10.10 (a) The relationship between nominal radar look angle θ_0, radar incident angle β and terrain slope angle α. (b) Given the nominal radar look angle θ_0 and the incident angle β, the ground range resolution R_y is decided by the terrain slope angle α. When the slope is perpendicular to the incident radar beam or $\beta = 0$, the range resolution R_y is indefinite.

where the superscripts represent the image pair for each coherence image.

Alternatively, we can state that the total ratio of coherence consists of a temporal ratio part and a spatial ratio part as

$$\eta = \eta_{temporal} \cdot \eta_{spatial} \qquad (10.29)$$

From the above equations, we have the following observations:

1 Obviously, the temporal ratio part always satisfies $\eta_{temporal} \leq 1$ because the temporal change is a cumulative process and $\rho_{temporal}^{13} \leq \rho_{temporal}^{12}$ is always true for the temporal separation $\Delta T^{13} > \Delta T^{12}$.

2 If the baseline $B_{\perp}^{13} \geq B_{\perp}^{12}$, then $\rho_{spatial}^{13} \leq \rho_{spatial}^{12}$ and thus the spatial ratio $\eta_{spatial} \leq 1$ and the total ratio $\eta \leq 1$ for all slopes.

3 For the case of $B_{\perp}^{13} < B_{\perp}^{12}$, then $\rho_{spatial}^{13} > \rho_{spatial}^{12}$ and the spatial ratio $\eta_{spatial} > 1$ in general. The difference between $\rho_{spatial}^{13}$ and $\rho_{spatial}^{12}$ will become significant when the terrain slope is nearly normal to the incident radar beam (i.e. the slope angle approaches the radar look angle) and the spatial ratio will be abnormally high $(\eta_{spatial} \gg 1)$ producing high total ratio coherence $\eta \gg 1$. However, for the areas other than direct radar-facing slopes, we would have $\rho_{spatial}^{13} \approx \rho_{spatial}^{12}$ and $\eta_{spatial} \approx 1$.

The item 3 above specifies the feasible working condition for the ratio coherence:

$$\eta = \frac{Coherence \ of \ large \ \Delta T \ and \ short \ B_{\perp}}{Coherence \ of \ small \ \Delta T \ and \ long \ B_{\perp}} \qquad (10.30)$$

Equation 10.30 specifies a ratio coherence image in which the numerator coherence image has a longer temporal separation and shorter baseline than the denominator coherence image. This ratio coherence image provides effective identification and separation of de-coherence features sourced from spatial and temporal de-correlation, as itemized below (Fig. 10.11):

1 Areas of total topographic de-correlation along the radar-facing slopes are highlighted as abnormally bright features because $\eta_{spatial} \gg 1$ and then $\eta \gg 1$.

2 The temporal de-correlation of random changes on land surface appears as dark features because $\eta_{temporal} < 1$ and $\eta_{spatial} \approx 1$ for areas not subject to severe topographic de-correlation and thus $\eta < 1$.

3 The stable areas, not subject to either temporal or spatial de-correlation, appear as a grey background where $\eta \approx 1$ as $\eta_{temporal} \approx 1$ and $\eta_{spatial} \approx 1$.

10.6 Fringe smoothing filter

A repeat-pass SAR interferogram is often noisy because of reduced coherence caused by temporal and spatial de-correlation, as shown in Fig. 10.12(a). The quality of an interferogram can be considerably improved if we can reduce noise by smoothing. Since the interferogram ϕ is a discontinuous periodic function wrapped in 2π-modulo [Fig. 10.13(a)], ordinary smoothing filters are not directly applicable for the noise reduction.

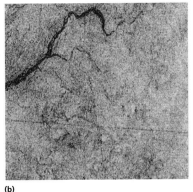

(a)   (b)   (c)

Fig. 10.11 Generation of a ratio coherence image: (a) *Coh1*: coherence image of 35 days and $B_{\perp} = 263$ m (short temporal separation with long baseline); (b) *Coh2*: coherence image of 350 days and $B_{\perp} = 106$ m (long temporal separation with short baseline); and (c) the ratio coherence image *Coh2/Coh1*: bright features represent topographic de-correlation on steep slopes facing the radar illumination, whilst dark features are detected random changes over a largely stable background in grey. (Lee and Liu 2001. Reproduced with permission of IEEE.)

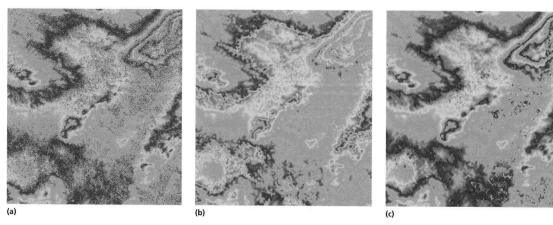

(a)   (b)   (c)

Fig. 10.12 Illustration of fringe smoothing filtering: (a) the original interferogram is quite noisy because of temporal and spatial de-correlation; (b) direct application of a 5 × 5 smoothing filter to the interferogram smears the fringe patterns, making the data no longer usable; and (c) after fringe smoothing filtering using a 5 × 5 smoothing filter, the interferogram fringes are well preserved, smoother and clearer than the original.

As illustrated in Fig. 10.12(b), direct application of smoothing filter to the interferogram will average the very high and very low values along the fringe edges, producing fake data and making the interferogram not usable. Here we introduce a simple phase fringe filtering technique (Wang *et al.* 2001), which has the following steps:

1 For the interferogram ϕ, the $\sin\phi$ and the $\cos\phi$ are continuous functions of ϕ, as shown in Fig. 10.13(b). Thus we convert the original interferogram into images of $\sin\phi$ and $\cos\phi$. All the trigonometric functions are wrapped in a cycle of π. Within the 2π cycle of ϕ, we can only retrieve angles in the range of $[0, \pi/2]$ and $[3\pi/2, 2\pi]$ from the $\sin\phi$, and $[0, \pi]$ from the $\cos\phi$. The combination of the $\sin\phi$ and $\cos\phi$, through a tangent function, allows us to retrieve the phase angle ϕ in its original 2π cycle.

2 A smoothing filter can then be applied to these two images of $\sin\phi$ and $\cos\phi$.

3 Retrieval of the filtered interferogram $\bar{\phi}$ from the smoothing filtered $\overline{\sin\phi}$ and $\overline{\cos\phi}$ is performed by $\bar{\phi} = \arctan\left(\dfrac{\overline{\sin\phi}}{\overline{\cos\phi}}\right)$. Here the signs of $\overline{\sin\phi}$ and $\overline{\cos\phi}$ dictate the quadrant of the smoothed phase angle $\bar{\phi}$. The smoothed phase angle $\bar{\phi}$ is within $0{\sim}\pi/2$ for $(+,+)$; $\pi/2{\sim}\pi$ for $(-,+)$; $\pi{\sim}3\pi/2$ for $(-,-)$; and $3\pi/2{\sim}2\pi$ for $(+,-)$.

The window size of the smoothing filter used must be small compared to the half-wavelength of $\sin\phi$ and $\cos\phi$ in the image of interferogram. Figure 10.12(c) is

a filtered interferogram with a 5 × 5 mean filter showing well-perservered fringe patterns and significantly reduced noise.

10.7 Summary

Several InSAR techniques are introduced in this chapter. It is very important to notice that different techniques apply to different applications, as outlined below:

• *InSAR interferogram is used for DEM generation.* The single-pass configuration of InSAR is preferred for the complete elimination of temporal de-correlation and to ensure high quality interferogram. The wider the perpendicular baseline $B_\perp$ is, the higher the elevation resolution. A wider $B_\perp$ introduces more severe spatial de-correlation, which degrades the quality of interferogram and then the DEM. Usually a $B_\perp$ of a few 10s to a few 100s of metres is used. The relationship between InSAR and topography is established by the geometry of the slightly differing view angles between the two observations. The elevation resolution of InSAR is, therefore – for C-band SAR, for example – no better than 10 m in general.

• *DInSAR is used for the measurement of terrain deformation.* This uses repeat-pass InSAR with at least one pair of cross-deformation event SAR acquisitions. The differential phase difference is directly translated from the deformation magnitude measured at 2π for half radar wavelength (e.g. 28 mm for ERS/ENVISAT C-band SAR). The technique is therefore very sensitive

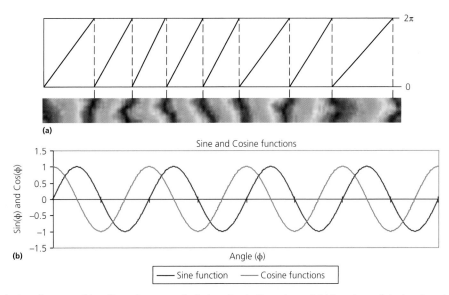

Fig. 10.13 (a) An interferogram ϕ is a discontinuous periodic function in 2π cycle; and (b) functions of $\sin\phi$ and $\cos\phi$ are also 2π periodic, but they are continuous.

to terrain deformation and can achieve millimetre-scale accuracies. This millimetre-scale accuracy is, however, an average deformation over an image pixel, which is, for ERS InSAR, about 25 m. A short $B_\perp$ is preferred for DInSAR to minimise fringes of topography. Ideally, a zero baseline InSAR configuration replaces DInSAR.

- *InSAR coherence technique is for random change detection.* It is very important to notice that this technique is for detection rather than measurement. A coherence technique must be based on repeat-pass InSAR. The random land surface changes are cumulative with time, and this reduces coherence in a coherence image derived from two SAR SLC images with temporal separation and can then be detected as low coherence features. As coherence is affected by both temporal and spatial (topographic) de-correlation, a short $B_\perp$ is preferred to minimise the spatial de-correlation, thus to achieve effective temporal change detection.

- *Ratio coherence technique is developed to separate spatial and temporal de-correlation.* This technique is defined as the ratio of a coherence image with long temporal separation and short $B_\perp$ divided by a coherence image with short temporal separation and long $B_\perp$. Such a ratio coherence image presents spatial de-correlation in bright tones and temporal de-correlation of random changes in dark tones on a grey background representing the stable land surface.

10.8 Questions

1 Describe the basic configuration of single path and repeat path InSAR, using diagrams.

2 Use a diagram to explain the principle of InSAR.

3 What are the differences between InSAR, DInSAR and coherence techniques and what are their main applications?

4 In comparison with the photogrammetric method using stereo imagery, discuss the advantages and disadvantages of using InSAR for DEM generation.

5 Select the only correct statement from the four below and explain the reasons for your choice:
  (a) InSAR technique can be used to produce DEMs of centimetre level accuracy of elevation.
  (b) DInSAR technique can be used to produce DEMs of centimetre level accuracy of elevation.
  (c) DInSAR technique can detect up to centimetre level (half radar wavelength) random changes of land surface.
  (d) SAR coherence image can detect up to centimetre level (half radar wavelength) random changes of land surface.

6 Describe the ratio coherence technique and explain the conditions of baseline and temporal separation necessary for this technique to be effective.

7 Why can ordinary smoothing filters not be applied directly to an interferogram image? Describe the procedure of fringe smoothing filtering.

CHAPTER 11

Sub-pixel technology and its applications*

A digital image is composed of pixels in lines and columns as a raster dataset, and therefore the minimum unit to contain image information is a pixel or a digital number (DN). When two images of the same scene are compared, they can be compared pixel by pixel; however, the pixel-based comparison may not be exactly accurate. For remote sensing satellite images, each pixel covers an area depending on the image spatial resolution, while the corresponding pixels in the two images in comparison usually do not overlap exactly covering the same area, as shown in Fig. 11.1(a) – there is a sub-pixel shift between the two corresponding pixels. If the imaging conditions of the two images are identical and there is considerable spatial variation in the images, then the two corresponding pixels may be recorded as slightly different DNs, but the difference of DNs does not really tell us the sub-pixel shift between them. Now, let's consider such a pair of corresponding pixels and their neighbourhoods; as all the pixels within the neighbourhood shift in the same way, they collectively formulate feature (or texture) shift as illustrated in Fig. 11.1(b), which is a 64 × 64 pixels patch of two images of the same scene displays in an RG (red and green) colour composite. The subtle red and green edges indicate the shifts between the two images. The image is enlarged by 10 times in order to show the sub-pixel shifts.

The sub-pixel technology has been developed to estimate and analyse image feature shifts down to sub-pixel magnitude for various applications, as discussed later in this chapter. There are many techniques developed in the last 10 years for image feature sub-pixel shift estimation in the computer vision field. These techniques can generally be categorized as image gradient-based methods (Horn & Schunck 1981; Lucas & Kanade 1981; Black & Anandan 1996; Brox *et al.* 2004), image domain correlation methods (Anandan 1989; Szeliski & Coughlan 1997; Lai & Vemuri 1998; Shimizu & Okutomi 2005; Heo *et al.* 2011), frequency domain phase correlation methods (Kuglin & Hines, 1975; Stone *et al.* 2001; Foroosh *et al.* 2002; Hoge 2003; Liu & Yan 2008; Ren *et al.* 2010; Liu *et al.* 2011).

In this chapter, we focus on one of the advanced sub-pixel techniques: phase correlation.

11.1 Phase correlation algorithm

Phase correlation (PC) is one of the most effective, accurate and robust techniques for sub-pixel image matching. Through areal correlation between two overlapped images, the shift between these two images can be identified at sub-pixel magnitude based on the spatial textures formulated by all the image pixels involved in the comparison.

PC is based on the well-known shift property of Fourier transform (FT): two functions related by a translational shift in the spatial coordinate frame result in a linear phase difference in the frequency domain of their FTs (Foroosh *et al.* 2002). Given two two-dimensional (2D) functions $g(x,y)$ and $h(x,y)$ representing two images related by a simple translational shift a in horizontal and b in vertical directions – the corresponding FTs are denoted as $G(u,v)$ and $H(u,v)$ – then,

$$H(u,v) = G(u,v)\exp\{-i(au + bv)\} \qquad (11.1)$$

In this formula, the phase shift is defined by $au + bv$. The PC is defined as the normalized cross-power

Image Processing and GIS for Remote Sensing: Techniques and Applications, Second Edition. Jian Guo Liu and Philippa J. Mason.
© 2016 John Wiley & Sons, Ltd. Published 2016 by John Wiley & Sons, Ltd.

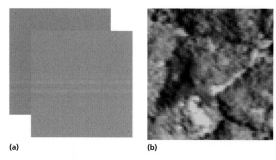

Fig. 11.1 (a) Sub-pixel shift between a pair of corresponding pixels in two images of the same scene. (b) Feature shift between two corresponding image neighbourhoods of 64 × 64 pixels displayed in an RG colour composite. The subtle red and green edges indicate the shifts between the two images. The image is enlarged by 10 times in order to show the sub-pixel shifts.

spectrum between G and H, which is a matrix of complex numbers:

$$Q(u,v) = \frac{G(u,v)H(u,v)^*}{\left|G(u,v)H(u,v)^*\right|} = \exp\{-i(au+bv)\} \quad (11.2)$$

If $G(u,v)$ and $H(u,v)$ are continuous functions, the inverse FT (IFT) of $Q(u,v)$ is a delta function:

$$IFT\big(Q(u,v)\big) = q(x,y) = \delta\big(x-a, y-b\big) \quad (11.3)$$

The peak of this delta function identifies the image shift (a, b) (Stiller & Konrad 1999; Hoge 2003; Balci & Foroosh 2005). However, as imagery data are discrete integer raster data, $G(u,v)$ and $H(u,v)$ are not smoothly continuous, and then eqn. 11.3 defines a Dirac delta function. The peak of this Dirac delta function identifies the nearest integer position of the true shift (a, b). Thus this straightforward FT-IFT solution can only resolve the image shift at integer pixel level.

Algorithms have been developed for direct solution in frequency domain without IFT, which can resolve the image shift at sub-pixel accuracy. As the magnitude of $Q(u,v)$ is normalized to 1, the only variable in eqn. 11.2 is the phase shift defined by $au + bv$ where a and b are the horizontal and vertical magnitudes of the image shift between $g(x,y)$ and $h(x,y)$. Obviously, if we can solve a and b accurately based on the PC matrix $Q(u,v)$, then the non-integer translation estimation at sub-pixel accuracy can be achieved in frequency domain. Two algorithms were then proposed: the SVD (singular vector decomposition) method (Hoge 2003) and the 2D fitting method (Balci & Foroosh 2005). The SVD method is based on complex matrix operations; it is robust to a relatively

large magnitude of shift between images and can tolerate poor correlation; however, it is relatively slow and becomes unstable for very small images. The robust 2D fitting method (Liu & Yan 2006) offers a much simpler and fast solution, and in the following, we explain the algorithm based on the 2D fitting method.

Actually, we can understand the data of PC matrix $Q(u,v)$ defined by eqn. 11.2 in a simple geometry. The phase shift angle in 11.2 is

$$c = au + bv \quad (11.4)$$

This equation is simply a 2D plane in u-v coordinates defined by coefficients a and b. Thus a seemingly complicated problem of complex numbers in frequency domain becomes a simple issue of finding the best 2D fitting of the phase shift angle data in $Q(u,v)$ to a plane defined by eqn. 11.4 in the Cartesian coordinates of u and v. This is to say that the image shift magnitudes a and b can be solved by the rank one (linear) approximation of PC matrix $Q(u,v)$. However, as angular data the phase shift c is wrapped by 2π-modulo in the direction defined by a and b. 2D unwrapping is therefore essential for the 2D fitting for c. There are many ready-to-use unwrapping algorithms.

As shown in Fig. 11.2, the image pair (a) and (b) of a Landsat 7 ETM+ scene image is generated by artificially shifting one image by 10 pixels to the right (positive) in horizontal and 13.33 pixels upwards (negative) in vertical directions to generate the other. Figure 11.2(c) and (d) are the power spectra of the FTs of the two images, and (e) presents the data of PC matrix $Q(u,v)$, which is the cross-power spectrum between the FT of the two images. We can see that the phase angle data of $Q(u,v)$ are in 2π cycle fringes, which are shown in three dimensions (3D) in Fig. 11.2(f). The density of these fringes is determined by the magnitude of the shift: $\sqrt{a^2+b^2}$, while the orientation by the ratio between a and b: b/a. After applying 2D unwrapping, we obtain a 2D plane as shown in Fig. 11.2(g), and here the gradient of the plane along u-axis is a while that along v-axis is b. The PC estimation of the shift is $a = 10.0037$ pixels and $b = -13.3439$ pixels in horizontal and vertical directions.

The example in Fig. 11.2 is to illustrate the principle of image PC, and thus identical images with an artificial shift are used. In reality, the two images for matching are not identical. There can be many de-correlation factors such as temporal changes, illumination variation, different spectral bands, and so forth. Figure 11.3(a) and (b) shows two ETM+ images of the same scene but

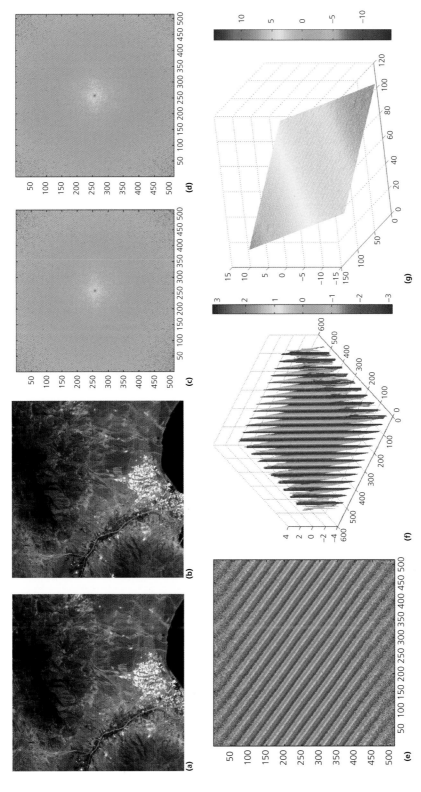

Fig. 11.2 (a) A 512 × 512 Landsat 7 ETM+ band 1 image. (b) The image is generated by artificially shifting image (a) 10 pixels to the right and 13.33 pixels upwards. (c) and (d) Power spectra of the FTs of image (a) and (b). (e) and (f) The PC matrix presented in 2D and 3D. (g) 3D view of the unwrapped PC matrix data, which form a plane.

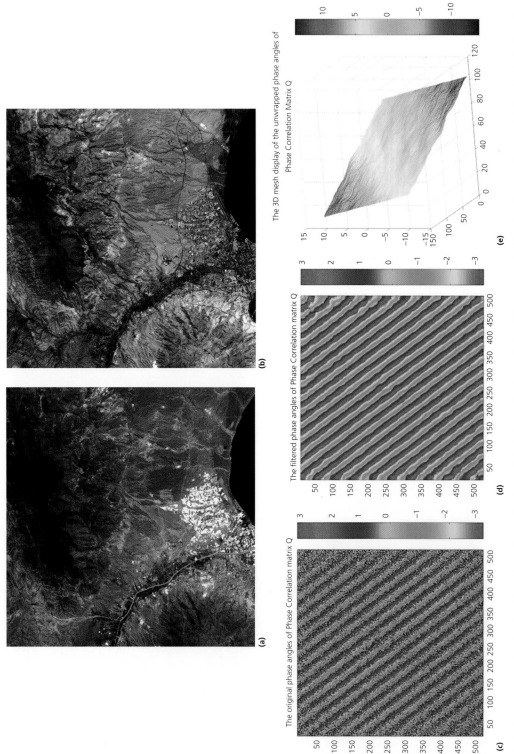

Fig. 11.3 Image (a) and (b) are Landsat 7 ETM+ image bands 1 and 5 with inter-band correlation of 0.69. Image (b) is artificially shifted to the right by 13.33 pixels and up by 10 pixels. (c) The PC matrix. (d) The PC matrix after applying the fringe smoothing filter (filter size = 15). (e) 3D view of the unwrapped PC matrix data, which form a slightly fluctuated plane.

different bands: band 1 (blue) and band 5 (short-wave infrared); the correlation between the two bands is 0.69. The band 5 image is artificially shifted 13.33 pixels to the right (positive) and 10 pixels upwards (negative) in comparison to the band 1 image. Figure 11.3(c), the cross-power spectrum of PC $Q(u,v)$ between the two images, looks rather noisy. As shown in Fig. 11.3(d), the noise can be quite effectively suppressed using the fringe smoothing filter introduced in § 10.6. The data of $c = au + bv$ in Fig. 11.3(e) do not form a perfect plane after unwrapping; it is slightly fluctuated. Such fluctuation can be much more significant if the correlation between the two images is low. However, using robust 2D fitting algorithms, a plane that optimally fit the unwrapped $Q(u,v)$ data can be found, and then a and b, the translational shift, can be determined at sub-pixel accuracy as $a = 13.1459$ and $b = -9.9819$.

In a similar way, PC can also be used to estimate the rotation and scale change between two images based on the rotation and scale properties of FT via coordinate transformations (Reddy & Chatterji 1996; Liu & Yan 2008).

For image rotation, consider image $f_2(x, y)$ is a replica of image $f_1(x, y)$ with rotation θ_0, then they are related by

$$f_2(x,y) = f_1(x\cos\theta_0 + y\sin\theta_0, -x\sin\theta_0 + y\cos\theta_0) \quad (11.5)$$

Given $F_1(\xi,\eta)$ and $F_2(\xi,\eta)$ representing the FT of f_1 and f_2, the Fourier rotation property shows that the FTs between f_1 and f_2 are related by

$$F_2(\xi,\eta) = F_1(\xi\cos\theta_0 + \eta\sin\theta_0 - \xi\sin\theta_0 + \eta\cos\theta_0) \quad (11.6)$$

If the frequency domain is presented in polar co-ordinates, then the rotation will be a shift on the axis corresponding to the angle. Using a polar co-ordinates system expressed in magnitude ρ and angle θ, we then have

$$F_2(\rho,\theta) = F_1(\rho,\theta - \theta_0) \quad (11.7)$$

The rotation can thus be found as a phase shift in the frequency domain that again can be determined by PC.

Similarly, if image f_2 is a replica of f_1 scaled by (a,b), then they are related by

$$f_2(x,y) = f_1(ax,by) \quad (11.8)$$

The Fourier scale property shows the FTs of f_1 and f_2 are related by

$$F_2(\xi,\eta) = \frac{1}{|ab|} F_1\left(\frac{\xi}{a},\frac{\eta}{b}\right) \quad (11.9)$$

Ignoring the multiplicative factor and taking logarithms,

$$F_2(\ln\xi,\ln\eta) = F_1(\ln\xi - \ln a, \ln\eta - \ln b) \quad (11.10)$$

Therefore, a change in scale can be determined based on a phase shift in the frequency domain presented in logarithmic coordinate units and thus again can be determined by PC.

By estimation of shift, rotation and scale difference between two images at sub-pixel accuracy, PC algorithm enables precise image orientation and matching, as illustrated in Fig. 11.4 and Table 11.1. Here, we need to emphasise that images for matching using PC algorithms must be taken under the same or at least very similar imaging geometry and largely overlapped. For instance, PC algorithms cannot match a nadir view image with an oblique view image. For remote sensing applications, most images are taken in nadir or near nadir view.

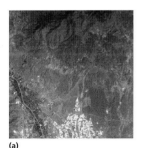 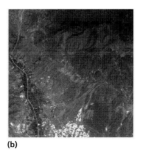 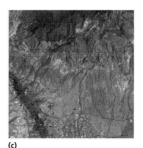 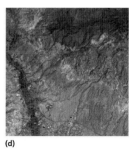

(a) (b) (c) (d)

Fig. 11.4 Image matching between two images of different spectral bands, shift and rotation. (a) and (c) are Landsat 7 ETM+ bands 1 (blue) and 5 (short wave infrared). (b) and (d) are generated from (a) and (b) by shifting 50 pixels to the right and 30 pixels downwards, and then rotation of 15°.

Table 11.1 PC estimations of translational shift, rotation and scale change between images in Fig. 11.4.

Images	Rotation	Scale	Shift	
			a	b
(a), (b)	−15.1146	0.9961	50.4805	30.3334
(c), (d)	−15.1370	0.9952	50.3802	30.3294
(a), (d)	−15.2454	0.9969	49.5674	29.6564

Image registration for alignment can therefore be done easily using PC in a matter of seconds. Mismatching between spectral bands is a common problem of many multi-spectral sensors. Images of different spectral bands may have a few to a few 10s of pixels shift in image row and column directions. This problem can be fixed automatically using PC image alignment.

Image registration for alignment is based on rigid shift, rotation and scaling, but there unavoidably exist irregular localised geometric variations between two images for many reasons, such as sensor optical differences, topography that results in slightly different view angles, ground objects change and motion. Thus image alignment does not achieve true image co-registration. PC scanning provides a way for pixel-wise disparity estimation that can be used for pixel-wise image co-registration, DTM (digital terrain model) generation, ground deformation analysis and target motion measurement.

PC is an image matching algorithm based on areal correlation; it does not provide measurement of correlation quality like the normalised cross-correlation (NCC). We therefore need to be careful that PC will give an estimation of the shift between two images that resemble each other, but we do not know if the estimation is reliable or not at all. One way to assess the PC quality is to calculate the NCC after the images are matched based on the PC estimated shift. A high NCC indicates reliable matching based on accurate shift estimation.

11.2 PC scanning for pixel-wise disparity estimation

11.2.1 Disparity estimation by PC scanning

Instead of applying PC to the whole scene of a pair of overlapped images, we can apply PC with a window-based scanning mechanism to estimate the feature shift (disparity) between the two images within the PC scanning window for every pixel. Figure 11.5 illustrates the major steps.

- Given two images for matching, set one as *Reference* image and the other *Target* image. If the pixel sizes of the two images are different, the *Target* image should be either oversampled or down sampled to the same pixel size of the *Reference* image.
- Apply the PC algorithm to a square region that roughly covers the most part of the overlapped image area between the *Reference* and the *Target* images. This will determine the frame shift, rotation and scale change between these two images. Then the *Target* image is aligned to the *Reference* image by standard interpolation routines.
- Scan the pair of aligned images using PC with an $n \times n$ calculation window from the first effective calculation position (the pixel at line $n/2$ and column $n/2$ of the top left corner of the overlapped image area) to the last effective calculation position (the pixel that is $n/2$ line and $n/2$ column to the bottom right corner of the overlapped image area). Thus the relative column and line shifts ($\delta x, \delta y$), the disparity, between any pair of correspondent pixels in the *Reference* and the *Target* images are determined. This process produces two disparity maps (images) of local feature shift in column and line directions, ΔX and ΔY.

The usual scanning window size is 16, 32 and 64. The window-based disparity estimation using PC is the average shift of image features of all the pixels in the scanning window. Thus a small window allows estimation of more localised disparity. However, with too few pixels within a small window, the FT may become unstable and the PC cross-power spectrum very noisy, and consequently, the disparity estimation not reliable. A large window ensures stable performance of PC at the cost that the disparity images are slightly smoothed. There is another catch; the size of the window must be sufficiently larger than the expected disparity between the two pixels for matching so that there are adequate corresponding pairs of pixels within the window for PC calculation. This means that the maximum ability of the disparity estimation is much smaller than the scanning window size for PC. According to our study, a 32×32 scanning window size generally achieves best performance in the disparity values not greater than 4 pixels.

In order to deal with large disparity estimation, a coarse-to-fine multi-resolution scheme can be employed as shown in Fig. 11.6. In this scheme, the pair of images

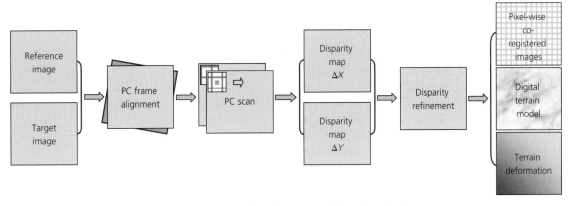

Fig. 11.5 Major steps of PC scanning process to generate disparity maps and the final application outputs.

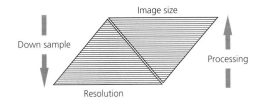

Fig. 11.6 The pyramid data processing structure from fine-coarse down sampling to coarse-fine multi-level PC scanning.

are down sampled to lower resolution in a fine-to-coarse sequence forming an image pyramid, and then PC is applied at each resolution level from coarse to fine. In this procedure, the same window size covers different sizes of image areas at each resolution level. For instance, a 16 × 16 window covers an area of 16 × 16 pixels of the original images. If the image is down sampled 10 times, the same window covers an area of 160 × 160 pixels of the original images. The disparities estimated at each level of the image pyramid are accumulated to achieve the final estimation. In this way, very large disparity can be estimated at sub-pixel accuracy better than 1/30th pixels.

11.2.2 The median shift propagation technique for disparity refinement

Pixel-wise disparity estimation via PC scan is based on the similar image textures in two images within the scanning window. If the image patch within the window is smooth without texture (featureless) or spectrally very different or subject to significant temporal changes, then there is no effective correlation and PC will fail to produce reliable estimation. Featureless is a general

problem for image matching techniques. Matching is a relative comparison, and if there is nothing to compare, such as between two images of clear blue sky, then there is no definition of matching. It is therefore necessary to assess the quality of the disparity estimation in every level of a coarse-to-fine pyramid processing structure to curb the propagation of errors from low resolution level to high resolution level. There are several general techniques and processing schemes to detect, eliminate and suppress the unreliable estimations of disparity (e.g. NCC and median filter). Here we introduce the median shift propagation (MSP) technique to amend the data gaps that result from masking off the bad disparity data (Liu & Yan 2008).

As shown in Fig. 11.7, the PC failed in featureless areas and areas of significant spectral differences between the *Reference* image [Fig. 11.7(a)] and the *Target* image [Fig. 11.7(b)]. Because of low correlation in these areas, the disparities estimated using PC are incorrect and of random direction and magnitude, as illustrated by feature shift vector representation of ΔX and ΔY [Fig. 11.7(c)].

As the magnitude of PC is unified, it does not give a direct measure of correlation quality. The correlation quality can, however, be reliably assessed based on the 2D fitting quality via the regression coefficient of determination (RCD) or NCC in a normalised value range [0,1]. A correlation quality assessment image can then be generated together with the disparity maps by 2D fitting PC scanning. Thus the low quality disparity data can be identified and masked off with a *Null* value from the disparity maps ΔX and ΔY by an RCD threshold, say at RCD < 0.75 [Fig. 11.7(d)].

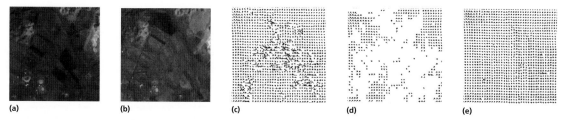

(a)   (b)   (c)   (d)   (e)

Fig. 11.7 Illustration of MSP technique. (a) The *Reference* image, ASTER band 3. (b) The *Target* image, ETM+ pan. The two images have significant spectral differences and several featureless areas. (c) The vector representation of ΔX and ΔY (re-sampled at 8-pixel interval and magnitude enlarged by four times for illustration) indicates obvious errors that result from the malfunction of PC in disparity estimation. (d) The areas of errors are masked off based on low quality of the 2D fitting, RCD < 0.75. (e) The gaps are smoothly filled using the MSP.

1.63	1.60	1.62	1.61	1.59	1.57	1.58
1.59	1.57	1.55	1.52	1.53	1.51	1.52
1.61	1.58					
1.56	1.54					
1.55	1.53					
1.52	1.51					
1.50	1.49					

(a)

1.63	1.60	1.62	1.61	1.59	1.57	1.58
1.59	1.57	1.55	1.52	1.53	1.51	1.52
1.61	1.58	*1.58*	*1.57*	*1.57*		
1.56	1.54	*1.56*	*1.55*	*1.55*		
1.55	1.53	*1.55*	*1.55*	*1.56*		
1.52	1.51					
1.50	1.49					

(b)

Fig. 11.8 Numerical explanation of the MSP: (a) the RCD masked shift image (e.g. ΔX) with good quality data in bold font and gaps; and (b) the numbers in italics in the central 3 × 3 box with bold frame are filled via the MSP from the existing quality data.

Now the problem is how to fill the gaps in the masked ΔX and ΔY images with shift values of reasonable quality. A simple and effective solution for the problem is the MSP as detailed below (Liu & Yan 2008):

- Scan the masked ΔX and ΔY images to search for a gap.
- Whenever a *Null* value is encountered, a median filtering mechanism in a calculation window (e.g. 5 × 5 pixels) is applied to the disparity images ΔX and ΔY. This process replaces the *Null* with the median of the values of non-*Null* value pixels in the calculation window.
- Continue the median filtering on the modified ΔX and ΔY, till all the gaps are filled to form a smooth optical flow, as in Fig. 11.7(e). The key difference between the MSP and the ordinary median filter is that instead of always applying the filter to the original ΔX and ΔY images, the filter is applied to the ΔX and ΔY images that are modified by the last filtering action. In such a way, the ΔX and ΔY images are updated continuously during the filtering process and the feature shifts are thus self-propagated from high quality data to fill the *Null* value gaps.

Though we borrowed the term 'median filter', the process is not filtering but self-propagation via median. As shown in Fig. 11.8(a), the 3 × 3 gap area in the box with a bold frame can be filled gradually via median propagation from the data in the top two lines and left two columns using a 5 × 5 processing window. The first cell (top left) in this 3 × 3 gap area is filled with the median, which is 1.58, of its surrounding 5 × 5 neighbourhoods. The second cell (top middle) is filled with the median of the non-*Null* value pixels in its surrounding 5 × 5 neighbourhoods, including the just filled number in the first cell, 1.58, and so on. As such, the 3 × 3 gap area is filled up as shown in Fig. 11.8(b). If there is an even number of values within the 5 × 5 neighbourhoods, the average of the two middle values is used.

11.3 Pixel-wise image co-registration

One application of PC scanning for pixel-wise disparity estimation is pixel-wise image co-registration.

From Chapter 9, we already know that image co-registration is an essential process for multi-temporal and multi-source image applications such as change detection. It also formulates an essential part of image geo-rectification. The 'rubber' warping based on a polynomial transformation model derived from ground control points (GCPs) is among the most popular techniques. With this classical approach, images can be quite accurately co-registered via a transform grid established from the polynomial model if the GCPs are of high position quality. However, co-registration in such a way is not achieved at pixel-to-pixel level. The registration error within a scene can vary from place to place depending on the local relative deformation between the images. The imaging geometry-related systematic deformation can be modelled and compensated, while the irregular deformation sourced from the quality of sensor optics and precision of platform status and so forth cannot be effectively corrected using the GCP-based approach.

A different approach for image co-registration is a forced pixel-to-pixel image co-registration (Lucas & Kanade 1981). Instead of using a transform grid determined by the coordinates transformation derived from GCPs, this approach achieves image co-registration pixel by pixel via imagery local spatial feature matching. Fundamentally different from image warping, the pixel-wise image co-registration is not geo-rectification. In other words, an image cannot be transformed to a given map projection system based on the map coordinates using this method, while this can be done using GCP-based techniques. The pixel-wise image co-registration can achieve image geo-rectification only when the reference image is geocoded.

11.3.1 Basic procedure of pixel-wise image co-registration using PC

Precise image feature matching has to be at the sub-pixel level as it is unlikely that the mismatch between two images is always at integer pixel position; a pixel in one image often corresponds to a decimal pixel position in the other. Obviously PC scanning for pixel-wise disparity estimation can be used for pixel-wise image co-registration in a similar procedure (Liu & Yan 2008):

- *Start.* Given two images for co-registration, set one as *Reference* and the other *Input* (*Target*), and then the *Output* (co-registered) image frame is set based on the *Reference* image. If the pixel sizes of the two images are different, the *Input* image should be either over-sampled or down sampled to the same pixel size of the *Reference* image.

- *Image alignment.* Apply PC to a square region roughly covering the most part of the overlapped image area between the *Reference* and the *Input* images to align the *Input* image to the *Reference* image.

- *PC scanning.* Carry out PC scan on the two aligned images with a small calculation window (e.g. 16 × 16, 32 × 32) in the pyramid processing structure to generate disparity maps ΔX and ΔY in which the relative column and row shifts (δx, δy) between any pair of correspondent pixels in the *Reference* and the *Input* images are recorded at the corresponding pixel position of the *Reference* image.

- *Disparity quality assessment.* The RCD image recording the correlation quality between every pair of corresponding pixels of *Reference* and *Input* images is also generated during the PC scanning. The low correlation areas in the ΔX and ΔY images are masked off with a threshold of RCD < 0.75.

- *Disparity refinement.* Apply the MSP technique to fill the gaps in the masked ΔX and ΔY images produced in the last step.

- *Output.* An output-to-input mapping procedure is employed to carry out a pixel-wise co-registration rectifying the *Input* image to precisely match the *Reference* image. This is a process to build an *Output* image based on the *Reference* pixel positions and *Input* pixel DNs. Specifically, at each integer pixel position of *Output* image, the pixel DN is drawn from its precise corresponding position in the *Input* image based on the disparities (δx, δy), where $\delta x \in \Delta X, \delta y \in \Delta Y$, between the two corresponding pixels in the *Input* and *Reference* images.

The above scheme registers the *Input* image to the *Reference* image pixel by pixel to formulate the *Output* (co-registered) image. With the sub-pixel resolution of (δx, δy) calculated via the PC, the *Output* image matches the *Reference* image at a sub-pixel accuracy at every pixel.

11.3.2 An example of pixel-wise image co-registration

Figure 11.9(a) and (b) are 1024 × 1024 sub-scenes of a Terra-1 ASTER band 3 image and a Landsat 7 ETM+ pan (panchromatic band) image. The two images were taken from different satellite platforms with different sensors. ETM+ is a two-way across-track scanner, while ASTER is an along-track push-broom scanner. Both images are of 15 m spatial resolution but in different spectral bands: 0.52~0.9 μm for the ETM+ pan and 0.76–0.86 μm for the ASTER band 3. There are shift, rotation and irregular geometrical variation between the two images, besides spectral differences. The images were chosen for the test for their rich variety in both spatial and spectral features in the scene.

The PC image alignment measured 0.838° rotation and 6.825 and 6.832 pixels shift in horizontal and vertical directions between the two images. The PC scan was then carried out to produce the ΔX and ΔY between the two aligned images. The variation of ΔX is –8.38~6.22 pixels and ΔY –1.61~1.27 pixels. The ETM+ pan image is then co-registered to the ASTER band 3 image pixel by pixel based on the disparity maps ΔX and ΔY. For visual exam, colour composites of ASTER band 3 in green and the ETM+ pan before and after co-registration in red were produced as shown in Fig. 11.9(c) and (d). Irregular mismatching between the two images was clearly shown in Fig. 11.9(c), while the crystal sharpness of Fig. 11.9(d) indicates very high quality of co-registration in every part of the image. Readers can zoom in on the image online to observe the registration quality at pixel level and there is no mismatching. The white box in Fig. 11.9(b) indicates the area of Fig. 11.7. In this area, PC failed initially, resulting in poor pixel-wise co-registration in Fig. 11.9(e). However, the MSP refinement amended the faulty disparity data and yielded high quality co-registration, as shown in Fig. 11.9(f). In this way the clear red and green patches reveal the spectral changes or differences between the two images, demonstrating the capability of the method for change detection.

11.3.3 Limitations

The PC pixel-wise image co-registration may fail if a large part of the images is of low correlation as the result of widespread spectral differences and featureless areas. In this case, manual GCP-based co-registration is more robust, as human eyes can pick up GCPs accurately based on comprehension of textures and spatial patterns.

As mentioned before, for the images with different spatial resolution, we can oversample or down sample the *Input* image to the same pixel size of the *Reference* image before the PC scanning to derive pixel-wise disparity maps. However, there is a limit for PC to be functional. Our experiments indicate that the PC pixel-wise co-registration method works the best for the images of the same or similar spatial resolution and it performs well till about 5 times spatial resolution difference. Certainly, for co-registration between a high spatial resolution image and a low spatial resolution image, the co-registration precision is of the sub-pixel accuracy of the lower resolution image, which can be several pixels of the higher resolution images if the resolution difference between the two is great.

11.3.4 Pixel-wise image co-registration-based SFIM pan-sharpen

In Chapter 6, we introduced the SFIM (smoothing filter-based intensity modulation) technique for image fusion (e.g. pan-sharpen). When the panchromatic images and multi-spectral images are from different sources, such as SPOT and Landsat, the precision of image co-registration is a major factor to affect the quality of SFIM pan-sharpened images. Even subtle misregistration between the panchromatic images and multi-spectral images can degrade the pan-sharpen quality in textual details. According to the discussion in the previous sub-sections, the PC-enabled pixel-wise image co-registration provides a solid solution for high quality SFIM pan-sharpen of multi-source images. Here we introduce the image fusion process with an example of using a 5 m resolution SPOT 5 panchromatic image to sharpen 30 m resolution Landsat 8 Operational Land Imager (OLI) multi-spectral images. The general procedure comprises the following steps.

1 Both Landsat multi-spectral and SPOT 5 pan images are rectified to the same map projection system (or raw) to make the images taken by different satellites frame aligned and registered.

2 Down sample the 5 m resolution SPOT pan image to 30 m resolution by bicubic interpolation.

3 Pixel-wise image co-registration via PC scanning using the Phase Correlation Image Analysis System (PCIAS). The 30 m resolution Landsat multi-spectral band images (*Input*) are co-registered to the down sampled 30 m resolution SPOT 5 pan (*Reference*). The PC window size is 16 × 16. The process makes the Landsat images comply with the map projection (if any) of the SPOT pan image and pixel-wise co-registration at 30 m pixel size.

4 Up sample the co-registered 30 m Landsat multi-spectral images to 5 m pixel size using bicubic interpolation.

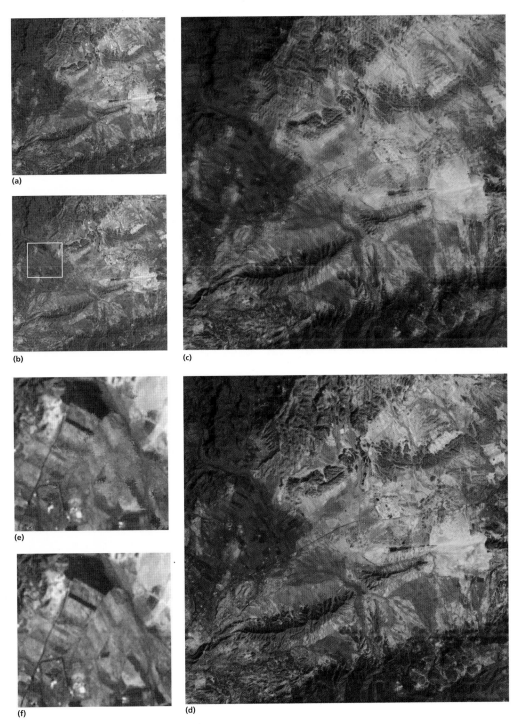

Fig. 11.9 (a) *Reference* image, ASTER band 3. (b) *Input* image, ETM+ pan. (c) RG colour composite of the original ETM+ pan in red and the ASTER band 3 in green. (d) The RG colour composite of co-registered ETM+ pan and the ASTER band 3. (e) The co-registered ETM+ pan image in the area denoted by the white box in (b) is initially contaminated by misregistration errors as the result of featureless and spectral differences. (f) After the MSP disparity refinement, the co-registration errors are effectively eliminated.

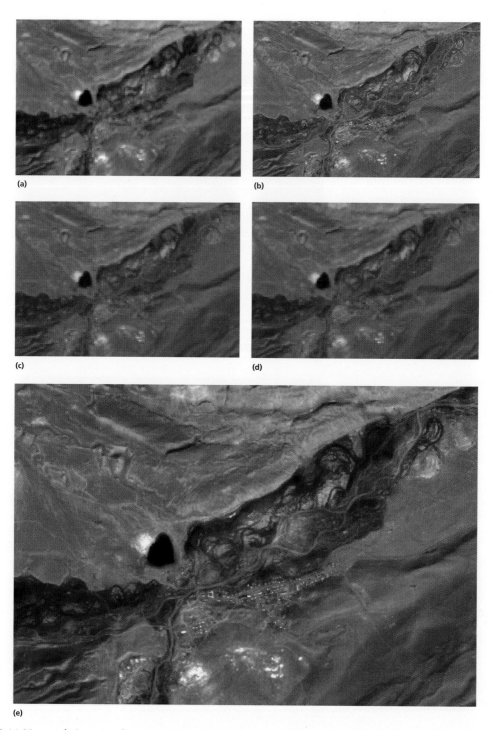

Fig. 11.10 (a) 30 m resolution true colour composite image of Landsat OLI bands 4-3-2 in RGB. (b) 5 m resolution SPOT 5 panchromatic band image. (c) Colour composite of OLI band 4 in red and SPOT 5 pan (down sampled to 30 m resolution) in green, before pixel-wise image co-registration. (d) The same colour composite as (c) but after pixel-wise image co-registration. (e) SFIM pan-sharpened true colour composite of Landsat OLI bands 4-3-2 in RGB using 5 m resolution SPOT 5 panchromatic band.

5 Apply the original 5 m resolution SPOT pan image to sharpen the pixel-wise co-registered 5 m pixel size Landsat multi-spectral images via SFIM with a 13 × 13 smoothing filter.

6 The final pan-sharpened fused Landsat multi-spectral images have 5 m texture resolution and the original spectral property is fully preserved.

It is desirable to use a higher resolution image as the *Reference* and down sample it to the pixel size of a lower resolution image (*Target*) for pixel-wise co-registration. Precise co-registration can be achieved only in this case.

Though PCIAS is capable of estimating pixel-wise disparity at sub-pixel accuracy, the pixel-wise image co-registration can only be pixel based. For fusion of images with very different spatial resolution, the pixel-wise image co-registration is necessary only when the irregular mismatching is greater than half of the low resolution pixel.

Figure 11.10 presents the SFIM process of a 30 m resolution Landsat 8 OLI bands 4-3-2 RGB true colour composite [Fig. 11.10(a)] pan-sharpened by a 5 m resolution SPOT 5 panchromatic image [Fig. 11.10(b)]. Though both images are rectified to comply with the same Universal Transverse Mercator (UTM) map projection, there are plenty of minor misregistrations showing as red and green edges in the RG colour composite of OLI band 4 in red and SPOT 5 pan (down sampled to 30 m resolution) in green in Fig. 11.10(c). After pixel-wise image co-registration using PC scan, the two images are precisely co-registered as shown in Fig. 11.10(d). Compare the SFIM pan-sharpened OLI 4-3-2 true colour composite [Fig. 11.10(e)] with the original colour composite [Fig. 11.10(a)]; the improvement of spatial textures is obvious and significant, and meanwhile, there is no observable colour distortion at all. You may view images covering a much larger area with greater spatial and spectral variation in full resolution by clicking the image links.

11.4 Very narrow-baseline stereo matching and 3D data generation

PC scanning enables generation of terrain 3D data (e.g. DTM) from stereo images with various baseline settings from conventional wide baseline to unconventional very narrow baseline. In this sub-section, we focus on unconventional very narrow-baseline stereo matching,

as it can only be done using sub-pixel technology (Morgan *et al.* 2010).

11.4.1 The principle of stereo vision

The stereo principle provides that when a 3D point is imaged from two distinct viewpoints it will project to a different location in each image. If it is possible to locate the corresponding points in both images, then the height of the 3D point can be established by a process of triangulation, provided that the imaging parameters and orientation of the camera/sensor are known.

For an ideal stereo geometry, shown in Fig. 11.11, with two cameras separated by a baseline B and at an altitude H above the datum, the height h of a 3D point can be determined by measuring the image distance between the projections of that point in the left and right images, known as the parallax or disparity d, as discussed before. Based on the similar triangles of this geometry, the following simple expressions can be derived (Delon & Rougé 2007; Igual *et al.* 2007):

First, the disparity in ground units D can be expressed as,

$$D = \frac{B}{(H-h)}h \qquad (11.11)$$

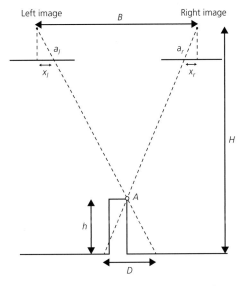

Fig. 11.11 The 3D point A is projected to the two 2D locations a_l and a_r in the left and right images. The relative distance between the image locations of these two corresponding points is the image disparity $d = (x_l - x_r)$ in pixels. Disparity, d, is directly proportional to the height h of point A. See eqns. 11.11 to 11.14 in the text.

Re-arranging,

$$\frac{D}{h} = \frac{B}{(H-h)} \tag{11.12}$$

D is simply related to the image disparity d by the system resolution or pixel size L (metres/pixel), and thus $D = dL$,

$$d = \frac{Bh}{(H-h)L} \tag{11.13}$$

Finally, if we make the assumption that H >> h, which is the case for aerial/satellite imagery, then 11.12 can be simplified to

$$d \approx \frac{B}{H}\frac{1}{L}h \tag{11.14}$$

From 11.13 it can be seen that the measurement precision of the height h is directly proportional to that of the disparity d and the baseline B (or B/H ratio) and inversely proportional to the pixel size L. In most real scenarios H and L will be fixed, so to obtain accurate height estimation, high disparity accuracy is required. The most convenient approach to achieving this is by maximising B or the B/H ratio. Consequently, large B/H ratios of ~0.6–1 (Mikhail *et al.* 2001) are conventionally chosen for aerial/satellite stereo systems.

11.4.2 Wide-baseline vs. narrow-baseline stereo

A wide-baseline or large B/H ratio generates a large viewing angle difference between an image pair, so all 3D scene objects display large relief distortion, which in turn increases the likelihood that taller objects will obscure nearby smaller ones. This will result in fewer corresponding points being visible in both images (Klette *et al.* 1998); obviously the disparity of a point can only be measured if clear correspondence can be achieved. This situation is known as *occlusion* and it is currently one of the outstanding problems in stereo vision (Brown *et al.* 2003). Occlusion is a particularly common problem for terrain with high local relief, such as in urban settings where taller buildings will obscure lower features.

Since the likelihood of occlusion is increased with a wide baseline, then it naturally follows that it may be reduced by narrowing the baseline. By allowing the $B/H \rightarrow 0$ (i.e. a very narrow baseline), occlusions will be minimised, because each image view becomes more geometrically similar, as illustrated in Fig. 11.12.

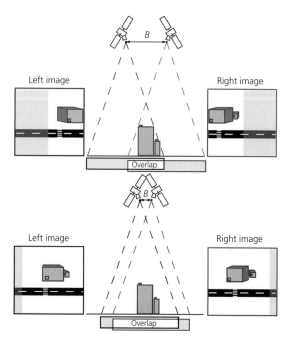

Fig. 11.12 (Top) A standard wide-baseline stereo configuration: two cameras, separated by a distance B, view two buildings from two different vantage points. In the left image the small building is entirely occluded by the taller building, making correspondence with the right image impossible in that region. (Bottom) By significantly narrowing B, both buildings are visible in both images because the imaging geometry is much more similar. Additionally, the useful overlapping image area is increased. However, the relief distortion and thus the disparity range are also significantly reduced.

However, when the baseline is significantly narrowed the magnitude of the disparities can be reduced to very small sub-pixel levels. Therefore, if we wish to make use of very narrow-baseline stereo whilst also maintaining adequate height accuracy, then an ability to measure sub-pixel disparities with very high precision is an essential requirement. The PC scanning for pixel-wise disparity estimation provides a solution.

11.4.3 Narrow-baseline stereo matching using PC

The narrow-baseline stereo matching procedure is essentially the same as the PC scanning for pixel-wise disparity estimation. Given a pair of stereo images of a scene of terrain, the pixel-wise disparity maps are actually the relative DTM or terrain 3D data that are not yet rectified to a map projection coordinates system but can be directly converted to DTM via standard procedure.

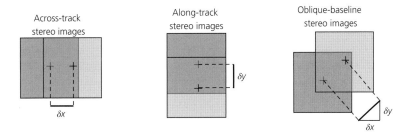

Fig. 11.13 The relationship between disparity and stereo orientation.

Depending on the baseline direction in relation to the orientation between the stereo images, either ΔX or ΔY or both provide terrain 3D data. As illustrated in Fig. 11.13, if the baseline is in the horizontal direction of the images, the case of across-track stereo, disparity map ΔX provides the topographic parallax data or DTM; if the baseline is in the vertical direction of the images, the case of along-track stereo, disparity map ΔY is the DTM; and if the baseline is oblique to image orientation, DTM can be derived from ΔX and ΔY as the magnitude of the pixel-wise disparity: $\sqrt{\delta x^2 + \delta y^2}$ in the direction of $arctg(\delta y / \delta x)$.

The main steps of an automated stereo matching algorithm are the same as the PC scanning for pixel-wise disparity estimation as outlined in the flow chart shown in Fig. 11.5. The procedure is applicable to both conventional wide-baseline images and unconventional narrow-baseline images, as described below.

- *Start.* Given a pair of stereo images, one is the *Left* (*Reference*) image and the other the *Right* (*Target*) image.
- *Image alignment.* Apply PC to a square region roughly covering the most part of the overlapped image area between the *Left* and the *Right* images to orientate the *Right* image to the *Left* image. Ideally we assume that the images have been acquired to comply epipolar geometry roughly.
- *PC scanning.* Carry out PC scan on the two aligned images with a small calculation window (e.g. 16×16, 32×32) in the pyramid processing structure to generate disparity maps ΔX and ΔY.
- *Output and refine.* As explained before, depending on the baseline direction in relation to the image pair orientation, the relative DTM is either ΔX or ΔY or can be derived from both. Some post-processing, such as MSP and filtering, can be applied to suppress error estimations of disparity. The final output is a refined relative DTM raster dataset (an image) that has the same spatial resolution as the input images because the processing is pixel-wise, and thus no further data interpolation is needed.

11.4.4 Accuracy assessment and application examples

11.4.4.1 Landscape model experiment

This experiment is to test the extreme of the capability of PC for very narrow-baseline stereo matching. A model landscape was created with a range of elevations up to 51 mm in height and subtle topographic details representing a continuous natural terrain surface [Fig. 11.14(a)]. Figure 11.14(b) shows the stereo image pair of the model that was imaged from nadir with baseline $B = 5$ mm in distance $H = 1500$ mm, giving a B/H ratio ~0.003 that is only about 1/200th of the conventional baseline B/H ratio ~0.6, providing a maximum scene disparity up to just 1.2 pixels.

Disparity maps were computed using the PC and the NCC integer pixel algorithms for comparison, as shown in Fig. 11.14(c) and (d) in 3D display with pseudo-colours. The NCC integer result shown in Fig. 11.14(c) can only provide disparities rounded to the nearest whole pixel in either 1 or 0 disparity value, as the total elevation range of the model is only 1.2 pixels. In contrast, the disparity map produced by PC method in Fig. 11.14(d) reveals subtle topographic detail, expressing the true topography shown in the picture of the model in 11.14(a). This example reinforces the point that with very low B/H ratio images, detailed and useful terrain information can be encoded in sub-pixel disparities but will be seemingly invisible to basic integer methods.

To give an indication of accuracy for the disparity estimation, the heights of 18 mm, evenly distributed control points measured from the model were converted to

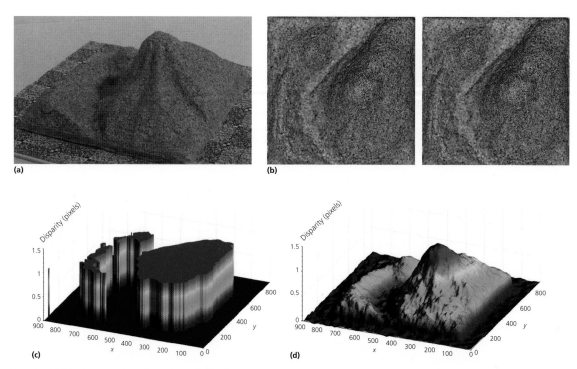

Fig. 11.14 (a) A photograph of the scale model landscape scene. (b) Stereo pair of the scale model of landscape with B/H ratio = 0.003. (c) The disparity maps computed by the NCC integer pixel algorithm using a 17 × 17 window; and (d) by PC using a 32 × 32 window, all presented in 3D sun-shaded views in pseudo-colours.

disparity values. These true values were then compared to corresponding areas within the disparity map computed from the PC. The RMS (root mean square) error is only 0.013 pixels, which is less than +/− 0.5 mm height error for an approximate height range of 0–51 mm.

11.4.4.2 Very narrow-baseline satellite image pair

A pair of Taiwanese Formosat-2 satellite images (Formosat-2 image ©2015 National Space Organization, Taiwan) with very narrow baseline, 1.1° view angle difference (B/H = 0.019) along track direction and nearly no angle difference along scanning direction was processed by PC scanning for DTM generation. Figure 11.15 presents the disparity estimation results from the 4652 × 3461 pixels sub-scenes of the pair. Figure 11.15(a) is the RG (red and green) colour composite of the two images; the spatial discrepancies between them are hardly discernible because the disparities between most corresponding pixels in the two images are less than a pixel. Figure 11.15(b) shows the disparity map ΔY along track direction and its histogram

in Fig. 11.15(c) illustrates the continuous data distribution from −12.6270 to 6.6884 pixels, evidencing the decimal disparities measured for an elevation range from 615 m to 1865 m. Figure 11.15(d) presents a 3D perspective view of the DTM in pseudo-colour, while Fig. 11.15(e) a 3D perspective view of the reference image draped on the DTM. The disparity map is a relative DTM that can be easily rectified to a geo-coded DTM via ground control points with known coordinates and elevation.

11.4.4.3 Mars DTM generation from standard stereo images

The coarse-to-fine pyramid processing structure enables PC scanning to cope with very large disparity range at sub-pixel accuracy. Thus it is capable of generating super-resolution DTMs from standard stereo images. A pair of HiRISE stereo images of Kasei Valles on Mars were processed to generate high resolution DTM, as shown in Fig. 11.16. The Mars surface has very high relief and with 1 m spatial resolution, the disparity range of this 4000 × 4000 pixels sub-scene is 186 pixels,

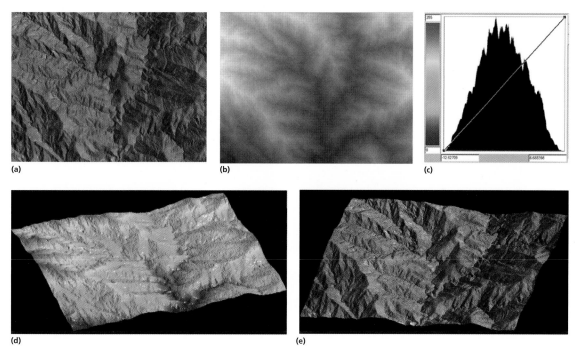

(a)

(b)

(c)

(d)

(e)

Fig. 11.15 Disparity estimation from a pair of 4652 × 3461 pixel sub-scenes of two Formosat-2 satellite images with 1.1° along-track view angle difference. (a) Red and green colour composite of the image pair. (b) Along-track disparity map ΔY (relative DTM) generated using PC. (c) The histogram of the disparity map. (d) 3D prospective view of the DTM in pseudo-colours. (e) 3D perspective view of the reference image draped on the DTM.

while for the whole image processed, the disparity range is greater than 500 pixels. This represents a serious challenge of large disparity estimation, but the coarse-to-fine multi-level PC in a pyramid processing structure coped well to generate 1 m spatial resolution DTM of the area. The data reveal steep slopes, deep valleys, impact craters and even the subtle topography of sand dunes in the valley.

11.4.4.4 DTM generation from multi-sensor images

The versatile ability of PC operating in a coarse-to-fine multi-level pyramid structure can cope with a wide range of disparities from several hundred pixels to a fraction of one pixel. This means that DTM or terrain 3D data can be generated from virtually any of a pair of overlapped images of similar spatial resolution.

Interestingly enough, the Landsat 7 ETM+ pan image and the Terra-1 ASTER band 3 image used for the pixel-wise image co-registration experiment in § 11.3 actually formulate a narrow-baseline stereo pair. While the disparity maps between the two images can be used

for pixel-wise image co-registration, the ΔX disparity map is actually a relative DTM of the terrain as shown in Fig. 11.17.

11.5 Ground motion/deformation detection and estimation

Disparity is fundamentally the relative motion between a pair of corresponding pixels in two images. For pixel-wise image co-registration, the relative motion is the result of irregular distortion between the two images. For stereo matching of terrain 3D data generation, the relative motion results from the projective shift of topography in the two images taken in different imaging angles. Disparity can certainly be the actual motion of ground or ground objects.

A strong earthquake can produce significant horizontal ground motion, especially if the earthquake is triggered by a strike-slip fault. While differential interferometric synthetic aperture radar (DInSAR) technique can measure the terrain surface deformation in LOS (line of sight) or slant range direction, the sub-pixel technology

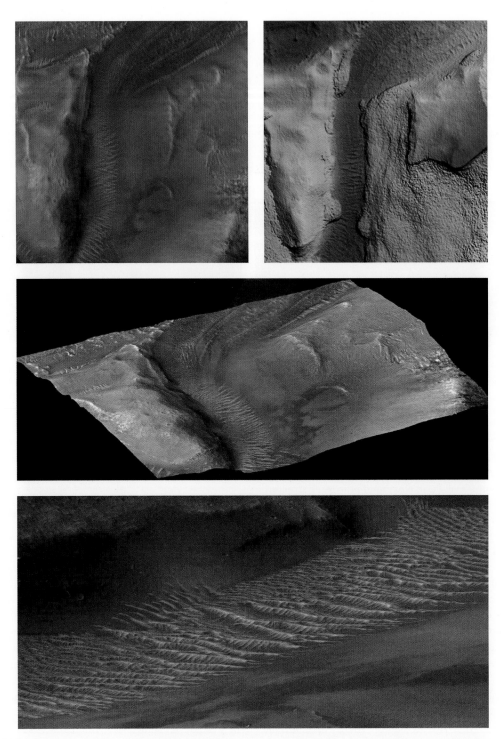

Fig. 11.16 Top left: a pair of 4000 × 4000 pixels stereo images of an area in Kasei Valles on Mars displayed in an RG colour composite with left image in red and right in green. Top right: the disparity map ΔX (relative DTM) displayed in pseudo-colours with sun shading. Middle: a perspective 3D view of the landscape of the Martian surface in Kasei Valles by draping the right image onto the DTM. Bottom: a closer look at the sand dune features in 3D perspective view.

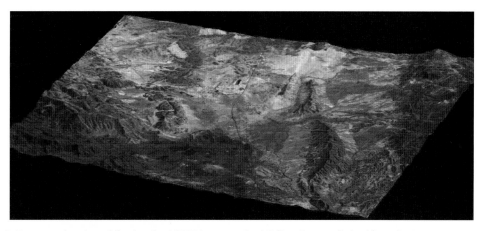

Fig. 11.17 A 3D perspective view of draping the ASTER image on the ΔX disparity map derived from the ETM pan/ASTER band 3 image pair used for pixel-wise image co-registration in § 11.3.

allows the measurement of ground horizontal motion in the image line and column directions as disparity maps, ΔX and ΔY, from either SAR intensity images or optical images. This approach has long been used for studying co-seismic deformation and is called the "pixel-offset" technique. Indeed, the capability of the technique depends on the sub-pixel accuracy. For instance, given a pair of across-event optical images with 5 m spatial resolution, 1/30th pixel accuracy of PC enables the measurement of horizontal motion as small as 16.7 cm. The accuracy is still low in comparison with DInSAR measurement based on half radar wavelength, but it is getting much closer.

Figure 11.18 shows (a) a pair of ALOS PALSAR intensity images, displayed in an RG colour composite, of an area of Beichuan County before and after the Mw 7.9 Wenchuan earthquake occurred on 12 May 2008 in Sichuan Province, China, and (b) and (c) the disparity maps in the SAR range direction and azimuth direction. The RG colour composite illustrates that these across-event images are nearly identical except in the areas where significant changes resulted from the major earthquake damages, such as river blockages and landslides shown in distinctive red and green colours. The disparity maps reveal considerable ground displacement in a belt along the fault line of the major seismic fault: Yingxiu-Beichuan fault, on its hanging wall, providing concrete evidence of ground motion. The ΔX map indicates, by high positive disparity values in Fig. 11.18(b), that the displacement is mainly in the ground range

direction toward east, and therefore the faulting in this area is a thrust. The effective disparity value range of the ΔX is −1.64~2 pixels and with 5 m nominal resolution; the maximum ground range displacement along the fault is about 10 m. Figure 11.18(d) is a smoothed ΔX map in pseudo-colours overlay on the pre-event SAR intensity image in grey scale to present the ground range displacement with topography.

The ground range is related to LOS by the sin function of SAR look angle θ, as

$$Range_{grnd} = \frac{LOS}{\sin \theta} \qquad (11.15)$$

It is important to notice that the ΔX map provides the data of ground range displacement but not true horizontal displacement in ground range direction. As explained by Fig. 10.6 in Chapter 10, the LOS displacement can result from the land surface deformation in either horizontal or vertical or both directions as the result of the SAR side-look ranging configuration. The same principle applies to the ground range displacement according to eqn. 11.15. On the other hand, the ΔY disparity map presents true horizontal motion in the SAR azimuth direction.

If we have good quality nadir view optical images of an area before and after a strong earthquake, then the disparity maps ΔX and ΔY between the two images may reveal the true horizontal displacement of co-seismic motion.

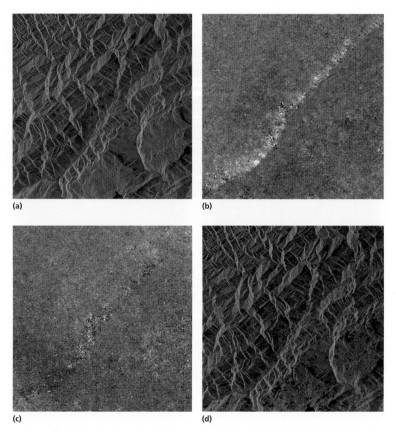

Fig. 11.18 Wenchuan 7.9 Mw earthquake across-event ALOS PALSAR images and disparity maps. (a) RG colour composite of ALOS PALSAR intensity images of Beichuan area before and after the Wenchuan 7.9 Mw earthquake, Sichuan Province, China. (b) and (c) Across-event disparity maps ΔX and ΔY. (d) Smoothed ΔX map in pseudo-colours overlay on the pre-event SAR intensity image in grey scale.

11.6 Summary

This chapter covers an advanced topic: sub-pixel technology. We first briefly explain the general principle of sub-pixel image shift and why useful information can be extracted from sub-pixel image matching. Then our discussion is focused on the PC sub-pixel matching method and its applications.

PC is one of the most robust and accurate methods for sub-pixel image matching. It is based on the well-known Fourier shift property to identify image feature shift at sub-pixel magnitude in image frequency domain via FT. By applying PC to two similar images of the same area, these images can be precisely aligned or frame registered.

PC can be performed with a window-based scanning mechanism to estimate the feature shift (disparity)

between the two images for every pixel to generate pixel-wise disparity maps in image column and line direction. The image pixel-wise disparity maps between the two images can result from many factors such as irregular discrepancy, the projective topographic distortion, terrain or ground objects motion, and so forth. As such, PC has the following major applications:

• *Pixel-wise image co-registration*. The disparity maps of two images for matching record the position shift between every pair of correspondent pixels and can therefore be used to force one image to be co-registered to the other pixel by pixel. This is fundamentally different from the image co-registration based on GCP-derived polynomial deformation model for warping transform, which cannot ensure the precise co-registration for every pixel. Here the sub-pixel accuracy of disparity estimation is vitally important because the true position discrepancy between any pair of correspondent pixels is very unlikely to be an exact integer number of pixels. The PC-derived sub-pixel disparity maps ensure high quality image co-registration at every image pixel regardless of irregular deformations between the two images, and therefore can achieve very high co-registration accuracy. The method is not, however, a geo-rectification process and so is not versatile for dealing with raster datasets of very different properties such as a satellite image and a map.

- *3D data generation*. Terrain 3D data (e.g. DTM) can be generated from stereo image pairs. Images forming a stereo pair are taken from different positions separated by a baseline. The same ground point of a particular height will be viewed in different angles from the two camera positions and thus recorded at different relative locations in the two images because of the projective topographic distortion. The disparity map between such a pair of stereo images in the baseline direction is therefore a relative DTM of the scene. Conventional stereo imaging requires a wide baseline to ensure that the disparity can be measured in adequate integer pixels for DTM generation. The ability of measuring disparity in the magnitude of a fraction (e.g. 1/30th) of a pixel enables generating DTM, or 3D data in general, from stereo image pairs with baseline as narrow as 1/200th of the conventional.

- *Motion/deformation detection and measurement*. Pixel-wise disparity is fundamentally the relative motion between the corresponding pixels in two images. If the terrain or ground objects moved or deformed between two images taken in nearly identical view angle before and after the motion, the disparity maps derived from these two images provide pixel-wise measurement of the motion/deformation. For instance, the earthquake-induced terrain deformation can be measured in such a way using the disparity maps derived from across-event image pairs; such a method is often called pixel-offset used in conjunction with DInSAR for earthquake deformation study. As the average co-seismic motion is usually small in comparison to the image resolution, the sub-pixel accuracy is essential for the method to be effective.

The above-described PC functionality and applications can all be performed by a software package: PCIAS (Phase Correlation Image Analysis System) that we have developed based on years of research (Liu *et al.* 2012). The PCIAS is a professional software package written in C++. Besides being a powerful PC engine, it also provides commonly used image processing and manipulation functions for users to carry out sub-pixel processing and analysis easily.

PART II
Geographical information systems

CHAPTER 12

Geographical information systems

12.1 Introduction

GIS requires no introduction these days since it has almost become a household term; it is still a massively expanding and developing technology, but most people are now aware that it affects almost every aspect of our daily lives. So many aspects of GIS are taken for granted, such as online route finding tools to help us plan our journeys on holidays and at work and to navigate by car, foot, bike or ski. Tools like Google Earth have transformed the public's perception of their environment and increased awareness of geospatial science. It is now considered that GIS is fast becoming part of the mass media, an extension to human thought and creative processes. Essentially, people think far more geographically than ever before. GIS has also become part of the GCSE school curriculum in the UK in the last few years. We are already seeing a marked increase in the number of young people embarking on geospatial careers.

Software tools become more sophisticated, easier to use, more customisable, more effective and faster. This is of course advantageous but we must still be wary of glossing over the basic operations and processes behind complex workflows, wizards and algorithms. The data we use are more voluminous, more efficiently organised, more effectively processed and more widely available. There is always the danger of not being aware of what happens behind a simple button click and of missing a mistake when it happens. In this respect, the roles of simple visualisation and human critique become more important, not less, in ensuring quality, reliability and relevance of results.

One of the principal tools in the hands of remote sensing scientists is image processing. It is not the only tool but is one that is fundamental in the visualisation of remotely sensed imagery. Visualisation is also a vital part of any digital analysis in GIS; written reports are important but rather impotent without the ability to actually see the results. A great many image processing procedures lie at the heart of GIS, mainly in raster GIS, but there are also parallels in vector GIS. We have tried to point out the links between the various aspects of GIS described here and their relevant image processing counterparts described in Part I wherever possible, as well as linking to case studies in Part III in which the techniques have been applied.

There are many excellent general texts on GIS that have already been published, as listed in the general references, so this part of our book is not intended as a general textbook but aims to summarise the techniques and principles that are useful in remote sensing applications within or connected with 'geoscientific' fields. We refer here to the many preparatory tasks as well as to the spatial analysis of data within this geoscientific context. Of course, that does not imply that the tasks dealt with here are the exclusive domain of the geosciences, but we stress that some of the more advanced tasks and techniques are perhaps not generally within the scope of GIS use by the general geographical public.

This seems a convenient point to clarify our definition of the acronym GIS in this book. In some instances it is taken to stand for *Geographic Information Systems* or *Geographical Information Science* and, in some places, *Geographic Information Systems and Science*; all of these are correct, but we will stick with *Geographic Information*

Image Processing and GIS for Remote Sensing: Techniques and Applications, Second Edition. Jian Guo Liu and Philippa J. Mason.
© 2016 John Wiley & Sons, Ltd. Published 2016 by John Wiley & Sons, Ltd.

Systems. The descriptions of processes and procedures in this part of the book are largely generic and do not therefore represent the practice of any particular software suite or the theory of any particular individual, but where certain aspects are relevant to specific GIS software packages or published research, these are stated.

12.2 Software tools

In the light of the conceptual overlap that is the subject of this book, it should also be understood that there are a great many other software tools, used in related disciplines, that have elements in common but do not generally qualify as GIS. Such tools involve spatially referenced information and perform similar operations on raster and vector objects, but we tend not to refer to them in the same breath because generally speaking they are in use within far more specific application areas; one of GIS's great strengths is its status as a general-purpose tool.

Such software suites include *ERMapper* (a sophisticated and totally transparent image processing engine); *Geosoft* (a raster processing suite containing tools for the processing of geophysical data and the production of maps, used largely by the mining/exploration industry); *Micromine* (which is a truly three-dimensional [3D] GIS package for managing, analysing and displaying sub-surface geological and geochemical data, largely in use within the mining/exploration industry); *Petrel* and *Landmark* (suites used almost entirely by the petroleum industry, for the processing, analysis and visualisation of surface and sub-surface raster and vector data); and *Surfer* (a sophisticated and generic toolset for gridding of vector data to produce raster grids). There are many more.

One of the limitations of conventional GIS in geosciences lies in the fundamental concept of the *layer*. Sub-surface geological phenomena exist as discrete features in one or more layers, but there is no way to adequately describe the relationships between these sub-surface layers or the way they may change location between layers at different depths in a conventional GIS. The Earth's surface is a conceptual boundary for GIS. For example, sub-surface horizon maps can be treated like any other spatial data layer, but features, such as faults, that intersect one layer cannot be made to intersect another layer at a slightly different location in a geologically meaningful way. This is partly because the

separation between layers is an arbitrary distance and is not deemed to be significant for the purposes of most GIS operations and procedures. Fortunately, however, there are other software suites that do allow for such concepts and provide more complete 3D functionality for geoscientific visualisation and analysis, such as *GeoVisionary, Move* and *Micromine*.

One of the biggest growth areas in GIS is in open source development using one of a growing number of tools, like Python, Pearl, R and MATLAB. Open source development is as old as GIS itself but has really expanded in recent years. The Open Source Geospatial Foundation (OSGeo, www.osgeo.org) is a not-for-profit organisation that was created to support collaborative development of open source geospatial software and to promote its use; it contains a wealth of valuable resources for anyone interested in GIS development. We have provided a list of GIS software packages and resource sites and the URLs of their websites in Appendix B; the list includes both proprietary and shareware (open source) tools.

12.3 GIS, cartography and thematic mapping

What is the relationship between GIS and a conventional cartographic map? Any map is of course a GIS in itself, that is, it is an analogue spatial database that requires perception and interpretation to extract the embedded information. Once on the paper it is, however, fixed and cannot be modified. A GIS display of a map, on the other hand, is inherently flexible. Unlike the conventional paper map, it does not require every piece of information to be visible at the same time. It can also change the depiction of a particular object according to the value of one of its attributes. Let's not forget, of course, that the cartographic map is still also a vital analogue output of our digital analysis!

The ability to selectively display information according to a particular objective is known as *Thematic Mapping*. Thematic maps are commonplace in atlases, textbooks and other published journal articles; standard topographic maps, for instance, are thematic. Thematic maps can be divided into two groups: qualitative and quantitative. The former show the location or distribution of nominal data and can be considered as 'general-purpose maps' focusing on specific themes such as you would find in an atlas. The latter show the variations or changing magnitudes of particular spatial phenomenon (such as population density, vegetation cover or CO_2 emissions) from place to place; they also illustrate data that can be processed and presented in various numerical scales – ordinal, interval or ratio. The thematic map is therefore a very basic component and may also be a product of remote sensing applications that involve image processing and/or GIS. The case studies described in Part III all involve production of some kind of thematic product, either as the end result or as an intermediary stage product of a broader analysis.

12.4 Standards, inter-operability and metadata

With data and software growing in volume and capability every day, so follows the increasing need to be transparent and rigorous in our recording of quality, provenance and validity, as a matter of best practice in GIS. The Open Geospatial Consortium (OGC) has been formed in relatively recent years as an internationally coordinated volunteer organisation (of which there are several hundred member companies, agencies and universities) that is responsible for the driving of standards in inter-operability and quality within the geospatial community (in its broadest sense). A wealth of information can be found on the OGC website, describing OpenGIS technical standards and specifications, model schemas and best practices for data, metadata and procedures (*www. opengeospatial.org/standards/*) within all aspects of GIS.

Many proprietary software suites now incorporate open application programming interfaces to allow users to customise and develop tools for their own working environment, both locally and for communication via wired and wireless internet. The current trend in software development in a growing international market is towards scalable and modular products, allowing users to customise the tools according to their own needs. A parallel trend is in the sharing of technological development, with highly specialised third-party modules from one product being incorporated (as 'plugins', for instance) into the main suite of another; GIS has now entered the world of 'plug and play'!

The improvement of inter-operability and standards is one of the great 'reliefs' as GIS comes of age. Moving information, integrating it, sharing it and recording its provenance and quality demands openness about its format and structure, how it was created and what has

been done to it subsequently. Being able to import data is vital but so too is the ability to export and share. Thankfully, there have been great advances in satisfying these needs and the trend in inter-operability and open standards is ongoing and entirely positive.

Metadata describe many aspects of geospatial information, especially the content, quality and provenance of the data. This information is vital to the understanding of the information by others who may use the data. The main uses are for general database organisation and for digital catalogues and internet servers. Such information is usually stored in a very standard form, such as *.xml*, and so can be created or edited using a standard text editor program. The recording of metadata seems such an easy and trivial thing that it is often overlooked or neglected, but with with the growing volumes of data at our fingertips it becomes more and more of a necessity. There are several basic categories of metadata that should be recorded, and this may seem obvious but it is often overlooked:

- general identification – the name and creator of the data, its general location, date of creation and any restrictions on its use;
- data quality – the general reliability and quality, stated accuracy levels, level of completeness and consistency and the source data;
- data organisation – the data model used to create and encode the information;
- spatial referencing – the coordinate system used to georeference the data, geographic datums used and any transformation parameters (relative to global standards) that may be specific to the geographic region ;
- attribute information – what kind of descriptive attribute system is used, any codes and schemas that the descriptive information conforms to;
- distribution – where the data were created, formats and other media types available, online availability, restrictions and costs;
- metadata referencing – when, where and by whom the data (and metadata) were compiled.

Metadata has a vital role in a world where digital analysis is commonplace and digital data are growing in volume all the time. The catch is that although metadata provides improved understanding and provenance tracking, it cannot prove the quality or the trustworthiness of the data.

12.5 GIS and the internet

With easy-to-use web browsers, GIS on the internet provides a much more dynamic tool than a static map display. Web-enabled GIS brings interactive query capabilities and data sharing to a much wider audience. It allows online data commerce and the retrieval of data and specialised services from remote servers, such as the Environmental Systems Research Institute Inc. (ESRI) online data server, the Geography Network (*www. geographynetwork.com*) and many others (numbers growing very fast!).

The development of languages and data models like GML, the Geography Markup Language (an extension of XML); VRML (superceded by X3D); and KML (Keyhole Markup Language) also make GIS far more accessible to the general, computer-literate public. KML, for instance, is a file structure for storage and display of geographic data, such as points, lines, polygons and images, in web browser applications such as Google Earth & Maps, MS Virtual Earth, ArcGIS Explorer, Adobe Photoshop and AutoCAD. It uses tags, attributes and nested elements and in the same way as standard HTML and XML files. KML files can be created in a simple text editor or in one of many script editing applications. More information on this and may others can be found at *www.opengeospatial.org/standards/kml*.

The massive benefit of such open standard tools is that they allow geographic information to be created and shared between users who do not have access to proprietary (costly) GIS software licenses, once again broadening the geographic user community.

CHAPTER 13
Data models and structures

13.1 Introducing spatial data in representing geographic features

Data that describes a part of the earth's surface or the features found on it can be described as *geographic* or *spatial*. Such data include not only cartographic and scientific data but also photographs, videos, land records, travel information, customer databases, property records, legal documents, and so on. We also use the term 'geographic features' or simply 'features' in reference to objects, located at the surface of the earth, whose positions have been measured and described.

Features may be naturally occurring objects (rivers, vegetation), or anthropogenic constructions (roads, pipelines, buildings) and classifications (counties, land parcels, political divisions). Conventional cartographic maps represent the real world using collections of points, lines and areas, with additional textual information to describe the features. Most GIS construct maps in a similar way but the features appearing on the map are stored as separate entities that have other intelligence stored with them as 'attributes'. It is also worth noting that the terms 'feature' and 'object' are commonly used interchangeably; here we have tried to stick to the term 'feature' when referring to a discrete entity usually in vector form, but there are times when both terms are used for clarity.

13.2 How are spatial data different from other digital data?

There are four main aspects that qualify data as being *spatial*. First, spatial data incorporates an explicit relationship between the geometric and attribute aspects of the information represented, so that both are always accessible. For instance, if some features are highlighted on a map display, the records containing the attributes of those features are also highlighted (automatically) in the file or associated attribute table. If one or more of those features are edited in some way in the map, those changes are also automatically updated in the table, and vice versa. There is therefore a dynamic link between a feature's geometry and its attributes.

Second, spatial data are referenced to known locations on the earth's surface, that is, they are 'georeferenced'. To ensure that a location is accurately recorded, spatial data must be referenced to a coordinate system, a unit of measurement and a map projection. When spatial data are displayed, they also have a specific scale just like on an analogue map, but in GIS this scale can be modified.

Spatial data also tend to be categorised according to the type of features they represent, that is, they are sometimes described as being 'feature-based'. For example, area features are stored separately from linear or point features and, in general, cannot co-exist in the same file structure.

Last, spatial data are often organized into different 'thematic' layers, one for each set of features or phenomena being recorded. For instance, streams, roads, rail and landuse could be stored as separate, themed 'layers', rather than as one large file. In the same way, within each 'theme' there may be sub-themes of the same feature type that can be usefully grouped together, such as different classes of roads and railways, all of which are linear features that belong to the theme 'transport'. This also makes data management, manipulation and analysis rather more effective.

Image Processing and GIS for Remote Sensing: Techniques and Applications, Second Edition. Jian Guo Liu and Philippa J. Mason.
© 2016 John Wiley & Sons, Ltd. Published 2016 by John Wiley & Sons, Ltd.

13.3 Attributes and measurement scales

Descriptive attributes can also be described as being *spatial* or *non-spatial,* though the difference between them may be subtle and ambiguous. The nature of the information stored, and thereby the scale of measurement to which it belongs, dictates what kind of processing or analysis can be performed upon it. *Measurement scales* describe how values are assigned to features represented in the GIS. The type of scale chosen is dictated by our intended use of the recorded information. There are five scales commonly used in GIS: *nominal, ordinal, interval ratio* and *cyclic,* and these are summarised, along with the numerical operators appropriate to each case, in Table 13.1.

Nominal or categorical scales include numerical values used as 'pointers' to real-world objects or qualitative descriptions held in attribute tables. *Ordinal* measures involve values ranked or ordered according to a relative scale and that generally have unequal intervals. Greater than or less than operators are therefore useful, but addition, subtraction, multiplication and division are not appropriate. One example could be multi-element geochemical data where element concentrations are given on a percentile scale and the intervals between classes are not constant but arbitrary. *Interval* measures are used to denote quantities like distances or ranges,

but in this case the intervals between the values are based on equal or regular units. There is, however, no true zero on an interval scale because the position of zero depends on the units of the quantity being described. Temperature scales are a good example of this since the position of zero temperature depends on the units of measurement, Fahrenheit or Celsius. *Ratio* measures are similar to interval scales and are often used for distances or quantities, but the zero value represents absolute zero, regardless of the units. *Cyclic* measures are a special case describing quantities that are measured on regular but circular or cyclic scales, such as aspect or azimuth directions of slopes, or flow directions, both of which are angular measures made with respect to north. Appropriate operators for cyclic measures are then any or all of the previously mentioned arithmetic and average operators.

13.4 Fundamental data structures

There are two basic types of structure used to represent the features or objects; these are *raster* and *vector* data, and as a consequence, there are different types of GIS software architecture, and different types of analysis, that have been designed in such a way as to be most effective with one or other type, such as Idrisi or ERDAS Imagine (raster) and MapInfo or ArcGIS (vector).

Rasters, images or grids consist of a regular array of digital numbers, or DNs, representing picture elements or pixels that have equal x and y dimensions. In raster form, point, line and area features are represented as individual pixels or groups thereof.

Vector or discrete data stores the geometric form and location of a particular feature (recorded as single points or vertices connected by arcs) separately from its attribute information describing what the feature represents, which is stored in a database file or table. Vector data typically resembles cartographic data.

The most basic units are the *point* (*vertex*) and *pixel,* which represent discrete geographic features of no or limited area, or which are too small to be depicted in any other way, such as well locations, geochemical sample points, towns or topographic spot heights. *Lines* (or *polylines*) are linear features consisting of vertices connected by arcs that do not in themselves represent area, such as roads, rivers, railways or elevation contours. *Areas* are closed polyline features that represent

Table 13.1 Measurement scales: methods for describing and operating on thematic information.

	Operators	Examples
Nominal	=, ≠ *and* mode	Categorical (class) identifiers (e.g. 1 = forest, 2 = pasture, 3 = urban)
Ordinal	<, ≤, ≥, > *and* median	Sequences of natural order, e.g. 1, 2, 3, 4 etc.
Interval	+, −, ×, ÷ *and* mean	Ranges between, and sections along, distance measures, e.g. temperature scales
Ratio	All the above	Distance measures, and/or subdivisions thereof, along lines and routes
Cyclic	All the above	Special measures, e.g. 360-degree bearings (azimuth), flow directions

the shape, area and location of homogeneous entities such as countries, land parcels, buildings, rock types or landuse categories. A *surface* describes a variable that has a value for every position within the extent of the dataset, such as elevation or rainfall intensity, and implies data of a 'continuous' nature. Surfaces are typically represented on conventional cartographic maps as a series of isolines or contours; within GIS there are other possibilities. Deciding how these features should be stored in the database, and represented on the map, depends on the nature of that information and the work it will be required to do.

All complex data structures in GIS stem from, and depend on, one or the other of these two basic structures, point and pixel. GIS operations and spatial analysis can be performed on either type of data. We shall now describe these structures in turn.

13.5 Raster data

A raster image is a graphic representation or description of an object that is typically produced by an optical or electronic device. Some common examples of image data include remotely sensed (satellite or airborne survey data), scanned data and digital photographs.

Raster data represents a regular grid or array of digital numbers, or pixels, where each has a value depending on how the image was captured and what it represents. For example, if the image is a remotely sensed satellite image, each pixel DN represents reflectance from a finite portion of the earth's surface; or in the case of a scanned document, each pixel represents a brightness value associated with a particular point on the original document.

In general, one important aspect of the basic raster data structure is that no additional attribute information is stored about the features it shows. The DN values in a single raster grid therefore represent the variance of one attribute, such as reflectance or emittance in a particular waveband from the earth's surface as in the case of remotely sensed data; a separate raster is required to represent each attribute that is used. This means that image databases generally occupy considerably more disk space than their vector counterparts. Any additional attribute information must be stored in a separate table that can then be linked or related to that raster – for example, a raster in which nominal integer values between 1 and 4 are stored, representing some phenomena (landuse

categories, in this case). These values/classes are linked via an attribute table to other descriptive textual information (two further attributes, 'land value' and 'lease expiry date'), such as is illustrated in Fig. 13.1.

For an attribute such as landuse, as in the example shown in Fig. 13.1, the discrete raster model involves data redundancy, and there is a better way to represent such categorical information.

13.5.1 Data quantisation and storage

The range of values that can be stored by image pixels depends on the quantisation level of the data, that is, the number of binary bits used to store the data. The more bits, the greater the range of possible values. For example, if 1-bit data is used, the number of unique values that can be expressed is 2^1, or 2. With eight bits, 2^8 or 256 unique values can be expressed, with 16 bits, that number is 2^{16} or 65,536 and so on. The most common image data quantization formats are 8-bit and 16-bit. Raster can also be stored in ASCII format, but this is the least efficient way of storing digital numbers and the file sizes get very large indeed. The binary quantisation level selected depends partly on the type of data being represented and what it is used for. For instance, remotely sensed data are commonly stored as 8-bit integer data, allowing 256 grey levels of image information to be stored; so that a three-band colour composite image provides three times 255 levels (8-bits) and hence 24-bit colour is produced. Very high resolution image datasets are now often stored on an 11-bit value range. Digital elevation data, on the other hand, may well represent height values that are in excess of 256 m above sea level. The value ranges of 8-bit, or 11-bit, will not allow this and so elevation data are usually stored as 16-bit data, as integers or real numbers (floating point values).

Image data can be organized in a number of ways and a number of standard formats are in common use. In many cases, the image data file contains a header record that stores information about the image such as the number of rows and columns in the image, the number of bits per pixel and the georeferencing information. Following the image header are the actual DNs of the image. Some formats contain only a single raster image, while others contain multiple images or bands. The file extension used usually reveals the method of storage of the image data, for example, band interleaved by line

Discrete raster

Associated attribute table

DN	Landuse	Value_per-ha	Lease_exp
1	Coniferous	30	2017
2	Deciduous	20	2020
3	Grassland	40	2017
4	Scrubland	10	2019

(a)

(b)

Fig. 13.1 (a) A discrete raster map containing values on a nominal scale, which are entirely arbitrary without reference to (b) the associated attribute table. The nominal codes 1–4 in this case depict landuse categories, and they are linked to two further attributes, land value and lease expiry date (both are on interval scales). Without such a table, separate rasters would be required to represent the values of the other two attributes.

has the extension *.bil*, band interleaved by pixel *.bip*, and band sequential *.bsq*.

There are other formats adopted on a national level, such as the National Imagery Transmission Format Standard (NITFS), which is a format for exchange, storage and transmission of digital image products defined by the US government. There are also several sub-types of NITF files and none should be confused with the National Transfer Format (NTF), which is a format administered by the British Standards Institution and is the standard adopted by the Ordnance Survey for their digital data.

The Geospatial Data Abstraction Library, or GDAL, is a translator library for both raster and vector geospatial data formats that is released under an open source license by OSGeo. A comprehensive and continually updated list of formats supported by GDAL, along with a list of software that uses GDAL, can be found at www.gdal.org. A detailed description of GDAL's raster data model is given at www.gdal.org/gdal_datamodel.html.

With the increasing availability of data, increasingly high resolution and increasing speed and computing power, so our capacity (and desire) to process large data volumes grows. In parallel with this has been the need to develop better methods of storage and compression. The goal of raster compression is then to reduce the amount of disk space consumed by the data file whilst retaining the maximum data quality. Newly developed image compression methods include *discrete wavelet transforms*, which are produced using an algorithm based on multi-resolution analysis. Such methods are much less 'lossy' than block-based (discrete cosine transformation) compression techniques such as JPEG. The advantage of wavelet compression is that it analyses the whole image and allows high compression ratios while maintaining good image quality and without distorting image histograms. These techniques are essential if large volumes of raw data are to be served via internet/intranet. The most well known include the following, and useful white papers can be found on the Open Geospatial Consortium (OGC) and relevant company websites:

- JPEG2000 – a new standard in image coding that uses bi-orthogonal wavelet compression. It produces better compression ratios than its predecessor, JPEG, and is almost 'lossless'.
- ECW or Enhanced Compressed Wavelet (proprietary format developed by Earth Resource Mapping Ltd, now owned by Intergraph, patent numbers US 6201897, US 6442298 and US 6633688) uses multiple scale and resolution levels to compress imagery, while maintaining a level of quality close to the original image. Compression ratios of between 10:1 to 20:1 (panchromatic images) and 30:1 to 50:1 (colour images) are commonly used; ratios of between 2:1 and 100:1 are achievable.
- MrSID or Multi-resolution Seamless Image Database (developed by Los Alamos National Laboratory and now marketed by LizardTech Corporation) – uses an algorithm very similar to ECW, with similar compression ratios and with similarly 'lossless' results, for serving over the Internet.

13.5.2 Spatial variability

The raster data model can represent discrete point, line and area features but its ability to do so is limited by the size of the pixel and by its regular grid-form nature. A point's value is assigned to and represented by the value at the nearest pixel. Similarly a linear feature would be represented by a series of adjacent pixels and an area would be shown as a group of adjacent pixels over an area that that most closely resembles the original shape of that area, as illustrated in Fig. 13.2.

The kind of information to be stored should be considered before choosing a model for storing it. Clearly points, lines and areas can be represented by pixels, but since the raster grid is by definition contiguous – that is, every possible position within the area of interest must be represented by a pixel value – we are forced to store values for all the areas not represented by significant features. So that in the case of the landuse map in Fig. 13.1, the information of value occurs at the boundaries between the classes. The distribution of scrubland, for instance, is such that it covers the area taken by 12 pixels; so its value (4 in this case) is stored 12 times. There is no further variation within this area and a similar situation exists for the other three classes. Such information can be considered to be of low spatial

variability and, if represented by a raster, means that many duplicate, and insignificant, values are stored in addition to the important ones where the class boundaries occur; this constitutes a degree of data redundancy and represents wasted storage space. In such cases, it would be better to choose the alternative, vector model of storage.

The raster is therefore most appropriately used where the data can be considered of relatively high spatial variability; that is, where every pixel has a unique and significant value. This applies to satellite imagery or photographs, or any case where we are interested the spatial variability of the attribute or phenomenon across an area, not just in discrete classes.

There are times during spatial analysis, for instance, when it becomes necessary to convert such maps and images of low spatial variability into raster form. Usually such files are intermediary and they are normally automatically removed at the end of the analysis.

13.5.3 Representing spatial relationships

Spatial relationships between the pixels of a raster are implicit in its regular data structure since there can be no gaps or holes in the grid. Any further

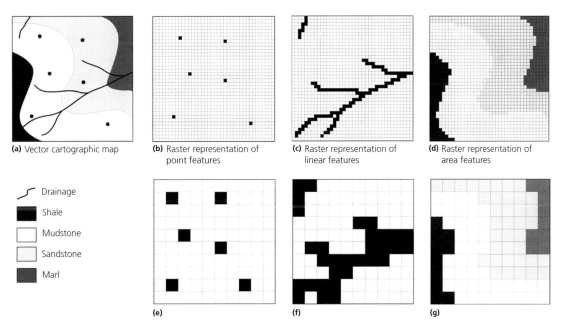

(a) Vector cartographic map

(b) Raster representation of point features

(c) Raster representation of linear features

(d) Raster representation of area features

Drainage

Shale

Mudstone

Sandstone

Marl

(e)

(f)

(g)

Fig. 13.2 (a) Vector features (point, line and area) representing a cartographic map and their raster equivalents depicted in grids of relatively high spatial resolution (b, c and d) and low spatial resolution (e, f and g).

explicit description of spatial relationships between pixels is unnecessary.

Each raster is referenced at the top left corner; its location is denoted by its row and column position and is usually given as 0,0. All other pixels are then identified by their position in the grid relative to the top-left. Each pixel in the raster grid has eight immediate neighbours (except those on the outside edges). In Fig. 13.3, the cell at position 2, 2 is located at 3 pixels along the *x* axis and 3 down on the *y* axis. Finding any one of eight neighbours simply requires adding or subtracting from the *x* and/or *y* pixel locations. For example, the value immediately to the left of (2, 2) is (2-1, 2 or 1,2).

Since the spatial relationship between pixels is implicit, the whole the raster is effectively georeferenced by this system; specifying the coordinate reference of the 0, 0 origin and the cell size in real-world distances enables rapid calculation of the real world locations of all other positions in the grid. The upper-left pixel being used as the origin or reference point for 'raster space' is in contrast to 'map space' where the lower-left corner is the geographical coordinate origin; this difference has an effect on the way raster images are georeferenced.

Another benefit of implicit spatial relationships is that spatial operations are readily facilitated. Processing can be carried out on each individual pixel value in isolation, between corresponding pixel positions in other layers, between a pixel value and its neighbour's values or between a zone of pixel values and zones on other layers. This is discussed in more detail in Chapter 14.

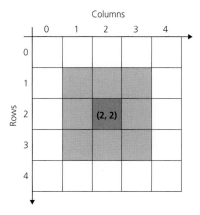

Fig. 13.3 The organisation of a simple raster grid, its origin and row and column reference system.

13.5.4 The effect of resolution

The accuracy of a cartographic map depends on the scale of that map. In the raster model the resolution, scale and hence accuracy, depend on the real-world area represented by each pixel or grid cell. The larger the area represented, the lower the spatial resolution of the data. The smaller the area covered by the pixel, the greater the spatial resolution and the more accurately and precisely features can be represented. The pixel can be thought of as the lower limit beyond which the raster becomes discrete. This is demonstrated in Fig. 13.2(b)–(g), where the boundaries of the lithological units in the geological map are most closely resembled by the raster grid whose spatial resolution is highest, that is, whose pixels are the smallest [Fig. 13.2(b)–(d)]. Problems arise at the boundaries of the classes; that is, where a class boundary passes through the area represented by a pixel, rules need to be applied to decided how that pixel will be encoded. This issue is discussed further in later in this chapter.

The raster data model may at first seem unappealing, within GIS, because of its apparent spatial inaccuracy, that is, the limitation of the sampling interval or spatial resolution, but it more than makes up for this in its convenience for analytical operations (being a two-dimensional [2D] array of numbers). This is especially true for any operations involving surfaces or overlay operations, and of course with remotely sensed images. With computer power becoming ever greater, we may have fewer concerns over the manageability of large, high resolution raster files. Providing we maintain sufficient spatial resolution to adequately describe the phenomenon of interest, that is, resolution should be as high as is necessary, then we should minimise problems related to accuracy.

13.5.5 Representing surface phenomena

Rasters are ideal for representing surfaces since a value, such as elevation, is recorded in each pixel and the representation is therefore continuously sampled across the area covered by the raster. Conceptually, we find it easiest to think of a surface, from which to generate a perspective view, as being elevation or topography but any raster can be used as a surface, with its pixel values (or DN) being used to generate 'height' within three-dimensional (3D) space (Fig. 13.4).

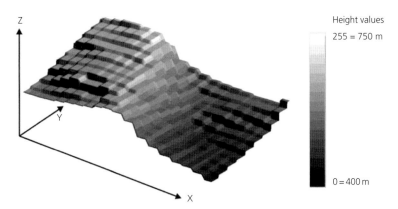

Fig. 13.4 A raster grid displayed, in perspective view, to illustrate the way that pixel values are used to denote height and to form a surface network. The pixel values can be colour coded on a relative scale, using a colour lookup table in which height values range from low to high, on a black to white scale in this case.

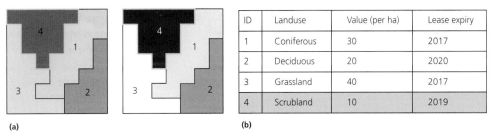

ID	Landuse	Value (per ha)	Lease expiry
1	Coniferous	30	2017
2	Deciduous	20	2020
3	Grassland	40	2017
4	Scrubland	10	2019

(a)   (b)

Fig. 13.5 (a) A discrete vector polygon file symbolised by the information in the two separate attributes of landuse and value; and (b) an integral attribute table, dynamically linked to the geometric information. Together they form a simple vector polygon file colour coded to represent (a) landuse type (nominal) and land value (ordinal); and (b) its integral attribute table with feature number 4 highlighted in both the map and in the attribute table.

The structure and use of raster surfaces are dealt with in more detail in Chapter 17. The input dataset representing the surface potentially contributes two pieces of information to this kind of perspective viewing. The first is the magnitude of the DN, which gives the height, and the second is given to the way the surface appears or is encoded visually, that is, the DN value is also mapped to colour in the display.

13.6 Vector data

In complete contrast, the vector model incorporates, discretely, both the geometry and location of geographic *features* and all the attribute information describing them. For convenience, the attribute information is generally viewed and accessed in tabular form. Each vector feature has a unique identifying code or number and then any number of numerical or textual attributes, such that the features can be symbolised according to any of those attributes. In Fig. 13.5 the simple vector map is shown where the individual land parcels are coded to represent two stored attributes in Fig. 13.5(a) and (b), respectively: landuse (nominal) and land value (ordinal); the alpha-numeric attributes used are shown in Fig. 13.5(c).

This association of geometry and tabular attribute is often referred to as a 'georelational' data structure and in this way, there is no limit to the attribute information that can be stored or linked to a particular feature object. This represents one very clear advantage over a raster's single attribute status. As with raster data, additional tables can also be associated with the primary vector feature attribute tables, to build further complexity to the information stored.

Tabular data represents a special form of vector data that can include almost any kind of information, whether or not it contains a geographic component, that is, tabular data are not necessarily spatial in nature. A table whose information includes and is referenced by coordinates can be displayed directly on a map. Those that do not must be linked to the other spatial data that do have coordinates before it can be displayed.

Vector data consist of a series of discrete features described by their geometry and spatial reference (coordinate positions) rather than graphically or in any regularly structured way. The vector model could be thought of as the antithesis of raster data in this respect, since it does not fill the space it occupies; that is, not every conceivable location is represented by a data point, only those locations where some feature of interest exists. If we were to choose the vector model to represent some phenomenon that varies continuously and regularly across a region, such that the dataset necessarily becomes so densely populated as to resemble a regular grid, then we would probably have chosen the wrong data model for those data.

13.6.1 Vector data models

Vector data, like vector graphics, involve the use of geometric primitives (points, lines and polygons) to represent features, all of which are based on mathematical expressions. A basic feature model for vector data description and function is described in detail on the GDAL website (http://www.gdal.org/ogr_arch.html).

This model includes descriptors for all aspects of vector data: geometry, spatial reference, feature and feature class definitions, layers and drivers.

13.6.2 Representing logical relationships through geometry and feature definition

While the term *topography* describes the precise physical location and shape of geographical objects, the term *topology* defines the logical relationships between those objects, so that data whose topological relationships are defined can be considered intelligently structured. The GIS can then determine where they are with respect to one another, as well as what they represent and where they are in absolute terms. The difference between these two concepts is illustrated in Fig. 13.6.

Looking at the topographic map in Fig. 13.6(a), it is an easy for us to interpret the relationships; that Regents Park completely encloses the Boating Lake and shares a boundary with the zoo but it is not as easy to express this digitally. The GIS can only understand these spatial relationships through topological constructs and rules.

There have been a number of vector data models developed over the past decades, which support topological relationships to varying degrees, or not at all. The representation, or not, of topology dictates the level of functionality that is achievable using those data. These models began with unstructured (or 'spaghetti')

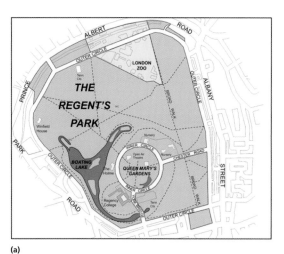

(a)

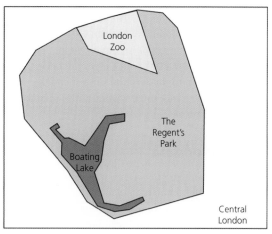

(b)

Fig. 13.6 (a) A topographic map showing the Regent's Park area in London; and (b) a topological map showing the locations of London Zoo and the Boating Lake that lies inside Regent's Park in London.

data, vertex dictionary, Dual Independent Map Encoding (DIME) and arc-node structure (formally known as POLYVRT). To understand the significance of topology it is useful to consider the evolution of these models.

13.6.2.1 Unstructured or 'spaghetti' data

At the simplest level, each vector location is recorded as a single *x, y* coordinate pair, representing a *point*. Points are either isolated or joined to form lines, when each is then termed a *vertex*. Lines are then recorded as an ordered series of vertices, and areas are delimited by a series of ordered line segments that enclose the area.

The information describing the features is stored as a simple file listing the *x, y* coordinate pairs of all the points comprising the feature, and a unique identifying character for each feature. Three simple vector features (area, line and point respectively) and their file format are shown in Fig. 13.7. Each feature is identified by a unique code (A, B or C in this case) and by the number of vertices it contains, followed by the coordinate pairs defining all the constituent vertex locations. The order of the listed points comprising the area and line features is significant; they must be listed according to the sequence of connection. Notice also that the area feature is different from the line feature only in that the first and last listed points are identical, indicating that the feature is closed.

In this 'spaghetti' form, vector data are stored without relational information; there is no mechanism to describe how the features relate to one another, that is, there is no topology. Any topological relationships must

be calculated from the coordinate positions of the features; at the time this was considered inefficient and computationally costly.

The arguments in favour of unstructured data are that their generation demands little effort (this is perhaps debatable) and that the plotting of large unstructured vector files is potentially faster than structured data. The disadvantages are that storage is inefficient, there may be redundant data and that relationships between objects must be calculated each time they are required.

13.6.2.2 Vertex dictionary

This structure is a minor modification of the 'spaghetti' model. It involves the use of two files to represent the map instead of one. Using the same map, and shown in Fig. 13.7(c) as an example, in the first file the coordinate pairs of all the vertices are stored as a list, with each pair being identified by a unique ID number or letter. The second file stores the information describing which coordinate pairs are used by each feature. This prevents duplication, since each coordinate pair is stored only once but it does not allow any facility to store the relationships between the features, that is, topology is still unsupported.

13.6.2.3 Dual Independent Map Encoding

This structure was developed by the US Bureau of the Census for managing their population databases; both street addresses and Universal Transverse Mercator (UTM) coordinates were assigned to each entity in the

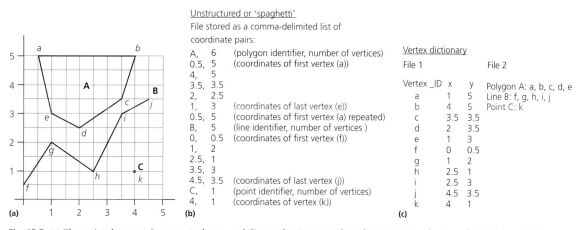

Fig. 13.7 (a) Three simple vector features (polygon, polyline and point) stored as: (b) unstructured or 'spaghetti' data and (c) vertex dictionary.

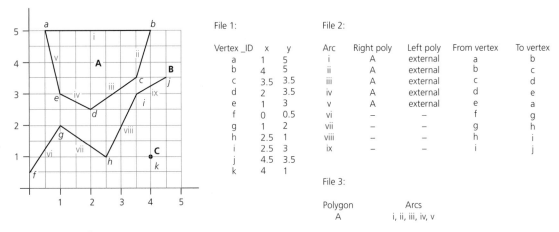

Fig. 13.8 The DIME data structure.

database. Here again, additional files (tables) are used to describe how the coordinate pairs are accessed and used (as shown in Fig. 13.8). This time the vertices and arcs are given unique codes and each arc is assigned a directional identifier to denote the direction in which the feature is constructed (the *from* and *to* vertices). In this way some topological functionality is supported, through the connectivity of features. Arcs that form polygons are listed in a third file.

The US Bureau of Census later developed this and released the TIGER/Line (or Topologically Integrated Geographic Encoding and Referencing) ASCII format in the 1990s. TIGER/Line incorporates a higher block-level so that a further level of hierarchy (with unique block ID numbers) to add complexity. TIGER/Line is still in use at present in the USA.

13.6.2.4 Arc-node structure

Here vector entities are stored separately but are linked using pointers. Nodes represent the start and end vertices of arcs. Arcs which share a node are connected. Polygons are defined by a series of connected arcs. A further concept of chains is also added, in which chains form collections of line segments with directional information.

When capturing the boundaries of two adjacent area features from a map, the double tracing of the boundary between them is inefficient and can lead to errors (either gaps or overlaps). The same is true for a point shared by a number of lines, because the point may be stored many times. A more efficient model for the storage of vector data, and one that supports topological relationships, is the 'arc-node' data structure (illustrated in Fig. 13.9).

We also have a potential problem in storing polygons that have holes or islands in them. A polygon that exists inside another polygon is often termed an *island* polygon (or hole) but if any of the previously described data models are used, we have no way of conveying that one lies inside the other, they merely have their location in common. There are ways to get around this with other models if, for whatever reason, the data must remain unstructured. If a value field is added to the polygon attribute table that denotes some level of priority, then this value can be used to control how the file is drawn or plotted on the screen or display; polygons with low priority attributes would be drawn first and those with high priority would be drawn last. In this way the smaller, high priority island polygons could be made to draw on top of the outer polygon that encloses it. If the polygon file in question contains lithological outcrop patterns in a geological map, for example, this task could get rather complicated with many levels of priority required to ensure the polygons are drawn in the correct sequence. Not an ideal solution!

A further level of complexity is therefore required to properly define such complex relationships and so some more definitions would be helpful at this point. A sequence of vertices forms a line where the first and last vertices may be referred to as *start* and *end* vertices, and these have special significance and confer the direction of capture. An *arc* is a line that, when linked with other arcs, forms a *polygon*. Arcs may be referred to as *edges* and sometimes as *chains*. A point where arcs terminate or connect is described as a *node*. Polygons are formed from an ordered

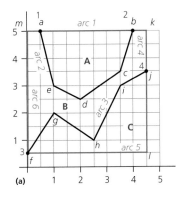

(a)

File 1: Coordinate pairs for all vertices and nodes			
Arc	From_node	To_node	Vertices
1	0.5, 5	4, 5	
2	4, 5	0.5, 5	3.5, 3.5 2, 2.5 1, 3
3	0, 0.5	4.5, 3.5	1, 2 2.5, 1 3.5, 3
4	4, 5	4.5, 3.5	4.5, 5
5	0, 0.5	4.5, 3.5	4.5, 0.5
6	0, 0.5	0.5, 5	0, 5

(b)

File 2: Arc topology				
Arc	From_node	To_node	Left_poly	Right_poly
1	a	b	external	A
2	b	a	B	A
3	f	j	B	C
4	b	j	external	B
5	f	j	C	external
6	f	a	external	B

(c)

File 3: Node topology	
Node	Arcs
1	1, 2, 6
2	1, 2, 4
3	3, 5, 6
4	3, 4, 5

(d)

File 4: Polygon topology	
Polygon	Arcs
A	1, 2
B	2, 3, 4, 6
C	3, 5

(e)

Fig. 13.9 A simple 'map' illustrating the concept of arc-node structure.

sequence of arcs and may be termed *simple* or *complex* depending on their relationship to other polygons. Where one polygon completely encloses another, the outer polygon is described as being complex. The outer polygon's boundary with the inner, *island* polygon may be referred to as an *inner ring* (of which there may be more than one) and its other, outer boundary is called its *outer ring*. So that one or more arcs form a *ring*.

This scheme is created from the moment of capture, when the features are 'digitised'. Each arc is digitised between nodes, in a consistent direction – that is, it has a start and end node – and given an identifying number, and at the same time, further attribute information is entered that describes the identity of the polygon features that exist to the left and to the right of the arc being captured. When all the arcs are captured and identified in this way, the topological relationships between them are calculated as the polygons are constructed. Usually some minor mistakes are made along the way but these can then be corrected and the topological construction

process repeated until the polygon map is complete. The same principle is then applied to line maps, for the construction of networks for example.

A scheme of tables is used to describe how each of the levels of information relates to each other. One stores the coordinate pairs describing each arc, including its *from* and *to* nodes, and any other vertices in between. Another table describes the arc topology, that is, which polygon lies to the left or to the right of each arc (on the basis of the stored direction of each arc). A third table describes the polygon topology, listing the arcs that comprise each polygon. The last table describes the node topology, which lists the arcs that are constructed from each node.

For example, consider the example used before (and shown in Fig. 13.10), this time slightly modified by the addition of lines forming a boundary around the polygon and line, forming new polygons (three in total, labelled A, B and C). Nodes are created where more than two lines intersect, arcs are created between the nodes, with vertices providing shape.

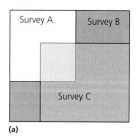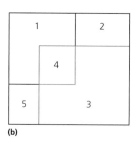

No	A	B	C	All
1	1	0	0	0
2	0	1	0	0
3	0	0	1	0
4	0	0	0	1
5	0	1	0	0

(a)   (b)   (c)

Fig. 13.10 (a) Map of the boundaries of three survey areas, carried out at different times. Notice that the areas overlap in some areas; this is permitted in 'spaghetti' data but not in arc-node structure. (b) The same survey maps after topological enforcement to create mutually exclusive polygonal areas. (c) The attribute table necessary to link the newly created polygons (1–5) to the original survey extents (A, B & C) (Bonham-Carter, 2002. Reproduced with permission from Elsevier.)

Since the development of *arc-node topology*, and in the context of superior computing power, a new generation of vector models has been formed that do not demand the rigorous construction of topologically correct data. The Environmental Systems Research Institute Inc. (ESRI) *shapefile* is a good example of this. In such cases, topological relationships are computed in memory, 'on-the-fly'. Conceptually, the shapefile lies somewhere between 'spaghetti' and arc-node structure. The shapefile achieves this by enforcing the organisation of data into separate types of shape file according to *feature class* type, for example, point, polyline, polygon, pointZ, pointM, polylineZ and so forth; in all there are 14 possible types and no two types can co-exist in the same file. The shape file actually consists of a minimum of three files (.shp, .shx and .dbf) to store coordinates, geometry and attributes, with a further file (.prj) to describe spatial reference (map projection). This hybrid vector format provides liberation from the requirement to rigorously capture and construct topologically correct data but does not replace it. Generally speaking, vector models tend to be application specific, that is, different application areas tend to have different demands of the data and so tend to adopt slightly different formats as standard, for example, TIGER/Line.

Many GIS employ a relational database management system (DBMS) to connect the attribute information to the geometric information. More recently, object-oriented databases have been developed to allow that discrete objects belong to discrete classes, and these may be given unique characteristics. Most modern GIS are hybrids of these, in which the GIS functionality is closely integrated with the relational management system. Such systems allow vector and raster data to be managed and used together. The object-oriented type also integrates a spatial query language to extend the hybrid model; the ArcGIS *geodatabase* is an example of this type of structure. Both ESRI *coverages* and *geodatabases* are relational structures storing vector data; and allowing geometry to be shared between attributes and vice versa. A *coverage* consists of a database directory holding the vector feature data and its attributes as a series of files of specific names and extensions, according to the type of coverage. Both types use rules to validate the integrity of data and to derive topological relationships between features. An ever growing list of common vector formats can be found on the GDAL website (www.gdal.org/ogr_formats.html).

The use of topological relationships, however they are defined, has several clear advantages including the fact that the data are stored more efficiently, enabling large data sets to be processed quickly. One important advantage is that analytical functions are facilitated through three major topological concepts, *connectivity*, *area definition* and *contiguity*.

13.6.2.5 Connectivity
Connectivity allows the identification of a pathway between two locations, between your home and the airport, along a bus, rail and/or underground system, for instance. Using the arc-node data structure, a route along an arc will be defined by two endpoints, the start or *from-node* and the finish or *to-node*. *Network connectivity* is then provided by an arc-node list, that identifies which nodes will be used as the *from* and *to* positions along an arc. All connected arcs are then identified by searching for node numbers in common. In Fig. 13.9, for example, it is possible to determine that arcs 1, 2 and 6 all intersect because

they share node 1. The GIS can then determine that it is possible to travel along arc 1 and turn onto arc 2 because they meet at node 1.

13.6.2.6 Area definition

This is the concept by which it is determined that the Boating Lake lies completely within Regents Park, that is, that it represents an island polygon inside it, as shown in Fig. 13.6.

13.6.2.7 Contiguity or adjacency

Contiguity, a related concept, allows the determination of adjacency between features. Two features can be considered adjacent if they share a boundary. Hence, the polygon representing London Zoo can be considered adjacent to Regents Park as shown in Fig. 13.6.

Remembering that the *from-node* and *to-node* define an arc's direction, so that the polygons on its left and right sides must also be known, *left-right topology* describes this relationship and therefore adjacency. In the arc-node data structure illustration in Fig. 13.9, polygon B lies to the left of arc 2, and polygon A lies to the right, so we know that polygons A and B are adjacent.

Notice the outer polygon that has no label here but is often called the *external* or *universe* polygon; it represents the world outside the area of interest and ensures that each arc always has a left and right side defined. So that the arc joining points *a* and *b*, when captured in that order, has polygon A to its right and the universe polygon to its left. From this logic, entered at the time of data capture, topological entities are constructed from the arcs, to provide a complex and contiguous dataset.

13.6.3 Extending the vector data model

Topology allows us to define areas and to describe *connectivity, area definition* and *adjacency* (or *contiguity*) but we may still need to add further complexity to the features we wish to describe. For instance, a feature may represent a composite of other features, so that a country could be modelled as the set of its counties, where the individual counties are also discrete and possibly geographically disparate features. Alternatively, a feature may change with time, and the historical tracking of the changes may be significant. For instance, a parcel of land might be subdivided and managed separately but the original shape, size and attribute information may also need to be retained. Other examples include naturally overlapping features of the same

type, such as the territories or habitats of several species, or the marketing catchments of competing supermarkets, or surveys conducted in different years as part of an exploration program (as illustrated in Fig. 13.10).

The 'spaghetti' model permits such area subdivision and/or overlap but cannot describe the relationships between the features. Arc-node topology can allow overlaps only by creating a new feature representing the area of overlap, and can only describe a feature's relationship with it's subdivisions by recording that information in the attribute table.

Several new vector structures have been developed by ESRI and incorporated into its ArcGIS technology. These support and enable complex relationships and are referred to as *regions, sections, routes* and *events*.

13.6.3.1 Regions

A region consists of a lose association of related polygons and allows the description of the relationships between them. In the same way that a polygon consists of a series of arcs and vertices, a series of polygons form a region. The only difference in the structure being that, unlike the vertices listed in a polygon, the polygons comprising the region may be listed in any order. As with points, lines and polygons, each region has a unique identifier. The polygons representing features within the region are independent, they may overlap and they do not necessarily cover the entire area represented by the region. So that the overlapping survey areas in Fig. 13.11 could simply be associated within a survey region, as related but separate entities, without having to create a new topologically correct polygon for the overlap area.

Normally, each feature would be represented by a separate polygon but the region structure allows that a single feature could consist of several polygons. For example, the islands around Great Britain might be

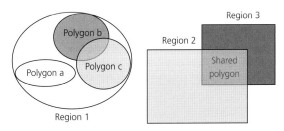

Fig. 13.11 Illustration of different types of *region*: associations of polygons and overlapping polygons that share a polygon.

stored as independent and unconnected polygons but could belong to the collective region of the United Kingdom and be given collective properties accordingly.

Constructing overlapping regions is rather similar to constructing polygons, that is, where regions overlap, they share a polygon in the same way that polygons share an arc where they meet, as shown in Fig. 13.11. The use of regions should assist data management since several different types of feature may be integrated into a single structure whilst the original characteristics remain unchanged.

13.6.3.2 Linear referencing

Routes, sections and events can be considered together since they tend not to exist on their own, and together they constitute a system of *linear referencing* as it is termed in ESRI's ArcGIS. The constructed route defines a new path along an existing linear feature or series of features, as illustrated in Fig. 13.12. If we use the same example of getting from home to the airport, your 'airport' route would consist of a starting and finishing location and a unique path, which follows existing roads, bus or railways but does not necessarily involve the entire lengths of those roads and railways. *Routes* may be circular, beginning and ending in the same place. They may be disconnected, such as one that passes through a tunnel and so is not visible at the surface. A further piece of information necessary for the description of the route, is the unit of measurement along the route. This could be almost any quantity and, for the example of a journey, the measure could be time or *distance*.

Sections describe particular portions of a route, such as where road works are in progress on a motorway, where speed limits are in place on a road, or where a portion of a pipeline is currently undergoing maintenance. Again,

starting and ending nodes of the section must be defined according to the particular measure along the route.

Similarly, *events* describe specific occurrences along a route, and they can be further subdivided into *point* and *linear events*. A *point event* describes the position of a point feature along a route, such as an accident on a section of motorway or a leak along a pipeline. The point event's position is described by a measure of, for instance, distance along the route. A *linear event* describes the extent of a linear feature along a route, such as a section of an oil pipeline that is under maintenance, and is rather similar in function to a section. A linear event is identified by measures denoting the positions where the event begins and ends along the route.

Route and event structures are of use in the description of application specific entities such as seismic lines and shot-point positions. Since conventional vector structures could not inherently describe the significance of discrete measurements along such structures. Along seismic lines the shot-points are the significant units of measurement but they are not necessarily regularly spaced or numbered along that line, so they do not necessarily denote distance along it or any predictable quantity. The use of routes and events becomes an elegant method of accommodating such intelligence within GIS since the route can be annotated with a measure that is independent of its inherent geometric properties.

13.6.4 Representing surfaces

The vector data model provides several options for surface representation: isolines (or contours), the *Triangulated Irregular Network*, or TIN, the *wireframe* and, although less commonly used, Thiessen polygons. Contours can only discretely describe the surfaces from

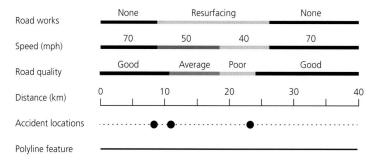

Fig. 13.12 Several routes representing different measures (linear and point events), created from and related to a pre-existing polyline feature representing, in this case, a road network.

which they were generated and so do not readily facilitate the calculation of further surface parameters, such as slope angle, or aspect (the facing direction of that slope); both of these are important for any kind of 'terrain' or surface analysis. The techniques involved in calculating contours are comprehensively covered in many other texts and so we shall skirt around this issue here.

13.6.4.1 The TIN surface model or tessellation

The TIN data model describes a three-dimensional surface composed of a series of irregularly shaped, linked but non-overlapping triangles. The TIN is also sometimes referred to as the *irregular triangular mesh* or *irregular triangular surface model*. The points that define the triangles can occur at any location, hence the irregular shapes. This method of surface description differs from the raster model in three ways. First, it is irregular in contrast with the regular spacing of the raster grid; second, the TIN allows the density of point spacing (and hence triangles) to be higher in areas of greater surface complexity (and requires fewer points in areas of low surface complexity); and last, also incorporates the topological relationships between the triangles.

The process of *Delaunay triangulation* (formulated by Boris Delaunay in 1934) is used to connect the input points to construct the triangular network. The triangles are constructed and arranged such that no point lies inside the *circumcircle* of any triangle (see Fig. 13.13). Delaunay triangulation maximises the smallest of the internal angles and so tends to produce 'fat' rather than 'thin' triangles.

As with all other vector structures, the basic components of the TIN model are the points or nodes, and these can be any set of mass points with which are stored a unique identifying number and one or more z values in their attribute table. Nodes are connected to their nearest neighbours by edges, according to the Delauney triangulation process. Left-right topology is associated with the edges to identify adjacent triangles. Triangles are constructed and break-lines can be incorporated to provide constraints on the surface.

The input mass points may be located anywhere. Of course the more closely spaced they are, the more closely the model will represent the actual surface. TINs are sometimes generated from raster elevation models, in which case the points are located according to an algorithm that determines the sampling ratio necessary to adequately describe the surface. Well-placed mass points occur at the main changes in the shape of the surface, such as ridges, valley floors, or at the tops and bottoms of cliffs. By connecting points along a ridge or cliff, a break-line in the surface can be defined. By way of a simple example, for an original set of mass points (as shown in Fig. 13.13(b) the resultant constructed TIN is formed with the input point elevations becoming the TIN node elevations [Fig. 13.13(c)]. If this TIN is found to have undersampled (and so aliased) a known topographic complexity, such as a valley, a break-line can be included, such as along a valley bottom. This then allows the generation of additional nodes at the intersection points with the existing triangles, and thereby further triangles are generated to better model the shape of the valley [Fig. 13.13(d)].

Once the TIN is constructed, the elevation of any position on its surface can be interpolated using the x, y, z coordinates of the bounding triangle's vertices. The slope and aspect angles of each triangle are also calculated during TIN construction, since these are constant for each one.

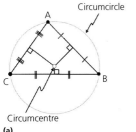 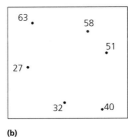 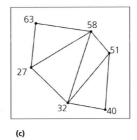 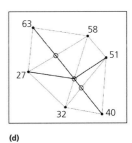

(a)   (b)   (c)   (d)

Fig. 13.13 (a) The Delauney triangle constructed from three points by derivation of the circumcircle and circumcentre; the position of the latter is given by intersection of the perpendicular bisectors from the three edges of the triangle. (b) Set of mass points; (c) the resulting TIN; and (d) the new set of triangles and nodes formed by addition of a break-line to the TIN.

Since elevation, slope and aspect are built into the structure of a TIN, they can very easily be displayed simply by symbolizing the TIN faces using these attributes. Four different edge types are stored within TINs; these may be referred to as *hard* (representing sharp breaks in slope or elevation), *soft* (representing gentle changes in slope), *outside* (generated automatically when the TIN is created, to constrain or close the outer triangles), and *regular edges* (all remaining edges other than the hard, soft or outside). A TIN surface can be displayed merely as edges, nodes, or faces, or as combinations of these features to enable the surface to be interpreted and any potential errors can be identified.

Another description of the surface can be given using Voronoi polygons (named after Georgy Voronoi). It can be said that the Delauney triangulation of a set of points is equivalent to the dual graph, or is the topological and geometric dual, of the Voronoi polygon tessellation (Whitney 1932). The Voronoi tessellation consists of a set of lines, within a plane, that divide the plane into the area closest to each of a set of points. The Voronoi polygons are formed by lines that fall at exactly half the distance between the mass points and are perpendicular to the Delauney triangle edges. Each Voronoi polygon is then assigned the *z* value of the point that lies at its centre. Once polygons are created, the neighbours of any point are defined as any other point whose polygon shares a boundary with that point. The relationship between the Voronoi polygons to the input points and Delauney triangles is shown in Fig. 13.14.

TINs allow rapid display and manipulation but have some limitations. The detail with which the surface morphology is represented depends on the number and density of the mass points and so the number of triangles. To represent a surface as well and as continuously as a raster grid, the point density would have to match or exceed the spatial resolution of the raster. Second, whilst TIN generation involves the automatic calculation of slope angle and aspect for each triangle, in the process of its generation, the calculation and representation of other surface morphological parameters, such as curvature, are rather more complex and generally best left in the realm of the raster. The derivation of surface parameters is dealt with in Chapter 17.

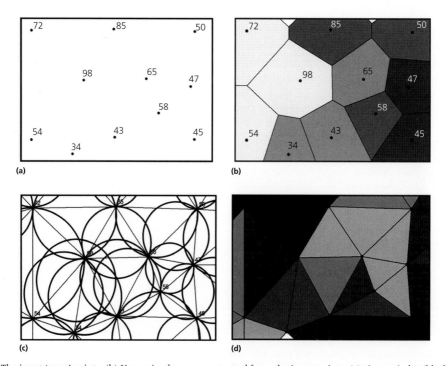

Fig. 13.14 (a) The input 'mass' points; (b) Voronoi polygons constructed from the input points; (c) circumcircles (black) constructed from the input points and Delauney triangles; and (d) the final TIN surface.

13.7 Data conversion between models and structures

There are sometimes circumstances when conversion between raster and vector formats is necessary for display and/or analysis. Data may have been captured in raster form through scanning, for instance, but may be needed for analysis in vector form (e.g. elevation contours needed to generate a surface, from a scanned paper topographic map). Data may have been digitised in vector form but subsequently needed in raster form for input to some multi-criteria analysis. In such cases it is necessary to convert between models and some consideration is required as to the optimum method, according to the stored attributes or the final intended use of the product. There are a number of processes that fall under this description, and these are summarised in Table 13.2.

13.7.1 Vector to raster conversion (rasterisation)

These processes begin with the identification of pixels that approximate significant points, and then pixels representing lines are found to connect those points. The locations of features are precisely defined within the vector coordinate space but the raster version can only approximate the original locations and so the level of approximation depends on the spatial resolution of the raster. The finer the resolution, the more closely the raster will represent the vector feature.

Many GIS programs require a 'blank' raster grid as a starting point for these vector-raster conversions where, for instance, every pixel value is zero or has a null or no data value. During the conversion, any pixels that correspond to vector features are then 'turned on'; that is, their values are assigned a numerical value to represent the vector feature.

13.7.1.1 Point to raster

For conversions between vector points and a discrete raster representation of the point data, there are several ways to assign a point's value to each pixel (as shown in Fig. 13.15). The first is to record the value of the unique identifier from each vector point. In this case, when more than one vector feature lies within the area of a single pixel, there is a further option to accept the value of either the first or the last point encountered since there may be more than one within the area of the pixel. Another is to record a value representing merely the presence of a point or points. The third choice is to record the most frequently occurring value. A further choice is

Table 13.2 Summary of general conversions between feature types (points, lines and areas), in vector/raster form.

Conversion Type	To Point/Pixel	To Line	To Polygon/Area
From point/pixel	Grid or lattice creation	Contouring, line scan conversion/ filling	Building topology, TIN, Thiessen polygons/ interpolation, dilation
From line From polygon/area	Vector intersection, line splitting Centre point derivation, vector intersection	Generalising, smoothing, thinning Area collapse, skeletonisation, erosion, thinning	Buffer generation, dilation Clipping, subdivision, merging

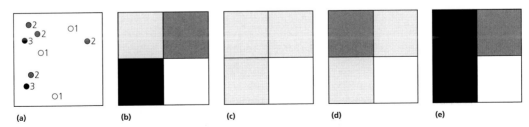

Fig. 13.15 (a) Input vector point map (showing attribute values); (b) to (e) are resulting rasters based on (b) the first point value encountered (1 in this case); (c) the presence of a point or points; (d) the most frequently occurring value (if there is no dominantly occurring value, then the lowest values is used); and (e) highest priority class rule (where attribute values 1-3 are used to denote increasing priority).

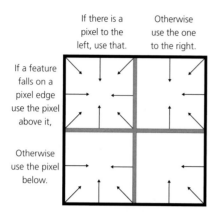

If there is a pixel to the left, use that.

Otherwise use the one to the right.

If a feature falls on a pixel edge use the pixel above it,

Otherwise use the pixel below.

Fig. 13.16 Boundary inclusion rules applied when a feature falls exactly on a pixel boundary. Arrows indicate the directional assignment of attribute values (modified after ESRI ArcGIS online knowledge base).

to record the sum of the unique identifying numbers of all vector points that fall with the area of the output pixel.

Point, line and polygon features can be converted to a raster using either textual or numerical attribute values. Only numbers are stored in the raster file; numbers in a value range that dictates how the raster data are quantised, as byte or integer data, for instance. So that if text fields are needed to describe the information in the output raster, an attribute table must also be used to relate each unique raster DN to its text descriptor. When pixels do not encounter a point, they are usually assigned a null (NoData) or zero value.

The above rules decide the value assigned to the pixel but further rules are required when points fall exactly on the boundary between pixels. These are used to determine which pixel will be assigned the appropriate point value. The scheme used within ESRI's ArcGIS is illustrated in Fig. 13.16, in which a kind of kernel and associated logical rules provide consistency by selecting which edge and which direction the value will be assigned.

Point to raster area conversions also include conversion to the 'continuous' raster model. This category generally implies interpolation or gridding, of which there are many different types. These processes are dealt with in Chapter 17 rather than here.

13.7.1.2 Polyline to raster

A typical line rasterising algorithm first finds a set of pixels that approximate locations of nodes. Then lines joining these nodes are approximated by adding new

pixels from one node to the next one, and so on, until the line is complete. As with points, the value assigned to each pixel when a line intersects it is determined by a series of rules. If intersected by more than one feature, the cell can be assigned the value of the first line it encounters, or merely the presence of a line (as with point conversions above), or of the line feature with the maximum length, or of the feature with the maximum combined length (if more than one feature with the same feature ID cross it), or of the feature that is given a higher priority feature ID (as shown in Fig. 13.17). Again, pixels that are not intersected by a line are assigned a null or NoData value. Should the features fall exactly on a pixel boundary, the same rules are applied to determine which pixel is assigned the line feature value, as illustrated in Fig. 13.16.

The rasterising process of a linear object initially produces a jagged line, of differing thickness along its length, and this effect is referred to as aliasing. This is visually unappealing and therefore undesirable but it can be corrected by anti-aliasing techniques such as smoothing. When rasterising a line or arc the objective is to approximate its shape as closely as possible; but, of course, the spatial resolution of the output raster has a significant effect on this.

13.7.1.3 Polygon to raster

The procedures used in rasterising polygons are sometimes referred to as *polygon scan conversion* algorithms. These processes begin with the establishment of pixel representations of points and lines that define the outline of the polygon. Once the outline is found, interior pixels are identified according to inclusion criteria; these determine which pixels that are close to the polygons edge should be included and which ones should be rejected. Then the pixels inside the polygon are assigned the polygon's identifying or attribute value. That value will be found from the pixel that intersects the polygon centre.

The inclusion criteria in this process may be one of the following, and their effects are illustrated in Fig. 13.18:

1 *Central point* rasterising where the pixel is assigned the value of the feature that lies at its centre.
2 *Dominant unit* or *largest share* rasterising where a pixel is assigned the value of the feature (or features) that occupies the largest proportion of that pixel.

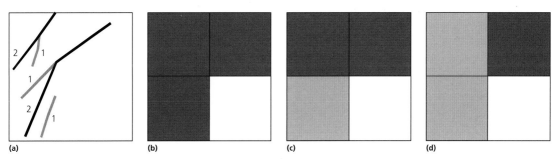

Fig. 13.17 (a) Input vector line map; (b)–(d) three different resulting raster versions based on a maximum length rule (or presence/absence rule) (b), maximum combined length rule (c), and a highest priority class rule (d), where the numbers indicate the priority attribute values.

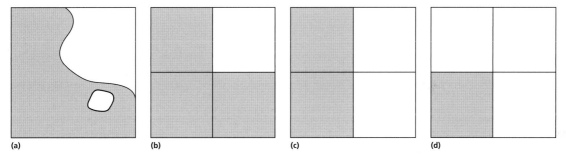

Fig. 13.18 (a) Input vector polygon map; (b)–(d) three different resulting raster versions based on a dominant share rule (b), a central point rule (c), and a most significant class rule (d).

3 *Most significant class* rasterising where priority can be given to a certain value or type of feature, such that if it is encountered anywhere within the area of a pixel, the pixel is assigned with its value.

When viewed at pixel level it can be seen that the inclusion criteria have quite different effects on the form of the raster version of the input vector feature.

Again, if the polygon features edge falls exactly on a pixel edge, special boundary rules are applied to determine which pixel is assigned the line feature value, as illustrated in Fig. 13.16.

13.7.2 Raster to vector conversion (vectorisation)

13.7.2.1 Raster to point

All non-zero cells are considered points and will become vector points with their identifiers equal to the DN value of the pixel. The input image should contain zeros except for the cells that are to be converted to points. The *x, y* position of the point is determined by the output point coordinates of the pixel centroid.

13.7.2.2 Raster to polyline

This process essentially traces the positions of any non-zero or non-null raster pixels to produce a vector polyline feature, summarised in Fig. 13.19. One general requirement is that all the other pixel values should be zero or a constant value. Unsurprisingly, the input cell size dictates the precision with which the output vertices are located. The higher the spatial resolution of the input raster, the more precisely located the vertices will be. The procedure is not simple, nor 'fool-proof' and generally involves several steps (and some editing) and is summarised as follows:

1 *Filling* – The image is first converted from a greyscale to a binary raster (through reclassification or thresholding), then any gaps in the features are filled by *dilation*.

2 *Thinning* – The line features are then *thinned, skeletonised* or *eroded*; that is, the edge pixels are removed in order to reduce the line features to an array or line of single but connected pixels.

3 *Vectorising* – The vertices are then created and defined at the centroids of pixels representing nodes, that is,

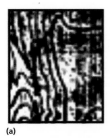   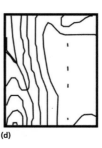 

(a)   (b)   (c)   (d)   (d)

Fig. 13.19 Illustration of polygon scan vectorisation procedures: (a) image improval by conversion to binary (bi-level) image; (b) thinning or skeletonisation process to reduce thickness of features to a single line of pixels; (c) vectorised lines representing complex areas; (d) lines collapsed but still including spurious line segments; and (e) the collapsed lines are smoothed or 'generalised' to correct pixelated appearance and line segments removed.

where there is a change in orientation of the feature. The lines are produced from any connected chains of pixels that have identical DN value. The resultant lines pass through the pixel centres. During this process, many small and superfluous vertices are often created and these must be removed. The vectors produced may also be complex and represent area instead of a true linear feature.

4 *Collapsing* – Complex features are then simplified by reducing the initial number of nodes, lines and polygons and, ideally, collapsing them to their centre lines. One commonly adopted method is that proposed by Douglas and Peuker (1974), which has subsequently been used and modified by many other authors.

5 *Smoothing* – The previous steps tend to create a jagged, pixelated line, producing an appearance that is rather unattractive to the eye; the vector features are then smoothed or generalised, to smooth this appearance and to remove unnecessary vertices. This smoothing may be achieved by reducing the number of vertices or using an averaging process (e.g. a 3-point or 5-point moving average).

13.7.2.3 Raster to polygon

This is the process of vectorising areas or regions from a raster. Unless the raster areas are all entirely discrete and have no shared boundaries, it is likely that the result will be quite complex. It is common therefore that this process leads to the generation of both a line and a polygon file, in addition to a point file representing the centres of the output polygons.

The polygon features are constructed from groups of connected pixels whose values are the same. The process begins by determining the intersection points of the area boundaries and then follows by generating lines at either external pixel centroids or the boundaries. A background polygon is also generated otherwise any isolated polygons produced will float in space. Again, such vectorisation procedures from raster images are usually followed by a smoothing or generalisation procedure, to correct the 'pixelated' appearance of the output vectors.

There are now a great many software suites available that provide a wealth of tools to perform these raster-vector conversions; some are proprietary and some are 'shareware', such as MATLAB (MathWorks), AutoCAD (Autodesk), R2V developed (Able Software Corp), Illustrator and Freehand.

13.8 Summary

Understanding the advantages and limitations of particular methods of representing data is key not only to effective storage and functionality but also to production of a reliable/accurate result. Knowing how each data structure is used to describe particular objects and what the limitations or drawbacks might be is also useful.

The ability to describe topological relationships is one significant advantage of the vector model. The ability to describe (numerically) the relationships between objects is very powerful indeed and has no real equivalent in raster processing. Spatial contiguity, connectivity and adjacency have to be inferred through the implicit relationship of one pixel to another, but there is no inherent intelligence such as is enabled through topology. This may not present too much of a disadvantage in geoscientific analyses, since we are often more interested in the relationship between different parameter values at one geographical position than in the relationship between objects that are geographically separated. In this sense, again it is the local or neighbourhood raster operations (discussed in Chapters 4 and 15) that gain our attention.

13.9 Questions

1 Why is it important to understand the scales or levels of data measurement when observing and recording information?

2 What are the essential differences between 'continuously' or 'regularly' sampled and 'discrete' data?

3 How should you decide on the most appropriate structure for a dataset?

4 What is an approriate level of accuracy when using or capturing digital data?

5 Why does the *shapefile* not include rigorous topological descriptions?

6 What are the differences between the topological vector model and the spaghetti vector model? What are the advantages and disadvantages of using each one?

7 How should more complex vector features (regions, routes and events) be organised?

8 What are the effects of varying spatial resolution on the raster representation of digital information and on accuracy?

9 What problems can arise during conversion between data models?

10 Do I need to invest in proprietary GIS software in order to work with geographic data?

CHAPTER 14

Defining a coordinate space

14.1 Introduction

Map projections and datums have been described very comprehensively by many other authors and we do not wish repeat or compete with these, but this topic is an extremely important one in the understanding of GIS construction and functionality, and as such cannot be ignored here. We have therefore attempted to overview the main principles and to concentrate on the practical applications, referring the reader, wherever possible, to other, more detailed texts.

To make GIS function, we must be able to assign 'coordinates' of time and location in a way that is generally understood. Calendar and temporal information can easily be included later as attributes or in metadata. Location is therefore the most important reference for data in a GIS. Several terms are commonly used to denote the positioning of objects; *georeference, geolocation, georegistration* and *geocoding* are all commonly used. The main requirement for a georeference is that it is unique, to avoid any confusion. Hence the address, or georeference, of the Royal School of Mines, Prince Consort Road, London SW7 2AZ, United Kingdom, only refers to one building; no other in the world has this specific address. Georeferencing must also be persistent through time, again to avoid both confusion and expense.

Every georeference also has an implication of resolution, that is, to a specific building or collection of buildings or a region. Many georeferencing systems are unique only within a limited area or domain, for example, within a city or county – for instance, there may be several towns called Boston, but only one is in Lincolnshire. Some georeferences are based on names

and others on measurements. The latter are known as 'metric' georeferences and they include latitude/longitude and other kinds of regular coordinate system; such metric coordinates are the more useful to us as geoscientists. Some coordinate systems involve combinations of metric measures and textual information. For instance, the six digits of a UK National Grid reference repeat every 100 km so that additional information (letters are used in this case) is required to achieve countrywide uniqueness. Metric systems provide infinitely fine resolution to enable accurate measurements. So how are systems for describing an object's location (and measurements associated with it) established? The answer requires a metaphorical step backward and consideration of the shape of the earth.

14.2 Datums and projections

The earth is a three-dimensional (3D) object, roughly oblately spherical in shape, and we need to represent that 3D shape in a two-dimensional (2D) environment, on paper or on a computer screen. This is the reason for the existence of a multitude of map projections – since you cannot project information without also distorting it, accurate measurements become potentially ambiguous. To achieve this projected 2D representation, two things need to be approximated: the shape of the earth and the transformations necessary to plot a location's position on the map.

14.2.1 Describing and measuring the earth
The system of latitude and longitude is considered the most comprehensive and globally constant method of description and is often referred to as the geographic

Image Processing and GIS for Remote Sensing: Techniques and Applications, Second Edition. Jian Guo Liu and Philippa J. Mason.
© 2016 John Wiley & Sons, Ltd. Published 2016 by John Wiley & Sons, Ltd.

system of coordinates, or *geodetic system*, and it is the root for all other systems. It is based on the earth's rotation about its centre of mass.

To define the centre of mass and thereby latitude and longitude (see Fig. 14.1), we must first define the earth's *axis* of rotation and the plane through the centre of mass perpendicular to the axis (the *equator*). Slices parallel to the axis but perpendicular to the plane of the equator are lines of constant *longitude*; these pass through the centre of mass and are sometimes also referred to as *great circles*. The slice through Greenwich defines zero degrees longitude and the angle between it and any other slice defines the angle of longitude. So that longitude then goes from 180 degrees west to 180 degrees east of Greenwich. A line of constant longitude is also called a *meridian*. Perpendicular to the meridians, slices perpendicular to the axis and passing through the earth but not through its centre are called *parallels*, also referred to as small circles, except for the equator (which is a great circle).

We also need to describe the shape of the earth, and the best approximation of this is the *ellipsoid of rotation* or *spheroid*. An ellipsoid is a type of quadric surface and is the 3D equivalent of an ellipse. It is defined, using x, y, z Cartesian coordinates by the following:

$$\frac{x^2}{a^2} + \frac{y^2}{b^2} + \frac{z^2}{c^2} = 1 \qquad (14.1)$$

The earth is not spherical but oblate and the difference between the ellipsoid or spheroid and a perfect sphere is defined by its *flattening (f)*, or its reduction in the shorter, *minor axis* relative to the *major axis*. *Eccentricity (e)* is a further phenomenon that describes how the shape of an ellipsoid deviates from a sphere (the eccentricity of a circle being zero). Flattening and eccentricity then have the following relationships:

$$f = \frac{(a-b)}{a} \quad \text{and} \quad e^2 = \frac{a^2 - b^2}{a^2} \quad \text{or} \quad e^2 = 2f - f^2 \ (14.2)$$

where a and b are the lengths of major and minor axes respectively (usually referred to as semi-axes or half

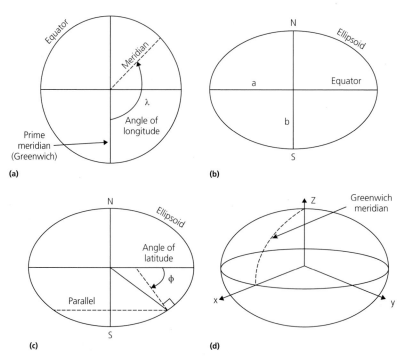

Fig. 14.1 Schematic representation of the earth, looking (a) down the pole, perpendicular to the equator; (b) and (c) perpendicular to the pole from the equator; and (d) obliquely at the earth. Illustrating the relationship between longitude and the meridians (a); between the equator and major and minor semi-axes (b); and between latitude and parallels (c); and the locations of the x, y and z axes forming Cartesian coordinates (d).

lengths of the axes). The actual flattening for the earth's case is about 1 part in 300. Some of the first ellipsoids to be established were not particularly accurate and were not actually centred on the earth's centre of mass. Fortunately, and rather ironically, the cold war, nuclear arms race and the need to target intercontinental missiles helped to drive the development of an international standard ellipsoid. The ellipsoid known as the *World Geodetic System of 1984* (or WGS84) is now accepted as this standard although many others are in use.

Latitude can now be defined as the angle between the equator and a line perpendicular to the ellipsoid, which ranges from 90 degrees north or south of the equator. Latitude is commonly given the Greek symbol phi (ϕ), and longitude lambda (λ). A line of constant latitude is known as a *parallel*. Parallels never meet since they are parallel to one another, whereas the meridians (lines of longitude) converge at the poles.

Longitude is more complex and only east-west measurements made at the equator are true. Away from the equator, where the line of latitude decrease in length, measures are increasingly shortened, by approximately the cosine of latitude. This means that at 30 degrees north (or south), shortening is *ca* 0.866, 0.707 at 45 degrees and 0.5 at 60 degrees. At 60 degrees north or south, one degree of longitude will represents 55 km ground distance.

14.2.2 Measuring height: The geoid

The true shape of the earth can be described as a surface that is perpendicular to the direction of gravity, or as an *equipotential surface*, in which there are fluctuations and irregularities according to variations in the density of the crust and mantle beneath. The spheroid or ellipsoid can therefore be thought of as a reasonable representation of the shape of the earth but not the true shape; this we refer to as the *geoid* (see Fig. 14.2), and it is defined

as an 'equipotential surface that most closely resembles mean sea level'. Mean sea level is used in this context since it responds everywhere to and is a perpendicular surface to gravity. In general, the differences between them (referred to as *separation*) are most great where undulations in terrain surface are of the greatest magnitude, but they are generally less than 1 m.

The significance of the geoid's variability is that it leads to different definitions of height from one place to another, since mean sea level also varies. Different countries may define slightly different equipotential surfaces as their reference. We should therefore be a little careful to distinguish between heights above geoid or spheroid. Fortunately the differences are small so that only highly precise engineering applications should be affected by them. For reference, *orthometric heights* and *spheroidal heights* are those defined with respect to the geoid and spheroid respectively. The variation in height from the geoid gives us topography.

14.2.3 Coordinate systems

Many different generic types of coordinates systems can be defined and the calculations necessary to move between them may sometimes be rather complex. In order of increasing complexity they can be thought of as follows. *Spherical coordinates* are formed using the simplest approximation of a spherical earth, where latitude is the angle north or south of the equatorial plane, longitude is the angle east or west of the prime meridian (Greenwich) and height is measured above or below the surface of the sphere. If high accuracy is not important, then this simple model may be sufficient for your purposes. *Spheroidal coordinates* are formed using a better approximation based on an ellipsoid or spheroid, as described in § 14.2.1, with coordinates of true latitude, longitude and height; this gives us the *geodetic coordinate* system. *Cartesian coordinates* involve values of

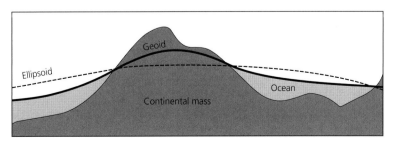

Fig. 14.2 Heights above spheroid and geoid.

x, y and z, as defined with their origin at the centre of a spheroid [see Fig. 14.1(d)]. The x and y axes lie in the equatorial plane, with x aligned with the Greenwich meridian, and z aligned with the polar axis. *Projection coordinates* are then defined using a simple set of x and y axes, where the curved surface of the earth is transformed onto a plane, the process of which causes distortions (discussed later).

Polar coordinates, generically, are those that are defined by distance and angle, with distance usually denoted r and angle θ. *Planar coordinates* refer to the representation of positions, as identified from polar coordinate positions, on a plane within which a set of orthogonal x, y axes is defined. The conversion between these polar and planar coordinates, for any particular datum, is relatively straightforward and the relationship between them is illustrated in Fig. 14.3. The following expression can be used to derive the distance (d) between two points a and b on an assumed spherical earth:

$$d(a,b) = R\arccos\left[\sin\phi_A\sin\phi_B + \cos\phi_A\cos\phi_B\cos(\lambda_A - \lambda_B)\right]$$
(14.3)

where R is the radius of the earth, λ is the longitude and ϕ the latitude. Generically, the x, y (planar) positions of the two points can be derived from the polar coordinates as follows:

$$x = r\sin\theta \quad \text{and} \quad y = r\cos\theta \qquad (14.4)$$

$$r = \sqrt{x^2 + y^2} \quad \text{and} \quad \theta = \tan^{-1}\left(\frac{y}{x}\right) \qquad (14.5)$$

where θ is measured clockwise from north. The Pythagorean distance between two points (a and b) can

then be found by the following, where the two points are located at (x_a, y_a and x_b, y_b):

$$d(a,b) = \sqrt{(x_a + x_b)^2 + (y_a + y_b)^2} \qquad (14.6)$$

14.2.4 Datums

A geodetic datum is a mathematical approximation of the earth's 3D surface and a reference from which other measurements are made. Every spheroid has a major axis and a minor axis, with the major axis being the longer of the two (as shown in Fig. 14.1), but is not in itself a *datum*. The missing information is a description of how and where the shape deviates from the earth's actual surface. This is provided by the definition of a *tie-point*, which is a known position on the earth's surface (or its interior, since the earth's centre of mass could be used), and its corresponding location on or within the ellipsoid.

Complications arise because datums may be global, regional or local, so that each is only accurate for a limited set of conditions. For a *global datum*, the tie point may well be the centre of mass of the earth meaning that the ellipsoid forms the best general approximation of the earth's shape, and that at any specific positions and accuracies may be quite poor. Such generalisations would be acceptable for datasets that are of very large or global extent. In contrast, a *local datum*, which uses a specific tie point somewhere on the surface, near the area of interest, would be used for a 'local' project or dataset. Within this area, the deviation of the ellipsoid from the actual surface will be minimal but at some distance from it may be considerable. This is the reason behind the development of the vast number of datums and projections worldwide. In practice, we chose a datum that is appropriate to our needs, according to the size and location of the area we are working with, to provide us with optimum measurement accuracy. Some common examples are given in Table 14.1.

It is worth noting that in many cases there may be several different versions of datum and ellipsoid under the same name, depending on when, where, by whom and for which purpose they were developed. The differences between them may seem insignificant at first glance but, in terms of calculated ground distances, can make very significant difference between measurements.

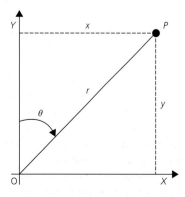

Fig. 14.3 The relationship between polar and planar coordinates in a single datum.

Table 14.1 Examples of geodetic datum and their descriptive parameters.

Datum	Spheroid	a	b	f	Tie Point
National Grid of Great Britain (OSGB)	Airy 1830	6377563	6356256.9	1/299.32	Herstmonceux
Pulkovo*	Krassovsky 1940	6378245	6356863	1/298.3	Pulkovo Observatory, Russia
M'poraloko 1951*	Clarke 1880	6378249.2	6356515.0	1/293.47	Libreville, Gabon

*Indicates that more than one variant exists for this datum

14.2.5 Geometric distortions and projection models

Since paper maps and geospatial databases are flat representations of data located on a curved surface, the *map projection* is an accepted means for fitting all or part of that curved surface to the flat surface or plane. This projection cannot be made without distortion of shape, area, distance, direction or scale. We would ideally like to preserve all these characteristics but we cannot, so we must choose which of those map characteristics should be represented accurately at the expense of the others, or whether to compromise on several characteristics.

There are probably 20 or 30 different types of map projections in common usage. These have been constructed to preserve one or other characteristic of geometry, as follows:

1 *Area* – many map projections try to preserve area, so that the projected region covers exactly the same area of the earth's surface no matter where it is placed on the map. To achieve this the map must distort scale, angles and shape.

2 *Shape* – there are two groups of projections that have either
   (a) *Conformal* property where the angles and the shapes of small features are preserved, that is, the scales in *x* and *y* are always equal (although large shapes will be distorted); or
   (b) *Equal-area* property where the areas measured on the map are always in the same proportion to the areas measured on the earth's surface but their shapes may be distorted.

3 *Scale* – no map projection shows scale correctly everywhere on the map, but for many projections there are one or more lines on the map where scale is correct.

4 *Distance* – some projections preserve neither angular nor area relationships but distances in certain directions are preserved.

5 *Angle* – although conformal projections preserve local angles, one class of projections (called *azimuthal projections*) preserve the easting and northing pair, so that angle and direction are preserved.

Scale factor (*k*) is useful to quantify the amount of distortion caused by projection and defined by the following:

$$k = \frac{\text{Projected distance}}{\text{Distance on the sphere}} \qquad (14.7)$$

The relationship will be different at every point on the map and in many cases will be different in each direction. It only applies to short distances. The ideal scale factor is 1, which represents no distortion at all; most scale factors approach but are less than 1.

Any projection can achieve one or other of these properties but none can preserve all, and the distortions that occur in each case are illustrated schematically in Fig. 14.4.

Once the property to be preserved has been decided, the next step is to transform the information using a 'projectable' or 'flattenable' surface. The transformation or projection is achieved using planar, cylindrical or cone shaped surfaces that touch the earth in one of a few ways; these form the basis for the three main groups of projection. Where the surface touches the earth at a point (for a plane), along a great circle (for a cylinder) or at a parallel (for a cone), projections of the *tangent* type are formed. Where the surface *cuts* the earth, rather than just touching it, between two parallels, a *secant* type of projection is formed (see Fig. 14.6). The conic is actually a general form, with azimuthal and cylindrical forms being special cases of the conic type.

In the planar type (such as a stereographic projection) where the surface is taken to be the tangent to one of the poles, the following relationship can be used to derive polar positions. All the equations given here assume a spherical earth for simplicity.

$$\theta = \lambda \quad \text{and} \quad r = 2\tan\left(\frac{\chi}{2}\right) \qquad (14.8)$$

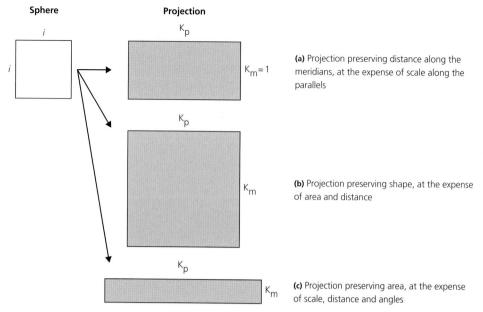

Fig. 14.4 Schematic illustration of the projection effects on a unit square of side length *i*, where K represents the scale along each projected side, and subscripts m and p represent meridian and parallel.

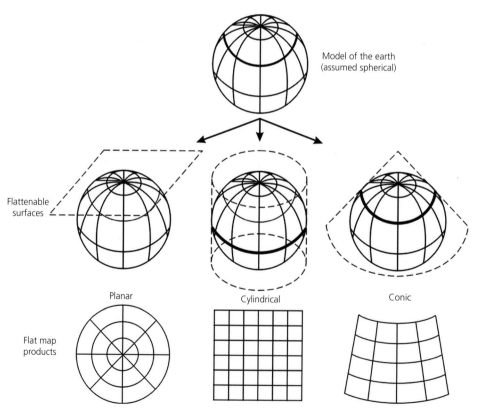

Fig. 14.5 The three main types of projection that are based on the tangent case: planar (left), cylindrical (centre) and conic (right), (Bonham-Carter, 2002. Reproduced with permission from Elsevier.)

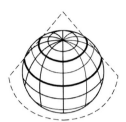

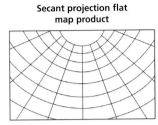

Flattenable surface
(two standard parallels are intersected by the cone)

Secant projection flat map product

Fig. 14.6 The conic projection when based on the secant case, where the conic surface intersects at two standard parallels (Bonham-Carter, 2002. Reproduced with permission from Elsevier.)

where χ represents the colatitude $(\chi = 90 - \phi)$; the resultant polar coordinates can then be converted to planar coordinates using 14.4. Of course, the plane could be a tangent to the earth at any point, not only at one of the poles.

For cylindrical projections, the axis of the cylinder may pass through the poles, for example, so that it touches the sphere at the equator (as in the Mercator). In this case, positions may be derived as follows:

$$x = \lambda \quad \text{and} \quad y = \log_e \tan\left(\frac{\pi}{4} + \frac{\phi}{2}\right) \quad (14.9)$$

For conic projections of the tangent type, the following can be used to derive positions, assuming one standard parallel at a colatitude of χ_0:

$$r = \theta = \lambda \cos(\chi_0) \quad \text{and} \quad \tan(\chi_0) + \tan(\chi + \chi_0) \quad (14.10)$$

14.2.6 Major map projections

Based on the projection models described so far, there are three broad categories of projection: equidistant, equal area and conformal.

(a) Cylindrical equidistant projection

In this projection the distances between one or two points and all other points on the map differ from the corresponding distances on the sphere by a constant scaling factor. The meridians are straight and parallel, and distances along them are undistorted. Scale along the equator is true but on other parallels is increasingly distorted towards the poles; shape and area are therefore increasingly distorted in this direction. An example is shown in

Fig. 14.7(a). The *plate carrée* is an example of this type of projection.

(b) Cylindrical equal-area projection

Here scale factor is a function of latitude so that, away from the equator, distances cannot be measured from a map with this type of projection. Shape distortion is extreme towards the poles. Scale factor is almost 1 on both meridians and parallels, as follows, but only near the equator.

$$k_m k_p = 1 \quad (14.11)$$

Scale factor along the parallels is given by $sec\phi$ and distortion along the meridians by $cos\phi$. An example is the *Lambert cylindrical equal-area* projection [e.g. Fig. 14.7(b)].

(c) Conformal: Mercator

In this case, the poles reach infinite size and distance from the equator, producing a very distinct looking map. The meridians are parallel and angles are preserved so these maps are acceptable for navigation [illustrated in Fig. 14.7(c) and Fig. 14.8]. The scale factor (k) at any point on the map is a function of latitude, as follows:

$$k_p = k_m = \sec\varphi \quad (14.12)$$

The distortion that occurs towards (and that is infinite at) the poles is a product of the way the Mercator is constructed, as illustrated in Fig. 14.8(a). This can be thought of by considering an area between two meridians, both at the equator and at a high latitude. The equatorial area appears square with sides of 1-degree length in both longitude and latitude. At high latitudes the area covers the same distance in latitude as at the equator and is still 1-degree wide but is narrower and covers a distance shown by x in Fig. 14.8, so the area is no longer square but

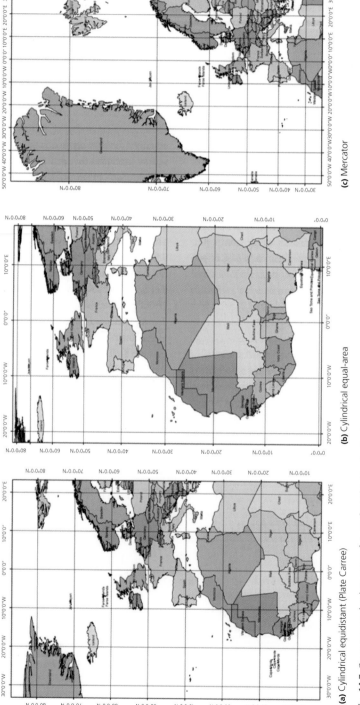

(a) Cylindrical equidistant (Plate Carree) **(b)** Cylindrical equal-area **(c)** Mercator

Fig. 14.7 Common projection examples, as described in the text.

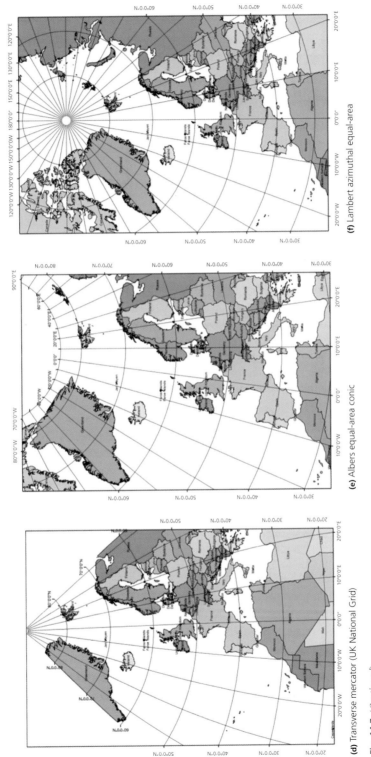

(f) Lambert azimuthal equal-area

(e) Albers equal-area conic

(d) Transverse mercator (UK National Grid)

Fig. **14.7** (*Continued*)

Sphere   Map

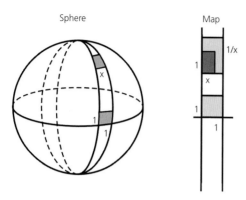

Fig. 14.8 The relationship between meridians and scale in the Mercator projection.

rectangular. On the projected map [Fig. 14.8(b)] the two meridians are shown, representing a difference of 1-degree longitude, and they are parallel, so when the rectangle is projected it must be enlarged by a factor 1/x to fit between them. This is the reason that areas are larger at higher latitudes and when this scaling is done repeatedly, from the equator northwards, the Mercator coordinate net is produced. On the earth, a 1-degree distance at the equator is about 111 km.

(d) Conformal: Transverse Mercator (TM)

This projection is a modification of the standard Mercator designed for areas away from equator. In this instance, the cylinder is rotated through 90° so that its axis is horizontal and its line of contact with the earth is no longer the equator but a meridian. The scale factor is same in any direction, and is defined by:

$$k = \sec \theta \qquad (14.13)$$

Where θ is equivalent to latitude (ϕ), except that it is the angular distance from the central meridian rather than from the equator. In this case the meridians are no longer parallel or straight (except the central meridian), and the angle made between the meridians and the central meridian (which is grid north) can be described as *convergence* (γ). For the sphere projection convergence is defined by the following:

$$\gamma = \delta\lambda = \sin \phi \qquad (14.14)$$

Then by turning the cylinder horizontally, the central meridian of a TM projection could be based on any line of longitude around the globe, so as to

be appropriate to any particular country or region. The British National Grid (OSGB) is a good example of this type of projection; the UK covers greater distance in the north-south direction than it does in the east-west direction, so to prevent north-south distortion (as would be produced by a Mercator), a TM with its central meridian at zero degrees longitude (i.e. Greenwich) is used [see Fig. 14.7(d)]. Further rules can then be applied to the TM to produce equal-area, equidistant or conformal projections.

(e) Conic projections

In these projections, the meridians are straight, have equal length and converge on a point that may or may not be a pole. The parallels are complete concentric circles about the centre of the projection. Such projections may be conformal, equidistant or equal-area. Examples include the *Albers equal-area conic* projection [illustrated in Fig. 14.7(e)].

(f) Planar (azimuthal) projections

Azimuthal projections preserve the azimuth or direction from a reference point (the centre of the map) to all other points, that is, angles are preserved at the expense of area and shape. In the polar case, the meridians are straight and the parallels are complete concentric circles. Scale is true only near the centre, with the map being circular in form. Such projections may be conformal, equidistant or equal-area. Examples include the *Lambert azimuthal equal-area* projection [Fig. 14.7(f)], or the *stereographic azimuthal* projection.

(g) Conformal: Universal Transverse Mercator (UTM)

A further modification of the Mercator allows the production of the UTM projection system (illustrated in Fig. 14.9). It is again projected on a cylinder tangent to a meridian (as in the TM) and, by repeatedly turning the cylinder, about its polar axis, the world can be divided into 60 east-west zones, each 6 degrees longitude in width. The projection is conformal so that shapes and angles within any small area will be preserved. This system was originally adopted for large-scale military maps for the world, but it is now a global standard and again is useful for mapping large areas that are oriented in a north-south direction.

Projected UTM grid coordinates are then established that are identical between zones. Separate grids are also established for both northern and southern

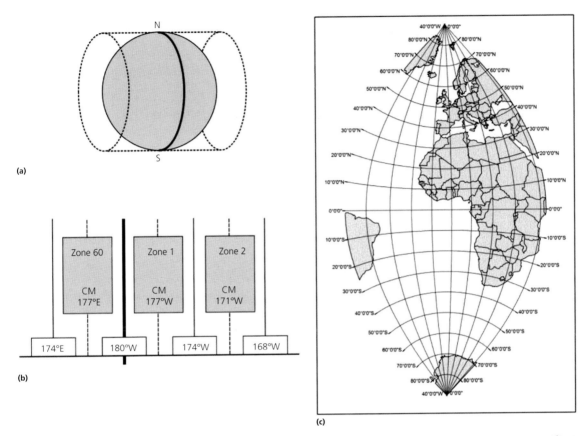

Fig. 14.9 The UTM system: (a) the cylindrical projection rotated about its axis; touching the earth along a meridian (any meridian) rather than the equator; (b) the arrangement of 6-degree zones and their central meridians, starting at 180°W (there are 60 zones in all); (c) the arrangement of parallels and meridians, in UTM zone 30, about the central meridian, which is in this case 0° at Greenwich; appropriate for the UK, western Europe and north-west Africa (separate projected grid coordinates are established for northern and southern halves of each UTM zone so that negative northings are never encountered).

halves of each UTM zone to ensure there are no negative northings in the southern hemisphere. Hence, when quoting a UTM grid reference, it is essential to state eastings, northings, zone number and the hemisphere (north or south) to ensure clarity.

14.2.7 Projection specification

Several other parameters are necessary to precisely define a particular projection. For example, the Mercator requires a *central meridian* to be specified, and this is given as the line of longitude on which the projection is centred (it is commonly expressed in radians). Coordinates defining a *false northing and easting* are also required, to position the origin of the projection. These are normally given in metres and are used to ensure

that negative values are never encountered. A *scale factor* is also given, as a number by which all easting and northing values must be multiplied, to force the map onto the page. A conic projection of the secant type requires the specification of the *first and second parallel* (as latitudes thereof). Some projections also require the origin of the projection to be given as a latitude and longitude coordinate pair; these are sometimes referred to as the *central meridian* and *central parallel*.

The *Geodetic* projection is a special type of map projection (the simplest possible map projection) where the easting value is exactly longitude, and the northing value is exactly latitude. Since it does not preserve angle, scale or shape, it is not generally used for cartographic purposes but is commonly used as the default

Table 14.2 Some examples of projected coordinates systems.

Projection	Central Meridian (Longitude)	Central Parallel (Latitude)	False Easting	False Northing	Scale Factor
OS GB	2°W	49°N	+400,000	−100,000	0.9996012
UTM $\phi > 0°$	Zonal	0°	+500,000	0	0.9996
UTM $\phi < 0°$	Zonal	0°	+500,000	+10,000,000	0.9996
GK TM zone11	63°E	0°	+500,000	0	1

option for recording simple coordinate positions since it is the only globally consistent system. The datum associated with this projection, to preserve its global applicability, is always the WGS84.

14.3 How coordinate information is stored and accessed

Vector data stores its coordinate information implicitly with each node position in real-world coordinates and these are used directly to plot positions and to re-project from one coordinate system to another. Raster data on the other hand has a regular local row and column number system for each pixel so that internally, the geometry and position of a position are implicit. Externally however, its world geographic reference must be explicitly stated; this requires the geographic location of the image origin and the ground distance represented by each pixel. A transformation is then performed, that converts local image coordinates to real-world coordinates for each pixel location using the geometric operations described in Chapter 9. This transformation information is also stored explicitly.

There are many binary image formats, such as IMG, BSQ, BIL, BIP, GeoTIFF and various ASCII grid formats, which are now accepted as standards. Some of these store the georeferencing information in the header portion of the actual image data file. Other image formats store the information in a separate ASCII text file; sometimes referred to as the *header* or *world file*, since it contains the real-world transformation information used by the image. Since these files are in ascii text format, they can be created or edited with any text editor. Most GIS/mapping/CAD software will detect and read this information automatically, if it is present. The image-to-world transformation is accessed each time an image is displayed and visualised. See also the

raster data model descriptions on GDAL (www.gdal.org/gdal_datamodel.html).

The contents of a world file, for a projected raster image with plane coordinates, will look something like this:

10.000	A The x scale (dimension of a pixel in x direction)
0.00000000000000	B Rotational term
0.00000000000000	C Rotational term
−10.000	D The negative of y scale (dimension of a pixel in y direction)
567110.113454530548	E Translational terms (centre coordinates of the top left pixel)
9415540.44549960335	F Translational terms (centre coordinates of the top left pixel)

Most GIS software, when this file is present, will perform an affine transformation of the following form:

$$x1 = Ax + Cy + E \quad \text{and} \quad y1 = Bx + Dy + F \qquad (14.15)$$

Where x1 and y1 are the calculated coordinates of a pixel on the map, x and y are the pixel column and row numbers in the image, A is the x scale and D is the negative of y scale (dimension of a pixel in map units in x and y directions), B and C are rotational terms, and E and F are translational terms (centre coordinates of the top left pixel). Notice that the y scale (D) is negative; this is because the origins of a geographic coordinate system and of a raster image are different. The geographic origin is usually in the lower-left corner whereas the origin of an image is the upper-left corner, so that y-coordinate values in the map increase from the origin upward and raster row values increase from the origin downwards.

Programs such as ERMapper, ArcInfo (ArcGIS), ENVI, ERDAS Imagine and PCI Geomatica contain routines to convert both image and vector data between projections and datums, while some programs only support re-projection of vector data.

14.4 Selecting appropriate coordinate systems

It is common to receive geospatial data created or acquired by someone else, and we frequently need to overlay datasets that are in different or unknown or unspecified coordinate systems. Establishing the data's projection and datum becomes of vital importance before beginning any work, and here again the use and upkeep of *metadata* is vital. The metadata should always contain information on the datum and projection of a geospatial dataset in addition to other information necessary to document its provenance. Such information is often found in the *header file* (or metadata) in the case of raster images, or failing that the dataset may have to be visually compared to another dataset of known datum and projection.

When creating a new dataset, or defining a new project, selection of the most appropriate map projection for any input data it is very important. Selection considerations include the extent of the project area (e.g. the world, a continent or a small region), its location (e.g. polar, mid-latitude or equatorial) and its predominant extent (e.g. circular, east-west axis, north-south axis or oblique axis). If there is a pre-existing base layer, such as a scaled map or georectified image, this may form the framework for all other data that is added or created within the project.

The UTM projection is nearly correct in every respect, for relatively small project areas, and is a very common choice. There are also some general *rules of thumb* that are useful for continental-scale, or smaller, regions in mid-latitude zones. For instance, for areas with a dominantly north-south axis, UTM will provide conformal accuracy. For those with an east-west axis, selection of a Lambert conformal projection will give conformal accuracy, or Albers equal area will preserve area. Areas with an oblique axis could be represented well by an oblique Mercator projection (for conformal accuracy); and for an area that has equal extent in all directions, a polar or stereographic projection will give conformal accuracy, or a Lambert azimuthal projection could be chosen for area preservation.

The standard Gauss-Kruger (GK) projection is sometimes also known as the Pulkovo 1942 Gauss-Kruger projection. Gauss-Kruger projections are implemented as a National Grid in Germany, referred to as the DHDN Gauss-Kruger, and are also commonly used in Russia and China. The GK projection is particularly suited for this part of the world since these countries occupy large continental masses of considerable east-west extent and relatively limited north-south extents. A zonal system, similar to that of the UTM but with zones of 3-degrees width instead of 6, is used for Russia and China, to ensure minimal distortion and maximum conformality across the continent.

14.5 Questions

1 Why is it important to establish a structural framework for the representation of digital data?
2 What properties of the spherical earth are affected by the use of map projections?
3 How will you decide on the most appropriate framework for a project?
4 What are the advantages and disadvantages of the UTM system of reference?
5 What happens to the area represented by a pixel when it is transformed between geodetic and projected coordinates, at the equator and at high latitudes?
6 How have you recorded your geographic location data (field localities) in the past? How accurately did you record their positions? What coordinate system did you choose? And how would you choose to record and display them now?

CHAPTER 15

Operations

15.1 Introducing operations on spatial data

It's probably fair to say that the average GIS suite contains far more functions than most people will ever need or be aware of! A long list of these, with some descriptions, would be useful but would make rather dull reading, and so it is helpful for the purposes of understanding to categorise them in some way.

The way processes are carried out depends on how the data are structured and stored, and since we have already described the fundamental differences between vector and raster data (in Chapter 13), we should begin to understand the implications of using one method of description over another. The difference between operations on raster and vector data could also be thought of as a dimensional one. Since one raster represents the variance of one attribute, operations affecting one raster pixel value occur in a 'vertical' dimension, through the stack of raster data attributes, or in a 'horizontal' dimension on one attribute only (Fig. 15.1), or a combination of these two. Operations on vector features attribute values occur in an n-dimensional space since the features' values are stored in an attribute table that has n number of attribute fields and the spatial extents of the input features are neither regular nor necessarily equal.

Operations could alternatively be grouped on the basis of being either 'spatial' or 'non-spatial'. Those falling into the non-spatial category could include reclassifications or statistical operations carried out within tables. Truly spatial operations could include neighbourhood processes such as convolution filtering (see also Chapter 4), or functions used to enhance the contrast of a raster image, since these involve the statistics of a region or of the whole image (see also Chapter 3). In these cases, the processing itself involves manipulation of data in a spatial context and produces results that reveal spatial patterns more clearly.

A further, rather useful hybrid classification of analytical operations could be made on the basis of the type of output, that is, map or table, as well as on whether spatial variables were involved or not, as summarised in Table 15.1. The simple re-assignment of values in one raster to another scheme of values in a new raster (i.e. reclassification) could be considered to produce map output but not necessarily involve spatial attributes; or at the opposite end of the spectrum, the calculation of spatial auto-correlation amongst point values to produce a variogram provides tabular output and involves the use of attributes with spatial qualities.

Since the objectives of many 'non-spatial' operations may also lead to and be part of wider spatial analyses, it is perhaps more useful to describe them as operations that are carried out on spatial data. Clearly there is a grey area here, and this is the reason for referring simply to operations that are performed upon data that is spatial in its nature rather than classifying the operations themselves.

Another perhaps more instructive way of classifying or grouping operations is in terms of the number of input data 'layers' involved; one, two or more (Fig. 15.1). In general, those operations applied to raster data are essentially the same as those applied in image processing, and here is the overlap between image processing and GIS. For instance, multi-layer raster data operations are essentially the same as local point operations that one would carry out on multi-spectral imagery except that the attributes carry different meaning.

Image Processing and GIS for Remote Sensing: Techniques and Applications, Second Edition. Jian Guo Liu and Philippa J. Mason.
© 2016 John Wiley & Sons, Ltd. Published 2016 by John Wiley & Sons, Ltd.

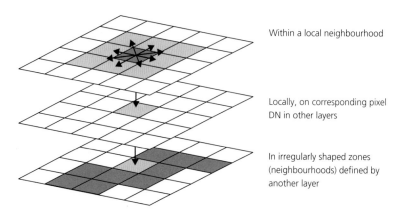

Within a local neighbourhood

Locally, on corresponding pixel
DN in other layers

In irregularly shaped zones
(neighbourhoods) defined by
another layer

Fig. 15.1 Stacked georeferenced rasters indicating multi-layer operations, both local, focal and neighbourhood.

Table 15.1 Operations categorised according to their spatial or non-spatial nature (after Bonham-Carter 2002).

Output	Spatial Attributes involved	
	Yes	**No (Not Necessarily)**
Map or image	Neighbourhood processing (filtering), zonal and focal operations, mathematical morphology	Reclassification, re-scaling (unary operations), overlay (binary operations), thresholding and density slicing
Tabular	Spatial auto-correlation and variograms	Various tabular statistics (aggregation, variety) and tabular modeling (calculation of new fields from existing ones), scatter graphs

15.2 Map algebra concepts

Map algebra is an informal and commonly used term for manipulating continuously sampled (i.e. raster) variables defined over a common area. It is also a term used to describe calculations within and between GIS data layers, according to some mathematical expression, to produce a new layer, and one that was first described and developed by Tomlin (1990). Map algebra can also be used to manipulate vector map layers, sometimes resulting in the production of a raster output. Although no new capabilities are brought to the GIS, map algebra provides an elegant way to describe operations on GIS datasets. It can be thought of simply as algebra applied to spatial data that, in the case of raster data, is facilitated by the fact that a raster is a georeferenced numerical array.

15.2.1 Working with Null data
An essential part of map algebra or spatial analysis is the coding of data in such a way as to eliminate certain areas from further contribution to the analysis. For instance, if

the existence of low grade land is a prerequisite for a site selection procedure, we then need to produce a layer in which areas of low grade land are coded distinctively so that all other areas can be removed. One possibility is to set the areas of low grade land to a value of 1 and the remaining areas as zero. Any processes involving multiplication, division or geometric mean that encounters the zero value will then also return a zero value and that location (pixel) will be removed from the analysis. The opposite is true if processing involves addition, subtraction or arithmetic mean calculations, since the zero value will survive through to the end of the process. The second possibility is to use a *Null* or *NoData* value instead of a zero. The Null is a special value that indicates that no information is associated with the pixel position, that is, there is no digital numerical value. In general, unlike zero, any expression will produce a Null value if any of the corresponding input pixels have Null values.

Many functions and expressions simply ignore Null values, however, and in some circumstances this may be useful, but it also means that a special kind of function must be used if we need to test for the presence of (or to assign) Null values in a dataset. For instance, within Environmental Systems Research Institute Inc.'s (ESRI) ArcGIS, the function *ISNULL* is used to test for the existence of Null values and will produce a value of 1 if Null, or zero if not. Using ERMapper's formula editor, Null values can easily be assigned, set to other values, made visible or hidden. Situations where the presence of Nulls is disadvantageous include instances where there are unknown gaps in the dataset, perhaps produced by measurement error or failure. Within map algebra, however, the Null value can be used to great advantage since it enables the selective removal or retention of values and locations during analysis.

15.2.2 Logical and conditional processing

These two types of process are quite similar and they provide a means of controlling function behaviour; allowing us to evaluate some criterion and to specify what happens next if the criterion is satisfied or not.

Logical processing describes the tracking of true and false values through a procedure. Normally, in map algebra, a non-zero value is considered to be a logical true, and zero, a logical false. Some operators and functions evaluate input cell values and return either logical true values (1) or logical false values (zero); for example, relational and Boolean operators. The return of a true or false value acts as a switch for one or other consequence within the procedure.

Conditional processing allows the same but with more flexibility. A particular action can be specified, according to the satisfaction of various conditions, that is, if the conditions are evaluated as true then one action is taken, and an alternative action is taken when the conditions are evaluated as false. The conventional *if-then-else* statement is a simple example of a conditional statement, such as

$$\text{if } i < 16 \text{ then } 1 \text{ else null} \quad \text{where } i = \text{input pixel dn}$$

Conditional processing is especially useful for creating analysis 'masks'. In Fig. 15.2, each input pixel value is tested for the condition of having a slope angle equal to or less than 15 degrees. If the value tests true, that is, slope angle is 15 degrees or less, a value of 1 is assigned to the output pixel. If it tests false, that is, exceeds 15 degrees, a Null value is assigned to the output pixel. The output could then be used as a mask to exclude areas of

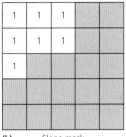

(a)   Slope angle (degrees)   **(b)**   Slope mask

Fig. 15.2 Logical test of slope angle data, for the condition of being no greater in value than 15 degrees: (a) slope angle raster; and (b) slope mask (pale grey blank cells indicate Null values).

steeper slopes and allow through all areas of gentle slopes, such as might be required in fulfilling the prescriptive criteria for a site selection exercise.

15.2.3 Other types of operator

Expressions can be evaluated using *arithmetic operators* (addition, subtraction, logarithmic, trigonometric) and performed on spatially coincident pixel digital number (DN) values within two or more input layers. Generally speaking, the order in which the input layers are listed denotes the precedence with which they are processed; the input or operator listed first is given top priority and is performed first, with decreasing priority from left to right.

Relational operators enable the construction of logical functions and tests by comparing two numbers and returning a true value (1) if the values are equal or false (0) if not. For example, this operator can be used to find locations within a single input layer with DN values representing a particular class of interest. These are particularly useful with discrete or categorical data.

Boolean operators, for example, AND, OR and NOT, also enable the logical functions and tests of set theory to be applied to spatial data. Like relational operators, Boolean operators also return true (1) and false (0) values. They are performed on two or more input layers to select or remove values and locations from the analysis. For example, to satisfy criteria within a slope stability model, Boolean operations could be used to identify all locations where values in one input representing slope are greater than 40 degrees AND where values in an aspect layer are between 030 and 180 degrees [as in Fig. 15.3(a)]. In this way a number of

Table 15.2 Summary of common arithmetic, relational, Boolean, power, logical and combinatorial operators used in map algebraic expressions.

Arithmetic		Relational (Return True/False)		Boolean (Return True/False)	
+	Addition	==, EQ	Equal	^, NOT	Logical complement
-	Subtraction	^=, <>, NE	Not equal	&, AND	Logical And
*	Multiplication	<, LT	Less than	|, OR	Logical Or
/	Division	<=, LE	Less than/equal to	!, XOR	Logical Xor
MOD	Modulus	>, GT	Greater than		
		>=, GE	Greater than/equal to		
Power		**Logical**		**Combinatorial**	
Sqrt	Square root	DIFF	Logical difference	CAND	Combinatorial And
Sqr	Square	IN {list}	Contained in list	COR	Combinatorial Or
Pow	Raised to a power	OVER	Replace	CXOR	Combinatorial Xor

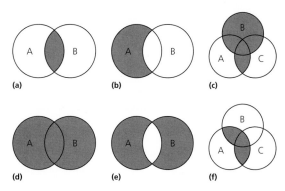

Fig. 15.3 Use of Boolean rules and set theory within map algebra; the circles represent the feature classes A, B and C, illustrating Boolean applied to geographic datasets (especially rasters) to extract or retain values: (a) A AND B (intersection or minimum); (b) A NOT B; (c) (A AND C) OR B; (d) A OR B (union or maximum); (e) A XOR B; and (f) A AND (B OR C).

selection criteria can be evaluated, and the areas that pass all the tests can be retained as being suitable or 'stable', as in the case of a slope stability model.

Logical comparisons of two inputs can also be useful, allowing the assignment of one or other of the input values under certain circumstances. For instance, for two inputs (A and B) *A DIFF B* assigns the value from A to the output pixel if the values are different, or a zero if they are the same. Alternatively, thre expression *A OVER B* assigns the value from A if a non-zero value exists, if not then the value from B is assigned to the output pixel.

Combinatorial operators find all the unique combinations of values among the attributes of multiple input rasters, and assign a unique value to each combination

in the output layer. The output attribute will contain fields and attributes from all the input layers.

All these operators can be used, with care, alone or sequentially, to selectively remove, test, process, retain or remove values (and locations) from datasets alone or from within a spatial analytical procedure.

15.3 Local operations

Local operations involve the production of an output value as a function of the value(s) at the corresponding locations in the input layer(s). As described in Chapter 3, these can be considered point operations when performed on raster data – that is, they operate on a pixel and its matching pixel position in other layers, as opposed to groups of neighbouring pixels, which are dealt with in § 15.4. They can be grouped into the following types that

- derive statistics from multiple input layers (e.g. mean, median, minority);
- combine multiple input layers;
- identify values that satisfy specified criteria;
- identify the number of occurrences that satisfy specified criteria;
- identify the position in an input list that satisfy a specified criteria.

All types of operator previously mentioned can be used in these contexts.

Commonly they are further subdivided according to the number of input layers involved at the start of the process. They include primary operations where nothing exists at the start, to n-ary where n number of layers may be involved and they are summarised in Table 15.3.

Table 15.3 Examples of local operations within groups based on the number of input layers.

Type	Function:	Examples
Primary	Creation of a layer from nothing	Rasters of constant value or containing randomly generated values
Unary	Conversion of units of measurement and as intermediary steps of spatial analysis	Re-scaling, negation, comparing or applying mathematical functions, reclassification, e.g [input1(int)]* −1
Binary	Operations on ordered pairs of numbers in matching pixels between layers	Arithmetic and logical combinations of rasters, e.g.[input1(int)]*[input2(float)]
N-ary	Comparison of local statistics between several rasters (many to one or many to many)	Change or variety detection, e.g. localstats(variety) {[input1(int)], [input2(int)], [input3(int)],}

15.3.1 Primary operations

This description refers primarily to operations used to generate a layer, conceptually, from nothing, for example, the creation of a raster of constant value, or of a blank (empty) raster, or one containing randomly generated numbers, such as could be used to test for error propagation through some analysis. An output pixel size, extent, data type and output DN value (either constant or random between set limits) must be specified for the creation of such a new layer.

15.3.2 Unary operations

These operations act on one layer to produce a new output layer and they include tasks such as *re-scaling*, *negation* and *reclassification*. Rescaling is especially useful in preparation for *multi-criteria analysis* where all the input layers should have consistent units and value range. For instance, in converting between byte data, with 0–255 value range, to a percent scale (0–100) or a range of between 0 and 1, and vice versa. Negation is used in a similar context, in modifying the value range of a dataset from being either entirely positive to entirely negative and vice versa. Reclassification is especially significant in data preparation for spatial analysis and so deserves rather more in depth description but all these activities can be and are commonly carried out in GIS and image processing systems.

15.3.2.1 Reclassification

This involves the process of re-assigning a value, a range of values, or a list of values in a raster to new output values, in a new output raster. For instance, if one class (or group or range of classes) is more important to us than the other classes, its original values can be assigned a specific value and all the others can be changed into a different (background) value. This process creates a discrete raster from either a continuous one or another discrete raster. Reclassification can also be applied to both vector and raster objects.

In the case of discrete raster data, a reclassification may be required to produce consistent units among a set of input raster images, in which case a one-to-one value change may be applied. The output raster would look no different, spatially, from the input, having the same number of classes, but the values would have changed.

Different classes or types of feature may be reclassified according to some criteria that are important to the overall analysis. During the reclassification process, weighting can be applied to the output values to give additional emphasis to the significant classes, whilst at the same time reducing the significance of other classes.

The example in Fig. 15.4(a) shows a discrete raster representation of a geological map in which nine lithological units are coded with values 1 to 9 and these have been labelled for the purposes of presentation, according to their name, rock type and ages. For the purposes of some analysis it may be necessary to simplify this lithological information, for example, according to the broad ages of the units – Pre-Cambrian, Palaeozoic and Mesozoic. The result of such a simplification is shown in Fig. 15.4(c); now the map has only three classes and it can be seen that the older rocks (Pre-Cambrian and Palaeozoic) are clustered in the south-western part of the area, with the younger rocks (Mesozoic) forming the majority of the area as an envelope around the older rocks. So the simplification of the seemingly quite complex lithological information shown in Fig. 15.4(a) has revealed spatial patterns in that information that are of significance and that were not immediately apparent beforehand.

Figure 15.4(c) shows a second reclassification of the original lithological map, this time on the basis of relative

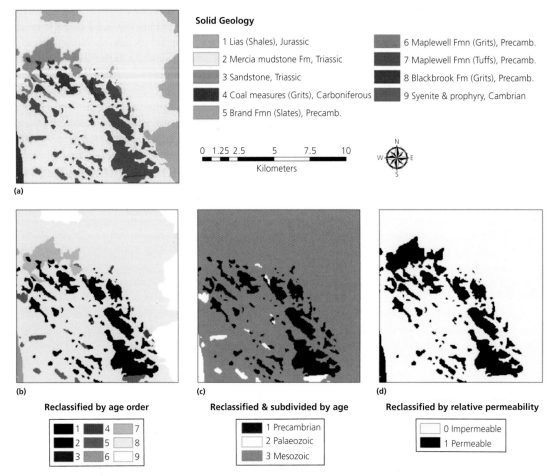

Fig. 15.4 (a) Discrete raster representation of a geological map, with nine classes representing different lithologies; (b) one-to-one reclassification by age order (1 representing the oldest, to 9 the youngest); (c) a reclassified and simplified version where the lithological classes have been grouped and re-coded into three broad age categories (Pre-Cambrian, Palaeozoic and Mesozoic); (d) a second reclassified version where the lithologies have been grouped according to their relative permeability, with 1 representing impermeable rocks and 0 permeable; such an image could be used as a mask.

permeability. The information is again simplified by reducing the number of classes to two, impermeable and permeable. Such a map might form a useful intermediary layer in an exercise to select land suitable for waste disposal but also illustrates that subjective judgements are involved at the early stage of data preparation. In the very act of simplifying information, we introduce bias and error to the analysis. We also have to accept the assumptions that the original classes are homogeneous and true representations everywhere on the map, which they may not be. In reality there is almost certainly heterogeneity within classes and the boundaries between the classes may not actually be as rigid as our classified

map suggests (these matters are discussed further in Chapter 17).

'Continuous' raster data can also be reclassified in the same way. The image in Fig. 15.5(a) shows a digital elevation model (DEM) with values ranging between 37 and 277, representing elevation in metres above sea level (for the same geographic area as shown in Fig. 15.4). Reclassification of this dataset into three classes of equal interval to show areas of low, medium and high altitude, produces the simplified image in Fig. 15.5(b). Comparison with Fig. 15.6(b) shows that the areas of high elevation coincide with the areas where older and permeable rocks outcrop at the surface

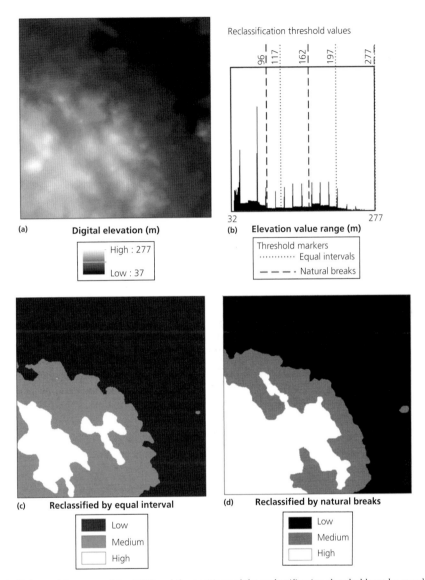

Fig. 15.5 (a) DEM; (b) image histogram of the DEM and the positions of the reclassification thresholds set by equal interval and natural break methods; (c) DEM reclassified into three equal interval classes; and (d) DEM reclassified by natural breaks in the histogram.

in the south-west of the area, again revealing spatial patterns not immediately evident in the original image. Reclassification of the DEM into three classes, this time with the classes defined according to the natural breaks in the image histogram (shown in Fig. 15.6) produces a slightly different result, Fig. 15.5(c). The high elevation areas are again in the south-west but the shape and distribution of those areas is slightly different. This demonstrates several things. Firstly that very different results

can be produced when we simplify data so that (and secondly) we should be careful in doing so, and thirdly, that the use of the image histogram is fundamental to the understanding of and sensible reclassification of continuous raster data. This issue is revisited later in Chapter 16.

Reclassification forms a very basic but important part of spatial analysis, in the preparation of data layers for combination, in the simplification of layer information

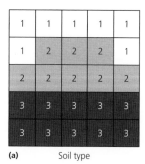

| (a) | Soil type | (b) | Average rainfall (m/yr) | (c) | Growing conditions |

Fig. 15.6 An example of a simple overlay operation involving two input rasters, (a) an integer raster representing soil classes (class 2, representing sandy loam, is considered optimum); (b) a floating point raster representing average rainfall, in metres per year (0.2 is considered optimum); and (c) the output raster derived by addition of (a) and (b) to produce a result representing conditions for a crop; a value of 2.2 (2 + 0.2), on this rather arbitrary scale, represents optimum growing conditions, and it can be seen that there are five pixel positions that satisfy this condition.

and especially when the layers have dissimilar value ranges. Reclassification is one or several methods of producing a common range among input data layers that hold values on different measurement scales.

Clear examples of the use of reclassification within case studies can be found in Chapters 21 and 22 (§ 21.3 and § 22.2).

15.3.3 Binary operations

This description refers to operations in which there are two input layers, leading to the production of a single output layer. *Overlay* refers to the combination of more than one layer of data, to create one new layer (using the standard operators described in § 15.3.4). The example shown in Fig. 15.6 illustrates how a layer representing average rainfall, and another representing soil type can be combined to produce a simple, qualitative map showing optimum growing conditions for a particular crop. Such operations are equivalent to the application of formulae to multi-band images, to generate ratios, differences and other inter-band indices (as described in Chapter 3), and as mentioned in relation to point operations on multi-spectral images, it is important to consider the value ranges of the input bands or layers, when combining their values arithmetically in some way. Just as image differencing requires some form of stretch applied to each input layer, to ensure that the real meaning of the differencing process is revealed in the output, here we must do the same. Either the inputs must be scaled to the same value range or if the inputs represent values on an absolute

measurement scale then those scales must have the same units.

The example shown in Fig. 15.6 represents two inputs with relative values on arbitrary nominal or ordinal [Fig. 15.6(a)] and interval [Fig. 15.6(b)] scales. The resultant values are also given on an interval scale and this is acceptable providing the range of potential output values is understood, having first understood the value ranges of the inputs, since they may mean nothing outside the scope of this simple exercise.

Another example could be the combination of two rasters as part of a *cost-weighted analysis* and possibly as part of a wider *least-cost pathway* exercise. The two input rasters may represent measures of *cost*, as produced through reclassification of, for instance, slope angle and land value; cost here being a measure of friction or the real cost of moving or operating across the area in question. These two cost rasters are then *aggregated* or *summed* to produce an output representing total cost for a particular area (Fig. 15.7).

15.3.4 N-ary operations

Here we deal with a potentially unlimited number of input layers to derive any of a series of standard statistical parameters, such as the mean, standard deviation, majority and variety. Ideally there should be a minimum of three layers involved but in many instances it is possible for the processes to be performed on single layers; the result may, however, be rather meaningless in that case. The more commonly used statistical operations and their functionalities are

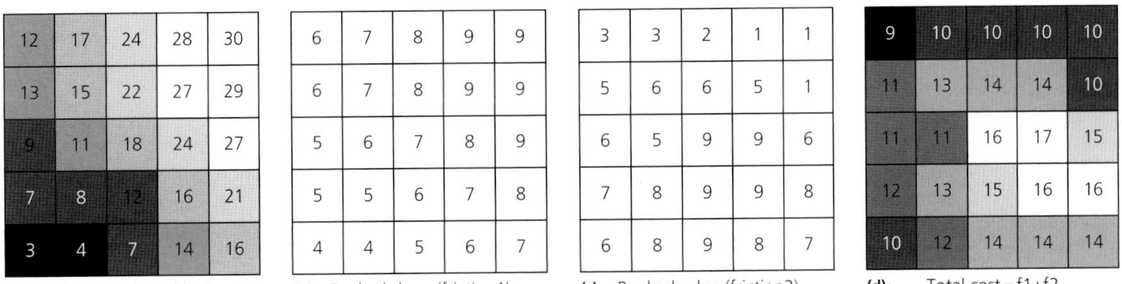

12	17	24	28	30
13	15	22	27	29
9	11	18	24	27
7	8	12	16	21
3	4	7	14	16

(a) Slope gradient (deg)

6	7	8	9	9
6	7	8	9	9
5	6	7	8	9
5	5	6	7	8
4	4	5	6	7

(b) Ranked slope (friction1)

3	3	2	1	1
5	6	6	5	1
6	5	9	9	6
7	8	9	9	8
6	8	9	8	7

(c) Ranked value (friction2)

9	10	10	10	10
11	13	14	14	10
11	11	16	17	15
12	13	15	16	16
10	12	14	14	14

(d) Total cost = f1+f2

Fig. 15.7 (a) Slope gradient in degrees; (b) ranked (reclassified) slope gradient constituting the first cost or friction input; (c) ranked land value (produced from a separate input landuse raster) representing the second cost or friction input; and (d) total cost raster produced by aggregation of the input friction rasters (f1 and f2). This total cost raster could then be used within a *cost-weighted distance analysis* exercise (see Fig. 15.9).

Table 15.4 Summary of local pixel statistical operations, their functionality and input/output data format.

Statistic	Input Format	Functionality	Data Type
Variety	Only rasters. If a number is input, it will be converted to a raster constant for that value	Reports the number of different DN values occurring in the input rasters	Output is integer
Mean		Reports the average DN value among the input rasters	Output is floating point
Standard deviation	Rasters, numbers and constants	Reports the standard deviation of the DN values among the input rasters	Output is floating point
Median		Reports the middle DN value among the input raster pixel values. With even number of inputs, the values are ranked and the middle two values are averaged. If inputs are all integer output will be truncated to integer	
Sum		Reports the total DN value among the input rasters	
Range		Reports the difference between maximum and minimum DN value among the input rasters	
Maximum	Only rasters. If a number is input, it will be converted to a raster constant for that value	Reports the highest DN value among the input rasters	If inputs are all integer, out will be integer, unless one is a float, then the output will be float
Minimum		Reports the lowest DN value among the input rasters	
Majority		Reports the DN value that occurs most frequently among the input rasters. If no clear majority, output = Null, e.g. if there are 3 inputs all with different values. If all inputs have equal value, output = input.	
Minority		Reports the DN value that occurs least frequently among the input rasters. If no clear minority, output = Null, e.g. if there are 3 inputs all with different values. If only 2 inputs, where different, output = null. If all inputs equal, output = input. If only 1 input, output = input.	

summarised in Table 15.4. As with the other local operations, these statistical parameters are point operations derived for each individual pixel position, from the values at corresponding pixel positions in all the layers, rather than from the values within each layer (as described in Chapter 2).

15.4 Neighbourhood operations

15.4.1 Local neighbourhood

These can be described as being incremental in their behaviour or operation. They work within a small neighbourhood of pixels (which in some circumstances can be user defined) to change the value of the pixel at the centre of that neighbourhood, based on the local neighbourhood statistics. The process is then repeated, or incremented, to the next pixel position along the row, and so on until the whole raster has been processed. It is equivalent to convolution filtering in the image or spatial domain, as described in some detail in Chapter 4 (§ 4.2). In image processing the process is used to quantify or enhance the spatial patterns or textures of a remotely sensed image, for instance. Here we are often dealing with data that is of implied three-dimensional character, for example, the gradient or curvature (Laplacian) of surface topography (see also Chapter 17) and if so, we're using the same process to quantify, describe or extract information relating to the morphology of the surface described by the DN in the local neighbourhood. Examples include calculations of slope angle, aspect and curvature, and mathematical morphology, that is, collapsing and expanding raster regions. These are described in more detail in Chapter 17. Otherwise these neighbourhood processes can be used to simplify or generalise discrete rasters.

15.4.1.1 Distance

Mapping distance allows the calculation of the proximity of any raster pixel to/from a set of target pixels, to determine the nearest or to gain a measure of cost in terms of distance. This is classified as neighbourhood processing since the value assigned to the output pixel is a function of its positional relation to another pixel. The input is a discrete raster image in which the target pixels are coded, probably with a value of 1 against a background of zero [as illustrated in Fig. 15.8(a)]. This input image may in itself be the product of an earlier reclassification.

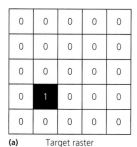
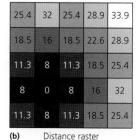

(a)   Target raster   **(b)**   Distance raster

Fig. 15.8 (a) An input discrete (binary) raster; and (b) the straight-line or Euclidean distance calculated from a single target or several targets are coded, to every other pixel in an input.

The simplest form of this operation involves the use of a *straight line distance* function, which calculates the *Euclidean distance* from every pixel to the target pixels [Fig. 15.8(b)]. Most GIS will also offer a spherical earth calculation as an alternative that does not use any geo-referencing (projection) information.

The output pixel values represent the Euclidean distance from the target pixel centres to every other pixel centre and are coded in the value units of the input raster, usually metres, so that the input raster will usually contain integers while the output will normally contain floating point numbers. The calculated distance raster may then be further *reclassified* for used as input to more complex *multi-criteria analysis* or used within a *cost-weighted distance analysis*.

15.4.1.2 Cost pathways

This moving window or kernel procedure is used to derive measures of *cost-weighted distance* and, from it, *cost-weighted direction* (these are referred to slightly differently depending on which software product you are using) as part of a *least-cost pathway* exercise. The cost weighted distance surface represents the aggregated cost of movement or construction as you move farther away from a source. The cost weighted direction surface represents opportunities and obstructions to the flow of cost-effective travel from any point in the study area back to that source.

Representation of 'cost' is given by an aggregation of a series of individual input cost layers, which have likely been produced by reclassification from their respective criterion layers, for example, gentle slopes generated from a slope layer, or low grade land from a landuse layer. In the example shown in Fig. 15.9, these are landuse, slope and lithology. Each is reclassified (coded) to reflect the cost, difficulty or prohibition of construction across a particular class. The coded costs are aggregated to derive a measure of *total cost* [Fig. 15.9(d)].

The cost-weighted distance function operates by evaluating each input pixel value in the *total cost* raster and comparing it to its neighbouring pixel values. The average cost between each pixel is multiplied by the distance between them [illustrated in Fig. 15.9(e)]. Cost-weighted direction is also generated from the total cost raster, where each pixel is given a value (from 1 to 8) using a direction encoded 3 × 3 kernel, which indicates the direction to the lowest cost pixel value from among its local neighbours [Fig. 15.9(f)]. These two rasters or

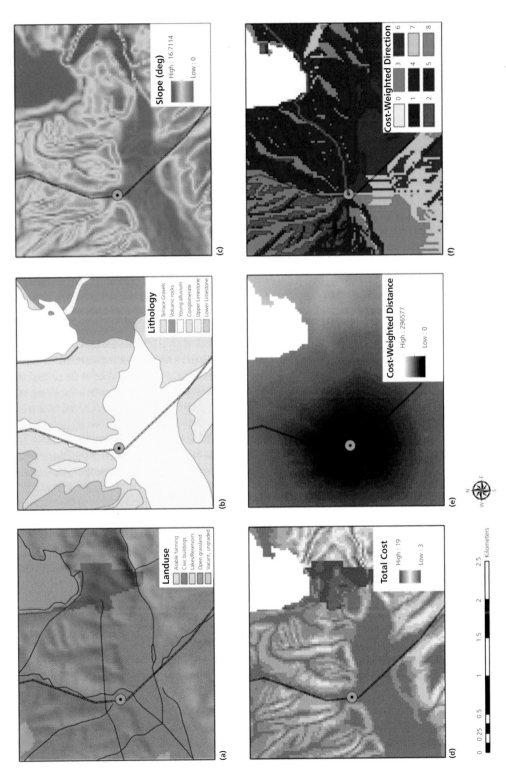

Fig. 15.9 Example of a least-cost pathway exercise in which the most efficient route for a pipeline is calculated from a source (shown here as an orange circle) to a destination (beyond the extents of the maps shown here). The inputs are (a) landuse; (b) lithology; (c) slope; and (d) total cost. The cost-weighted distance (e), increasing away from the source and cost-weighted direction (f), showing corridors of differing azimuth, are then used to derive the least-cost pathway, shown as a magenta line in (f).

surfaces are then combined to derive the *least-cost pathway* or route across the raster, to the target.

15.4.1.3 Mathematical morphology

Mathematical morphology (MM) can be thought of as the combination of conditional processing and convolution filtering. As a concept was first developed by Matheron (1975) and subsequently by many others. It describes the spatial expansion and shrinking of objects through neighbourhood processing and extends the concept of filtering. Such changes include *erosion* or *shrinking*, *dilation* or *expansion*, *opening* and *closing* of raster images. The size and shape of the neighbourhoods used are controlled by *structuring elements or* kernels that may also be of varying size and form. At its simplest, the kernel is a set of values passed across a binary raster image, whose status (1 or zero, 'on' or 'off') is changed according to agreement with the values in the kernel. The processing may not be reversible; for instance, after eroding such an image, using the erosion kernel, it is generally not possible to return the binary image to its original shape through use of a dilation kernel. Several different kinds of structuring kernels can be used, including those that are square, in addition to one-dimensional, hexagonal, circular and irregular in shape.

MM can be applied to vector point, line and area features but more often involves raster data, commonly discrete rasters and sometimes continuous raster surfaces, such as DEMs. In the latter case it can be used to find and correct for errors or extreme values (high or low) in those surfaces. Here we concentrate on the mechanism of the operations involved. It has also been used in mineral prospectivity mapping, to generate evidence maps, and in the processing of rock thin-section images, to find and extract mineral grain boundaries. The method also has applications in raster topology and networks, in addition to pattern recognition, image texture analysis and terrain analysis. These and other related methods have also been developed for edge feature extraction and image segmentation, for example, the Canny edge detector and OCR text recognition (Parker 1997).

To illustrate the effects consider a simple binary raster image showing two classes, as illustrated in Fig. 15.10. The values in the raster of the two classes are 1 (inner, dark grey class) and 0 (surrounding, white class); the image consists of a grid of ones and zeros. This input raster is processed using a series of 3 × 3 structuring elements or kernels (k), which consist of the values 1 and Null (rather than 0).

The kernels are passed incrementally over the raster image, changing the central pixel each time, according to the pattern of its neighbouring values. The incremental neighbourhood operation is therefore similar to spatial filtering but with conditional rather than arithmetic rules controlling the modification of the central value.

A simple dilation operation involves the growth, or expansion, of an object and can be described by:

$$o = i \oplus k \quad or \quad \delta k(i) \quad (15.1)$$

where o is the output binary raster, i is the input binary raster and k is the kernel that is centred on a pixel at i, and δ indicates a dilation. The Minkowski summation of sets ($a \oplus b$) refers to all the pixels in a and b, in which $\oplus$

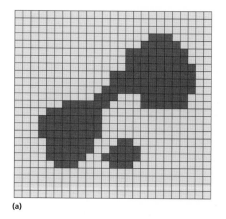

Structuring kernals k_1, k_2 and k_3

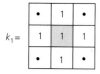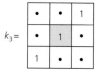

(a)　　　　(b)

Fig. 15.10 (a) Simple binary raster image (*i*); and (b) three structuring kernels (k_1, k_2 and k_3), the effects of which are illustrated in Figs. 15.12, 15.13 and 15.14. The black dots in the kernels represent Null values.

is the vector sum, and a belongs to set b, and b belongs to set a (Minkowski 1911). The Minkowski effect is where one shape is grown by the shape of another. The values of i are compared with the corresponding values in the kernel k, and are modified as follows: the value in o is assigned a value of 1 if the central value of i equals 1, or if any of the other values in k match their corresponding values in i. If they differ, the resultant value in o will be zero. The result of this is to leave the inner values as they are and to modify the surrounding outer values by the morphology of the kernel. The effect of a dilation, using kernel k_1, is to add a rim of pixels around the inner shapes, and in doing so the two shapes in the binary image are joined into one, both having been dilated, as in Fig. 15.11(b). If the output o_1 is then dilated again using k_1, then a second rim of pixels is turned on and so on. It can be seen that by this process, the features are merged into one. Using these conditional rules, the effect of a dilation can be considered equivalent to a maximum operation. Dilation is commonly used to create a map or image that reflects proximity to or distance from a feature or object, such as distance from road networks, or proximity to major faults. These distance or 'buffer' maps often form an important part of multi-layer spatial analysis, such as in the modelling of mineral prospectivity, where proximity to particular phenomenon is considered a significant and favourable condition.

A simple erosion operation $(a \ominus b)$ has the opposite effect, where $\ominus$ is a vector subtraction, that is, it involves the shrinking of an object using the Minkowski subtraction, and is described by:

$$o = i \ominus k \quad or \quad \varepsilon k(i) \qquad (15.2)$$

where ε indicates an erosion. The values in o are compared with those in k and if they are the same, then the pixel is 'turned off', that is, the value in o will be set to zero, or left unchanged if they are not the same. The effect of this, using kernel k_1, is the removal of a rim of value 1 (grey) pixels from the edges of the feature shown in Fig. 15.11(b) to produce that shown in Fig. 15.11(d). Using these conditional rules, the effect of an erosion operation can be considered equivalent to a minimum operation. Notice that the output, o_3, which is the product of the sequential dilation of i, then erosion of o_1, results in the amalgamation of the two original objects, and that the subsequent shrinking produces a generalised object that covers approximately the area of the original; an effect known as *closing* (dilation followed by erosion). Notice also that the repeated erosion of o_3 will not restore the appearance of the two original features in i.

In Fig. 15.12(b), an erosion operation is performed on the original i, removing one rim of pixels, causes the feature to be subdivided into two. When this is followed by a dilation, the result is to restore the two features to more or less their original size and shape except that the main feature has been split into two. This splitting is known as an *opening* (erosion followed by dilation) and is shown in Fig. 15.12(c).

$$Opening, \ \gamma k(i) = \delta k\left[\varepsilon k(i)\right] \quad and$$
$$Closing, \ \varphi k(i) = \varepsilon k\left[\delta k(i)\right] \qquad (15.3)$$

Again, repeated dilations of the features after opening, will not restore the features to their appearance in i.

Closing can be used to generalise objects and to reduce the complexity of features in a raster, such as

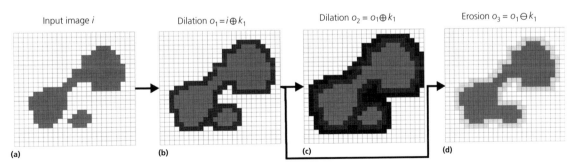

Fig. 15.11 Dilation, erosion and closing: (a) the original image i; (b) dilation of i using k_1 to produce o_1; (c) dilation of o_1 also using k_1 to produce o_2; and (d) erosion of o_1 using k_1 to produce o_3. Notice that o_3 cannot be derived from i by a simple dilation using k_1; the two objects are joined and this effect is referred to as *closing*. The pixels added by dilation are shown with dark grey tones and those pixels lost through erosion are shown with pale grey tones.

where a cluster of small features all representing the same class are dilated into one region representing that class and then eroded to reduce the features to approximately the same area as before but with reduced complexity. Opening can be used to perform a kind of sharpening or to add detail or complexity to the image.

Dilation and erosion operations can also be carried out anisotropically, that is, they can be applied by inequal amounts and in specific directions. Such directional operations are often relevant in geological applications where there is some kind of structural or directional control on the phenomenon of interest.

For example, the effect of kernel k_2 on i is shown in Fig. 15.13(a), where the effect is a westward shift of the features by one pixel. The effect of kernel k_3 is to cause dilation in the NW-SE directions, resulting in an elongation of the feature [Fig. 15.13(b)].

To consider the effect of MM on continuous raster data, we can simply take the simple binary image (*i*)

shown in Figs. 15.10 to 15.13 to represent a density slice through a raster surface, such as an elevation model, in which case the darker class would represent the geographical extent of areas exceeding a certain elevation value. Figure 15.14(a) shows the binary image and a line of profile [Fig. 15.14(b)] across a theoretical surface that could be represented by image (*i*). The effect of simple dilation and erosion of the surface is shown in Fig. 15.14(c); it can be seen that dilations would have the effect of filling pits or holes, and broadening peaks in the surface, while erosions reduce the peaks or spikes, and widen depressions. Such techniques could therefore be used to correct for errors in generated surfaces, such as DEMs, except that the dilations and erosions affect all other areas of the surface too, that is, the parts that do not need correcting. Such artefacts and errors in DEMs cannot be properly corrected by merely smoothing either, since the entire DEM will also be smoothed and so degraded. The use of median filters to smooth whilst retaining edge features has been proposed but

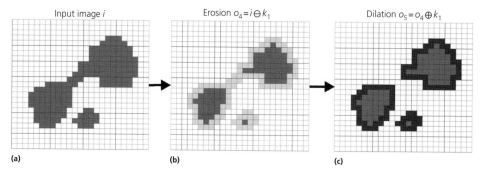

Fig. 15.12 Erosion, dilation and opening: (a) the original image i; (b) erosion of i using k_1 to produce o_4; (c) subsequent dilation of o_4, using k_1, to produce o_5. Note the initial erosion splits the main object into two smaller ones and that the subsequent dilation does not restore the object to its original shape, an effect referred to as *opening*.

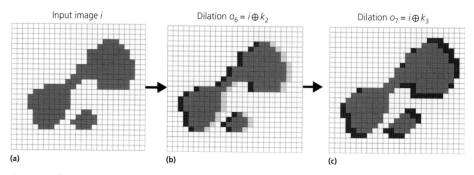

Fig. 15.13 Anisotropic effects: (a) dilation of i using k_2 to produce o_6 (b), causing a westward shift of the object; and (c) dilation of i using k_3, producing an elongation in the NE-SW direction on o_7.

again, this is also undesirable for the same reason. A modification of the MM technique has been proposed, known as *morphological reconstruction* (Vincent 1993), for the correction of DEM errors. In this case, the original image is used as a *mask* and the dilations and erosions are performed iteratively on a second version of the same image (*marker* image) until stability between mask and marker images is reached.

This process, using a *marker* and a *mask*, instead of an image and a structuring kernel, repeats until stability is reached, that is, when the image is fully reconstructed and no longer changes, when the holes or pits are corrected (Fig. 15.15). Since morphological reconstruction is based on repeated dilations, rather than directly modifying the surface morphology, it works by controlling connectivity between areas. The

marker could simply be created by making a copy of the mask and either subtracting or adding a constant value. The error-affected raster image is then used as the mask, and the marker (which is derived from it) is dilated or eroded repeatedly until it is constrained by the mask, that is, until there is no change between the two, and the process then stops. By subtracting a constant from the marker and repeatedly dilating it, extreme peaks can be removed; whereas by adding a constant and repeatedly eroding the marker, extreme pits would be removed. The extreme values are effectively reduced in magnitude, relative to the entire image value range, the in the reconstructed marker image. This technique can be (and has been) used selectively to remove undesirable extreme values from DEMs.

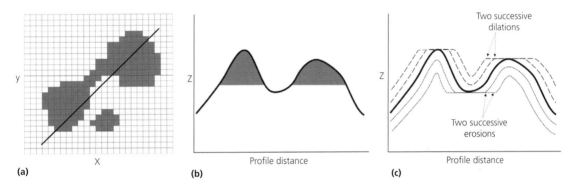

Fig. 15.14 (a) The original input image with the position of a profile line marked; (b) the theoretical cross-sectional profile with the shaded area representing the geographical extent of the darker class along the line shown in (a); and (c) the effect on the profile of dilations and erosions of that surface.

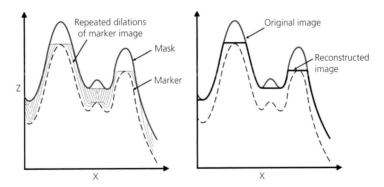

Fig. 15.15 Mechanism of morphological reconstruction of an image, as illustrated by a profile across the image. (a) In this case, by repeated dilations of the marker until it is constrained by the mask image; and (b) the extreme peaks are reduced in magnitude in the reconstructed image.

15.4.2 Extended neighbourhood

The term *'extended neighbourhood'* is used to describe operations whose effects are constrained by the geometry of a feature in one layer and performed on the attributes of another layer. These extended neighbourhood operations can be further described as *focal* and *zonal*. If for instance slope angles must be extracted from within a corridor along a road or river, the corridor is defined from one layer and then used to constrain the extent of the DEM from which the slope angle is then calculated (illustrated in Fig. 15.16).

15.4.2.1 Focal operations

Focal operations are used for generating corridors and buffers around features. Focal operations are those that derive each new value as a function of the existing values, distances, and/or directions of neighbouring (but not necessarily adjacent) locations. The relationships may be defined by such variables as Euclidean distance, travel cost, engineering cost or inter-visibility. Such operations could involve measurement of the distance between each pixel (or point) position, and a target feature(s). A buffer can then be created by reclassification of the output 'distance' layer.

This allows specific values to be set for the original target features, with the buffer zones and for the areas beyond the buffers. In this way, it is possible to establish the approximate proximity of objects using a buffer. Buffer zones can also be used as masks to identify all features that lie within a particular distance of another feature. Buffers can be set at a specified distance or at a distance set by an attribute. Since the buffer is a reclassification of the distance parameter, multiple buffer rings can also be easily generated. Buffers are therefore particularly useful for constraining the activities of spatial analysis. Dilation, as described previously, is just one method of creating a buffer.

15.4.2.2 Zonal operations

Zonal operations involve the use of the spatial characteristics of a zone or region defined on one layer, to operate on the attribute(s) of a second layer or layers. The zonal areas may be regularly or irregularly shaped. This process falls into the binary operations category since zonal operations most commonly involve two layers. An example is given in Fig. 15.17 where zonal statistics are calculated from an input layer representing the density of forest growth, within the spatial limits defined by a second survey boundary layer, to provide an output representing, in this case, the average forest density within each survey unit. Notice that the two raster inputs contain integer values but that the output values are floating point numbers, as is always the case with mean calculations.

15.5 Vector equivalents to raster map algebra

Map-algebra operations can be performed on vector data too. The operators behave slightly differently because of the nature of vector data but in many cases are used to achieve the similar results:

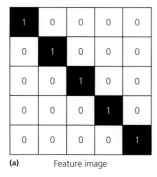 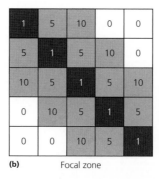 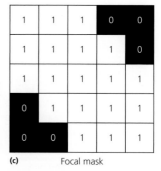

| **(a)** Feature image | **(b)** Focal zone | **(c)** Focal mask |

Fig. 15.16 Focal statistics: (a) a binary image representing a linear target feature (coded with a value of 1 for the feature and 0 for the background); (b) a 10 m focal zone mask created around the linear feature, where each pixel is coded with a value representing its distance from the feature (assuming that the pixel size is 5 m × 5 m), areas beyond 10 m from the feature remain coded as zero; (c) binary focal mask produced by thresholding the values in (b). This has a similar effect to a dilation followed by a reclassification, to produce a distance buffer.

15.5.1 Buffers

Zones calculated as the Euclidean distance from existing vector features, such as roads, are referred to as *buffers*. These are calculated at constant distance from the feature or at distances dictated by attribute values, and each zone will be the same width around the feature (illustrated in Fig. 15.18). The earth's curvature is not taken into account, so that the zones will be at the same width regardless of the coordinate system. Negative distance values can be used, and these will cause a reduction in the size of the input feature. Buffers can also be generated on only one side of input features (should this be appropriate). The input layer in this case is a vector feature but the output may be a polygon file or raster. The same buffering operation can also be applied to raster data by first calculating the Euclidean distance and then reclassifying the output to exclude distances within or beyond specified thresholds; the output will always be a raster in this case. Buffering in

this way can be considered as the vector equivalent of conditional logic combined with raster dilation.

15.5.2 Dissolve

When boundaries exist between adjacent polygon or line features, they can be removed or *dissolved* because they have the same or similar values for a particular attribute (illustrated in Fig. 15.19). Such as in a geological map where adjacent lithological units with similar or identical descriptions can sensibly be joined into one; the boundaries between them are removed by this process and the classes merged into one. Complications in the vector case arise if the features' attribute tables contain other attributes (besides the one of interest being merged) which differ across the boundary; choices must be made about how those other attributes should appear in the output dissolved layer. This is equivalent to merging raster classes through reclassification, or raster generalisation/simplification.

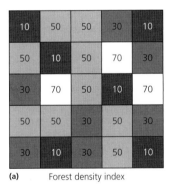

(a)   Forest density index

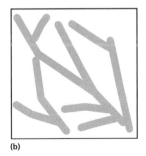

(b)   Survey boundaries

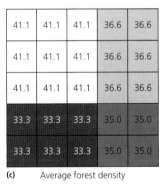

(c)   Average forest density

Fig. 15.17 Zonal statistics; (a) forest density integer image; b) survey boundaries (integer) image; and (c) the result of zonal statistics (in this case a zonal mean) for the same area. Note that this statistical operation returns a non-integer result.

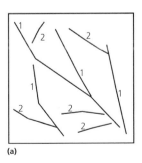

(a)

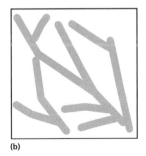

(b)

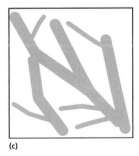

(c)

Fig. 15.18 (a) Simple vector line feature map, labelled with attribute values (1 and 2); (b) output with buffers of constant distance; (c) output map with buffers of distance defined by the attribute values shown in (a) (features with attribute value 1 having buffers twice the distance of those of features with attribute value 2).

15.5.3 Clipping

The geometry of a feature layer can be used as a mask to selectively extract a portion of another layer; the input layer is thereby *clipped* to the extents of the mask (as illustrated in Fig. 15.20). The feature layer to be clipped may contain point, line or polygon features, but the feature being used as a mask must have area, that is, it will always be a polygon. The output feature attribute table will contain only the fields and values of the extracted portion of the input vector map, as the attributes of the mask layer are not combined. Clipping is equivalent to a binary raster zonal operation, where the pixels inside or outside the region are set as Null, using a second layer to define the region or mask.

15.5.4 Intersection

If two feature layers are to be integrated while preserving only these features that lie within the spatial extent of both layers, an *intersection* can be performed (as illustrated in Fig. 15.21). This is similar to the clip operation except that the two input layers are not necessarily of the same feature type. The input layers could be point, line and/or polygon, so that the output features could also be point, line and/or polygon in nature. New vertices need to be created to produce the new output polygons, lines and points, through a process called *cracking*. Unlike the clip operation, the output

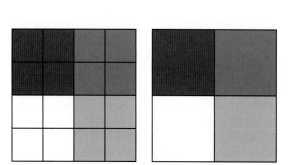

(a)   Input vector map   **(b)**   Simplified map

Fig. 15.19 (a) Vector polygon features; and (b) the dissolved and simplified output map.

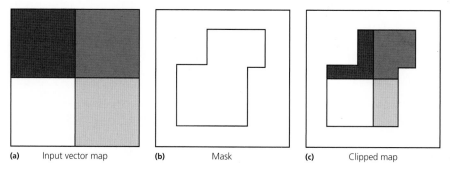

(a)   Input vector map   **(b)**   Mask   **(c)**   Clipped map

Fig. 15.20 (a) Vector polygon clipping, using an input vector layer from which an area will be extracted; (b) the vector feature whose geometric properties will be used as the mask; and (c) the output clipped vector feature.

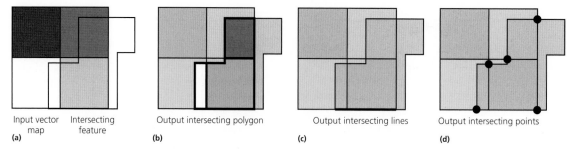

Input vector   Intersecting   Output intersecting polygon   Output intersecting lines   Output intersecting points
map   feature
(a)   **(b)**   **(c)**   **(d)**

Fig. 15.21 Intersection operation between two overlapping polygon features (a); the output intersecting bold polygon (b), which covers the extent and geometry of the area that the two inputs have in common; the intersecting (bold) line (c) and points (d) shared by both polygons. The output attribute table contains only those fields and values that exist over the common area, line and points.

attribute table contains fields and values from both input layers, over the intersecting feature/area. In the case of two intersecting polygons, intersection is equivalent to a Boolean operation using a logical AND (Min) operator between two overlapping raster images.

When two input overlapping feature layers are required to be integrated such that the new output feature layer contains all the geometric features and attributes of two input layers, the *union* operation can be used (as illustrated in Fig. 15.22). Since vector feature layers can contain only points, or only lines or only polygons, here the inputs must be of the same type but the number of inputs is not limited to two. Again, new vertices will be created through *cracking*. This is similar to the intersection operation but the output will have the overall extent of the input layers. New, minor polygons are created wherever polygons overlap. The attribute table of the output layer contains attribute fields of both the input layers, though some of the entries may be blank. In the polygon case, it is equivalent to a binary raster operation using a logical OR (Max) operator between two overlapping images.

15.6 Automating GIS functions

Automation makes map algebraic operations easier (once automated, a process can be run many times), faster (tools can be executed in sequence much faster than can be accomplished manually) and more reliable (performing multiple tasks manually create chances for error - once a process is configured correctly, a computer can usually be trusted to perform it every time).

There are many options for automating GIS processes, even if you are not an accomplished programmer. Many GIS include a macro-buidling facility, such as Model Builder in ArcGIS (an example is shown in Fig. 17.13), or Idrisi's batch mode script tool. Many GIS now include a scripting interface, such as for Python, JScript, Perl & R, which are relatievly easy to learn. There is of course Matlab, allowing the development of algorithms, models and applications independently of any proprietary software. In fact there are so many open source GIS packages, and repositories of code for specific and clever tools, freely available now that the sky is the limit.

15.7 Summary

The overlap with image processing is perhaps most obvious in this chapter. The processing of the DN values in various bands of a multi-spectral image is analogous with that of raster grids in a GIS using map algebra. Local, focal, zonal, incremental and global operations in raster GIS are synonymous with those of image processing even though the objective may be different. The use of conditional statements is another parallel and represents the first step in the development of a more complex spatial analysis, for example, in decision making. The use of the geometric properties of one layer to control the limits of operations on another is a minor departure since this is less commonly required in image processing, but it is perfectly possible using 'regions of interest' for which statistics have been derived (Chapter 3). Regions are spatially defined within the coordinate space of the image, the extents of which are recorded in association with the raster image information (header). Statistics can be calculated globally and for the

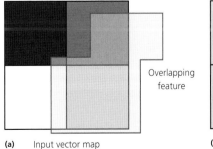
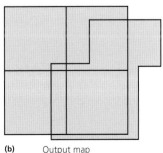

Overlapping feature

(a)   Input vector map   **(b)**   Output map

Fig. 15.22 (a) Vector polygon union operation where two polygon features overlap; and (b) the output object covers the extent and geometry of both inputs. The output attribute table also contains the attribute fields and values of both input features.

region and these can be manipulated on any of the bands of the image. These are in essence zonal operations.

This chapter is focused on operations that assume raster inputs. Many of these operations have vector equivalents, and in some circumstances it could be argued that they could be carried out more effectively using vector data. The diversity of raster-based operations is, however, testament to their significance in the processing of continuously sampled data whose spatial variability is significant. This is especially the case in geoscientific applications, where we are deeply concerned with the way variables change from one location to another as well as the spatial relationships between the variables.

Map algebra has a major role to play in multi-criteria and multi-objective problems, by linking together these simple processes and procedures to prepare data and build complex models and so to tackle difficult spatial problems, which are discussed further in Chapter 19.

15.8 Questions

1 With respect to the nature of the classes being represented, what assumptions are made during spatial operations on categorical (discrete) rasters?

2 What are the considerations for scaling (preparing) data for input to spatial analysis? And what are the effects of using those scales?

3 How should you decide on the threshold values for reclassification schemes?

4 Why is it important to understand the nature of the input recorded data when applying local statistical operations?

5 What are the practical applications of mathematical morphology?

6 For further consideration beyond this chapter:
How do these individual operations combine and contribute to more complex spatial analytical models?
Are discrete and continuous rasters treated differently within spatial analyses?

CHAPTER 16

Extracting information from point data: Geostatistics

16.1 Introduction

The data that we have at our disposal are never complete; we either have the wrong kind or insufficient or only partial coverage! Naturally, we seek ways to predict the values between, or to extrapolate beyond the limits of, our data; indeed, therein also lies the role of multi-criteria spatial analysis, but we shall deal with that in Chapter 19.

This chapter deals with two topics: gaining a better understanding the data and dealing with incomplete data. If we understand the nature and meaning of the sample data we do have, the better our chance of producing a reliable prediction of the unknowns. After all, one of the most important messages of this book is that producing an impressive result is not enough; if it cannot be explained or understood, it is absolutely meaningless.

This chapter therefore covers the subject of *geostatistics*, a term first coined with "trepidation" by Hart (1954) and first used in mineral resource evaluation in an attempt to predict the potential economic value of a mineral deposit from limited sample data by George Matheron and Daniel Krige (1951). Such techniques have subsequently been applied to many disciplines other than the geosciences. The many and varied uses of geostatistics include, for example, the description and summary of spatial data attributes, simplifying complex spatial patterns, inferring the characteristics of a larger population on the basis of a sample, estimating the probability of an outcome and establishing how closely a predicted spatial pattern matches an actual one. Geostatistics is concerned with the description of patterns in spatial data; each known data point has a geographic location and a value, and the connection between them is exploited to help predict values at the unknown locations. There are many truly comprehensive accounts of geostatistical methods that are listed in the general references and further reading section. We aim only to give an overview of the main issues and methods involved in extracting and exploiting statistical data and in getting over the problem of incomplete data.

Our early qualitative questions about the nature of processes and phenomena have quickly developed from 'what?' and 'where?' into more quantitative questions, such as 'how much?' or 'to what degree?' and 'how sure am I that the result is true or representative?' This touches on the issue of uncertainty in data and analysis (this is discussed in more detail in the next chapter). Asking 'why?' is rather more tricky for GIS to tackle since it requires the unravelling of causative links between phenomena, and this is a dangerously speculative area.

In dealing with the estimation of unknown values from known ones, this chapter also overlaps with topics in Chapter 9 (§ 9.2).

16.2 Understanding the data

We should never underestimate the importance of understanding our data, how they were collected, how reliable and accurate are their geographic positions, what area of ground they represent and do their values represent one or more statistically independent populations; all these (and more) should be considered when thinking about how to process data and interpret the result.

Image Processing and GIS for Remote Sensing: Techniques and Applications, Second Edition. Jian Guo Liu and Philippa J. Mason.
© 2016 John Wiley & Sons, Ltd. Published 2016 by John Wiley & Sons, Ltd.

16.2.1 Histograms

A vital tool in the understanding of data, and one that should be our first port of call, is the *histogram*. It shows us the count of data points falling into various ranges, that is, the frequency distribution. The histogram shows us the general shape and spread, symmetry of distribution and modality of the data, and should reveal any outliers.

No interpolation method should be used without full understanding of the implications and effects on the result. If any interpolation method is applied to data that comprises more than one statistically independent population, then the result is flawed. If quantitative predictions are to be based on calculations from datasets that contain more than one population, the actual values could be considerably less than the predicted values. For instance, in the case of predicted output of a precious metal from an actively producing mine, predictions of total metal production based on the concentrations measured at discrete sample points within the mine may significantly overestimate production if the existence of statistically (and geologically) independent populations within the sample data is not recognised.

Here again the data histogram becomes a vital tool in understanding the problem. It should always be carefully examined before hand, to establish how the data are composed. If the data are not normally distributed about the mean but skewed, there must be a reason for it, and it will be necessary to consider the existence of minor populations as the reason (illustrated in Fig. 16.1). Skewness in the data histogram could indicate a sampling, measurement or processing problem, or it could point to some real but unknown pattern in the data; either way, it should be investigated before proceeding. There are two important messages here, first the possibility of erroneous numerical predictions from the data, and second the fact that interpolating such data across its multiple populations could produce a result that is invalid.

16.2.2 Spatial auto-correlation

The simplest method of estimating values at unknown positions from known sample values might be to average them but this is sensible only if the values are independent of their location. Normally, however, a variable defined continuously over an area does not change greatly from one place to another and so we can expect the unknown values to be related to those at nearby known points. This behaviour is described by Tobler's *First Law of Geography* (Tobler 1970), which states that "everything is related to everything else, but near things are more related than distant things". The formal property that describes this is known as *spatial auto-correlation (SAC)*. Correlation represents the degree to which two variables or types of variable are related; spatial auto-correlation represents the degree to which that correlation changes with distance. In the context of a raster image, this can be likened to making a copy of an image and overlaying the first one by the copy precisely; the two should be exactly correlated. If the copy image is then shifted by one pixel relative to the other and the correlation between them is examined again, they should still be very highly correlated. Continuing this process, shifting by one pixel and recalculating the correlation, should lead to a point where the two images have been shifted so far that they are almost uncorrelated. If the collective results of this process are examined and plotted, a measure of SAC is produced. Understanding spatial auto-correlation is very useful since it can reveal and describe systematic patterns in the data that may not otherwise be obvious and may in turn reveal some underlying control

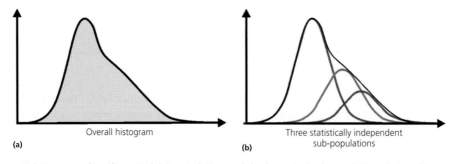

(a) Overall histogram

(b) Three statistically independent sub-populations

Fig. 16.1 Schematic histograms of (a) theoretical data population; and (b) theoretical sub-populations that could exist within the dataset.

on variation. *SAC* is important because it provides a measure of correlation in the dataset, and because it tests the assumption of randomness in the data. Patterns that are truly random exhibit no spatial auto-correlation. Here we recall the auto-correlation matrix that characterizes the SAC in every direction.

Both positive and negative auto-correlation exist and are opposites of one another. *Positive spatial auto-correlation* occurs where near values or areas are alike and *negative spatial auto-correlation* where near values or areas are dissimilar, whilst zero indicates no correlation between the two.

In general, two assumptions are made of *spatially auto-correlated errors* ('error' here does not mean a mistake but simply variation about the mean or trend). The first is that the average error will be zero because the positive and negative fluctuations around the trend will cancel one another out. The second is that the precise locations of the errors are not significant, only their relative positions with respect to each other, a relationship known as *stationarity*.

16.2.3 Variograms

The *variogram* or more commonly the *semi-variogram* (half the variogram) is the main measure of similarity within a data population and is a principle tool of geostatistics. It is a statistical function that describes the decreasing correlation between pairs of sample values as separation between them increases. Other tools such as *correlograms* and *covariance* functions are also used but these are all very closely related to the variogram. A *covariance cloud* reveals the auto-correlation between pairs of data points, where each point in the cloud also represents a pair of points in the dataset. The cloud is generated by plotting the distance between the points against the squared difference between their values for every conceivable pair; points that are close together should also be close in value. As distance between points increases, the likelihood of correlation between the point values decreases. The form of the cloud and the function fitted to it comprise the semi-variogram. The semi-variogram $z(d)$ is a function describing the degree of spatial dependence of a variable or process, and in general is defined by the following:

$$z(d) = \frac{1}{2n} \sum_{i=1}^{n} \{z(x_i) - z(x_i + d)\}^2 \qquad (16.1)$$

where n is the number of sample pairs (observations of value z separated by distance d being evaluated, and x_1 and x_2 are the positions of the points being compared. An idealised (theoretical) semi-variogram is a function defined by the relationship between the semi-variance $z(d)$ and distance d between sample points, as in in Fig. 16.2, and is used to describe data populations. There

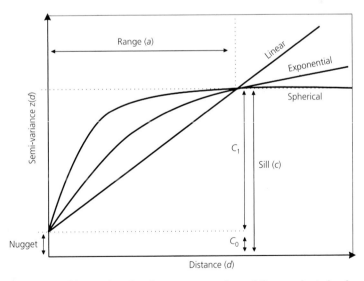

Fig. 16.2 The form of several theoretical forms of semi-variograms commonly used (linear, spherical and exponential, although there are several others), showing the relationship between the function, its sill (c, where $c = c_0 + c_1$) and the distance (d) at which the sill is reached (a). Pairs of points that plot in the lower left hand corner of the semi-variogram are close to one another spatially and have similar values; the opposite is true of points near the top right corner of the semi-variogram.

are several forms of theoretical semi-variogram, which are fitted to the actual data and used to predict unknowns, and these can be defined as follows and illustrated schematically in Fig. 16.2.

A *linear* variogram, $z_l(d)$, is a special case and rather rare in nature, since it never reaches a sill, and is described by the following:

$$z_l(d) = c_0 + pd \qquad (16.2)$$

where c_0 represents the nugget effect (which is random), p represents the gradient of the function, which is constant in the linear case, and d is the distance or *lag*. A *spherical* variogram function, $z_s(d)$, can be used to fit data that reach a distinct sill. When $z(d)$ becomes constant, that is, the sill roughly equals the calculated sample variance,

$$z_s(d) = c_0 + c_1 \left(\frac{3d}{2a} - \frac{1}{2}\left(\frac{d}{a}\right)^3 \right) \quad \text{when} \quad 0 \leq d \leq a$$

or $\qquad\qquad\qquad\qquad\qquad\qquad (16.3)$

$$z(d) = c_0 + c_1 \quad \text{when} \quad d \geq a$$

Here a is the distance to the sill or *range*. If there is a only a gradual approach to the sill, an *exponential* variogram, $z_e(d)$, provides a good fit, and can be described by the following:

$$z_e(d) = c_0 + c_1 \left(1 - \exp(-d/a) \right) \qquad (16.4)$$

The spherical and exponential variograms can be referred to as *transitive* forms because the correlation varies with distance or lag (d). Variogram forms that have no sill can be described as *non-transitive*, and the linear form falls into this case.

As illustrated in Fig. 16.3, in a semi-variogram, pairs of points that plot in the lower left hand corner are close one another spatially and have highly correlated values, and the opposite is true of points that plot near the upper right corner. The values of a pair of points that plot in the upper right of the semi-variogram (on or above the sill) can be considered to be uncorrelated. Like the histogram, the semi-variogram is useful for detecting *outliers* or pairs of points that have erroneous values, such as two closely adjacent points with wildly differing values.

16.2.4 Underlying trends and natural barriers

The *trend* can be thought of as an underlying control on the overall pattern of the data values. If for instance the variable being predicted is elevation, the trend might be the regional slope. In soil or sediment multi-element geochemical data the trend could represent slow ongoing slope processes in moving debris (and elements) down-slope under gravity. In airborne pollutants it could represent the prevailing wind direction. Such underlying trends may well affect the distribution of values and failure to consider them could produce misleading results. If the trend is not constant but variable, such as in the case of elevation data covering a sizeable area of terrain with, for example, a valley running across it, the function fitted to the data should allow for that. The assumption in this instance would then be that the mean is variable, and importantly that there is likely to be more than one statistically independent population present. Some methods of value estimation or prediction cannot make such allowances.

Other phenomena can also affect the data value distributions. Physical, geographic barriers that exist in the landscape, such as cliffs or rivers, present a particular challenge when describing a surface numerically because the values on either side of the barrier may be drastically different. Elevation values change suddenly and radically near the edge of a cliff but the known values at the bottom of the cliff cannot be used to accurately estimate values at the top of the cliff. If natural barriers are known to exist in the data population then it would be advantageous to use a method of value extraction that can selectively use values on one side or the other. Many interpolators smooth over these differences by averaging values on either sides of the barrier. The *inverse distance weighted (IDW)* method generally allows the inclusion of barriers to constrain the interpolation to one side. In most cases, such separate populations must be accommodated by partitioning the data before interpolation.

16.3 Interpolation

Regardless of the quantity in question – for example, rainfall intensity, pollution concentrations or elevation values – it is impossible or at best impractical, to measure such phenomena at every conceivable location

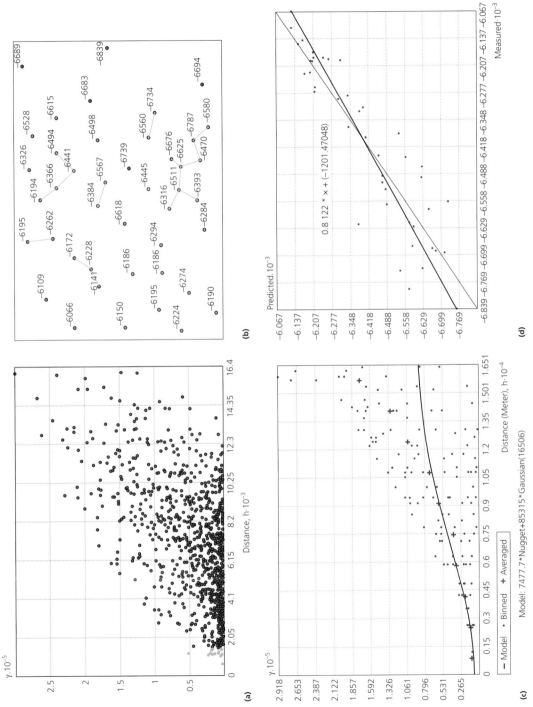

Fig. 16.3 (a) Non-binned semi-variance plot for the set of points (the y axis represents semi-variance $z(d)$ and the x axis the distance between points, as in Fig. 16.2); and (b) the map of those points showing the relative positions (as cyan coloured links) of the point pairs highlighted (also in cyan) in (a). (c) Semi-variogram of the binned point pairs (red points), the averages of the binned point pairs (blue points), and the model (Gaussian in this case) fitted to them; and (d) cross-validation plot of measured points against predicted values, with the regression line (blue) representing the prediction and a reference line (grey) representing parity between the two – these should be as close together as possible. The point values in this case represent depth below sea level to a stratigraphic horizon and reveal a gentle sloping surface; these values are used in the examples of interpolation types later on.

within an area. You can, however, obtain a sample of measurements from selected locations within that area, and from those samples, make predictions about the values over the entire area. *Interpolation* is a process by which such predictions are made.

The process begins with a set of sample points containing numerical measurements recorded at specific locations. Spatial auto-correlation is assumed so that an unknown value can be estimated from the neighbourhood of values. The aim is then to create a surface that models the sampled phenomenon so that the predicted values resemble the actual ones as closely as possible. Adjustments to the surface can be made by limiting the size of the sample used and controlling the influence that the neighbourhood of sample points has on the estimated values.

Interpolation can then be described as the process of estimating a value at a location (x, y) from irregularly spaced, assumed or measured values at other locations $(x1...xn, y1...yn)$, to produce a regularly or continuously sampled grid. It is possible to interpolate a surface from a very small number of sample points but more sample points will give a better result. Ideally, sample points should be well-distributed throughout the area. If there are some rapidly changing phenomena, then denser sampling may be needed.

16.3.1 Selecting sample size

This is an important step, as it controls the neighbourhood statistics from which the interpolated values will be estimated. Most interpolation methods allow control over the number of sample points used, in some way or other. For example, if you limit your sample by number, to five points for example, for every location the

interpolated value will be estimated from the five nearest points. The distance between each sample point varies according to the distribution of the points and, as we have said, this distance is important. Using many points will slow the process down but will mean that the distances between the points are smaller so variation between them will be lower and the result should be more accurate. Using fewer points will make the process faster, and sufficient points are likely to be found but the prediction may not really represent the statistics of the neighbourhood. The sample size can also be controlled by use of a *search radius* or by defining the minimum number of points to be used. The number of sample points found will depend on the density and regularity of the point distribution.

Two common approaches to sample selection are the *fixed distance* and the *nearest K neighbours* (where K is a specified number) methods. A *fixed search radius* will use only the samples contained within the specified radial distance of the unknown value, regardless of how large or small that number that might be. The *K nearest neighbour* method uses a *variable search radius*, which expands until the K neighbouring points are found. The fixed-distance technique, shown in Fig. 16.4(a), using a distance equal to the radius of the circles shown, would interpolate the value at point B using four neighbouring samples but would find only one sample to interpolate the value at point A. If instead, a variable radius were used [Fig. 16.4(b)], then the search around point A would have to expand considerably before four neighbours are found.

The fixed distance approach may fail to find any sample points and the interpolator will fail to estimate a value within an area of low density sampling. This is useful only in that it will reveal areas where there is

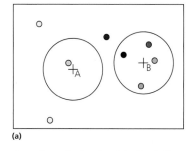
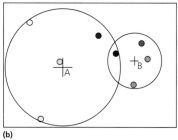

(a)   (b)

Fig. 16.4 Neighbourhood search methods using (a) a fixed distance; and (b) variable distance to find the k-nearest neighbours, to estimate two points located at A and B.

insufficient sampling but the resultant 'holes' in the interpolated surface are rather undesirable. The K-nearest-neighbours approach, in contrast, will always find sample points but they may be so far away [as in Fig. 16.4(b)] as to be unrelated to one another and so the predicted result may be misleading.

Once a set of neighbours is found, the interpolator must combine their values to produce the estimate. Clearly, the choice of method used will depend on the data, how they were collected, the desired characteristics of the output surface, and the nature of the decisions or analyses that will be performed with the resulting grid.

16.3.2 Interpolation methods

There are two broad classes of interpolator: *deterministic* and *stochastic*. A *deterministic* process is one where at any specific known instant, there is only one possible outcome. In general terms, a *stochastic* process, on the other hand, exhibits probabilistic behaviour, that is, it can be considered the opposite of deterministic, so that for one known condition there are many possible outcomes, some of which will be more likely than others. Deterministic interpolators create surfaces based either on the degree of similarity between sample values (as in the Inverse Distance Weighted or *IDW* method) or on the degree of smoothing (as with *radial basis functions*). Stochastic interpolators are based on the degree of auto-correlation between every possible combination of points in the input dataset. It is generally considered that, in situations where data is plentiful, stochastic interpolation methods are superior. A summary of different interpolation methods is shown in Table 16.1.

16.3.3 Deterministic interpolators

The majority of deterministic interpolators are polynomial in form, and of varying degrees of complexity. The general form of a polynomial function is as follows:

$$f(x) = a_n x^n + a_{n-1} x^{n-1} + \ldots + a_1 x^1 + a_0 x^0 \quad (16.5)$$

where x is the input value, the number of terms is variable and each term consists of two factors [a real number coefficient (a) and a non-negative integer power (n)]. The degree or order of a polynomial function is given by the highest value of n.

Global polynomial interpolators use a polynomial function to construct a very simple surface from all the sampled point values, that is, no neighbourhood is specified, and so smoothes over all local variations (as in Fig. 16.5(a)). First-, second-, third- or fourth-order polynomials, and so on, can be used to represent surfaces of increasing complexity. A smooth plane is created with a first-order, a surface with one bend or fold is made by a second-order, and one with two bends or folds by a third-order polynomial. Since the surface is relatively rigid it will not honour the data, that is, it will not necessarily pass through all the data point values. Global polynomial interpolators are often referred to as *inexact* interpolators for this reason.

For a set of sample points representing surface elevations, shown in Fig. 16.3(b), the result of interpolating using a first-order global polynomial is shown in Fig. 16.6(a).

If a specific neighbourhood is then selected, the result becomes a *local polynomial interpolator* (Fig. 16.5(b)). *Local polynomial* interpolation creates a surface using functions unique to a sample neighbourhood. By controlling of the number of points, the shape of the neighbourhood and the location of the points within

Table 16.1 A selection of interpolators compared where SAC - Spatial Autocorrelation, RBF - Radial Basis Functions and IDW - Inverse Distance Weighted.

Class	Type	Uses SAC	No. of Variables	Honours the Data?	Surface Type	Pros	Cons	Assumptions
Deterministic	Polynomial (global)	No	1	No	Prediction only	Simple	Too simple	None
	RBF			Yes		Simple		
	IDW			No		Simple & barriers		
Stochastic	Kriging	Yes	1	Variable	Prediction probability quantile and standard error	Maximum flexibility and allows for trends		Stationarity
	Co-kriging		2+	Variable				

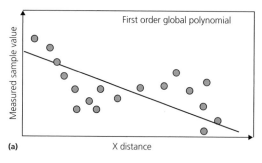 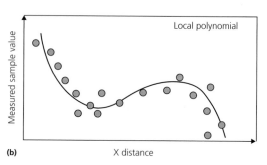

Fig. 16.5 (a) A first-order global polynomial surface profile – the planar surface (black line) does not pass through the sampled points and reflects only the gross scale pattern of the data; and (b) a local polynomial surface profile – the surface is no longer planar but has flexure. It still does not necessarily pass through all the data points.

the neighbourhood (i.e. the sector configuration), even more control is enabled. In this way the interpolation can be made to behave in a more (or less) local manner. The process is a little like convolution filtering, in that a function is fitted to the values in a neighbourhood to derive an estimated value for that unknown location and neighbourhood. The interpolator then shifts to the next unknown location and the process is repeated until a grid of estimated values is built up. Fig. 16.6(b) shows the result of local polynomial interpolation for the same group of points, as shown in Fig. 16.3. As with the global interpolator, selection of first, second, third-order, and so on, polynomial functions allow more complexity to be allowed for in the predicted surface except that these are fitted within the local neighbourhood. Hence if the neighbourhood size is increased to the point where it includes all the data points, the result will be equivalent to a global polynomial interpolator.

By adding more orders to the polynomial function, it can be made to fit almost any data distribution but if the data are not that complex, then why bother going to such effort when a simpler one would be quicker and more appropriate. If you have to work so hard to make the function fit data, that extra effort may not really provide much additional information; and perhaps this is telling you something about the data you have overlooked, such the presence of as an underlying trend and/or more than one population. Generally speaking, and for these reasons, the first and second orders of the polynomial are considered to be the most indicative and significant in fitting to the data and they are effectively estimates of the first- and second-order trends of the data.

Radial basis functions (RBF) and splines – a surface created by a spline is generated using a piece-wise function and can be thought of as a flexible membrane (rather than a rigid plane), stretched between the sample points, of which the total curvature is kept to a minimum but is variable. Variable weights are used to control the flexibility and curvature of the interpolated surface. The surface passes through every sampled point, and so spline functions can be described as being *exact* interpolators. The stretching effect is useful as it allows predicted values to be estimated above the maximum or below the minimum sampled values, so that highs and lows can be predicted when they are known to exist but are not represented in the sample data.

When sample points are very close together and have extreme differences in value, spline interpolation may be ineffective because it involves slope calculations and honours the data. High frequency (sudden) changes in value, such as caused by a cliff face, faults or other naturally occurring barrier, are not represented well by a smooth-curving surface. In such cases, another method may be more effective.

Several types of spline interpolator can be found in most GIS suites: *tension, thin-plate (minimum curvature), regularized, multi-quadratic* and *inverse multi-quadratic* splines. A *tension* spline is flatter and more rigid than a regularized spline of the same sample points - it forces the estimated values to stay closer to the sampled values (Fig. 16.7). The behaviour of a *regularized* spline is more flexible and elastic in character (it has greater curvature). Interpolated surfaces generated from the points in Fig. 16.3 are shown in Fig. 16.8.

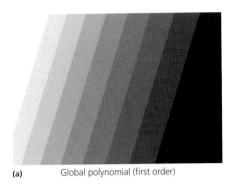

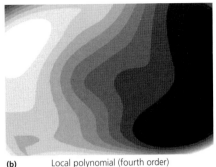

(a) Global polynomial (first order)   (b) Local polynomial (fourth order)

Fig. 16.6 Surfaces constructed from the sample points shown in Fig. 16.3(b), using (a) global polynomial; and (b) local polynomial functions.

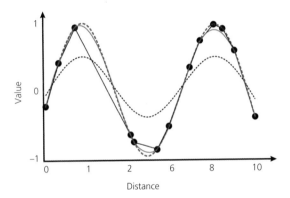

Fig. 16.7 Profile view through a theoretical surface constructed with a quadratic spline interpolator (red dashed line), smoothing or regularised spline (green) and a highly smoothed spline (black dashed), as compared to a linear surface (blue) from a series of z values (black dots).

The spline interpolation process can also be weighted. The simplest form of radial basis function is a weighted linear inverse distance function as follows:

$$z_p = \sum_{i=1}^{n} \lambda_i z_i \qquad (16.6)$$

Where z_p is the estimated value of the interpolated surface at a point p, and λ_i are the data weights.

The form of a tension spline can be expressed by the following:

$$\varphi(r) = \ln(cr/2) + l_0(cr) + \gamma \qquad (16.7)$$

Where $\varphi(r)$ is the radial basis function used, r is the distance between the point and the sample, c is a smoothing parameter, $l_0(\ )$ a modified Bessel function and γ is Euler's constant ($\gamma = 0.577$). The modified Bessel function is given by:

$$l_0(cr) = \sum_{i=0}^{\infty} \frac{(-1)^i (cr/2)^{2i}}{(i!)^2} \qquad (16.8)$$

The general form of a regularised spline can be described by the following:

$$\varphi(r) = \ln(cr/2)^2 + E_1(cr)^2 + \gamma \qquad (16.9)$$

where $E_1(\ )$ is an exponential integral function given by:

$$E_1(x) = \int_1^{\infty} \frac{e^{-tx}}{t} dt \qquad (16.10)$$

A multi-quadratic spline is defined as

$$\varphi(r) = \sqrt{r^2 + c^2} \qquad (16.11)$$

and a thin-plate spline by the following:

$$\varphi(r) = c^2 r^2 \ln(cr) \qquad (16.12)$$

In the regularised type the predicted surface becomes increasingly smooth as the weight value increases. With the tension type, increasing the weight produces a more rigid surface, eventually approaching a linear interpolation between sample point values. Splines cannot assess prediction error, cannot allow for SAC and do not involve any assumptions about the stationarity of the data.

A minimum curvature spline interpolation was used for the 'gridding' of multi-element geochemical sample point in Greenland (for which a smooth surface was required), within a multi-criteria evaluation for mineral prospectivity in § 23.1.

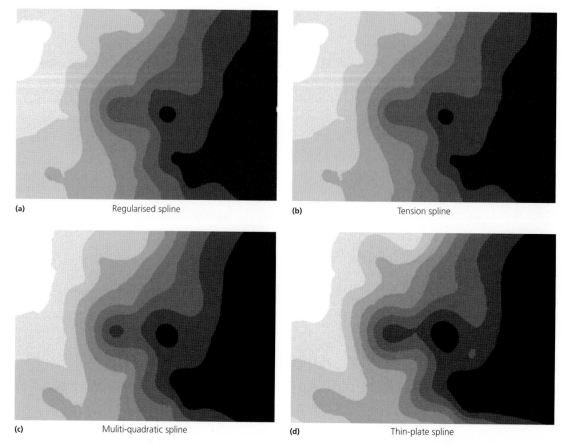

(a)   Regularised spline   (b)   Tension spline

(c)   Muliti-quadratic spline   (d)   Thin-plate spline

Fig. 16.8 Surfaces constructed from the same set of sample points shown in Fig. 16.3(b), using (a) regularised; (b) tension; (c) multi-quadratic; and (d) thin-plate splines. The difference between these are subtle, with surfaces in (c) and (d) being noticeably smoother than those in (a) and (b).

16.3.3.1 Inverse distance weighted average

The *IDW* is a localised interpolator that predicts values through averaging, as its name suggests, but that allows variable *weighting* of the averages according to the distances between the points, using a power setting. The weights are exponents of distance and are largest at zero distance from a location and decrease as the distance increases. For a position x, and for i to n data points with z known values, the unknown weighted average $(z(x))$ is derived as follows:

$$z(x) = \frac{\sum_i w_i z_i}{\sum_i w_i} \qquad (16.13)$$

Reducing the weight or power produces a more averaged prediction because distant sample points become more and more influential until all sample points have equal influence [Fig. 16.9(a)]. Increased weight means that the predicted values become more localised and less averaged, but the influence of the sample point decreases more rapidly with distance [Fig. 16.9(b)]. The weights are commonly derived as the inverse square of distance, so that the weight of a point drops by a factor of 4 as the distance to a point increases by a factor of 2. For the sample points shown in Fig. 16.4, surfaces interpolated by IDW, with low and higher power setting, are shown in Fig. 16.9.

Since IDW is an averaging technique, it cannot make estimates above the maximum or below the minimum sample values, and as a result the predicted surface will not pass through the sample points, so it can be referred to as an *inexact* interpolator. In a surface representing

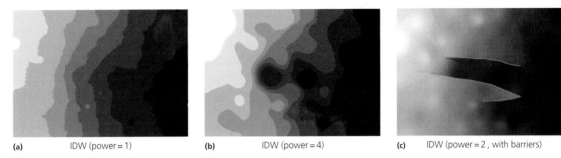

(a)   IDW (power = 1)   (b)   IDW (power = 4)   (c)   IDW (power = 2 , with barriers)

Fig. 16.9 Surfaces constructed from the same set of sample points shown in Fig. 16.3(b), using (a) IDW with a low weight setting; (b) IDW with high weight setting; and (c) using a moderate weight and breaklines (green) where rapidly changing values are known to exist. The interpolated grids in (a) and (b) were produced using ArcGIS's Geostatistical Analysist and (c) using ArcGIS's Spatial Analyst; the 'bullseye' pattern is visible in all three grids.

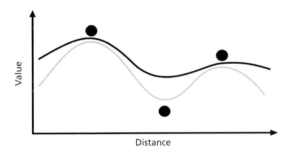

Fig. 16.10 Profile view of a theoretical surface constructed with an IDW interpolator: the black line represents the surface generated with a lower weight setting, the grey line represents that produced with a higher weight setting [compare with Fig. 16.9(a) and (b)].

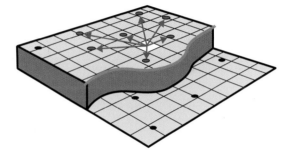

Fig. 16.11 Schematic representation of sample points lying across an abrupt change in values, such as might be caused by a cliff within elevation data. The IDW interpolator allows the incorporation of vector breaklines (green line) to constrain the interpolation process, so that only points on one side (red dots) are used to estimate the output value (white pixel).

elevation, for example, this has the effect of flattening peaks and valleys unless their high and low points are actual sample points (see Fig. 16.10).

One advantage of this method is that barriers can often be incorporated to restrict the predictions, geographically, if structures are known to exist that affect the shape of the surface. Sample points on one side of a barrier are excluded from the interpolation even if they are near to the prediction location (as illustrated in Fig. 16.11). In this way the IDW is prevented from averaging across significant structures.

Like most interpolators, IDW is most effective with densely and evenly spaced sample points. It cannot account for any directional trends in the data, and so the interpolated surface will average across any trend rather than preserve it. It is perhaps useful as a 'first attempt' when little is known about any complexities that may exist. IDW does not involve the assessment of prediction errors or allow for SAC either, and the weighting tends to produce 'bullseyes' around sample point locations that are not real and you may not like the look of.

16.3.4 Stochastic interpolators
16.3.4.1 Kriging

The kriging method was first developed by Georges Matheron, based on the work of Daniel Krige (1951). It is a method of estimation based on the *trend* and *variability* from that trend. Variability refers to the random errors about the trend or mean, but 'error' does not imply a mistake merely a fluctuation from the trend, and 'random' implies that the fluctuation is unknown, is not systematic and could be positive or negative. Kriging may be considered exact or inexact depending on the presence of measurement error or not. Kriging incorporates the principles of probability and prediction, and like IDW, is a weighted average technique except that a surface produced by kriging may exceed the value range of the sample points yet may still not actually passing through the points. Various statistical models can be chosen from, to produce four map outputs (or surfaces) from the kriging process. These include the interpolated surface (the prediction), the standard prediction errors (variance), probability (that the prediction exceeds a threshold) and quantile (for any given probability).

Simplistically, all forms of kriging are based on the following relationship:

$$z_{xy} = \mu_{xy} + \varepsilon_{xy} \qquad (16.14)$$

where z_{xy} is the predicted surface variable (at location xy), μ_{xy} is the deterministic mean or trend of the data and ε_{xy} is the spatially auto-correlated error associated with the prediction (see Fig. 16.12).

The general form of kriging can be defined as follows:

$$\hat{f}\left(z_{xy}^o\right) = \sum_{i=1}^{n} w_i f\left(z_{xy}^i\right) \qquad (16.15)$$

where the function determines the output prediction value of a location so that $\hat{f}\left(z_{xy}^o\right)$ is the predicted output value and is a weighted linear combination of the input values (ranging from i to n), and w_i refers to the weight for the ith input value.

So the predicted surface value at any position is a function of the trend and the deviation from that trend (Fig. 16.12). The differences between the different forms of kriging can be explained in reference to this relationship. *Ordinary kriging* assumes an unknown but constant mean, that is, $\mu_{xy} = \mu$ at all locations, so there is no underlying trend to the data, and that the sample values are random (spatially auto-correlated) errors about the unknown mean [Fig. 16.13(a), and an example is shown in 16.14(a)]. In situations where there is a trend and the mean is no longer constant, the trend is represented as a linear or non-linear regression; this is the basis of *universal kriging*, which assumes a varying but still unknown mean and that the sample values are random (spatially auto-correlated) errors about the mean [Fig. 16.13(b)]. In contrast, *simple kriging* assumes that the mean is known in advance and that it may be constant or variable [Fig. 16.14(b)]. *Indicator kriging* involves the use of other transformations,

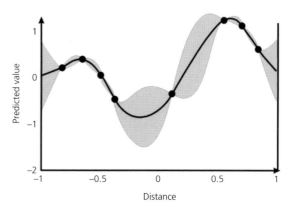

Fig. 16.12 Schematic two-dimensional illustration of interpolation by kriging. The black line indicates the interpolated (predicted) surface that follows the means of the normally distributed confidence intervals (grey areas) or variance about the mean. It can be seen that where the distance between sample points (black dots) is great, the confidence interval is correspondingly wide.

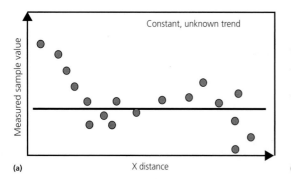 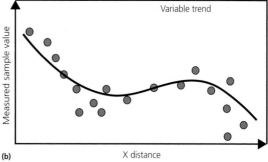

Fig. 16.13 Ordinary and universal kriging (illustrating constant mean and varying mean). (a) Ordinary kriging a constant mean (no trend) and that the mean value is not known in advance; and (b) universal kriging in which there is a trend in the data, but the terms of its function are not known.

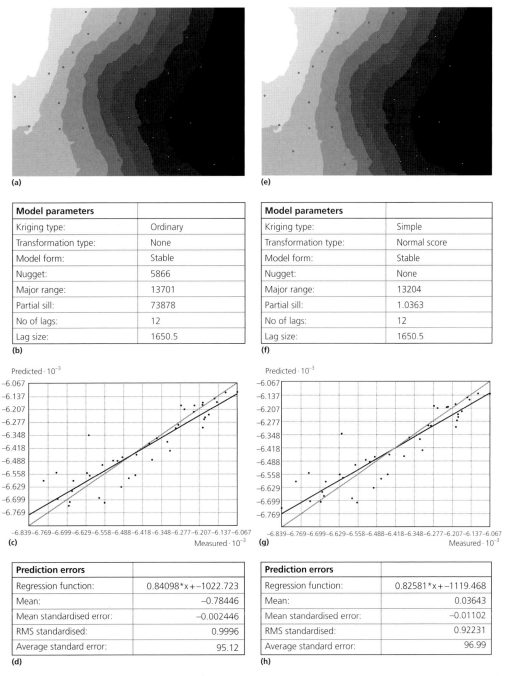

(a)

Model parameters	
Kriging type:	Ordinary
Transformation type:	None
Model form:	Stable
Nugget:	5866
Major range:	13701
Partial sill:	73878
No of lags:	12
Lag size:	1650.5

(b)

Predicted · 10^{-3}

(c)   Measured · 10^{-3}

Prediction errors	
Regression function:	0.84098*x + −1022.723
Mean:	−0.78446
Mean standardised error:	−0.002446
RMS standardised:	0.9996
Average standard error:	95.12

(d)

(e)

Model parameters	
Kriging type:	Simple
Transformation type:	Normal score
Model form:	Stable
Nugget:	None
Major range:	13204
Partial sill:	1.0363
No of lags:	12
Lag size:	1650.5

(f)

Predicted · 10^{-3}

(g)   Measured · 10^{-3}

Prediction errors	
Regression function:	0.82581*x + −1119.468
Mean:	0.03643
Mean standardised error:	−0.01102
RMS standardised:	0.92231
Average standard error:	96.99

(h)

Fig. 16.14 Interpolation result for the same point dataset using ordinary and simple kriging, with sample point positions indicated as dots. Predicted surfaces, model parameters, cross-validation plots and prediction errors are shown in (a)–(d) for ordinary kriging and in (e)–(h) for simple kriging. The predictions results are very similar but, interestingly, it can be seen from the cross-validation and prediction errors that the ordinary kriging result is slightly better. The bold regression line is slightly closer to the faint ideal line in (c) than it is in (g); the mean standardised error should be as low as possible; and the root mean square (RMS) standardised should be as close to 1 as possible.

$f(z_{xy})$, applied to the predicted value rather than the sample value, such that the predicted values are signed (0 or 1) representing the probability that the surface value will exceed or fall below a specified threshold; 1 if the value is above the threshold, or 0 if below. This may be useful if predicting values on which rigid decisions will be made, such as whether chemical substances are in high enough concentration to warrant the area being classified as contaminated and hazardous or not. It can be thought of as a combination of kriging with reclassification, and forms an area of overlap with multi-criteria evaluation. *Disjunctive kriging* forms a development of this approach in which a series of possible transformations is searched to predict the function of $f(z_{xy})$.

Kriging assumes stationarity in the data and, in some methods, that the data are normally distributed.

16.3.4.2 Co-Kriging

Where kriging involves interpolation of a single variable, *co-kriging* involves the simultaneous interpolation of more than one variable. As a result, *co-kriging* allows the derivation of *cross-correlation* as well as SAC and is given as a minor modification of eqn. 16.14.

$$z_{xy}^{i} = \mu_{xy}^{i} + \varepsilon_{ixy}^{i} \qquad (16.16)$$

In this way, different trends and SAC can be considered for each of the *i* variables. This may be useful if you do not have equal number sample points for all variables and need to share values; the prediction can be made from the values of both variables and from the correlation between them. For instance, if you have multi-element geochemical data for samples collected by different ground sampling strategies (rock in-situ samples, rock transport, stream sediment etc.) but not many samples for any single collection method, it may be useful to interpolate the concentrations from all available sample types for a particular element. Care must be taken, however, to ensure that the combination of sample points is actually conceptually meaningful, that is, that the two sample types being combined represent the same or comparable phenomena; they may not. Alternatively, if particular associations of elements are indicators for some phenomena of interest, co-kriging using those elements and evaluating the relationships between them (and their error patterns) may be very revealing. Here again is an overlap with multi-criteria evaluation and decision making.

All the forms of kriging product are also produced by co-kriging, creating *ordinary co-kriging, universal co-kriging, simple co-kriging, indicator co-kriging* and *disjunctive co-kriging*.

You will find many more, and more detailed, explanations and examples of kriging in the literature. Some early and key texts include Agterberg (1974), Cressie (1990), Krige (1951) and Matheron (1963), though there are many more applications and developments.

16.4 Summary

The important aspects to take away from this chapter include the importance of knowing your data from the start and understanding its make-up and provenance, so to make the best choice of interpolator. Understanding how the data were collected, recorded and measured point to what they represent or, importantly, what they do not represent, 'on the ground'. This is essential to the understanding of any statistics derived from the data. Realising the existence of populations within the data is important when interpolating since to treat the data as a single population when in fact it is several could produce meaningless results and could mean underestimation of calculations or forecasted quantities. Similarly, determining how points are selected for the interpolation process can have significant effects on the validity of the result.

In terms of control over the interpolation process, RBF or splines can be considered more flexible than the IDW and less flexible than kriging, but it is the distribution and quantity of your data that should dictate the kind of interpolation you use rather than the convenience of the tools. In general, when data are plentiful, geostatistical methods give better results and, unlike the simpler methods, do not treat noise as part of the data.

This chapter also touches on the conversion of vector point data into a raster representation about filling space; that is, estimating values where none exist. There are also instances where the concepts here have elements in common with decision making, multi-criteria evaluation and spatial analysis.

16.5 Questions

1 What are the best methods of sample selection? What are the potential effects of the methods used?

2 What are the advantages and potential dangers of using interpolation?

3 Under what circumstances might you decide it was not appropriate to use interpolation?

4 When are spline interpolation methods most and least useful?

5 Why would you need to include barriers in weighted interpolation methods?

6 What is the difference between kriging and other weighted methods of interpolation? When is kriging not likely to give better results than any other types?

7 Why is it important, in kriging, to have a thorough understanding of the variance (and semi-variance) of the data?

8 Why is it worth using interpolated data (think about explaining this to a non-technical decision maker)?

9 Consider some applications and decide which interpolation methods would be most appropriate – such as assessing the probability of geochemical contamination from regularly spaced soil samples, or estimating production of a mineral commodity using geochemical data derived from various samples of rock and sediment collected from several levels within a mine.

CHAPTER 17

Representing and exploiting surfaces

17.1 Introduction

A *surface* models a phenomenon that varies continuously across an area, such as elevation. Since the phenomenon could represent precipitation, temperature, magnetic susceptibility or any variable, a more general term could be a *statistical surface* because the surface describes the statistical representation of the magnitude of that variable. Surfaces provide the 'height' information, or *z-values*, necessary both for spatial analysis and for three-dimensional (3D) visualisation, either in the form of raster digital number (DN) values or the nodes of a Triangulated Irregular Network (TIN).

This chapter concentrates on the use of raster data since its structure lends itself to terrain or surface analysis, allowing the description and quantification of terrain morphology and the extraction of surface parameters in a more uniform, regular manner than from a vector surface. This chapter also deals with the visualisation of information that has an implied 3D quality within a two-dimensional (2D) environment on a map, the visualisation of information within a simulated 3D environment and the exploitation of surface data to derive parameters that quantify the 3D environment and are useful within the broader scope of spatial analysis. When talking about raster data visualisation, image processing techniques are embedded by default.

17.2 Sources and uses of surface data

Methods of surface description using raster and vector models have been mentioned earlier in Chapter 12. Primary sources of surface data include point surveys, *photogrammetry, phase correlation* (see Chapter 11) from stereo imagery or air photography, *interferometry* from radar imagery, and *altimetry*. Surfaces can also be produced by the digital capture of contours from analogue maps (and secondary conversion to a surface) and by interpolation from survey points. We shall not dwell on the use of contours (as these are familiar concepts that have been dealt with in many other texts) and interpolation from point data has already been dealt with in Chapter 16. The many processes of digital elevation model (DEM) generation are the subject of a great breadth of research and are covered in great detail in other texts. The technical details of such procedures are not within the scope of this book.

17.2.1 Digital elevation models

Digital elevation model or *DEM* is a term used widely to describe a representation of a continuously sampled surface representing ground surface height above a datum. A DEM generally represents the uppermost level of a surface feature, including vegetation canopies and buildings; that is, it does not necessarily represent the ground surface level of the earth. If this is required, the DEM must be modified to remove any such building and tree canopy heights. And when this is achieved, the product may be described as a *bare earth model, digital terrain model* (DTM) or *digital surface model* (DSM). The term *bare earth* can refer to either the DEM or the extracted contours, from which the effects of objects such as buildings and tree canopies have been removed, leaving only ground surface elevation values. The production of the DTM from the DEM requires either post-processing to correct to a *bare earth model*, or

Image Processing and GIS for Remote Sensing: Techniques and Applications, Second Edition. Jian Guo Liu and Philippa J. Mason.
© 2016 John Wiley & Sons, Ltd. Published 2016 by John Wiley & Sons, Ltd.

calculation from the raw acquired elevation data, as in the case of laser altimetry where the collected data represent complex information containing the uppermost surface and ground-level elevations (and any other objects in between). For the sake of simplicity in this chapter we shall stick to the acronym 'DEM' as a generic term in reference to all forms of digital surface data.

A DEM is a digital raster data file consisting of terrain elevations for ground positions at regularly spaced horizontal intervals. DEMs may be used aesthetically or analytically, that is, they can be used in combination with digital images and vectors to create visually pleasing and dramatic graphics or for the calculation of various surface parameters such as terrain slope, aspect or profiles.

One potentially misleading issue relates to the term 'continuous' which is frequently used to describe raster data. It is more correct to describe the model as being *continuously or regularly sampled* at discrete intervals rather than truly continuous in nature. Its ability to represent a surface depends on its spatial resolution and the complexity of the ground surface being represented. It is widely accepted that all natural surfaces are fractal in nature, so that at any particular scale, there will always be more detail than we can observe. We must therefore accept that the resolution of the data implies the level of detail that we can work with. The most important factors to be considered prior to the use of DEMs therefore include the planimetric and altimetric accuracies of the source data, the quality and quantity of both the source data and ground control data, the level of terrain complexity, the output spatial resolution as well as the algorithm used to generate the DEM.

17.2.1.1 Photogrammetry

Photogrammetry is the conventional process of obtaining reliable 3D measurements of physical objects and environment from measurements made from two or more photographs or images (Wolf & Dewitt 2000). The photographs or images must have been acquired from different positions with sufficient overlaps, that is, stereoscopically. Two forms of photogrammetry can be identified: metric and interpretative. The former refers to the quantitative measurement and analysis of objects for the purpose of calculating dimensions, including elevation and volume. The latter refers to the more qualitative interpretation and identification of objects and structures through analogue stereoscopy, with the aim of better understanding their relationships with their surroundings.

Elevation data can be derived photogrammetrically from a number of readily available data sources (airborne or spaceborne) such as stereo air photography, or from the many satellite sensors that can aquire stereo imagery, such as ASTER, Ikonos, SPOT, ALOS, GeoEye, Worldview & SkySat. The height accuracy of models generated this way depends mainly on the *base to height ratio* (B/H) and the accuracy of the parallax approximations. The sensor specifications of these instruments is given in Appendix A.

Also included in this category is the well known Global Aster DEM or *GDEM*, which was produced jointly by the Ministry of Economy, Trade, and Industry (METI) of Japan and the United States National Aeronautics and Space Administration (NASA). GDEM coverage spans the area between 83 degress North and 83 degrees South (some 99% of the earth's land surface). GDEM has been produced from hundreds of thousands of stero-pairs, thus reducing artifacts and improving accuracies; it is gridded at 30 m and released in 1 × 1 degree tiles. It is currently in the second version of its release (GDEM v2, October 2011).

The resolution of DEMs derived from aerial photography can vary greatly according to the flight height, camera quality and imaging configuration. Air photography is now collected using high resolution digital cameras, and with onboard Global Positioning System (GPS) devices to georeference the acquired data and inertial navigational units (to record and subsequently correct for the roll, pitch and yaw of the platform). *Unmanned airborne vehicles* (UAVs) are also now increasingly being used to map large areas at very high resolutions, and in stereo.

17.2.1.2 Laser Altimetry (LiDAR)

Surfaces can be generated from laser altimetry or *Light Detection and Ranging (LiDAR)* data, which may be acquired from satellite or airborne platforms. Airborne LiDAR has somewhat revolutionised the acquisition of high accuracy DEM data for large scale mapping applications.

A LiDAR system transmits pulses of light that reflect off the terrain and other ground objects. The receipt of laser pulses is continuous and so the first and last returned pulses can be extracted to differentiate between canopy elevations and true ground or *bare earth* elevations. The return travel time of a laser beam is measured, from the source instrument to the target and back; the distance is then computed (using the known speed of light) to give the height of the surveyed ground position. An airborne LiDAR system typically consists of a laser scanning instrument, a GPS and an inertial navigational unit. Airborne LiDAR derives height elevations with accuracies of between 10 and 15 cm (altimetric) and 15 and 30 cm (planimetric).

With the same ranging principle, airborne radar altimeters and barometric altimeters are also used to map terrain. The radar altimeter height can then be subtracted from the barometric altimeter height to give surface elevation with respect to sea level.

Some LiDAR datasets are available as off-the-shelf products, from satellite sources such as ACE, a 1 km (30 arc seconds) DEM, globally available from De Montfort University (see online resources). These are generally low resolution products aimed at small scale, regional applications. Airborne LiDAR surveys are normally bespoke and relatively expensive to commission and are therefore not likely to be freely available in the foreseeable future.

17.2.1.3 Synthetic aperture radar (SAR) interferometry

The theoretical basis for interferometric derivation of elevation from SAR data has already been described in some detail in Chapter 10, so we have no need to dwell on this here. InSAR DEMs can be generated from SAR imagery acquired by the ERS, Envisat, Radarsat and ALOS, TanDEM-X, TerrSAR and Sentinel-1. The unique advantage of InSAR DEM generation is its all weather capability. It can penetrate clouds under which conditions all optical sensor-based technology cannot operate.

One very well-known and widely used such InSAR DEM dataset was produced during the Shuttle Radar Topographic Mission (SRTM). SRTM was acquired during the year 2000, onboard the Space Shuttle Endeavour, when topographic data of roughly 80% of the earth's land surface (between latitudes 60 degrees north and 60 degrees south) was generated and then gridded at 90 m (a higher resolution product was, at the time, only available for the USA). The absolute planimetric and altimetric accuracies of the 90 m DEMs are 20 m and 16 m respectively. The SRTM product is ideal for regional mapping, typically at scales of between 100,000 to 150,000. Toward the end of 2014 it was decided that the 1 arc second (30 m) product would be released for the entire globe.

17.2.1.4 Phase-correlation-based DEM generation

The novel technique of surface generation from unconventional and narrow baseline stereo imagery using phase correlation, and without ground control data, is described in detail in Chapter 11.

17.2.2 Vector surfaces and objects

As described in Chapter 13, Triangulated Irregular Networks, or TINs, represent surfaces using a set of non-overlapping triangles that border one another and vary in size and form. TINs are created from input points with x, y coordinates and z-values, which become the triangle vertices (nodes). The vertices are connected by lines to form the triangle boundaries (edges), as illustrated in Fig. 17.1. The final product is a continuous surface of triangles, made of nodes and edges. This allows a TIN to preserve the precision of the input data while simultaneously modeling the values between known points. Any precisely known locations such as spot heights at mountain peaks, or road junctions, can be described and added as input points, to become new TIN nodes. TINs are typically used for high-precision modelling, in small areas, such as within engineering applications, where they are favoured because they honour the data.

A related and similar vector structure is the *wireframe* which is often used to describe surfaces and volumes [Fig. 17.1(c)].

17.2.2.1 3D vector features

Capitalising on the TIN method of height representation, ordinary 2D vector features (points, lines or polygons) can be displayed in a 3D environment, on, above or beneath a surface. 3D display requires height information but vector features may or may not have such information in their attribute tables. 2D vector features can be visualised in 3D space by exploiting the height information of other data layers or by projecting according to some numerical attribute (see § 17.3.2). The elevations are then calculated for the x and y positions of the vector feature's vertices and used to project them vertically. These heights, exploited from other layers are generally assumed to define the base level or ground height of the object. A feature may of course store further attributes

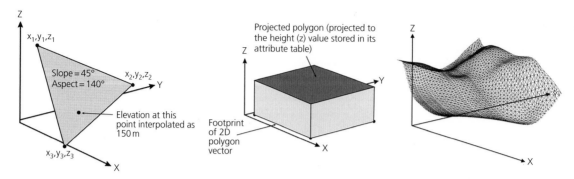

Fig. 17.1 (a) Individual triangular face of a TIN, defined by the 3D coordinate positions of the three irregularly spaced points. The slope and aspect of each face are constant for each triangular face and are calculated when the TIN is generated. The elevation of any position is then interpolated from its position with respect to the points and edges of the triangular face. (b) 3D shape formed from the projection of a vector polygon according to a height (z) value stored in its attributes table; these values could represent either the base level elevation of the object or the object's height above ground level or any other quantity described in the attribute table; (c) a wireframe constructed from x, y & z coordinates of a regularly spaced point file.

that represent the height of the object above ground level, such as building height, or some other quantity.

A feature that contains, in addition its x,y coordinates, one or more z-values as part of its geometry, is referred to as a *3D vector feature*. At their simplest, 3D points have one z-value; 3D lines and polygons have a z-value for each vertex defining the object. Actual or complex generic buildings and other far more complex structures can then be modeled by extending this concept.

The VRML (Virtual Reality Modelling Language) was developed as a standard file format for storing and displaying 3D vector graphics. This has been superceded by X3D, which is now an ISO standard; both formats use standard XML syntax. Google has its own version, called O3D, for creating 3D graphcs inside web browsers. The development of 3D vector models and vector topology is certainly a very active area of development.

17.2.3 Uses of surface data

Surface data have a great many potential uses in a many application areas. They are commonly used as relief maps to convey 3D information within the 2D environment and they are essential for 3D visualisation. DEMs are also required for the ortho-rectification of optical images and for making terrain corrections to radar and gravity survey data. They also form a valuable data source for the calculation of descriptive surface parameters, for flow-modeling (of water and mass movements), for *geomor-phometrics*, or geomorphological terrain analysis, and for the calculation of other engineering, hydrological and hydrogeological indices.

Elevation models are used in applications over a growing range of scales, from global, meso and topo (conventional mapping scales) to micro and nano scales (very small scale measurements), and within a diverse range of disciplines - meteorological, geological and geomorphological, engineering, biological and architec-tural. At the micro and nano scales, close-range photo-grammetry is now a rapidly developing science. Examples in these fields include the micro-scale analysis of the terrain of stream beds, to reveal the physical char-acteristics of habitats occupied by small organisms such as fish and crustaceans; and analysis of rock fracture surfaces for modelling fluid flow or of the frictional properties affecting rock strength.

17.3 Visualising surfaces

GIS visualisation tools, in both two and three dimen-sions, rely on the ability to share and integrate data, models and interpretations. The simplest form of visu-alisation involves the display of 2D images with conven-tional cartographic symbols. For instance the geological map can be recreated using conventional geological symbology, which can easily be incorporated within the GIS, through the use of special fonts. Thus structural information, for instance, can be presented to give the appearance of a published geological map.

GIS also bridges the gap between 2D and 3D display and analysis. This is especially useful in geosciences, because depth is such a fundamental consideration and the integration of sub-surface data has become an essential part of any digital mapping technology.

Visual exploration and interrogation, in several dimensions, facilitates enhanced understanding of structures and relationships. Virtual field visits become a route activity and are vital in assisting in logistical planning and to improve understanding prior to setting foot in the field, thus saving valuable time and reducing risks. All GIS software provide 2D and 3D tools for the manipulation of surfaces, images and maps in pseudo-3D space and for the mathematical derivation of other products, such as slope angle, aspect or azimuth, curvature, line-of-sight, watersheds and catchments. Examples of software providing excellent 3D manipulation, viewing and analytical capabilities are ERDAS Imagine, ER Mapper, ArcGIS 3D Analyst ArcScene & ArcGlobe, MapInfo Vertical Mapper, Geomatica Fly, Virtalis' Geovisionary, Move and Micromine.

17.3.1 Visualising in two dimensions

Appreciation of a truly 3D physical landscape on a conventional 2D (flat) maps requires some level of interpretation and imagination. Cartography has traditionally made use of a range of visual symbols to show height information and create the illusion of an undulating surface: elevation contours, spot height symbols, hill shading and cliff and slope symbols. GIS allows much more through the simulation of the 3D environment but we often still need to use 2D output to convey the results of our efforts. Again GIS cleverly provides us with the ability to visualise 3D quantities and objects within a 2D medium, through the use of *contours, shaded relief* or *hillshades* and *artificial illumination*.

17.3.1.1 Contours

Contour lines are the more familiar and mathematically more precise way of representing surface information but they are not visually powerful. A contour is a line connecting points of equal surface value. Contour lines reveal the rate of change in values across an area for spatially continuous phenomena. Where the lines are closer together, the change in values is more rapid. They are drawn at a specified interval, which represent the interval is simply the change in z value between the contour lines. For example, a contour map of precipitation with a contour interval of 10 mm would have contour lines at 10, 20, 30, and so on. Each point on a particular contour line has the same value, while a point between two contour lines has a value that is between the values of the lines on either side of it. The interval

determines the number of lines that will be on a map and the distance between them. The smaller the interval, the more lines will be created on the map. A base contour may also be specified, as a starting point; this is not necessarily the minimum contour, but refers to a starting point from which contour values may go both above and below, based on the contour interval. For example, the base contour may be set to 0 and the interval may be set to 10. The resulting contour values would be −20, −10, 0, 10, 20, and 30. Watson (1992) provides a comprehensive description of the concepts of all forms of surface modeling, including algorithms for contouring and surface generation.

17.3.1.2 Shaded relief or hillshading

Hillshading is a technique used to create a realistic 2D view of terrain by simulating light falling on a surface from a given direction, and the shadows this creates. It is often used to produce visually appealing maps that are easier to interpret. Used as a background, hillshades provide a relief over which both raster data or vector data can be displayed.

There are several types of hillshading: *slope shading* where tonal intensity is proportional to the angle of slope (e.g. the steeper the slope the darker the tone); *oblique light shading* where the pattern of light and dark on the surface is determined by a simulated oblique light source; and *combined shading*, which represents the combination of these two types.

Contours and hillshading are quite often used together since they complement one another: hillshading provides a qualitative impression of the terrain, while contours show quantitative height information but only at discrete locations.

Oblique light shading involves the simulation of the oblique illumination of a surface by defining a position (angle and height) for an artificial parallel light source and calculating a brightness value for each position, based on its orientation (slope and aspect) relative to the light source. The surface illumination is estimated as an 8-bit value in the range between 0 and 255, based on a given compass direction relative to the sun and an altitude above the horizon. *Analytical oblique light hillshading* estimates the brightness based on the angle between the selected *illumination direction* (vector) and the *surface normal vector* (see Fig. 17.2).

Conceptually, the illumination vector is defined by two angles, an angular attitude relative to north, or azimuth (given as a compass direction between 0° and

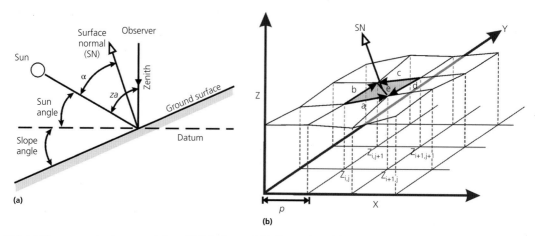

Fig. 17.2 (a) Slope geometry for computation of hillshades; and (b) illustration of the SN vector for a raster surface. SN is perpendicular to the average cross-product (e) of ab and cd, which approximates the surface area occupied by the pixel, depending on the resolution [(a) Baston *et al.* 1975. Reproduced with permission from US Geological Survey and (b) Corripio, 2003. Reproduced with permission from Taylor and Francis.]

360°) and an altitude (given as a horizontal angle between 0° and 90°). The surface normal (SN) is a vector perpendicular to a surface, as defined in the raster case, by a pixel and its closest 8 neighbouring grid cells (as illustrated in Fig. 17.2). The hillshade DN value assigned to the output pixel will be proportional to the cosine of the 3D angle between the SN and the illumination vector [shown as α in Fig. 17.2(a)]; a technique originally suggested by Wiechel (1878). So for slopes almost normal to the illumination direction, the angle will be very small, the cosine value large, and the estimated brightness proportionally high.

There are a number of ways of calculating this from a raster surface. The altitude and azimuth of the illumination source and the slope and aspect angles of the pixel being evaluated are needed. From these parameters, *hillshade* (h) can be calculated as follows (where all angles are calculated in radians):

$$h = 255\left(\cos(za)\left(\cos(slope)\right)\right.$$
$$\left. +\left(\sin(za)\sin(slope)\cos(az - aspect)\right)\right) \quad (17.1)$$

where *za* is the zenith angle, and *az* the azimuth angle. Altitude is normally expressed in degrees above the horizontal but the formula requires the angle to be defined from the vertical, that is, the *zenith angle (za)*, which is measured between the overhead zenith and the illumination direction [see Fig. 17.2(a)]. The zenith angle represents the 90° complement of altitude (90– altitude). The azimuthal angle of the illumination must

be changed from its compass bearing to a mathematical unit (*maz*), that is, the *maz* $= 360° - az + 90°$, and if the *az* angle is greater than or equal to 360 degrees, then *maz* $- 360$. Note that eqn. 17.1 is essentially the same as eqn. 3.27 introduced in § 3.7.1 for true sun illumination.

The *x*, *y*, *z* values of the vectors along the side of the central pixel, a, b, c and d, in Fig. 17.2(b) are given by the following:

$$\begin{aligned}
a &= (p, 0, \delta z_a), & \text{where } \delta z_a &= z_{i+1,j} - z_{i,j} \\
b &= (0, p, \delta z_b), & \text{where } \delta z_b &= z_{i,j+1} - z_{i,j} \\
c &= (-p, 0, \delta z_c), & \text{where } \delta z_c &= z_{i,j+1} - z_{i+1,j+1} \\
d &= (0, -p, \delta z_d), & \text{where } \delta z_d &= z_{i+1,j} - z_{i+1,j+1}
\end{aligned} \quad (17.2)$$

The vector normal to the surface in the central pixel in Fig. 17.2(b) is then defined by the following:

$$SN = \frac{a \times b}{2} + \frac{c \times d}{2} = \frac{1}{2}\begin{vmatrix} i & j & k \\ p & 0 & \delta z_a \\ 0 & p & \delta z_b \end{vmatrix} + \frac{1}{2}\begin{vmatrix} i & j & k \\ -p & 0 & \delta z_c \\ 0 & -p & \delta z_d \end{vmatrix} \quad (17.3)$$

Simplifying eqn. 17.3 we have the cell orientation defined by the heights of the central pixel corner points:

$$SN = \begin{pmatrix} 0.5p\left(z_{i,j} - z_{i+1,j} + z_{i,j+1} - z_{i+1,j+1}\right) \\ 0.5p\left(z_{i,j} + z_{i+1,j} - z_{i,j+1} - z_{i+1,j+1}\right) \\ p^2 \end{pmatrix} \quad (17.4)$$

Once *SN* is derived, the 3D angle between the SN and the illumination vector [shown as α in Fig. 17.2(a)] can

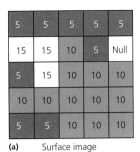

(a)		Surface image		

Fig. 17.3 Simple raster example: (a) the input surface; and (b) the calculated hillshade, with 135° as the azimuth and 45° as the altitude. The bold line in (b) encloses the nine central pixels for which hillshade values can be calculated using the moving 3 × 3 window; null values are generated around the edges and adjacent to any nulls in the input raster, hence the two null values inside the bold line.

be calculated and the DN value assigned to the output pixel will be proportional to the cos(α). A simple example is shown in Fig. 17.3.

The values returned by hillshading may be considered a relative measure of the intensity of incident light on a slope. Such measures could be useful for many applications, such as selecting suitable sites for particular agricultural practices or slopes suitable for ski resorts. Hillshading of DEMs can also accentuate faults and other geological structures and facilitate geological interpretation. Convolution gradient filtering can also be used to identify linear features in remotely sensed imagery but the result may not be as visually striking as a shaded relief image derived from a DEM.

When creating a cartographic hillshade for visualisation, convention dictates that the light source is placed in the north-west (upper left) quadrant of the map, so to cast a shadow at the bottom-right of the object. The eye tends to see objects better when the shadow is cast at the bottom of the view; placing the light source elsewhere creates a visual effect that makes hills look like hollows. Everyone's perception is slightly different, however, and examples are showin in Fig. 17.4 so you can judge for yourself.

17.3.2 Visualising in three dimensions

Visualisation in 'pseudo-3D' requires a height to convey the third dimension, whether it is topography or some other attribute. In fact several things are needed: the definition of the 3D coordinate space, a viewing perspective, a vertical exaggeration (VE) to control depth and (optionally) a simulated light source, in addition to the

datasets being visualised. Certain parameters necessary for visualisation in 3D are connected to the data (*base or ground elevations*) and while others are temporary or virtual and control the environment of visualisation (*artificial illumination* and *vertical exaggeration*).

17.3.2.1 Basal or ground elevations
These refer to the basal or ground level values used to display raster images or vector objects within the 3D space. They are needed to place an object correctly according to the z scale of the display. The height values can be derived from one of several sources: the values of the layer being displayed (the nodes of a TIN or the pixels DN values of DEM); the values stored in a different TIN or raster layer that covers the same geographic area; or last, from a value or expression (this would produce elevation as a fixed value or as a function of another attribute). Fig. 17.5 illustrates the principle and effects of this.

17.3.2.2 Vertical exaggeration
This refers to a kind of relative scaling used within 3D views to make subtle surface features (and any objects on the surface) more visible. This scaling applies to the environment of visualisation rather than to the datasets being visualized, that is, it is a temporary visual effect produced by multiplying the height values in the display by a constant factor. A VE of 2 multiplies all heights by 2 and exaggerates the vertical scale, whereas an exaggeration of 0.1 will actually suppresses the vertical scale, and so on, as shown in Fig. 17.6.

VE has two main uses: to emphasize subtle changes in elevation on a surface that is relatively flat or has great extent; and secondly to force the x, y units into proportion with the z units, if these represent different quantities or units.

17.3.2.3 Projection of 2D vector objects into 3D space
Vector projection or 'extrusion' represents the 3D projection of 2D vector features. For example, an extruded point becomes a vertical line; an extruded line becomes a vertical plane or wall; an extruded polygon becomes a block. In contrast to the base heights, extrusion controls upper elevation of features and the simulation of 3D objects using 2D map features. Since they are simulations, feature extrusions can be said to produce *geotypical* or generic representations of physical objects, rather than *geospecific* ones (i.e. actual objects).

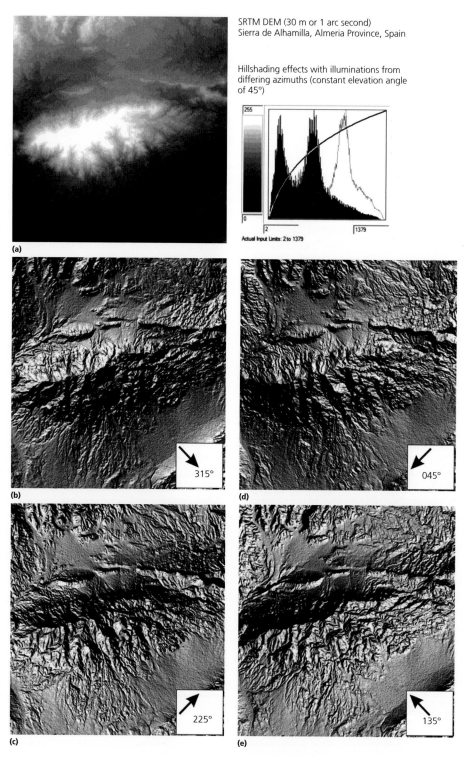

SRTM DEM (30 m or 1 arc second)
Sierra de Alhamilla, Almeria Province, Spain

Hillshading effects with illuminations from
differing azimuths (constant elevation angle
of 45°)

Actual Input Limits: 2 to 1379

Fig. 17.4 Hillshading examples for a portion of the 30 m SRTM DEM and its histogram (a) with the same illumination (elevation angle) of 45° but differing azimuth angles: (b) 045, (c) 135, (d) 225 and (e) 315 degrees.

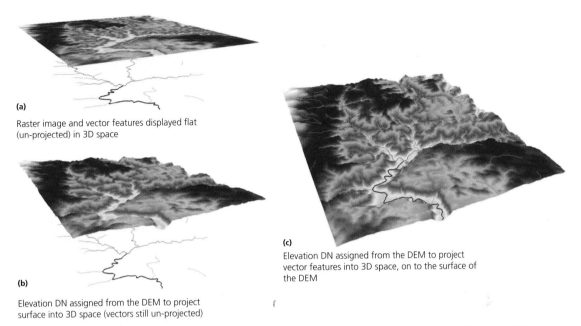

(a)
Raster image and vector features displayed flat (un-projected) in 3D space

(b)
Elevation DN assigned from the DEM to project surface into 3D space (vectors still un-projected)

(c)
Elevation DN assigned from the DEM to project vector features into 3D space, on to the surface of the DEM

Fig. 17.5 Schematic illustration of the effect of assigning basal elevations to a raster surface and vectors in 3D space: (a) flat raster and vector features; (b) surface extruded but vectors are still flat; and (c) both surface and vectors have height and the vectors now plot on the surface.

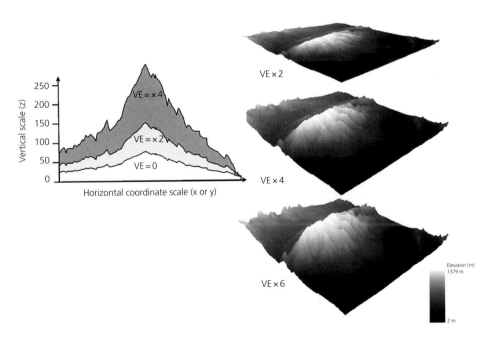

Fig. 17.6 (a) Schematic illustration of VE in 3D environment. Three surface elevation profiles are shown, with no exaggeration (lowest profile), and with VE factors of ×2 and ×4 (middle and upper profiles respectively); and (b) perspective view of the SRTM shown in in Fig. 17.4, with VE of ×2, ×4 and ×6.

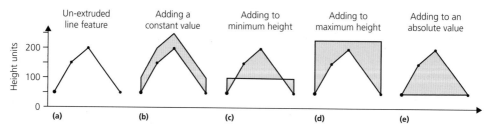

Fig. 17.7 Illustration of the mechanisms of extruding 2D vector objects into the 3D perspective environment, for the purposes of visualisation.

Features can be extruded by a variety of methods, as shown in Fig. 17.7. A simple line feature is shown whose heights, when added to by various methods, are altered to simulate other features. By adding a constant value, the line feature is extruded upward by a constant value [50 units, as shown in Fig. 17.7(b)]; this can be applied to points and lines only. A second form of extrusion is formed when a value is added to the minimum or maximum height of the feature (base or top), and all other vertices are extruded to the same absolute value, whether up or down [as shown in Fig. 17.7(c) and (d)]; this can be applied to lines and polygons only. Last, the vertices can be extruded to a specified absolute value, whether above or below the original values, as shown in Fig. 17.7(e); this can be applied to points, lines and polygons.

17.3.2.4 Artificial illumination

This also applies to the environment of visualisation rather than to the datasets themselves and is calculated by the same principle as the estimation of oblique light hillshading. Every 3D display has a theoretical light source, the position of which controls the lighting and shading of the display. The pseudo-illumination geometry is defined by azimuth and altitude angle settings. Again azimuth is a compass direction, measured clockwise in degrees from 0 (due north) to 360 (also due north). Altitude is the angle, measured in degrees from 0° to 90°, between the light source and the horizon. An altitude of 0° means the light source is level with the horizon; an altitude of 90° means it is directly overhead.

17.4 Extracting surface parameters

Surface or terrain parameters, such as slope angle (gradient) and orientation (aspect), are important controls on a number of natural processes, such as rainfall runoff and erosion, and incident solar radiation upon slopes. There are many published texts describing different methods of calculating these parameters from DEMs. As described in detail in Chapter 4 and briefly in Chapter 15, the calculation of slope gradient, aspect and curvature, from a raster surface are essentially neighbourhood operations or point spread functions, involving the use of a convolution kernel that is passed over the raster, to produce a new set of values that describe the variance of each parameter and the morphology of that surface. Here we provide a summary of the parameters and the more common of the methods for their calculation, referring the read to further texts where appropriate. Each parameter is illustrated here using a simple 5 × 5 raster surface.

17.4.1 Slope: Gradient and aspect

A slope is defined by a plane tangent to the surface, as modelled by a DEM at a point and it confers the angle of inclination (steepness) of that part of the surface. While typically applied to topography, slope may be useful in analysing other parameters, e.g. for a surface of rainfall intensity, showing where intensity is changing and how quickly (steeper 'slopes' indicate values that are changing faster).

Slope has two component parameters: a quantity (*gradient*) and a direction (*aspect*). Gradient (*g*) is defined as the maximum rate of change in altitude, and aspect (*a*) represents the compass direction, or azimuth, of this maximum rate of change. For the geoscientist this is equivalent to dip and dip direction of an inclined bedding or structural surface. More analytically, slope gradient at a point is the first derivative of elevation (*z*) with respect to the surface slope, where *g* is the maximum angle, and the direction or bearing of that angle is the *aspect*. Since gradient has direction, it is a vector product. At the same time the first derivative at a point can be defined as the slope (angular coefficient or trigonometric tangent) of the tangent to the function at that particular point.

The general mathematical concept and calculation of gradient for raster data have already been introduced in chapter 4 (see § 4.4, eqns. 4.12, 4.14, 4.15) in relation to high pass filters. Here we address the same parameter with direct relevance to surfaces and in particular to DEM data.

17.4.1.1 Gradient

Slope gradient may be expressed as either degrees or percent; the former are commonly used in scientific applications, while the latter is more commonly adopted in transport, engineering and other practical applications. *Percent* gradient is calculated by dividing the elevation difference (known as the rise) between two points, by the distance between them (known as the run), and then multiplying the result by 100. The *degree* of gradient is derived from the geometric relationship between the rise and run, as sides of a right-angle triangle; the angle opposite the rise. Since degree of slope is equal to the tangent of the fraction of rise over run, it can also be calculated as the arctangent of rise over run. Measures of slope in degrees can approach 90° but measures in percent can approach infinity, for instance in the case of a vertical cliff. An example is shown in Fig. 17.8.

So for a raster grid, gradient is calculated, on a pixel-by-pixel basis, within a moving 3 × 3 window, as the maximum rate of change in values between each pixel and its neighbours. The gradient (g) is calculated using a second order finite difference algorithm based on the four nearest neighbours:

$$\tan g = \sqrt{\left[\left(\delta z / \delta x\right)^2 + \left(\delta z / \delta y\right)^2\right]} \qquad (17.5)$$

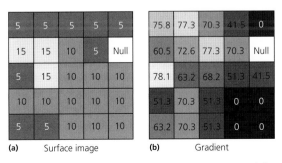

(a)   Surface image   **(b)**   Gradient

Fig. 17.8 Simple raster example: (a) surface (integers); and (b) gradient (floating point).

Such measures are given in degrees or percent, according to taste:

$$g^{degree} = \text{atan}\left(\frac{g^{percent}}{100}\right) \qquad (17.6)$$

$$g^{percent} = \tan\left(g^{degree}\right) \times 100 \qquad (17.7)$$

Common slope procedures involve calculation from the pixel values immediately above, below, to the left and to the right of the central pixel, but not the corner (diagonal) pixel values, and in such cases, it referred to as the *rook's case* because it resembles the way the rook moves on a chess board. Whilst elevation models are commonly stored as integer data (normally 16 bit), the output from a slope calculation will always be a real number, that is, floating point.

Since g is usually calculated in radians, conversion to degrees is given by:

$$g^{degree} = \text{atan}\left(g^{radian}\right) \times \left(\frac{180}{\pi}\right) \qquad (17.8)$$

See § 21.4, § 22.1 and § 22.3 for examples of the use of slope (gradient) within GIS case studies in Part III.

17.4.1.2 Aspect (or azimuth)

The companion component of gradient, *aspect (a)* identifies the *down-slope orientation* or *direction of gradient*, measured with respect to North. When calculated from surface elevation (topography), it is usually referred to as 'aspect'; in reference to quantities other than topographic variation, the term 'azimuth' tends to be used.

The pixel DN values in a raster representing aspect are compass directions or bearings, in degrees, measured in a clockwise direction from 0° to 360°, where North has a value of 0°, East 90°, South 180° and West 270°. An example is shown in Fig. 17.9.

Aspect is calculated in the same 'rook's case' manner, as in the calculation of gradient, as follows:

$$\tan a = \frac{\left(\delta z / \delta x\right)}{\left(\delta z / \delta y\right)} \qquad (17.9)$$

Any pixels calculated as having a zero slope, that is, representing areas that are 'flat', are given a special aspect value, usually −1, to indicate that they have no aspect direction.

$$\text{if}\left(\delta z_x = 0 \ AND \ \delta z_y = 0\right), \quad a = -1 \qquad (17.10)$$

An example of the use of aspect within a slope stability hazard assessment can be found in § 22.3 in Part III.

17.4.2 Curvature

Curvature (*c*) represents the *rate of change in* surface orientation of a variable across an area. It is calculated from a surface (raster or vector), such as elevation, and describes the convexity or concavity of that surface. Referring to Fig. 17.2 it can be considered as a measure of the variation in the SN vector across the image or map, and is therefore the first derivative of the SN vector, and the second derivative of position on a surface (with respect to the changing rate of gradient *g*, see also § 4.4).

Several measures of curvature are recognised: *profile (down-slope) curvature* and *cross-sectional (plan) curvature*, which are orthogonal to one another; and *total curvature* as a summation of profile curvature and cross-sectional curvature. Profile curvature is parallel to the direction of maximum gradient (or aspect) while cross-sectional curvature is perpendicular to it. Essentially, total curvature is the Laplacian, as further explained later.

Geomorphological forms can be discriminated in digital images by their curvature forms, examples of which are illustrated schematically in Fig. 17.10, e.g. ridges, are convex in cross-section and valleys are concave in cross-section (where gradient and laplacian may be variable in both cases), whereas planar slopes have zero cross-sectional curvatures since gradient is constant so curvature is zero. Peaks are convex in both cross-profile and cross-section (i.e. in all directions) and the reverse is true for pits, which are concave in all directions. If areas are flat or slopes are planar, curvature is zero.

As the summation of the changing rate of gradients in x and y directions for a 2D dataset, the Laplacian represents the curvature at a point without direction, that is, total curvature, and it can be derived by the application of a Laplacian convolution filter to a raster surface (see § 4.4). From differentiation theory, a negative Laplacian indicates a convex surface whilst a positive one a concave. If, however, we recall the principle of the Laplacian filter in § 4.4, we know that the conventional Laplacian filter in image processing gives the reverse so that a positive Laplacian indicates convexity and negative one concavity.

The Laplacian, as a scalar, is composed of components in both x and y directions, we can decompose it in the aspect, or gradient, direction and in the direction perpendicular to aspect, by a second partial differentiation of elevation (z) in these two directions. We denote these as $c_{profile}$ and c_{cross}, and they represent the profile curvature and cross-sectional curvature as we introduced earlier. For simplicity, we denote the Laplacian as c_{total}. There is a convention, in some GIS software packages, that the three measures of curvature are calculated in such a way as that positive values for c_{total}, or c_{cross}, indicate <u>upwardly convex</u> surface, whereas a positive $c_{profile}$

5	5	5	5	5
15	15	10	5	Null
5	15	10	10	10
10	10	10	10	10
5	5	10	10	10

198.4	171.9	153.4	135.0	–1.0
135.0	168.7	135.0	153.4	Null
293.2	341.6	90.0	180.0	225.0
–1.0	333.4	–1.0	–1.0	–1.0
341.6	333.4	270.0	–1.0	–1.0

(a)   Surface image   **(b)**   Aspect

Fig. 17.9 Simple raster example: (a) surface (integer); and (b) aspect (floating point).

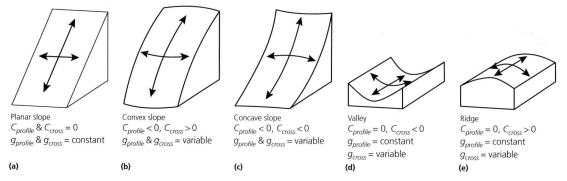

Planar slope
$C_{profile}$ & $C_{cross} = 0$
$g_{profile}$ & g_{cross} = constant

(a)

Convex slope
$C_{profile} < 0$, $C_{cross} > 0$
$g_{profile}$ & g_{cross} = variable

(b)

Concave slope
$C_{profile} < 0$, $C_{cross} < 0$
$g_{profile}$ & g_{cross} = variable

(c)

Valley
$C_{profile} = 0$, $C_{cross} < 0$
$g_{profile}$ = constant
g_{cross} = variable

(d)

Ridge
$C_{profile} = 0$, $C_{cross} > 0$
$g_{profile}$ = constant
g_{cross} = variable

(e)

Fig. 17.10 Some common schematic slope forms and the relationships between curvature and gradient: (a) planar slope; (b) convex slope; (c) concave slope; (d) channel; and (e) ridge, where curvature may be positive (concave) or negative (convex).

indicates that <u>upwardly concave</u> surface; the illustrations in Fig. 17.10 follow this convention. A zero value for any of these indicates no curvature. Other packages give all forms of curvature with positive values to indicate convexity and negative for concavity. Regardless of any convention, Laplacian, using a convolution kernel, is the simplest and reliable method of deriving curvature from raster data. A simple example is shown in Fig. 17.11, with the Laplacian result in Fig. 17.11(b).

Mathematically, true curvature along a curve in direction α is a function of both first and second derivatives defined as below:

$$c_{true} = \frac{\left(\partial^2 z / \partial \alpha^2\right)}{\left(1 + \left(\partial z / \partial \alpha\right)^2\right)^{3/2}} \qquad (17.11)$$

The estimation of curvature, as carried out in many proprietary software suites, including ESRI's ArcGIS and RiverTools, follows the method first formulated by Zevenbergen and Thorne (1987). This method estimates curvature using a second order polynomial surface (a parabolic surface) fitted to the values in a 3 × 3 window centred at x and y; the surface is constructed in the manner illustrated by the block diagram in Fig. 17.12(a), using parameters in Fig. 17.12(b), and is of the general form as shown here:

$$c = Ax^2y^2 + Bx^2y + Cxy^2 + Dx^2 + Ey^2 + Fxy + Gx + Hy + I \qquad (17.12)$$

Where c is the curvature function at position x and y as illustrated in Fig. 17.12(a). Coefficients A to I are derived

as in Fig. 17.12(b). The methods used in Idrisi and Landserf also use polynomials but of slightly different form.

Referring to the values in block diagram in Fig. 17.12(c), the term $c_{profile}$ can be estimated by the following (parameter D) in Fig.17.12(b):

$$c_{profile} = \frac{200\left(DG^2 + EH^2 + FGH\right)}{\left(G^2 + H^2\right)} \qquad (17.13)$$

and c_{cross} as follows (parameter E) in Fig. 17.10(b):

$$c_{cross} = \frac{-200\left(DH^2 + EG^2 - FGH\right)}{\left(G^2 + H^2\right)} \qquad (17.14)$$

Mean curvature, c_{total} is derived from D and E as in Fig. 17.12(b) as follows:

$$c_{total} = -200(D + E) \qquad (17.15)$$

which for the values shown in Fig. 17.13(c), and $p = 10$, gives $c_{profile} = 1.4$, $c_{cross} = -4.6$ and $c_{total} = -6$, indicating that the terrain within the 3 × 3 window represents a channel whose bed is slightly concave along its length, and that the overall window curvature is concave.

Several methods employ polynomial fitting for curvature calculation but there are some concerns with this method. Firstly, the polynomial surface may not necessarily pass exactly through all 9 elevation points (z1...z9). Secondly, if the complexity of the actual surface (represented by the 3 × 3 grid in Fig. 17.9) is low (nearly planar), then the coefficients A to F will be zero. It must be stressed that this method does not represent

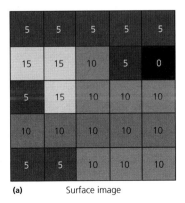

(a)   Surface image

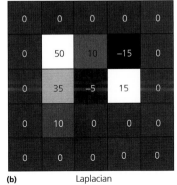

(b)   Laplacian

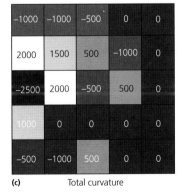

(c)   Total curvature

Fig. 17.11 Simple raster example: (a) surface; (b) the Laplacian, as representing total curvature (as would be calculated using 3 × 3 kernel, the outer rim of pixels always having zero value because of the kernel size); and (c) total curvature as calculated using the polynomial fitting method (as in Fig. 17.11). Since both (b) and (c) represent total curvature, positive values indicate convexity and negative values indicate concavity. Both input and output are integer values.

the calculation of true curvature, merely directional estimates of it; Laplacian is the only true measure.

From an applied viewpoint, curvature could be used to describe the geomorphological characteristics of a drainage basin in an effort to understand erosion and runoff processes. The gradient affects the overall rate of movement downslope, aspect defines the flow direction and profile curvature determines the acceleration (or deceleration) of flow. Profile curvature has an effect on surface erosion and deposition and as such is a very useful parameter in a variety of applications. Cross-sectional curvature affects convergence and divergence of flow, into and out of drainage basins and so can be used to estimate potential recharge and lag times in basin throughput (see also § 17.4.3).

The shape of the surface can be evaluated to identify various categories of geomorphology, such as ridges, peaks, channels and pits. These are summarized in Table 17.1, and illustrated in the 3D perspective terrain views shown in Fig. 17.13.

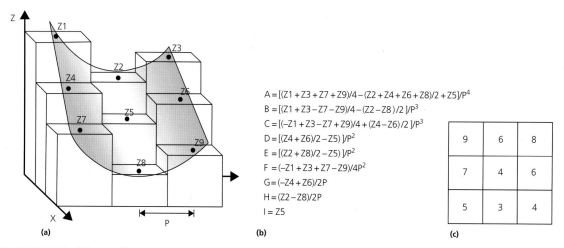

$$A = [(Z1 + Z3 + Z7 + Z9)/4 - (Z2 + Z4 + Z6 + Z8)/2 + Z5]/P^4$$
$$B = [(Z1 + Z3 - Z7 - Z9)/4 - (Z2 - Z8)/2]/P^3$$
$$C = [(-Z1 + Z3 - Z7 + Z9)/4 + (Z4 - Z6)/2]/P^3$$
$$D = [(Z4 + Z6)/2 - Z5]/P^2$$
$$E = [(Z2 + Z8)/2 - Z5]/P^2$$
$$F = (-Z1 + Z3 + Z7 - Z9)/4P^2$$
$$G = (-Z4 + Z6)/2P$$
$$H = (Z2 - Z8)/2P$$
$$I = Z5$$

9	6	8
7	4	6
5	3	4

(a)   (b)   (c)

Fig. 17.12 (a) Pixel diagram illustrating the relation of raster elevations to a conceptual curved, channel-like (concave) surface (modified after Zevenbergen & Thorne 1987); (b) equations used to derive the various directional coefficients of curvature from the surface in (a) where P is the increment between two pixels in either x or y direction; and (c) sample elevation z values representing the feature shown in (a).

Table 17.1 Geomorphological features, their surface characteristics and the pixel curvature formula relationships where positive values of $c_{profile}$ indicate upward concavity, and the opposite is true for c_{total} and c_{cross}.

Geomorphological Feature	Surface Characteristics	Second Derivatives: Profile and Plan Curvature
Peak	Point that lies on a local convexity in all directions (all neighbours lower).	$c_{profile} > 0$, $c_{cross} < 0$
Pit	Point that lies in a local concavity in all directions (all neighbours higher).	$c_{profile} < 0$, $c_{cross} > 0$
Ridge	Point that lies on a local convexity that is orthogonal to a line with no convexity/concavity.	$c_{profile} \approx 0$, $c_{cross} < 0$
Channel	Point that lies in a local concavity that is orthogonal to a line with no concavity/convexity.	$c_{profile} \approx 0$, $c_{cross} > 0$
Pass	Point that lies on a local convexity that is orthogonal to a local concavity (saddle).	$c_{profile} < 0$, $c_{cross} < 0$
Plane	Points that do not lie on any surface concavity or convexity (flat or planar inclined).	$c_{profile} = 0$, $c_{cross} = 0$

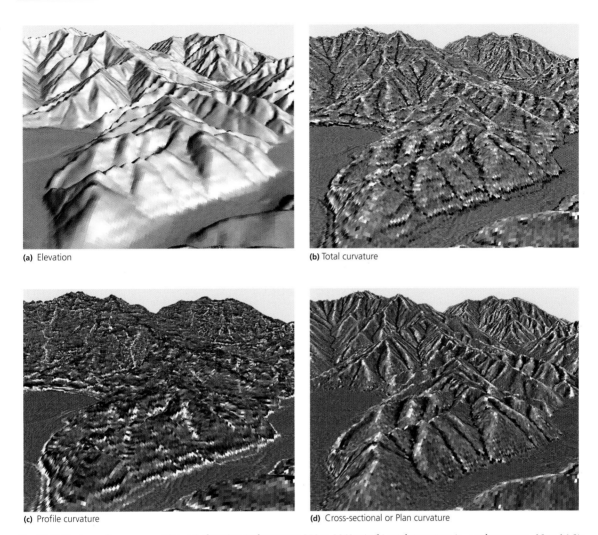

(a) Elevation

(b) Total curvature

(c) Profile curvature

(d) Cross-sectional or Plan curvature

Fig. 17.13 Raster surfaces representing: (a) elevation (value range: 350 to 1041 m); (b) total curvature (c_{total}; value range: −10 to 16.8); (c) profile curvature ($c_{profile}$; value range: −8 to 7.3); and (d) cross-sectional or plan curvature (c_{cross}; value range: −7 to 9.6). A greyscale colour lookup table with low values in black and high values in white is used in each case. As calculated using ArcGIS's Spatial Analyst curvature calculator, where positive values of $c_{profile}$ indicate upward concavity, whereas the opposite is true for c_{total} and c_{cross}.

17.4.3 Surface topology: Drainage networks and watersheds

Several other important parameters can be extracted from surfaces, and in particularly 'topographic' DEMs, and these relate to the connection between geomorphological features and the fluvial processes that produce them, hence the term *surface topology*. An increasing number of tools are now available for the extraction of stream networks, drainage basins, watersheds and flow grids from raw elevation data. These combine calculations of aspect and curvature to establish the flow direction and amount of water across a surface, by defining connectivity between drainage channels and to construct catchments and flow networks. This represents an active area of research and one that we cannot do justice to here.

The extraction of drainage networks has been investigated by many authors and these methods now appear in many proprietary software suites (e.g. RiverTools, ArcGIS's hydrological toolbox and Idrisi). There are many tricky issues associated with automated drainage network extraction from DEMs and it has been suggested that these are caused by scale dependence in the methods and a failure to accommodate the fractal natures of both elevation and drainage. Some algorithms begin with the calculation of *flow direction* or *flow routing*, then creation of drainage basins, from their outlet points, from which to progressively 'chase' upwards to the drainage divides (e.g. River Tools). Potential problems occur where there are pits in the DEM surface since these tend to stop the 'drainage chasing' algorithms and, in any case, these should be corrected first. Any pits in the DEM must be filled to create a '*depressionless*' DEM and this is used to derive *flow direction* and *flow accumulation* (to give a measure of flow through each pixel); the latter is then thresholded to retain pixels with the highest accumulation values that represent likely channels or streams. The extracted streams combined with the calculated watersheds between areas of drainage, enables the drainage network to be derived (e.g. as in ArcGIS and illustrated in Figs. 17.14, 17.15 and

17.16). Such a process is ideal for automation using either a macro or model builder or a script; an example of an ArcToolbox 'model' for extracting strem networks is shown in Fig. 17.14(b). Geomorphological features, including drainage networks can also be extracted by skeletonisation of the DEM; see also Chapter 15.

17.4.4 Viewshed

A *viewshed* represents an area or areas that are visible from a static vantage point. In the context of surface data, the viewshed identifies the pixels of an input raster (or positions on a TIN) that can be seen from one or more vantage points (or lines). The output product is a raster in which every pixel is assigned a value indicating the number of vantage points from which a pixel is visible. *Visibility,* in this respect, refers to the *line of sight*, but this 'sight' could also refer to the transmission of other signals, such as radio and microwaves. The viewshed is sometimes referred to as a 2D *isovist* (see online resources). Viewsheds are used in a variety of applications such as for the siting of infrastructure, waste disposal or landfill sites,

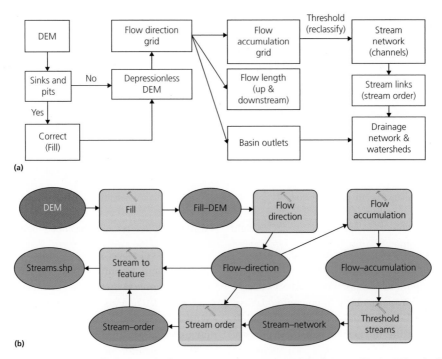

Fig. 17.14 (a) Schematic and generic illustration of a drainage network extraction method (as used in ArcGIS and other software); and (b) a stream network extraction 'algorithm' created using ArcToolbox's Model Builder (blue, input; orange, process; green, outputs).

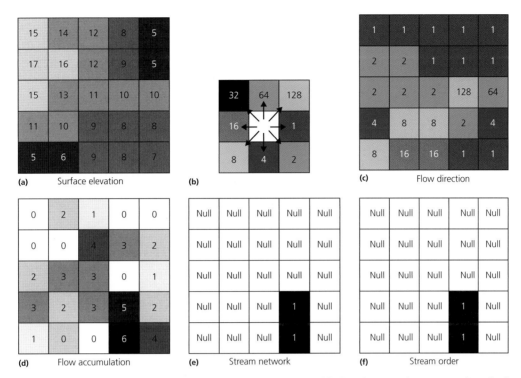

Fig. 17.15 Derivation of flow direction (or flow routing) and accumulation: (a) filled or 'depressionless' input surface; (b) direction encoding kernel; (c) flow direction grid (each pixel is coded according to the direction in which flow would pass from it, according to the values in the kernel, always moving to the lowest of the adjacent pixels); and (d) flow accumulation grid (the number of pixels that flow into each pixel); (e) stream network (via application of a threshold to the flow accumulation grid, values above which are selected to become the channels); and (f) stream network pixels are assigned numeric codes according to Strahler stream order for conversion to vector stream polylines.

and to select sites for mobile telephone transmission towers to avoid gaps in reception.

In an example with one vantage point, each pixel that can be seen from the vantage point is assigned a value of 1, while all pixels that cannot be seen from the vantage point are assigned a value of 0. The output is therefore a typically binary image.

The simplest viewshed calculations assume that light travels in straight lines in a Euclidean manner, that is, the earth is not curved and no refraction of light occurs, and that there are no restrictions on the distance and directions of view. This assumption is acceptable over short distances (of several kilometres) but corrections for the earth's curvature and optical refraction by the earth's atmosphere are necessary for accurate results over longer distances.

To describe the concept and framework of a viewshed, several controlling parameters can be defined. These are

the surface elevations of the vantage points, the limits of the horizontal angle to within which the viewshed will be calculated (given as two azimuths), the upper and lower angles, and the inner and outer radii (minimum and maximum distances from the vantage point), limiting the search distance within which the viewshed will be calculated from each vantage point, the vertical distance (if necessary) to be added to the vantage points, and finally the vertical distance (if necessary) to be added to the elevation of a vantage point as it is considered for visibility. These are illustrated in Fig. 17.17 and an example is shown in Fig. 17.18.

A modification of this could be used to model a *sound-shed*, that is, to identify areas of ground where noise can and cannot be detected, for example, from military installations or road traffic. Such soundshed analysis could then prove useful in designing sound barriers around potentially noise 'polluting' activities.

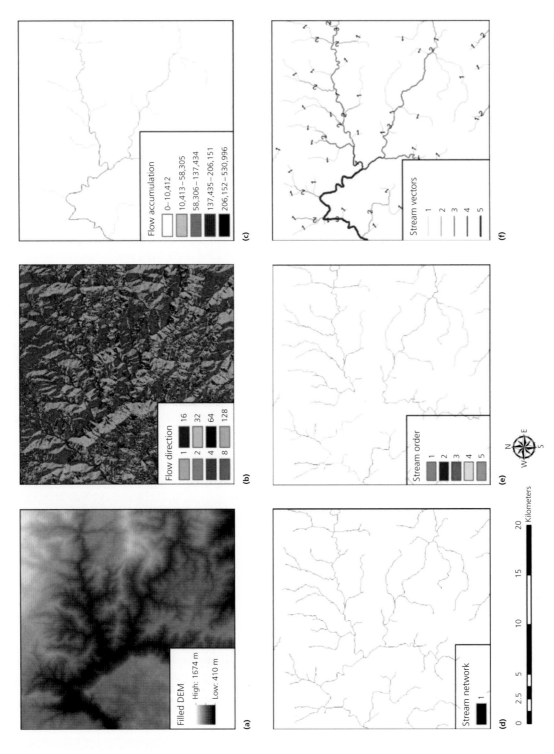

Fig. 17.16 Drainage extraction steps described in Figs. 17.14 and 17.15 above, illustrated using a real DEM example: (a) filled DEM; (b) flow direction calculated from (a); (c) flow accumulation; (d) stream network; (e) Strahler stream order; and (f) exported vector stream features labelled by stream order.

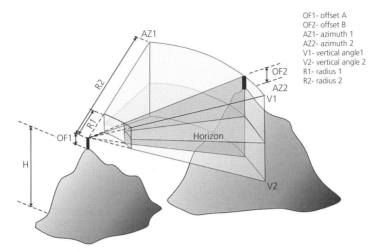

OF1- offset A
OF2- offset B
AZ1- azimuth 1
AZ2- azimuth 2
V1- vertical angle1
V2- vertical angle 2
R1- radius 1
R2- radius 2

Fig. 17.17 Schematic illustration of parameters defining the viewshed: elevations of the vantage point(s) (H), the limits of the horizontal angle to within which the viewshed will be calculated (azimuths 1 and 2), the upper and lower view angles (V1 and V2), and the inner and outer radii (R1 and R2), limiting the search distance within which the viewshed will be calculated from each vantage point, the vertical distance (OF1 and OF2) to be added to the vantage points (Reproduced with permission from ESRI Online Knowledge Base.)

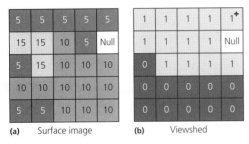

(a) Surface image   (b) Viewshed

Fig. 17.18 Simple raster example: (a) surface; and (b) viewshed (with simple vantage point indicated by a black cross, to which there are no added offset or specified angles or radii); both input and output are integer values. Viewshed values of 1 or 0 indicate that the position is either visible or invisible (respectively) from the vantage point.

17.4.5 Calculating volume

There may be many instances where we would like to estimate the volume of a quantity, as well as area or some other statistic. This may be very useful in the estimation of necessary costs associated with particular activities, such as for engineering or construction purposes, e.g. how much soils and rock need to be excavated to construct a cutting for a railway line, or the expected volume of a reservoir. Such calculations are usually referred to as *cut and fill analysis*. Here we extract the change between two surfaces (usually raster). These input surfaces might

represent the same location but recorded at different times. The calculation is made simple using raster because of the constant area of the pixel and the height given by the pixel DN, so that multiplying the two gives the volume occupied by that pixel. So that for a 100 × 100 raster, of 10 m spatial resolution (i.e. an area of 100 sq m), where each pixel DN represents a height of 2 m (above datum), the volume of the raster would be 100 m × 2 m × 10000 pixels, that is, 2 million cubic metres. If for each pixel we calculate the volume, as a local operation, then one volume raster can be subtracted or added to another to derive change. Depending on whether the change between two rasters is negative or positive, we have *cut* and *fill* statistics respectively.

17.5 Summary

It is worth noting once more the effect of the fractal nature of surface phenomena processes; a commonly overlooked phenomenon and one that has been commented on by many authors. Gradient, aspect and curvature are all phenomena that vary at different scales of topography and so scales of observation. Measuring any of these phenomena in the field yields very different results according to the distance over which they are being measured. It has also been shown that the effect of spatial resolution on their calculation from raster imagery can have a profound effect on the result.

Nevertheless, surfaces are clearly a valuable data source from which a great variety of descriptive morphological (geomorphometric) information can be extracted, and for many different applications. Image processing techniques and GIS-based tools overlap significantly in this chapter, though they are applied here with different intentions to those applied to multispectral images.

17.6 Questions

1 What are the differences between continuously sampled and discrete statistical surfaces?

2 For any particular phenomenon you are working with, which type of surface description should you choose and why?

3 What are DEMs and why are they so powerful for geoscientific use?

4 What effect does spatial resolution have on the estimation of slope gradient from a raster surface?

5 How can parameters like slope and aspect be used to derive neighbourhoods? How could such neighbourhoods then be used?

6 What other applications could there be for surface parameters such as gradient, aspect and curvature, that is, in application to surface data other than topography?

7 What other geomorphological parameters could we extract?

CHAPTER 18

Decision support and uncertainty

18.1 Introduction

Uncertainty in GIS is inevitable and arises for many reasons: the data we have at our disposal are never complete, our knowledge and understanding of a problem is flawed or limited, because of natural variation, because of measurement error or that the information is out of date. Albert Einstein is famously quoted as having stated that 'as far as the laws of mathematics refer to reality, they are not certain; as far as they are certain, they do not refer to reality'. We cannot get away from it, nor ignore its existence; we must therefore learn to live with uncertainty and try to reduce it.

Whilst we realise that we may not be able to tackle directly the causes or sources of uncertainty, we can recognise its existence and attempt to quantify it, track its course through any analysis and estimate or predict its probable effect on an outcome. The more advanced and involved our spatial analysis becomes, the more input factors are aggregated, the greater effect any potential errors and uncertainties are likely to have. We can also simulate or model potential outcomes and scenarios, varying the input parameters and the effects of errors and uncertainties as we go. These processes act as a form of quality control or validation for both data and analysis. In this way we improve our understanding of the problem and the potential reliability of a result, by more closely defining the limits of its applicability. Some of the key publications covering this subject include Goodchild and Gopal (1989); Burrough and Frank (1996); Burrough and McDonnell (1998); and Foody and Atkinson (2003).

Three key concepts that require further definition in this context are *decision support*, types of *uncertainty*, and *risk* and *hazard*. We will attempt to explain what we mean by these terms and how they are relevant to GIS. This chapter attempts to describe some of the surrounding issues, causes of and potential solutions to the problem of uncertainty.

18.2 Decision support

A *spatial decision support system* (SDSS) can be thought of as a knowledge-based information system that supports decision making or, more simply, a mechanism bringing variables together to enable better decisions to be made. An SDSS could involve a system designed to assist managers and/or engineers where the task at hand may be complex and where the aim is to facilitate skilled judgement. An SDSS could also be used to assist in problems that are of a poorly understood nature, or where data are incomplete, or where there are variables of unknown significance involved. Indeed there are many definitions because the SDSS is used in many, very different contexts.

A *decision* should be based on a level of acceptable risk and on a degree of confidence (error and uncertainty) in the available data. A decision may also imply the need for a quantitative prediction that demands the evaluation of the influential criteria to which the decision rules will be applied. A good decision may lead to a bad outcome (and vice versa) but if good decisions are continually attempted, then good outcomes are likely to occur more frequently (Ross 2004). Such decision making can be sub-divided

Image Processing and GIS for Remote Sensing: Techniques and Applications, Second Edition. Jian Guo Liu and Philippa J. Mason.
© 2016 John Wiley & Sons, Ltd. Published 2016 by John Wiley & Sons, Ltd.

according to the situations in which the decisions are made, as follows:

- *Deterministic decision making* is made when the 'controls' on the problem and the data are understood with some degree of certainty, so too are the relationships between each decision and the outcome. In such cases, categorical classes, rules and thresholds can be applied.
- *Probabilistic decision making* – here the surrounding environment, relationships and outcomes are uncertain or unknown (to some degree). In general, this approach treats uncertainty as 'randomness' but this is not always the case, especially not in the natural environment. Since a probabilistic approach produces only a true or false result, no degree of uncertainty can be accommodated unless it is considered as a separate and distinct state.
- *Fuzzy decision making* – this approach deals with uncertainties that are related to natural variation, imprecision, lack of understanding or insufficient data (or all these). Such ambiguities can be accounted for by allowing that classes can exist in varying degrees and amounts rather than as one of two end-member states (true or false), so that an infinite number of further states, representing the increasing *possibility* of being true, can be accommodated.

Probability and *possibility* form opposing but complementary concepts that co-exist within *Dempster-Shafer Theory* (Eastman 1997), described in Chapter 19.

Examples of applications in which the SDSS are frequently used might include, for example, a classification of locations in an area according to their estimated suitability for a pipeline route; or for a landfill site, for toxic waste disposal, or for a hazard assessment. Within these and other applications, the function of the SDSS is to help decision maker(s) to identify areas where there are unacceptable levels of risk associated with various predictive outcomes, so that they can then select appropriate courses of action.

18.3 Uncertainty

In general terms, uncertainty can be considered as an indeterminacy of spatial, attribute or temporal information of some kind. It can be reduced by acquiring more information and/or improving the quality of that information. There will be few cases where it can be removed altogether so it needs to be reduced to a level tolerable to the decision maker. Methods for reducing uncertainty include defining and standardising technical procedures, improving education and training (to improve awareness), collecting data more rigorously, increasing spatial/temporal data resolution during data collection, field checking of observations, better data processing methods & models and developing an understanding of error propagation in the algorithms used.

Assumptions must be made in all spatial analyses where any kind of 'unknowns' or uncertainties need to be dealt with. Examples of assumptions include that soil and rock classes (or any classes) are homogeneous across the area they represent; that slope angles classified as being stable are stable at every location characterised by that stable value; that classifications made at the time of data collection have not changed since then; or that geological boundaries are rigid and their positions are certain, everywhere. Just as uncertainties are unavoidable so too are these assumptions. There are methods we can employ to quantify these uncertainties, and so limit the effect of the assumptions, such as allowing for the gradational nature of natural and artificial boundaries and for establishing threshold values for classification and standardisation. Uncertainties are many and complex, and the underlying rule, once again, is to know the limitations of and to understand the effects on the data from the start.

Conditions of 'certainty', in contrast, could include situations where there is only one 'state of nature' or where any 'state of nature' that exists has only one effect on the outcome or only one outcome. Clearly such definitive, certain and simplistic states are rare or unlikely in nature but they are useful concepts from which to consider more realistic possibilities. There may be cases where an element of certainty may be acceptable, perhaps with respect to either data availability or finance since both are often in short supply. Such shortages often lead to compromise, when some areas of uncertainty may have to be ignored.

Uncertainties, specifically within spatial analysis, may be related to the validity of the information itself, to the potential effects of the phenomena or to the handling of the information. There are three types: *criterion*, *threshold* and *decision rule* uncertainties.

18.3.1 Criterion uncertainty
Criterion uncertainty arises from errors in original measurement, identification and/or data quality (perhaps during data collection by a third party). Broadly

speaking, criterion uncertainty may be considered to be related to measurement (primary data collection) or may be conceptual (interpretative). It includes *locational, attribute value, attribute class separation* and *attribute boundary* uncertainties and they may not be correctable. In such cases an important step is to estimate or record the potential errors for the benefit of future users, and for your own liability. Measurement errors may derive directly from instruments with limited precision or as a result of user error, observer bias, mismatches in data collected by different individuals, sampling errors or poor sampling. Understanding the target and objective are vital in designing the sampling strategy. Repeated sampling can often correct for such potential errors but the expense of doing this may be prohibitive.

Criteria prepared for spatial analysis inherently include some uncertainty (error) since they have probably been rescaled or converted to discrete rasters from continuously sampled ones, or they may be reclassified and/or generalised; they are thus the product of subjective decision making. Errors in spatial data are usually considered to be normally distributed and identifying them requires some ground truth information to allow comparison of the differences between measured and observed, such as through the root mean square (RMS) error calculation. One thing is common to all, the more criteria that are involved in the decision making, the greater the uncertainty will be. In this situation, one could restrict the analysis to a simplistic combination of a few criteria but the simplistic solution will still incorporate uncertainties, it will merely ignore them since it cannot account for them. In the end there are two choices where data or criterion uncertinaty is concerned: to reject the data (in favour of better data) or accept them and work around the uncertainties they contain. In many cases, the latter is the only course of action since there are usually no 'better' data.

18.3.2 Threshold uncertainty

There are two principal causes of uncertainties in this category. First, the classes we chose to define are generally heterogeneous, that is, we chose homogeneous classes for simplicity and convenience. The concept of 'possibility' could therefore be very useful when attempting to define class boundaries in natural phenomena, that is, when deciding whether an object belongs to one class or another. Second, the boundaries between natural

phenomena of any kind are rarely rigid or Boolean in character because again we define arbitrary classes for our own convenience (they may not exist in reality). By treating such divisions less rigidly, we can 'blur' the boundaries between them, thus allowing further potential states to exist, that is, in addition to 'suitable' or 'unsuitable', or prospective and non-prospective, stable and unstable and so on. Values can then be incorporated that represent the increasing likelihood of belonging to a class or state, as illustrated in Fig. 18.1.

A great deal of research has been carried out into the use of multi-source datasets for mineral prospectivity mapping. Many agree that identifying areas of very low or very high prospectivity has not been difficult but uncertainty always arises in defining areas of intermediate prospectivity, which then require further analysis and interpretation. In such cases prospectivity (suitability) would be much better treated as a continuous phenomenon, representing a measure of confidence in an outcome. The prospectivity map is then a tool to reduce risk associated with exploration investment and expense.

18.3.3 Decision rule uncertainty

This is subtly different from threshold uncertainty and refers to the way in which we apply thresholds to particular criteria to denote values of significance. With firm and reliable evidence about some phenomenon we may be able to apply 'hard' deterministic decision rules confidently. For instance, in cases where some prescriptive law governs the analytical selections we make, hard decision rules are applied so that the result complies categorically with that law. Conversely, where we have incomplete data and must rely on circumstantial

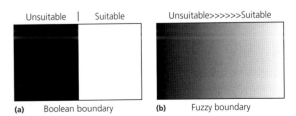

(a) Boolean boundary   **(b)** Fuzzy boundary

Fig. 18.1 Different boundary types between class thresholds: (a) crisp (Boolean) threshold producing two categorical classes of suitability: unsuitable and suitable; and (b) a fuzzy threshold representing increasing probability of membership to the class 'suitable'.

evidence then we should find a way to apply 'soft', probabilistic or fuzzy decision rules, ones in which a certain degree of confidence/uncertainty is attached to each result.

This type of uncertainty arises because of subjective judgements made by the decision maker in the selection of, ranking and weighting of criteria, in the method of criteria combination and in the choice of alternative courses of action. Uncertainties of this type are the most difficult to quantify, since they are not simply caused by mistakes or imprecision, and their correctness may never be precisely known.

18.4 Risk and hazard

The existence of uncertainties and errors in both data and spatial analysis introduces an element of risk. *Risk* can be thought of not only as the prediction of the likelihood of a (potentially unwelcome) event occurring but also of the chance of getting that prediction wrong. It is therefore normal for such spatial analysis to form part of a risk assessment framework of some kind, an example of which is illustrated in Fig. 18.2.

There are many definitions of risk and there is a tendency for the terms risk and hazard, risk and uncertainty, or risk and probability, to be used interchangeably but in all cases, this may because of a misunderstanding of their meanings. In many senses, *hazard* represents the probability of the occurrence of an unwelcome event and *risk* is that probability modulated by the economic value of the losses caused by event. To this end, Varnes (1984) defined the following relationship in an attempt to separate the related terms of risk and hazard as well as *vulnerability*, which represents a measure of the economic value (i.e. damage and cost, in both economic and human senses), as follows:

$$Risk = Hazard \times Vulnerability$$

Risk then represents the expected degree of loss, in terms of probability & cost, as caused by an event. Within the context of uncertainty, risk can be described by three phases of activity: (a) the prediction of one or more adverse events (scenarios) which have unexpected or unknown outcomes; (b) the probability of those predicted events occurring; and (c) their actual consequences.

There are several different kinds of risk (operational, financial, management, etc.) and consequentlty, different ways of dealing with it, whether that be to try to reduce it or only to define and quantify it. In defining risk, we should also consider what is an *acceptable level of risk,* since this is a quantity that varies with perception. One popular method is that proposed by Melchers (initially in 1993 and later published in 2000), which is known as the *as low as reasonably practicable* or *ALARP principle.* This represents the minimum limit below which risk can be practically ignored. This is of course subjective and there are several alternative definitions, such as the *lower limit beyond which further risk reduction cannot be justified.* Common responses to risk associated with uncertainty are to reduce it, either by *risk retention* (bearing the consequences) or *risk transfer* (insurance). However, all potential outcomes might not be insurable!

18.5 Dealing with uncertainty in GIS-based spatial analysis

There are a number of tools we can employ to quantify, track and test for ambiguity within our data and analytical methods. These include *error assessment, fuzzy membership functions, multi-criteria evaluation, error propagation* and *sensitivity analysis.*

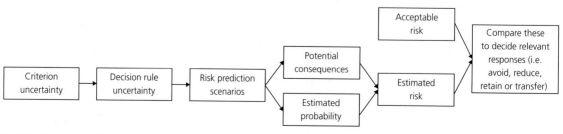

Fig. 18.2 Generalised risk assessment framework.

18.5.1 Error assessment (criterion uncertainty)

18.5.1.1 Error assessment

Errors can be defined as the deviation of a data value from the considered or measured 'true' or 'accurate' value. Other terms that need defining here include *precision* and *accuracy* since these vary within some interval according to errors in the data. Precision can be described as the level of measurement 'exactness', and is usually limited by the instrument and/or method, whereas accuracy may be thought of as the degree to which information matches true or accepted values. The standard deviation is usually taken as a measure of accuracy and is normally stated with reference to an interval, that is, ± a given value. High precision does not indicate high accuracy nor does high accuracy imply high precision.

Errors are considered either *systematic* or *random*. Random errors occur when repeated measurements do not agree and tend to be normally distributed about the mean. Systematic errors tend to be either positively or negatively skewed, indicating that they are of common source and of similar sign and value, such as an instrument error. Errors may be of position or attribute, and their source could be measurement (imprecision or instrument malfunction) or interpretation (conceptual). Ground-based sampling is usually taken to provide data that represent 'true' values. The errors are then identified by comparison with the 'true' values at the sample locations. For these locations, the *root mean square errors* (RMSE) can be calculated, as follows:

$$RMS = \left[\frac{\sum_i \left(x_i - x_{it} \right)^2}{n-1} \right]^{0.5} \qquad (18.1)$$

where x_i is the measured value, x_{it} is the true value and n is the number of measurements (and hence error values). The RMS is commonly used for error description in GIS for continuously sampled raster data. In classified or categorical data, errors are usually described using a *confusion matrix* (see § 8.5.2), which is constructed by *cross-tabulation* of observed (true) and mapped (estimated) values. This type of error assessment is commonly applied to classifications made from remotely sensed images, and has been the driving force for a great deal of work on validation within remote sensing applications. In such cases, the errors may come from any of a great many sources: misregistration of images, sensor properties, classification errors, ground truth data errors,

class definitions, pixel impurities and more. Such errors are commonly irregularly distributed (spatially) but the confusion matrix technique does not reveal this pattern. Alternatively a geostatistical approach could be used to model the geographic variation of accuracy within the results. Accuracy assessment in classifications from remotely sensed data are therefore not trivial but are now considered to be fundamental to all forms of thematic mapping. Despite this there is still seems to be no universally accepted standard method of making such assessments; refer back to § 8.5.2.

18.5.2 Fuzzy membership (threshold and decision rule uncertainty)

Fuzzy logic is a term first used in 1965 by Lofti Zadeh in his work on fuzzy set theory, and it deals with reasoning that is approximate or partial rather than exact or fixed. It is commonly be applied in GIS in one of two basic ways: either through the combination of Boolean type maps, using fuzzy rules, to yield a fuzzy output map or through the use of fuzzy membership functions as a tool for rescaling the data (or both). Fuzzy membership incorporates not only a measure of a phenomenon that exists on the ground but also of some level of knowledge (confidence) about that phenomenon.

Fuzzy membership or fuzzy sets provide an elegant solution to the problem of threshold and decision rule uncertainty by allowing 'soft' thresholds and decisions to be made. Fuzzy membership breaks the requirement of 'total membership'; instead of just two states of belonging for a class, a fuzzy variable can have one of an infinite number of states ranging from 0 (non-membership) to 1 (complete membership) and the values in between represent the increasing possibility of membership. A simple linear fuzzy set is defined by the following:

$$\mu(x) = \begin{cases} 0 & x < a \\ \dfrac{x-a}{b-a} & a < x < b \\ 1 & x > b \end{cases} \qquad (18.2)$$

where a and b are upper and lower threshold values of x defining the significant limits for the fuzzy set. Each value of x is associated with a value of $\mu_{(x)}$ and ordered pairs [x, $\mu_{(x)}$] that together comprise the fuzzy set.

The fuzzy set membership function can be most readily appreciated with reference to a simple linear function but it may also be sigmoidal or J-shaped, and monotonic or symmetric in form (Fig. 18.3). The threshold values, which define it, will depend on the phenomenon and desired outcome of the operation; that is, the threshold values applied to each membership function reflect their significance on the result. The resultant fuzzy set layers can then be combined in a number of ways, for example using Boolean logic and fuzzy algebra or 'set theory'.

An illustration of multi-criteria evaluation applied to hazard assessment, using fuzzy scaled inputs is described in a research case study, in § 22.3.

18.5.3 Multi-criteria decision making (decision rule uncertainty)

In this context we introduce the concept of *multiple criteria evaluation* or *multi-criteria decision making*, a process by which the most suitable areas for a particular objective are identified, using a variety of evidence layers, and any conflicts between objectives are resolved. The risks arising from decision rule uncertainty here are reduced through the integration of multiple evidence

layers representing the various contributory processes and quantities. This process allows the input criteria to be handled in different ways according to their desired function within the analysis. This is further enhanced by the evaluation of individual criterion significance, the ranking of the criteria and the assignment of weights to give certain criteria more influence inside the model. The criteria being evaluated are generally one of two types, *constraints* that are inherently Boolean and limit the area within which the phenomenon is feasible, or *factors* that are variable and are measured on a relative scale representing a variable degree of likelihood for the occurrence of the phenomenon. Decision rules, applied to these criteria, are commonly based on a linear combination of factors together, along with the constraints, producing an *index of suitability* or *favourability*. These topics are discussed further in the next chapter.

It is worth noting here the difference between *multi-attribute decision making* and *multi-objective decision making*, since both fall into this category of activities, and further to this, the possibility of involving either *individual* and *multiple decision making*. Multi-criteria

1. Linear membership functions

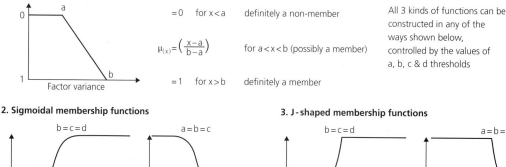

$=0$ for $x < a$ definitely a non-member

$\mu_{(x)} = \left(\dfrac{x-a}{b-a}\right)$ for $a < x < b$ (possibly a member)

$=1$ for $x > b$ definitely a member

All 3 kinds of functions can be constructed in any of the ways shown below, controlled by the values of a, b, c & d thresholds

2. Sigmoidal membership functions   **3. J-shaped membership functions**

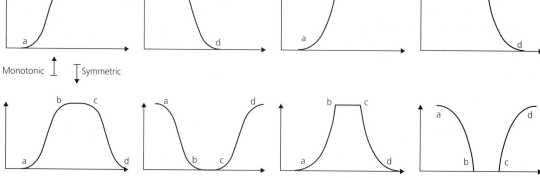

Fig. 18.3 Fuzzy set membership functions: linear, sigmoidal, J-shaped, monotonic and symmetric.

decision making is often used to cover both types, multi-attribute and multi-objective. The *attribute* is a measured quantity or quality of a phenomenon, whereas an *objective* infers the desired state of a system being analysed or simulated. The difference between *individual* and *multiple* decision making is not really about the number of people making the decisions but about the number of objectives being satisfied. For instance, an individual decision making process with one single goal may involve a group of several people or just one, and a multiple decision making process, involving a coalition of participating individuals may be competitive or independent, depending on the nature of the topic. Group decision making is common in the public sector especially since stakeholder involvement is often mandatory and there may be many points of view to be considered.

Commonly used multiple criteria combination methods that incorporate measures of uncertainty include weighted factors in linear combination (WLC), weights of evidence modelling (Bayesian probability), vectorial fuzzy modeling and Dempster-Shafer analysis (belief and plausibility). These will be described in more detail in the next chapter.

18.5.4 Error propagation and sensitivity analysis (decision rule uncertainty)

18.5.4.1 Error propagation

This refers to the process of determining the expected variability in the result as produced either by known errors, or errors deliberately introduced, to the input dataset. There are several popular methods of error propagation, the Monte Carlo simulation (which is probably the more widely used) and the analytical method or Taylor's series error propagation analysis.

18.5.4.2 Monte Carlo simulation

The Monte Carlo method is one of a group of rather computer intensive general methods for assessing the impact of statistical errors on the results of functions. For any particular variable, z, which is a function of various inputs $i_1...i_{n'}$ the idea is to determine the error associated with the estimate value of z, and the contributions of each input to the error. The variable is considered to have a normal probability distribution function with known mean and variance, and stationarity is assumed. The function is calculated iteratively, to generate values

of z many times, so to derive the average resulting z value and its standard deviation.

This method is commonly used to assess the errors associated with the calculation of surface parameters from a DEM, such as slope and aspect. In this context, an error dataset with $\mu = 0$, variance σ^2 is produced for a standard deviation of ±1 m for instance; this is added to the DEM. Slope, for instance, is then calculated many times – for example, 100, to produce 100 slightly different results. These are combined to produce an average slope image. Division of the standard deviation by the average slope then gives a measure of the relative error, both in magnitude and distribution.

18.5.4.3 Analytical method

This method uses a mathematical (polynomial type) function to describe the way errors are translated through a particular decision rule. If a function has continuous derivatives (up to the n+1th order) it can be expanded and if continued expansion causes the function to converge, then it is known as a Taylor's Series, which is infinitely differentiable. Error propagation of this type involves evaluating the effect of known errors on the function, where only the lowest order terms are considered important since any pattern can be simulated if a function of high enough order is used. These methods are described in detail by Heuvelink *et al.* 1989; Goodchild and Gopal 1989; and Burrough and McDonell, 1998.

18.5.4.4 Sensitivity analysis

This process revolves around the idea that the attribute values and their weights, within spatial analysis, are the most important aspect since they comprise the most subjective part of the analysis. The scaling and weighting of datasets involves interpretation and perceptive judgments, which introduce 'error'; if the rank order of inputs changes greatly as the weights are modified, then the latter should be re-evaluated. If the order does not change, then the model can be considered quite robust. Detailed descriptions of sensitivity analysis methods are provided by several standard texts.

The main difference between this and error propagation methods is that sensitivity analysis requires prior understanding of data and errors, whereas error propagation is a process by which errors are introduced to the analysis. Sensitivity analysis involves a number of methods aimed at determining the degree to which the result is affect by changes in the input data and decision rules (and weights); it is a *measure of the robustness* of a model.

18.5.5 Result validation (decision rule uncertainty)

Multi-criteria analysis is often carried out with little consideration of the meaning, relevance or quality of the output solution or the effect of potential errors it contains. The multi-criteria evaluation procedures discussed here and in the next chapter provide a means to allow for, quantify and reduce certain varieties of uncertainty, but it is in the final stages more than any other that the validity of the result should be questioned and tested. This often means some kind of 'blind' test to determine the validity of the output suitability map, and to do this requires some reliable 'ground truth' information. The ground truth information could come from a physical ground test but this is also likely to be subjective. What is really needed is an objective measure of the effectiveness of the result; revealing how predictive it is.

There are several methods for *cross-validating* or estimating the success of a particular result using some *training data*. This means the use of a partitioned dataset, where one part is used in the analysis and the other retained for confirming or testing of the result. The ground truth data are usually provided by some known occurrence data, such as known landslide locations or mineral occurrences. A widely used approach is the confusion matrix, which we already encountered in Chapter 8 (§ 8.5.2).

Several other error measures are used that are based on a *pairwise comparison* approach, such as the *Kappa statistic*, which describes the agreement between measured and observed spatial patterns, on a scale between 0 and 1, and is described formally in § 8.5.2 and simply as follows:

$$\kappa = \frac{p(a) - p(e)}{1 - p(e)} \qquad (18.3)$$

where $p(a)$ represents the relative observed agreement among the input values and $p(e)$ represents the probability that any agreement is caused by chance. If there is complete agreement then $\kappa = 1$, if the opposite is true and there is no agreement then $\kappa \leq 0$. The Cohen's Kappa statistic is only applicable in cases where two inputs are compared. A variation of this can be used to consider multiple inputs. In § 8.5.2, we have already presented the Kappa coefficient derived from confusion matrix for classification accuracy assessment with an example of multiple classes.

Other methods include the *holdout* or *test-set*, *k-fold* and *leave-one-out cross-validation* (LOOCV). The *holdout* or *test-set* method involves a random selection of, for example, 30% of the data, which is kept back from the analysis to act as a ground truth. If data are not plentiful this may be unacceptable; that is, we may not wish to waste 30% of the input data. The *k-fold* cross-validation method is slightly less wasteful and involves the random division of the dataset into *k* subsets, and a regression is performed *k* times on the subsets. Each time one subset is used as the test set and the other *k*-1 subsets are used as the training data, then the average error from the *k* regressions is derived. The *leave-one-out* method involves the iterative removal of one test data point; this is equivalent to the k-fold method taken to the extreme, where *k* = *n* (*n* being the number of test data points). This last method is useful if you have an independent set of very few ground truth data points. The entire analysis can be run iteratively, each time with one test data point removed, and the success of prediction examined each time. In this way, prediction curves can be constructed to give an idea of the effectiveness of the model and data used.

18.6 Summary

Uncertainty is a very active area of research within spatial analysis and remote sensing since the volume of data and our access to them are both growing. Clearly, what we do as geospatial scientists with our digital data in GIS is fraught with dangers and vague possibilities. These problems are often, alarmingly, overlooked. They cannot be avoided, but there are things we can do to minimise the risks and to allow for the uncertainties and errors. Whether we are becoming more or less critical of data quality and reliability is a moot point, but it is certain that we should continue developing tools and understanding to keep pace with these trends.

18.7 Questions

1 Why is it important to quantify uncertainty in GIS?

2 What are the main types of uncertainty and how do they affect the analysis?

3 What developments should GIS software provide in future to help deal with uncertainty?

4 List some examples of phenomena that cannot be realistically described by rigid (Boolean) functions and describe some fuzzy alternatives for each.

5 Why is error tracking important?

6 What is the difference between probability and possibility, and how do these two concepts help us in spatial analysis?

7 Why is validation important in multi-criteria evaluation problems?

8 What can standards and benchmarks contribute?

9 How significant is metadata in this context and why?

CHAPTER 19

Complex problems and multi-criterion evaluation

19.1 Introduction

This branch of GIS activity is sometimes referred to as *advanced spatial analysis*, *multi-criterion evaluation* or *multi-criterion decision analysis* (MCDA) and deals with multiple conflicting criteria that need assessment and evaluation to facilitate decision making. Multi-criterion evaluation is a discipline that has grown considerably over the last 10–15 years, resulting in what is now a well-established body of research on this topic. There is even an International Society on Multiple Criterion Decision Making, though their activities are rather broader than the use of MCDA in GIS.

Any procedure that uses spatial data to satisfy a particular request could be described as 'spatial analysis' and often is. This chapter deals with the procedures by which we deal with complex geospatial problems to which there may be many potentially unknown contributing processes and pieces of evidence. These procedures incorporate the conversion of real-world problems into a set of abstract quantities and operations, the accommodation of the vagaries of the natural environment, and of 'unknown' quantities, to produce a realistic and practical solution to the original problem. The terms *criterion* and *factor* tend to be used interchangeably in this chapter in reference to the multiple input layers; however, *factor* is used only in reference to continuously sampled (variable) data, whereas *criterion* is used in a more general sense for both categorical and variable inputs.

Data generated from modern-day surveys and exploration campaigns are not only diverse but voluminous. Sophisticated topographic, geological, geochemical, remote sensing, geophysical (high resolution ground and airborne) surveys not only make the analysis more quantitative (hopefully) but also make interpretation more difficult. A successful result lies in effective processing of the data, extraction of the relevant criteria and integration of these criteria to a single 'suitability' map or index.

Over the past decade or so, many techniques have evolved to exploit large datasets and construct maps that illustrate, for example, how mineral potential or prospectivity changes over an area (Knox-Robinson & Wyborn 1997; Chung *et al.* 2002; Chung & Keating 2002), or how slope instability (or vulnerability to slope failure) varies across an area (Wadge 1988; van Westen, 1993; Mason & Rosenbaum 2002; Chung & Fabbri 2003, 2005; Liu *et al.* 2004a to name but a few), or how rapid surface soil erosion can be discriminated from other kinds of small-scale surface change (Liu *et al.* 2004b). These kinds of analyses (often referred to collectively as 'modelling' even though it really isn't) demand the abstraction of reality, that is, the representation of physical properties numerically, and the application of statistical approaches to accommodate natural variations. Unsurprisingly, use of the simplest methods greatly outnumbers that of the more complex procedures.

In this chapter, we describe the main approaches and point to other more detailed texts, where appropriate. Generally speaking, and whatever the application area, the steps involved follow a similar path, always beginning with the definition of the problem, through data preparation to the production of a result and its validation, followed by some recommendations for action 'on the ground'. These phases are illustrated in Fig. 19.1.

Image Processing and GIS for Remote Sensing: Techniques and Applications, Second Edition. Jian Guo Liu and Philippa J. Mason.
© 2016 John Wiley & Sons, Ltd. Published 2016 by John Wiley & Sons, Ltd.

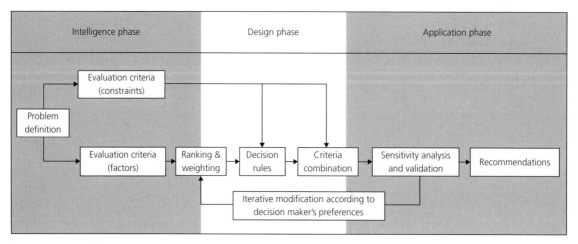

Fig. 19.1 Conceptual framework for multi-criteria spatial analysis.

19.2 Different approaches and models

There are a number of different approaches to multi-criteria decision making and analysis, with the aim of estimating *suitability* or *favourability* across a region, some of which pre-date GIS. They are often divided into two broad categories: *knowledge-driven* (conceptual) and *data-driven*, and of the latter, there are two further kinds. The first is *empirical* and tries to identify significant spatial relationships, and the second uses *artificial intelligence and neural networks* to objectively recognise patterns in the data.

19.2.1 Knowledge-driven (conceptual)

This approach generally involves consideration of a specific model for a particular type or case. The 'model' is then broken down to identify its significant contributing criteria on the basis of knowledge and evidence, perhaps based on other examples of the same type elsewhere. It is a method commonly used for hazard mapping (but sometimes also mineral prospecticity mapping on a regional scale). In the case of landslide hazards, decisions are made about, for instance, the particular angle above which slopes become unstable in that case, or the particular rock type that is susceptible to failure in that case. A database is then constructed that contains data appropriate to the description of all the criteria considered relevant. The datasets are then scaled and coded to suit the knowledge for each criterion. The criteria are then agregated combined in such a way as to identify areas of potential hazard or suitability. *Dempster-Shafer Theory* (DST) (§ 19.5.4) and *Analytical Hierarchy Process* (AHP) (§ 19.5.5) fall into this category.

19.2.2 Data-driven (empirical)

This approach is commonly applied to mineral prospectivity mapping, but in this case it is the dataset statistics of the particular case that drive the processing. The statistical characteristics of known examples or occurrences are evaluated, at the pixel level, and these pixels are used as 'seeds' or 'fingerprints' to classify other locations by their similarity to the seed or fingerprint locations. In this type of intuitive approach to prospectivity mapping, the aim is to identify areas that are statistically similar to the known mineral deposits or occurrences, but that have not yet been found nor undergone any systematic exploration. Exploration resources can then be focused on these more likely areas, thus reducing the risk of discovering nothing.

Often MCDM will include elements of both knowledge-driven and data-driven approaches at different stages, and both normally involve three basic steps; the identification of spatial relationships, the quantification of those relationships as layers and then the integration of the layers. Research in this area tends to focus on the third step (layer combination) and, as a result, there are an increasing number of techniques available, ranging from the very simple (*Boolean* set theory) to the increasingly complex *algebraic, weights of evidence* and *fuzzy-logic* methods.

19.2.3 Data-driven (neural network)

In contrast, the application of neural networks and other data-mining techniques involve the process of 'learning' and 'pattern' recognition from the data. Unlike the knowledge driven and empirical data-driven approaches, the neural-network approach evaluates all inputs simultaneously by comparison with a *training dataset*. This approach is particularly suited to very large data volumes where the number of significant factors and possible combinations is potentially huge, so the individual handling of layers and spatial relationships becomes impractical. Success relies heavily on the *training* or *learning* process and so, once trained for a particular case, the system must be re-trained before it can be applied to any other case. In general, this approach is very demanding computationally and generally far less commonly applied than the previous two and so we will not attempt to cover it in this book.

19.3 Evaluation criteria

Criteria are usually evaluated within a hierarchical structure, some consideration of which is necessary before the data can be prepared. The overall objective and any subordinate objectives should be identified (see Fig. 19.2) before the requisite criteria may be identified to achieve those objectives. Once the hierarchy of objectives and criteria is established, each criterion should be represented by a map or layer in the database.

Each chosen criterion should ideally be independent and unambiguous. Each must also be represented in common units otherwise, when the layers are combined, the value units of the result will be meaningless.

If we consider, as an example, the problem of slope stability assessment as a principal objective, beneath which are several sub-objectives (as illustrated in Fig. 19.2). The latter may represent intermediary steps in achieving the principal objective or valuable end-products in their own right. Each of the sub-objectives may involve multiple input criteria, for example three layers representing the variable mechanical properties of rocks and soils, groundwater levels and slope gradient may be required as layers to represent *conditioning engineering factors*; these could also be combined to produce as a *factor of safety* geotechnical measure. All the coded criteria are then combined to produce the overall slope stability hazard assessment map. This map may then in turn be used to give predictions of the potential environmental states, for example, the area is stable, stable only in the short term or presently unstable.

An alternative example could be in the evaluation of mineral exploration prospectivity (as the principal objective) where the sub-objectives for the same geographic area could be evaluation of potential for several commodities of economic value, such as gold, nickel and base metals, or for different deposit model types, such as epithermal gold, skarns and porphyry copper. Each sub-objective has a set of specific input criteria and

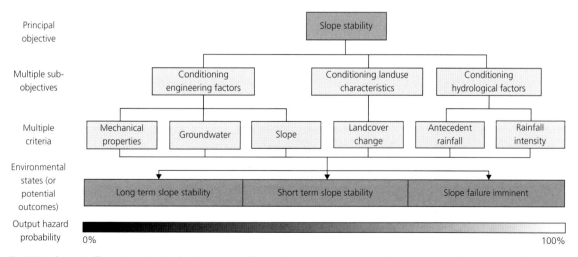

Fig. 19.2 Schematic illustration of a single-objective, multi-criteria evaluation system; with several sub-objectives under the main objective heading of slope stability analysis; and with several possible environmental states, defined as classes of hazard.

whilst these may be used by more than one sub-objective, they may be prepared, reclassified and rescaled uniquely for each sub-objective. In each case, the end result is a map showing the variation in prospectivity (or suitability for exploration) for a given commodity or mineralisation type over a given area. Thus we can see the distinction between *multi-objective* and *multi-criterion decision making* (MCDM). The identification of both the significant input criteria and the various sub-objectives are therefore important steps. Following this, appropriate data must be identified and prepared, to represent these criteria.

In choosing the criteria, there are two problematic situations: too many criteria so that the decision making process becomes too complicated to understand, or too few criteria, causing an over-simplification of the problem. Poor understanding may cause the former and the latter is usually caused by data shortage.

Assuming that data shortage is not a problem, there are a number of methods for selecting criteria, such as by researching past cases or by conducting a survey of opinions. As mentioned in the previous chapter, input criteria are generally of two types: *continuous* (later referred to as *factors*) and *thematic* (later referred to as *constraints* in some situations). The continuous type represent spatially variable, continuously sampled phenomena, often containing values on interval, ratio and cycle scales, such as those described in Chapter 12. The thematic type comprise representations of discretely sampled phenomena, usually containing values on nominal or ordinal measurement scale (as described in Chapter 13). The input criteria must be prepared and scaled, using operations and procedures described in Chapter 15, in such a way as to contribute correctly towards the end result.

19.4 Deriving weighting coefficients

In establishing the structure of the model, it generally becomes clear that some criteria play a more significant role than others in leading to the outcome. Having identified and prepared the input criteria, the next step is to assess and quantify their relative significances. To achieve this, the criteria must be ordered and a mechanism identified to describe their order numerically. Deciding on order and weight is perhaps the most difficult aspect of multi-criteria evaluation problems,

and it commonly requires discussion, field verification and iterative modification.

There are many weight derivation approaches and these differ in their complexity, accuracy and method. Weights should not be considered as simple indicators of criterion significance because they should allow for changes in the range of factor values as well in as the significance of each factor. The reason for this is that a factor weight could give an artificially small or large effect on an outcome simply by increasing or decreasing its range of values, for example, weights in the range of 1–1000 will have a far greater effect than those in the range 1–10. Weights applied to the criteria should always sum to 1, so that:

$$\sum w_i = 1 \qquad (19.1)$$

where there are *n* criteria and where the weights (*w*) range from $i = 1, 2, 3 \dots n$ Most multi-objective and multi-criteria evaluation procedures and decision making processes involve the combination of a series of input variables, and it is highly likely that these inputs will contribute to the outcome to varying degrees. If their significance in contributing to the outcome is not equal, then some means of quantifying those inequal contributions is necessary. Many types of weighting procedure have been proposed to allow this, and these include *rating, ranking, pairwise comparison* and *trade-off analysis*, some of which are more popular than others.

19.4.1 Rating

Rating involves the assignment of values on relative scales of significance, for example 0 to 10 or 100. One popular method, referred to as *point allocation*, involves the identification of a number of points or scores among the input criteria. For instance if a 0–100 scale is chosen and there are three input criteria, a value (*score*) of 50 out of 100 could be assigned to the most significant criterion, 40 to the next and 10 to the last. The resultant weights would then become 0.5, 0.4 and 0.1. Alternatively, in the *ratio estimation* method scores are assigned to the most and least significant criteria, the latter is then used as the reference from which all other ratio scores are calculated. Criteria are ranked and scored on a relative scale, as in the point allocation method, and then the score of the least significant one is divided by each other criterion score to give a weight, which is then normalised by the sum of weights. The process is repeated for the second least significant, and

Table 19.1 Resultant weights derived using rank sum, rank reciprocal and rank exponent methods above, for a set of (k) criteria numbered 1 to 4.

		Rank Sum		Rank Reciprocal		Rank Exponent	
k	Rank	Weight	Normalised Weight	Reciprocal Weight	Normalised Weight	Weight	Normalised Weight
(k)	(r)	($n-r_j+1$)		($1/r_j$)		($n-r_j+1$)^p, $p=0.8$	
1	3	2	0.200	0.333	0.160	1.741	0.213
2	4	1	0.100	0.250	0.120	1.000	0.122
3	1	4	0.400	1.000	0.480	3.031	0.371
4	2	3	0.300	0.500	0.240	2.408	0.294
		10	1.000	2.083	1.000	8.181	1.000

third least significant, and so on until all criteria have calculated ratio weight values. The result is then a measure of the difference between lowest and highest values for a particular criterion in comparison with those of the first (least significant) criterion.

19.4.2 Ranking

Here the criteria are first arranged in *rank order* according to their considered relative significance in affecting the outcome. The weights are then derived by one of a number of popular methods, summarised as follows. In all methods, the weights are normalised by the sum of the weights for all criteria.

1 *Rank sum*: this refers to the normalised summation of weights derived for each criterion, as follows:

$$w_i = \frac{n-r_j+1}{\sum(n-r_k+1)} \qquad (19.2)$$

Where w_i is the normalised weight for the ith criterion, n is the number of criteria being evaluated, r_j is the rank position of the ith criterion, and the criteria being evaluated (k) range from $k = 1, 2, 3 ...n$).

2 *Rank reciprocal*: this involves calculation of the weight reciprocals, normalised by the sum of weights:

$$w_i = \frac{1/r_j}{\sum(1/r_k)} \qquad (19.3)$$

3 *Rank exponent*: in this case, a 'most significant' criterion is identified and a variable is set to represent it.

It is then used as a power with which to multiply the normalisation:

$$w_i = \frac{(n-r_j+1)^p}{\sum(n-r_k+1)^p} \qquad (19.4)$$

A weight is specified for the most significant criterion (p). The value of p is then solved iteratively and the weights are derived. The higher the value of p, the more sharply the values of the normalised weights rise with increasing significance. If $p = 0$ the weights will be equal in value; if $p = 1$ the result is equivalent to the rank sum of weights. So this method allows a certain amount of control over or 'trade-off' between the weights.

These three methods involve only relative weight 'approximation' so that the larger the number of criteria, the less appropriate the method becomes. They are therefore considered acceptable for cases with few input criteria and are illustrated in Table 19.1.

19.4.3 Pairwise comparison

The *pairwise comparison matrix* (*PCM*) method was created and developed by Saaty (1980) for use within the *Analytical Hierarchy Process* (described in § 19.5.5). The method has in the past received criticism for its disassociation from the real measured or reference scales of the input criteria. It is therefore vital that the input criteria are normalised correctly and to common scales before combination. The PCM method is, however, flexible, easy to understand (since only two criteria are considered at a time) and appropriate for collective and iterative discussions of weighting. This method is incorporated into the decision support section of the *Idrisi* software suite (Eastman 1993). There are 3 steps involved, and these are summarised as follows and illustrated using slope stability assessment as an example:

1 Construction of the PCM: The matrix is constructed where every input criterion is compared to every other and is given a score representing its significance in contributing to the outcome. The values in the matrix are assigned from a relative scale of importance between 1 (equal importance) and 9 (extreme importance). Reciprocal values can be used to indicate the reverse relationship, that is, 1/9 indicating that one factor is extremely less important than another. The values on the diagonal are always one, where identical criteria are compared, and the values in the upper-right of the matrix are reciprocals of those in the lower left. The assigned value scale is described in Table 19.2 and an example is given in Table 19.3.

2 Derivation of weights: The weights are produced from the principal eigenvectors of the PCM and can be derived using the following method: the values in each column of the matrix are summed, to give column marginal totals (*ct*). A second matrix is then generated by dividing each matrix value by its column marginal total. These values are then averaged across the rows to derive the weight for each criterion.

3 Calculation of a consistency ratio, within the matrix, ensures that the logical relationships between the criteria are represented fairly. The value of this ratio should be as low as possible, indicating that the relative comparisons have been made sensibly. This process involves the calculation of several component parameters as follows:

(a) *Weighted sum vector* (WSV): where the first weight (*w*) is multiplied by the first column total (*ct*) in the matrix, the second by the second column and so on. These values are then summed over the rows to give the WSV

$$WSV = \sum_{i=1}^{n} w \; ct \qquad (19.5)$$

(b) *Consistency vector* (CV): here the WSV is divided by the criterion weights

(c) *Average consistency vector* (λ) is then calculated for all the criteria

(d) *Consistency index* (CI): since there are always inconsistencies within the matrix, λ is always greater than or equal to the number of input

Table 19.2 Table of significance estimations based on a nine-point scale (after Saaty 1980); reciprocal values can also be used.

Significance	Value	Significance	Value
Extreme importance	9	Moderate to strong importance	4
Very to extreme importance	8	Moderate importance	3
Very strong importance	7	Equal to moderate importance	2
Strong to very strong importance	6	Equal importance	1
Strong importance	5		

Table 19.3 PCM, used to assess relative factor significance, in contributing to slope instability, and to calculate criterion weights as shown in Table 19.4. The table should be read from the left, along the rows, so that *slope* is considered the most significant and *distance from drainage* the least significant.

PCM Step 1	Slope	Aspect	Factor of Safety	Distance from Drainage
Slope	1	3	5	7
Aspect	1/3	1	1	7
Factor of safety	1/5	1	1	7
Distance from drainage	1/7	1/7	1/7	1
Marginal totals	1.68	5.14	7.14	22.00

Table 19.4 Second table generated from the column totals of those in Table 19.3, to derive the weights.

PCM Step 2	Slope	Aspect	FS	Distance from Drainage	Weight
Slope	0.597	0.583	0.700	0.318	0.550
Aspect	0.199	0.194	0.140	0.318	0.213
Factor of safety	0.119	0.194	0.140	0.318	0.193
Dist. from drainage	0.085	0.028	0.020	0.045	0.045
Marginal totals	1.000	1.000	1.000	1.000	1.00

Table 19.5 Derivation of the Consistency Index (CI) using the weights derived in Table 19.4.

PCM Step 3	Weighted Sum Vector (WSV)	Consistency Vector (CV)
Slope	(0.55)*(1) + (0.213)*(3) + (0.193)*(5) + (0.045)*(7) = 2.47	4.49
Aspect	(0.55)*(0.33) + (0.213)*(1) + (0.193)*(1) + (0.045)*(7) = 0.90	4.23
Factor of safety	(0.55)*(0.2) + (0.213)*(1) + (0.193)*(1) + (0.045)*(7) = 0.83	4.29
Dist. from drainage	(0.55)*(0.143) + (0.213)*(0.143) + (0.193)*(0.143) + (0.045)*(1) = 0.18	4.06

criteria $(\lambda \geq n)$ for any reciprocal matrix. The closer the value of λ to n, $(\lambda = n$ in an ideal case) the more consistent the matrix is. So $\lambda - n$ represents a good measure of consistency and CI as the unbiased estimation of the average difference give a good judgement of consistency:

$$CI = \frac{\lambda - n}{n - 1} \qquad (19.6)$$

If we consider the example of slope stability, where there are four input criteria considered to have varying degrees of influence in causing slope failure for an area. A PCM could be constructed and used to derive the criterion weights, using the method described, as illustrated in Tables 19.3 and 19.4.

Using these scores and weights, the weighted sum vector and consistency vector can be derived for each criterion, as shown in Table 19.5.

For this example, with four criteria contributing to slope instability, the average consistency vector (λ) is

$$\lambda = \frac{4.49 + 4.23 + 4.29 + 4.06}{4} = 4.27 \qquad (19.7)$$

and so the calculated value of CI is

$$CI = \frac{4.268 - 4}{4 - 1} = 0.099 \qquad (19.8)$$

A value of 0.099 would be considered to represent acceptable consistency within the PCM.

19.5 Multi-criterion combination methods

Multi-criterion evaluation (MCE) is a process in which multiple scaled and weighted layers are aggregated to yield a single output map or *index of evaluation*. Often this is a map showing the suitability of area for a particular activity. It could also be a hazard or prospectivity map or some other variable that is a function of multiple criteria. Several methods are described here, in order of complexity: *Boolean combination, index-overlay, algebraic combination, weights of evidence modelling (Bayesian probability), Dempster-Shafer Theory, weight linear factors in combination (WLC)* – otherwise known as the *Analytical Hierarchy Process, fuzzy logic* and *vectorial fuzzy modelling*. Weights of evidence modelling, WLC, AHP, vectorial fuzzy modelling and DST can all be considered as providing *fuzzy measures* since all allow uncertainty to be incorporated in some way, either directly through the use of fuzzy membership sets, or through probability functions or some other gradational quantities.

19.5.1 Boolean logical combination

This represents the simplest possible method of factor combination. Each spatial relationship is identified and prepared as a map or image where every location has two possible conditions: *suitable* or *unsuitable*. One or more of the standard arithmetic operators is used to combine the spatial relationship factors into a single map. The Boolean *AND combinatorial operator* is very commonly used, and retains only those areas that are suitable in all input factors. Alternatively, the *combinatorial Boolean OR* can also be used, which represents the conceptual opposite of AND and will always result in more or larger areas being categorised as suitable. The former represents the '*risk averse*' or conservative of the two methods, and the latter is the more '*risk taking*' or liberal of the two. This method is simple, conceptually and computationally, but is in many cases rather rigid and over-simplistic, since there is no allowance for gradational quantities or for other forms of uncertainty.

19.5.2 Index-overlay and algebraic combination

Here criteria are still categorical but they may comprise more than two discrete levels of suitability. These are usually represented as ordinal scale numbers, so that a location with a value of 2 is more suitable than a location with a value of 1 but is not twice as suitable. The resultant suitability map is constructed by the summation of all the input factors; the higher the number, the more suitable the location. This approach is also simple and effective but also has some drawbacks. Since the classes of the adopted criteria are subjective and they behave like weights so that a factor divided into 10 levels will have greater effect on the result than one divided into only 3 levels. Ideally therefore, the input datasets should be scaled to the same number of classes. The result is also unconstrained such that an increase in the number of input criteria causes an increase the range of values in the suitability map.

The criteria are combined using simple summation, or arithmetic or geometric mean operators, according to the desired level of conservatism in the result. Use of the arithmetic mean is more liberal in that all criteria and locations pass through to the result even if a zero is encountered. The geometric mean can be considered more conservative since any zero value causes that location to be selectively removed from the result. These different operators should be used selectively to combine input criteria in different decision making situations. For instance, where there is considerable decision rule confidence, a more risk averse geometric mean method might be applicable. Conversely, if there is plentiful data but a great deal of decision rule uncertainty, the more

risk taking arithmetic mean may be more appropriate; allowing all values and positions through to the end result. A research case study, in which an index-overlay combination method based on the geometric mean is used for landslide hazard assessment, is shown in § 22.2.

The index-overlay method can be modified and improved upon by replacing the ordinal scale numbers with ratio scale numbers, so that a value of 2 means the location is twice as suitable as a location with a value of 1; this removes the need for further scaling. Criteria combination can then proceed in the same way by summation or arithmetic and geometric mean.

19.5.3 Weights of evidence modelling based on Bayesian probability theory

One of the most widely used statistical, multi-criteria analysis techniques is the *weights of evidence method*, which is based on *Bayesian* statistics. Here the quantitative spatial relationships between datasets representing significant criteria (input evidence) and known occurrences (outcomes) are analysed using Bayesian weights of evidence probability analysis. Predictor maps and layers are used as input evidence. The products are layers representing the estimated probability of occurrence of a particular phenomenon (according to a hypothesis) and of the uncertainty of the probability estimates. This involves the calculation of the likelihood of specific values occurring or being exceeded, such as the likelihood of a pixel slope angle value exceeding a certain threshold, as part of a slope stability assessment.

Bayesian probability allows us to combine new evidence about a hypothesis on the basis of prior knowledge or evidence. This allows us to evaluate the likelihood of a hypothesis being true using one of more pieces of evidence. Bayes' theorem is given as

$$p(h|e) = \frac{p(e|h) \times p(h)}{\sum_n p(e|h_n) \times p(h_n)} \qquad (19.9)$$

where h represents the hypothesis and there are n possible, mutually exclusive, statistically independent outcomes; e is the evidence (some kind of observation or measurement); $p(h)$, the *prior probability*, represents the probability of the hypothesis being true regardless of any new evidence; whereas $p(e|h)$ represents the probability of the evidence occurring given that the hypothesis is true, that is, the *conditional probability*. $P(h|e)$, the *posterior probability*, is the probability of the hypothesis being true given the evidence.

One assumption considered here is the independence, or lack thereof, between the input evidence layers. If two evidence layers, e_1 and e_2, are statistically independent, then the implied probabilities of their presence are:

$$p(e_1|e_2) = p(e_1) \quad and \quad p(e_2|e_1) = p(e_2) \quad (19.10)$$

that is, the conditional probability of the presence of e_1 is independent of the presence of e_2, and vice versa. If, however, the two variables are *conditionally independent* with respect to a third layer, t, then the following relationship exists:

$$p(e_1 \cap e_2|t) = p(e_1|t)p(e_2|t) \quad (19.11)$$

If e_1 and e_2 are binary evidence layers for an area and t represents known target occurrences in that area, then the following allows us to estimate the number of targets that might occur in the area of overlap between e_1 and e_2 (i.e. where both are present):

$$n(e_1 \cap e_2 \cap t) = \frac{n(e_1 \cap t)\, n(e_2 \cap t)}{n(t)} \quad (19.12)$$

where the predicted number of target occurrences in the overlap equals the number of target occurrences in e_1 times the number of target occurrences in e_2, divided by the total number of target occurrences, if the two variables are conditionally independent.

If the total estimated (predicted) number of targets is larger than the actual number of occurrences, then the conditional independence can be considered to be 'violated' and the input variables being compared should be checked.

When input layers are being prepared at the start of the MCE process, the binary evidence layers should be compared, for example in a pairwise fashion, to test for conditional independence. If necessary, problematic layers should be combined to reduce the conditional independence effect or they should be removed.

These descriptions seem rather abstract in themselves, so we could consider a simple example, with one target occurrence dataset and one evidence layer, to better explain the principle. If we take mineral exploration as the example, and a sample dataset of 100 samples, represented as a raster of 10 x 10 pixels. Of the 100 samples, there are five gold occurrences. So there are two possible mutually exclusive outcomes, containing gold or not containing gold, that is, $n = 2$.

From the number of occurrances and samples, it appears that the probability of finding gold is 0.05 and of not finding gold, is 0.95; this is the prior probability of finding gold in this area. The new evidence introduced in this case is a layer representing a geophysical parameter (such as total magnetic field). This layer contains an area of anomalously high values and so a reclassified evidence layer, containing only two classes (anomalous or not anomalous), is introduced. It is found that four of the five gold occurrences lie within the area characterised as anomalous, so it seems that the chances of encountering gold are much higher given the presence of the geophysical anomaly since 0.8 of pixels that contain gold, are also on the anomaly. Of the 100 pixels, 95 contain no gold but 25 of these are also geographically coincident with the anomaly, so that 0.275 of pixels with no gold occur in the anomaly. So using these values, the probability of *finding gold given the presence of the anomaly*, or the posterior probability, will therefore be:

$$p(h|e) = \frac{0.80 \times 0.05}{(0.80 \times 0.05) + (0.275 \times 0.95)}$$

$$= \frac{0.04}{0.04 + 0.261} = 0.133 \; i.e. > 0.05, \; the \; p(h) \quad (19.13)$$

And in a similar way, the posterior probability of finding gold where there is no measured anomaly will be:

$$p(h|e) = \frac{0.20 \times 0.05}{(0.20 \times 0.05) + (0.725 \times 0.95)}$$

$$= \frac{0.01}{0.01 + 0.689} = 0.0143 \; i.e. < 0.05, \; the \; p(h) \quad (19.14)$$

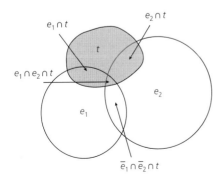

Fig. 19.3 Schematic illustration of the effect of conditional independence.

So after the introduction of new evidence (in this case, the presence of the geophysical anomaly), the posterior probability of finding gold is considerably greater than without that evidence; 0.133 as opposed to 0.05. Therefore the introduction of various new pieces of evidence (representing significant criteria) to the analysis increases our chances of finding the target, or of making good decisions about where to explore.

One limitation in this simple method is that gradational outcomes are not permitted, that is, only two states of nature can exist: that the pixel either contains gold or does not contain gold. The outcomes are said to be mutually exclusive. No account of the variable amount of gold present can be made. Similarly, the evidence layer contains two classes, anomalous or not anomalous, meaning that a subjective judgement has been made in the reclassification of this input evidence. In this simple example, if we wish to introduce gradational (continuous) values, they must be included as separate pieces of evidence or separate outcomes. This can of course be achieved; it simply makes the calculations more lengthy. Another drawback of this method is that you need to know all possible outcomes at the start, which may be unrealistic.

A method of factor combination based on Bayesian probability is shown in § 23.1. This case, set in southeast Greenland involves prospectivity mapping for a number of commodities, in which there is reliable but limited data available and reasonably good understanding about the deposit models.

19.5.4 Belief and Dempster-Shafer Theory

DST is a knowledge-driven approach based on *belief functions* and *plausible reasoning*, and is used to combine separate pieces of evidence to calculate the probability of an event or occurrence. The theory owes its name to work by Shafer (1976) in extending the Bayesian theory of statistical inference (Dempster 1967, 1968). The mathematical derivations are also dealt with in great detail in a number of other texts.

A limitation of many of the approaches described so far is the assumption (requirement) that all the input factors must contribute positively towards the outcome, that is, that high input values are correlated with suitability. As a consequence, there is no way to consider evidence that seems to be contradictory to the main hypothesis. DST allows the estimation of the likelihood of *suitability*, or *unsuitability*, in addition to estimates of *plausibility* and *belief* for any hypothesis being evaluated. In traditional Bayesian probability the absence of supporting evidence for a hypothesis is automatically assumed to support the alternative hypothesis, that is, unsuitability. The DS method is unique in allowing that 'ignorance' does not necessarily support that alternative hypothesis and in doing so provides a method for describing 'grey' areas and thus allowing for some uncertainty.

DST introduces six quantities: *basic probability assignment, ignorance, belief, disbelief, plausibility* and a *belief interval*. It also provides estimates of the confidence levels of the probabilities assigned to the various outcomes. The degree to which evidence supports a hypothesis is known as *belief*, the degree to which the evidence does not contradict that hypothesis is known as *plausibility*, and the difference between them is referred to as the *belief interval*; the latter serves as a measure of uncertainty about a particular hypothesis. Belief is always less than or equal to plausibility. Belief in a hypothesis is the sum of the probability 'masses' of all subsets of the hypothesis. Plausibility is therefore an upper limit on the possibility that the hypothesis could happen, that is, it 'could possibly happen' up to that value, because there is only so much evidence that contradicts that hypothesis. Plausibility represents 1 minus the sum of the probability masses of all sets whose intersection with the hypothesis is empty or, in other words, the sum of the masses of all sets whose intersection with the hypothesis is *not* empty. A *degree of belief* or *mass* is represented as a *belief function* rather than a probability distribution. Probabilities are therefore assigned to *sets* of possibilities rather than to single, definitive occurrences and this incorporates a measure of uncertainty.

Using the previous simple example, considering the hypothesis that a pixel position contains no gold, we may have evidence suggesting that the pixel area contains no gold, with a confidence of 0.5, but the evidence contrary to that hypothesis (i.e. pixel does contain gold) only has a confidence of 0.2. So for our hypothesis we have a belief of 0.5 (lower limit) and a plausibility of 0.8 (upper limit). The remaining mass of 0.3 (the gap between the 0.5 and 0.2) represents the probability that the pixel may or may not contain gold; this interval represents the level of uncertainty caused by a lack of the evidence for the hypothesis.

Table 19.6 Probability masses associated with the hypothesis that a pixel position contains no gold, under DST.

Hypothesis	Probability	Belief	Plausibility
Null (no solution):	0	0	0
Gold:	0.2	0.2	0.5
No gold:	0.5	0.5	0.8
Either (gold or no gold):	0.3	1.0	1.0

The null hypothesis is zero by definition and this represents 'no solution' to the problem. The mutually exclusive hypotheses 'Gold' and 'No gold' have probabilities of 0.2 and 0.5, respectively. The universal 'hypothesis 'Either' represents the assumption that the pixel contains something (gold or not), forms the remainder so that the sum of the probability masses is 1. The belief value for 'Either' consists of the sum of all three probability masses (Either, Gold, and No gold) because 'Gold' and 'No gold' are subsets of 'Either', whereas the belief values for 'Gold' and 'No gold' hypotheses are equal to their individual probability masses since they have no subsets. So that the plausibility of 'Gold' occurring is equal to the sum of probability masses for 'Gold' and 'Either'; the 'no gold' plausibility is equal to the sum of probability masses of 'No gold' and 'Either'; and the 'Either' plausibility is equal to the sum of probability masses of 'Gold', 'No gold' and 'Either'. The hypothesis 'Either' must always have 100% belief and plausibility, and so acts as a 'checksum' on the result.

Evidence layers are then brought together using *Dempster's rule of combination*, which is a generalisation of Bayesian theorem. The combination involves the summation of two input probability masses and normalisation to 1. If we use a simple example with two input datasets (*i* and *j*):

$$e_c = \frac{(e_i e_j) + (e_i u_j) + (e_j u_i)}{\beta} \quad (19.15)$$

$$d_c = \frac{(d_i d_j) + (d_i u_j) + (d_j u_i)}{\beta} \quad (19.16)$$

$$u_c = \frac{u_i u_j}{\beta} \quad (19.17)$$

where e = evidence (or belief) for input evidence datasets i and j, d = disbelief, u = uncertainty (which equals plausibility minus belief), e_c = the combined evidence,

d_c = the combined disbelief, u_c = the combined uncertainty and β = the normalisation factor. Where β is derived as follows:

$$\beta = 1 - e_i d_j - d_i e_j \quad (19.18)$$

This method tends to emphasise the agreement between input evidence and ignore all conflicts between them through the normalisation factor. The latter ensures that evidence (belief), disbelief and uncertainty always equal one ($e + d + u = 1$). Where conflicts between input evidence are known to be of significant magnitude, DST could produce meaningless results and an alternative method should be sought.

19.5.5 Weighted factors in linear combination (WLC)

This method, sometimes referred to as the *Analytical Hierarchy Process* (Saaty 1980), is a technique that allows consideration of both qualitative and quantitative aspects of decisions. It is popular for its simplicity, since it reduces complex decisions to a series of pairwise comparisons (via the use of a PCM) and then combines the results. The AHP method rather importantly accepts that:

- Certain criteria are more important than others;
- Criteria have intermediate values of suitability (i.e. they do not need to be classed rigidly as 'suitable' or 'unsuitable').

Criteria can also be coded to behave differently. Those acting as Boolean *constraints* can be used to selectively remove or 'zero-out' locations and regions. Others, which are continuous or variable in nature, are referred to as *factors*. The factors should be combined in a way that reflects the two points above and there are three important issues surrounding this combination: determination of the relative importance of each criterion; the standardisation or normalisation of each factor (since each must be on consistent scale and must contribute in the same direction towards suitability); and the method of factor combination.

It is assumed to be unlikely that all factors will have equal effect on the outcome. The assessment of their relative importance involves derivation of a series of weighting coefficients. These Factor Weights, control the effect that each factor has on the outcome. The relative significance of each factor, its influence on the other factors and on the outcome need to be compared. This involves the ordering of the factors into a hierarchy of significance, and assessment of their degree of influence on the outcome and on each other.

The WLC method allows a measure of suitability in one factor to be compensated for in another factor(s) through the use of weights. The general approach of this method is as follows:

1 Identify the criteria (decide which criteria are factors and which are constraints).
2 Produce an image or coverage representing each criteria.
3 Standardise or scale each factor image (for instance, using fuzzy functions), reclassifying to either a real number scale (0–1), byte scale (0–255) or percentage scale (0–100). All must be scaled to the same range and in the same direction, so that they each contribute either positively or negatively (generally the former), towards the outcome.
4 Derive weighting coefficients that describe the relative importance of each factor (by one of the methods already described, usually by the PCM method).
5 Linearly combine the factor weights with the standardised factors and the constraints (usually by aggregation) to produce the 'suitability' map.

The consequence of using an aggregation method is that all candidate pixels entering the model pass through to the end. At no point does an encountered zero value cause termination of the model for that pixel position. So whilst the factors are weighted in terms of their significance, there is total *trade-off* between each factor value encountered, that is, a low score in one factor can be compensated for by a very high score in another.

The WLC method is popular but can in some instances be considered too 'liberal' in its handling of the data in the system, since it involves equal ranking of the weighted factors and allows full trade-off between them. Its factor aggregation method can be likened to a parallel connection system, which allows all input criteria to survive to the end (the likelihood of the occurrence of a zero is low). It is also possible that the relationships between the input factors are not linear, in which case a more complex model will be required. In many cases, this parallel system may be appropriate, but in others a harsher, more risk-averse system may be better, one that enables certain factors or combinations of factors to be eliminated completely from the system, rather like a sequential connection system. Factor combination via calculation of the geometric mean

(as opposed to the arithmetic mean) represents such a system in which the occurrence of a zero rating terminates the system and eliminates that location from the analysis.

An illustration of this method applied using fuzzy scaled inputs (see also § 16.5.6) is described in a research case study, in § 22.3. The index-overlay combination introduced in § 19.5.2 is the very simplest case of WLC.

19.5.5.1 Ordered weighted average (OWA)

This represents a refinement of the WLC method, where the degree of trade-off between factors is controlled by a second set of *order weights*. With full control over the size and distribution of the order weights, the amount of *risk taking* and *degree of trade-off* (or *substitutability*) can be varied. Trade-off represents the degree to which a low score in one criterion can be compensated for by a higher score in another. The order weights define the rank ordering of factors for any pixel, that is, they are not combined in the same sequence everywhere.

In this way, the degree to which factors can pass through the system is also controllable. Using one arrangement of order weights, pixels may be eliminated in some areas but permitted through in others. Using another arrangement, all pixels may be permitted through (equivalent to WLC).

After the first set of *factor weights* have been applied, the results are ranked from low to high (in terms of their calculated 'suitability' value). The factor with the lowest suitability score is then assigned the first order weight and so on up to the factor with the highest suitability being assigned the highest order weight. The relative skew to either end of the order weights determines the level of *risk*, and the degree to which the order weights are evenly distributed across the factor positions determines the amount of trade-off (see Table 19.7).

For example, consider three factors *a*, *b* and *c*, to which we apply weights of 0.6, 0.3, 0.1, on the basis of their rank order (the order weights sum to 1.0). At one location the factors are ranked *cba*, from lowest to highest, the weighted combination would be $0.6c + 0.3b + 0.1a$. If at another location however, the factors are ranked *bac*, the weighted combination would be $0.6b + 0.3a + 0.1c$. A low score in one factor can therefore be compensated for by a high score in another, that is, there is trade-off between factors.

Table 19.7 Order weights for a five-factor example used by OWA.

Rank	1st	2nd	3rd	4th	5th	Description
Order weights:	1	0	0	0	0	Risk averse, no trade-off
Order weights:	0	0	0	0	1	Risk taking, no trade-off
Order weights:	0	0	1	0	0	Average risk, no trade-off
Order weights:	0.2	0.2	0.2	0.2	0.2	Average risk, full trade-off

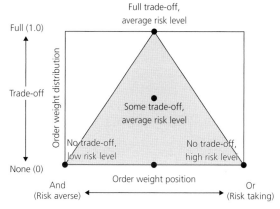

Fig. 19.4 Decision strategy space in order weighted averaging (Saaty, 1990. Reproduced with permission of RWS Publications.)

Two parameters, *ANDORness* and *TRADEOFF*, are used to characterize the nature of an OWA operation:

$$TRADEOFF = 1 - \sqrt{\frac{n \sum \left(w_i - 1/n \right)^2}{n-1}} \qquad (19.19)$$

$$ORness = 1 - ANDness \qquad (19.20)$$

$$ANDness = \left(1 \big/ (n-1) \right) \sum \left((n-i) w_i \right) \qquad (19.21)$$

where n is the total number of factors, i is the order of factors, and w_i is the weight for the factor of the ith order. From the equation, ANDness or ORness is governed by the amount of skew in the order weights and trade off is controlled by the degree of dispersion in the order weights.

For a *risk averse* or conservative result, greater order weight is assigned to the factors nearest the minimum value. For a *risk taking* or liberal result, full weighting is given to the maximum suitability score. If full weight is given to the factor with minimum suitability score and zero to all other positions, then the result will resemble that produced by the Boolean Min(AND) combination of factors and will represent no trade-off between factors.

If full weight is given to the maximum suitability score, then the result resembles Boolean Max(OR). If all order weights are equal fractions of 1, then full trade off is allowed and the result is equivalent to the WLC.

Order weights control the position of the aggregation operator on a continuum between the extremes of MIN and MAX, as well as the degree of trade-off. Examples are shown in Table 19.7 and illustrated conceptually by the decision strategy space shown in Fig. 19.4.

19.5.6 Fuzzy logic

Although the Bayesian and algebraic methods can be adapted to accommodate the combination of spatial continuously sampled data, there is a branch of mathematics, *fuzzy logic*, which is well suited for this purpose. Fuzzy logic represents a 'super-set' of Boolean logic, and deals with variables that incorporate some uncertainty or 'fuzziness'. Since fuzzy membership removes the requirements of 'total membership' of a particular class, it provides an ideal way to allow for the *possibility* of a variable being suitable or unsuitable. We have already described how this can be used to incorporate uncertainty by criterion scaling in the previous chapter.

The threshold values that define the fuzzy set become the input parameters when preparing each variable in our spatial analysis. The values are chosen based on prior knowledge and understanding of the data and decision rules. The type of membership function we choose will depend on the way the phenomenon contributes to the outcome, that is, positively or negatively, monotonically or symmetrically. The threshold values are applied according to the user's understanding and 'ground' knowledge of the phenomenon, to best reflect their significance on the result. An underlying assumption at this point is that the relationship between the input factors and the outcome is linear, but this may not necessarily be so.

Each input factor is scaled according to the chosen fuzzy membership function in preparation for factor combination. All the factors must be scaled in the same direction, that is, they must all contribute to the outcome in the same way, either positively or negatively. In this way, the locations that represent the most desirable characteristics are all coded with either very high values or very low values according to choice. An illustration of the WLC method applied using fuzzy scaled inputs is described in a research case study, in § 22.3.

The resultant fuzzy factor layers are then combined in one of a number of ways. The simplest option is to via simple set theory using map algebra, such as using Boolean logical intersection (AND) and logical union (OR) operators.

A series of fuzzy operators have subsequently been developed around set theory, for the combination of scaled (fuzzified) input factors: *fuzzy AND, fuzzy OR, fuzzy algebraic product, fuzzy algebraic sum* and a *fuzzy gamma function*.

$$\text{Fuzzy AND} \quad \mu_C = \min\left(\mu_1, \mu_2, \mu_3 \ldots \mu_n\right) \quad (19.22)$$

$$\text{Fuzzy OR} \quad \mu_C = \max\left(\mu_1, \mu_2, \mu_3 \ldots \mu_n\right) \quad (19.23)$$

$$\text{Fuzzy NOT} \quad \bar{\mu} = 1 - \mu \quad (19.24)$$

where μ_c represents the combined fuzzy membership function of n individual fuzzy inputs. These operate in much the same manner as the Boolean versions. The AND operator produces the most conservative result, producing low values, and allows only areas that are characterised by favourable conditions in all input layers to survive to the end result. In contrast, the OR operator produces the most liberal or 'risk taking' result and is suitable when it is desirable to allow any favourable evidence survive to be reflected in the end result.

This simple combination may not be considered suitable for the combination of multiple datasets because it is possible that extremely high and low values can propagate through to the final result. Two operators have been developed to overcome this problem: the *fuzzy algebraic product* (*FAP*) and the *fuzzy algebraic sum* (*FAS*). The FAP is the combined product of all the input values or 'fuzzy factors' in the following way:

$$\mu_C = \prod_{i=1}^{n} \mu_i \quad (19.25)$$

where μ_c represents the *FAP* fuzzy membership function for the n-th input factor. Since the values being combined are all fractions of 1, the values in the final result are always smaller than the lowest contributing

value in any layer. The function can be considered 'decreasive' for this reason. The *FAS* is not a true sum and is derived by the following:

$$\mu_C = 1 - \prod_{i=1}^{n}\left(1 - \mu_i\right) \quad (19.26)$$

Here the reverse is true, that is, the resulting value will always be larger than the largest contributing value in any layer but is limited by the maximum value of 1. This function therefore has the opposite effect to the FAP and is considered 'increasive'. Two pieces of input evidence that favour the result would reinforce one another in this method. Worth noting here is that the output value is partly affected by the number of input datasets, the more datasets involved, the greater the resulting value. FAP and FAS can also be combined into a single operation, called a *Gamma function*, which is calculated as follows:

$$\mu_c = \left[FAS\right]^{\gamma}\left[FAP\right]^{1-\gamma} = \left[1 - \prod_{i=1}^{n}\left(1 - \mu_i\right)\right]^{\gamma}\left[\prod_{i=1}^{n}\mu_i\right]^{1-\gamma}$$
$$(19.27)$$

where the gamma parameter (γ) varies between 0 and 1. When a value of 0 is chosen, the result is equivalent to the *FAP*. When gamma is 1.0 the result is equivalent to the *FAS*. A gamma value somewhere between provides a compromise between increasive and decreasive tendencies of the two separate functions. This method allows for uncertainty to be incorporated and allows all the input factors to contribute to the final result but has the drawback that all the input factors are treated equally. They must also contribute in the same direction toward the outcome.

In this way, pieces of evidence can be combined sequentially, in a series of carefully designed steps, rather than in one simultaneous operation. This gives more control over the final outcome and allows the different input layers to be treated differently, according to the understanding of the layer's contribution to the outcome.

19.5.7 Vectorial fuzzy modelling

In an attempt to improve on the above, the *vectorial fuzzy logic* method has been developed (Knox-Robinson 2000) in the mapping of mineral prospectivity or suitability. The fuzzy vector is defined by two values, calculated *prospectivity* (the fuzzy vector angle) and *confidence* (the fuzzy vector magnitude), the latter as a measure of similarity between input factors. Using the vectorial fuzzy logic method, null data and incomplete knowledge can be incorporated into the multi-criterion analysis. The 'confidence' value actually performs several functions: it represents confidence in the suitability value, the importance of each factor relative to others; and it allows null values to be used. The combination of the two values involves calculating a vector for each spatial relationship factor. The combined lengths and directions of each vector provide the aggregate suitability. The closer in value the inputs are, the longer the resultant combined vector (c_c) and the higher the confidence level. Confidence is a relative measure of consistency throughout the multi-criteria dataset for any particular location. The two values are derived as follows:

Fuzzy prospectivity:

$$\mu_c = \left(2/\pi\right)\arctan\left[\frac{\sum_{i=1}^{n}c_i\sin\left(\frac{\pi\mu_i}{2}\right)}{\sum_{i=1}^{n}c_i\cos\left(\frac{\pi\mu_i}{2}\right)}\right] \quad (19.28)$$

Fuzzy confidence:

$$c_c = \sqrt{\left(\sum_{i=1}^{n}c_i\sin\left(\frac{\pi\mu_i}{2}\right)\right)^2 + \left(\sum_{i=1}^{n}c_i\cos\left(\frac{\pi\mu_i}{2}\right)\right)^2} \quad (19.29)$$

where μ_i is the fuzzy suitability value for the ith factor input layer $\left(0 \leq \mu_i \leq 1\right)$. Fig. 19.5 shows the concept of variable suitability (prospectivity) represented as a vector using this method. In Fig. 19.5a the vector quantity of suitability (μ) is given as the direction of the vector and the confidence level (c) of that suitability

value is represented by the vector's length. In Fig. 19.5(b) two examples illustrate vectors of equal suitability, that is, constant direction, but of differing confidence levels, or different lengths. The vector (μ_2) is longer and therefore represents a greater level of confidence in the suitability value, and so the corresponding pixel would have a greater influence on the final output value.

In Fig. 19.6a, two input criteria with different suitability values but equal confidence levels provide a combined suitability value of 0.5 (the average μ value) and a confidence level of 1.41 (derived simply by Pythagorean geometry); the confidence level is dropped since the two input suitability scores are conflicting. In Fig. 19.6(b) the two inputs have identical suitability and confidence level values, so they combine to give the same suitability score of 0.5 but with double the confidence level, that is, $c_c = 2.0$; from this result we could be more confident that the result is more representative of the true suitability at that position.

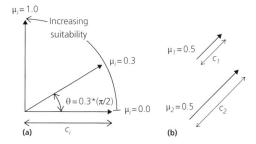

Fig. 19.5 Illustration of variable prospectivity derived using the vectorial fuzzy modelling method (Knox-Robinson, 2000. Reproduced with permission of Taylor and Francis.)

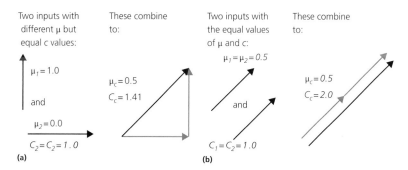

Fig. 19.6 Examples of the combination of two 'fuzzy' vectors: (a) two input criteria fuzzy vectors with equal confidence but different suitability (prospectivity); and in (b) two inputs with identical suitability and confidence levels (modified from Knox-Robinson 2000).

In this method all input factors must contribute positively towards the outcome. The existence or possibility of trade-off between input factors is considered undesirable or irrelevant in this method. Its use is therefore most appropriate when there is considerable understanding about the influence of each piece of evidence in leading to the outcome, that is, where there is considerable decision rule confidence but perhaps some data uncertainties.

19.6 Summary

Here we have summarised some of the better-known methods of criteria combination; it is not an exhaustive list and there are many variations and, regardless of the level of complexity involved, they all have strengths and weaknesses. It may be desirable, and is often appropriate, to use more than one of the criteria combination methods described here, that is, to use a mixture of fuzzy and non-fuzzy operators within the same model. Certainly none of these methods can be considered correct or incorrect when it comes to integrating spatial datasets or combining multiple criteria. The important issue is that each method is appropriate to a particular set of circumstances and objectives that should be carefully considered before choosing that method.

It is important to note, however, that multi-criteria analysis is often carried out with little consideration of the meaning, relevance or quality of the output solution or the effect of potential errors it contains. Some methods for cross-validation, error propagation and sensitivity analysis have already been described (in Chapters 16 and 18), and they are especially relevant here too. The object of spatial analysis in these circumstances is to predict conditions beyond the location or times where information is available. The multi-criteria evaluation procedures discussed here provide a means to allow for, quantify and reduce certain varieties of uncertainty in achieving these predictions, but not all types. So it is in the final stages of such analysis, more than any other, that the validity of the 'prediction' should be questioned and tested. The paradox is that there is always a solution but there is never a perfect solution, and we always want to know exactly how good a solution is whilst the answer will forever be fuzzy.

19.7 Questions

1 What are the assumptions made in choosing a weighting method?
2 Why is it important to know the measurement scale of the input criteria?
3 What is the difference between order weights and factor weights?
4 Which weighting method is the most appropriate for a group decision making process?
5 Why are certain weighting methods subject to uncertainty?
6 What are the chief differences between a simple Boolean model and one that incorporates uncertainty?

PART III
Remote sensing applications

In Parts I and II of this book, we learnt the essential image processing and GIS techniques. Here we will demonstrate, step by step, with examples, how these techniques can be used effectively in remote sensing applications. Although many case studies are drawn from our own research projects on earth science and the terrestrial environment, it is important to stress the generic sense of these examples in terms of concept and methodology for wider applications. From this viewpoint, our aim is not to provide rigid recipes for fixed problems but to provide guidance on 'how to think' and 'how to approach'. When first presented with a new project, beginners and students may feel a little lost, rather as in the Chinese saying *the tiger wants to eat the sky but doesn't know where to bite*. We will be satisfied if this part serves as a catalyst to get you on the right track for your own remote sensing application project.

Image Processing and GIS for Remote Sensing: Techniques and Applications, Second Edition. Jian Guo Liu and Philippa J. Mason.
© 2016 John Wiley & Sons, Ltd. Published 2016 by John Wiley & Sons, Ltd.

CHAPTER 20

Image processing and GIS operation strategy

In this chapter, we describe how the processing, interpretation and analysis of multi-source image and map data (in general, raster datasets) should be approached to produce thematic maps for a typical project. Following the discussion of basic strategy, a simple example of digital geological mapping, based on processing and interpretation of Landsat 7 ETM+ imagery, is presented to demonstrate the workflow from image processing to map composition.

We suggest the following *rules of thumb* as general guidance for operational strategy:

- *Purpose*: The aims and objectives of the project should be the driving force behind the image processing and multi-source data manipulation. In other words, it should be application driven rather than data processing driven. This is different from algorithm development, which may be triggered by application requirements but is focused on the technical part of data processing and its effectiveness; application examples serve as demonstrators of the algorithm.

- *Keep things simple*: Recall Fig. P.1 in Part I; it is not only true for image processing but serves as good advice for any application project. Nowadays image processing and GIS software packages are highly functional and are supported by ever-increasing computing power. It is far too easy to be dragged into a complicated 'computer game' rather than focus on the central theme of the project and to produce the required result in the simplest and most effective way. For learning purposes, we encourage students to experiment with all the relevant processing techniques, while in a real operational case the simplest, cheapest and most effective method is the best choice. Simplicity is beauty.

- *From simple to complex*: Keeping things simple does not necessarily mean they can be achieved in a simple way. If simple image processing and GIS techniques were adequate for all applications, more complicated and advanced techniques would never be developed! Some key information can only be enhanced, analysed and abstracted using complex algorithms and methodology. Starting at the simple end, with general image enhancement techniques, will allow us to better comprehend the scope and true nature of the task. Only then should complex techniques be configured and employed to reveal the specific diagnostic features of the intended targets and to extract the most crucial information.

- *Reality checks*: After you have performed some processing and produced an exciting-looking result, always ask yourself if the result is realistic; does it make sense? Performing this kind of 'reality check' could involve correlating the result with simpler images (e.g. colour composites) or other forms of analysis and/or published information (if available). Such information may in itself be insufficient, out of date, at too gross a scale or geographically incomplete but, when compared with your result, may collectively point to your having produced something useful and realistic. If all scant pieces of evidence point in the same direction, you should be on the right track!

- *Relationship between analysis and visualisation*: A remote sensing application project normally begins with image visualisation, and its final results are often in the form of maps and images, for which visualisation is again necessary. As a two-dimensional (2D) array of numbers, a digital image can easily be numerically analysed. The results of the analysis are not necessarily

Image Processing and GIS for Remote Sensing: Techniques and Applications, Second Edition. Jian Guo Liu and Philippa J. Mason.
© 2016 John Wiley & Sons, Ltd. Published 2016 by John Wiley & Sons, Ltd.

raster datasets, but they can always be visualised in one way or another. In general, we are far more able to comprehend complicated information graphically than numerically. For remote sensing applications, visualising and interpreting the results of every stage of image processing and GIS analysis are essential to help assess your progress and to decide on the next step toward the final goal of the project.

- *Thinking in three dimensions (3D):* Remote sensing deals with nothing but images, and an image is a 2D representation of 3D phenomena. For centuries, we have tried every possible approach to presenting the 3D earth on a 2D plane – that is, projected maps – and therein lies the science and engineering of geomatics and geodesy. Now, thanks to the development of computer graphics, moving between 2D and 3D is much easier. The digital form of topographic maps, the digital elevation model (DEM), is in itself a 2D raster dataset representing the 3D topography of the earth surface. Using the powerful 3D graphical functions in modern image processing and GIS software packages, we can easily simulate the 3D environment using 2D data, by draping multi-spectral satellite colour composite images over a 3D perspective view of a DEM (as described in Chapter 17). Thinking and viewing in 3D can make some tricky information in 2D images suddenly obvious and clearly understandable. In fact, it may reveal information that was completely unknown from 2D observation. For instance, a low angle reverse fault appears to be a curved line in an image, depending on its intersection relationship with topographic slopes. The 3D thinking and visualisation make you realise that it is a low angle, planar surface rather than a steeply dipping curved one.

20.1 General image processing strategy

Image processing is almost always the first step of any remote sensing application project but it is often given greater significance than it deserves. In fact, one of the main objectives of image processing is to optimise visualisation of particular thematic dataset. Visual interpretation is therefore essential. Thematic maps are the most important products of remotely sensed imagery, and they are derived either by visual interpretation or image segmentation (computerised classification). Thus far, broadband multi-spectral and synthetic aperture

radar (SAR) images are the most commonly used datasets. The image processing strategy proposed in this section is most relevant to these types of data, and its goal is the effective discrimination of different spectral and spatial objects. We use the word discrimination advisedly in this context; in general, it is only possible to differentiate between rocks, soils and mineral groups using broadband image data, rather than identify them.

In contrast, the processing of hyperspectral image data is to achieve spectral target identification, to species level in the case of rock-forming minerals, and thus has a different strategy. Many people make the mistake either of thinking that hyper-resolution is the answer to all problems, or of being put off investing in such technology at all because they do not understand its role or are suspicious of its acclaimed capability. A hyperspectral dataset is acquired using an imaging spectrometer or hyperspectral sensor, which is a remote sensing instrument that combines the spatial presentation of an imaging sensor with the analytical capabilities of a spectrometer. Such a sensor system may have up to several hundred narrow bands, with a spectral resolution on the order of 10 nm or narrower. Imaging spectrometers produce a near complete spectrum for every pixel of the image, thus allowing the specific identification of materials rather than merely the discrimination between them. A hyperspectral dataset truly provides a data volume or cube. Here it is more important to analyse the spectral signature of each pixel than to perform general image enhancement. The processing methodology and strategy are therefore different from broad band image processing in many aspects, although the enhancement for image visualisation is still important. Considering that hyperspectral remote sensing is a broad and important topic on its own, covering data processing to application development, in this book we have decided to discuss it only briefly and to focus instead on broad band multi-spectral remote sensing.

When you begin a project, you should think along the following lines and, broadly speaking, in the following order:

1 What is the application theme and overall objective of the project?
2 What kind thematic information do I need to extract from remotely sensed images?
3 At what scale do I need to work? That is, what is the geographic extent and what level of spatial or spectral detail is required within that area?

4 What types of image data are required and what is the availability?

5 What is my approach and methodology for image/GIS processing and analysis?

6 How do I present my results (interpretation and map composition)?

7 Who will need to use and understand the results (technical, managerial or layperson)?

Once these steps have been completed and the data have been acquired, the generation of thematic maps from remote sensing image data is generally carried out in three stages:

1 Data preparation.

2 Processing for general visualisation and thematic information extraction.

3 Analysis, interpretation and finally map composition.

In the following sections, we describe the thematic mapping procedure in a linear sequence for clarity. In reality, the image processing and image interpretation are dynamically integrated. The interpretation of the results of one stage of image processing may lead to the image processing and data analysis of the next stage. Often, you may feel you have reached the end, after producing a wonderful image; the subsequent interpretation of that image may then spur you on to explore something further or something completely different. A thematic map derived from remotely sensed image may be used alone or as an input layer for further GIS analysis, as outlined in § 20.2.

20.1.1 Preparation of basic working dataset

20.1.1.1 Data collection

At the stage of sensor development, sensor configuration (spectral bands, spatial resolution, radiometric quantization etc.) is decided based on wide consultation of application sectors, in attempt to provide capabilities that meet actual requirements, subject to the readiness of the technology of course. The data collection is, however, often dictated by what is available, or what the budget will allow, rather than what is actually required. As a result, remotely sensed image datasets are generally aimed at a broad range of application fields and so may not be able to satisfy the most specifics needs of some cases. Within the context of these constraints we should therefore always ask the following questions: What is the purpose of the job? And which dataset will be the most cost effective or will provide the most information relevant to the

purpose of the job? In many ways, image processing aims to enhance and extract thematic information that is not directly sensed or distinctively recorded in any one single image. Sometimes, it is a matter of detective work!

At these early stages of choosing and preparing the dataset, it is also important to consider the most appropriate mapping scales for particular datasets, or rather to choose the most appropriate data to suite the mapping requirements of the task being undertaken. If working at a country-wide or regional scale, it would be rather unwise to select very high resolution (VHR) data for the work since the costs of doing so might be prohibitive and it would generate huge data volumes that may be unworkable or provide a level of detail that is just unnecessary at that stage. On the other hand if the first regional scale work is likely to be followed by more detailed second phase, of the same geographic extent, it may then be necessary to acquire VHR for the entire area at the start. This is, however, not the normal way of doing things for the majority of application cases. Commonly, the regional scale work is followed by more detailed studies in selected areas, in which case you would then acquire higher resolution imagery for those selected areas. The appropriate mapping scale of data is dictated largely by the spatial resolution and partly by the swath width or footprint of the dataset. A summary of common remotely sensed datasets (spaceborne and airborne, and medium to VHR), and the mapping scales at which they are commonly used, is shown in Table 20.1. Table 20.2 presents a summary of the current

Table 20.1 Remotely sensed (EO) datasets and their appropriate mapping scales.

Dataset	Spatial Resolution	Swath	Mapping Scales
Airborne optical	<10–50 cm	Variable	<1:10,000
Airborne hyperspectral	2–10 m	Variable	1:10,000
VHR satellite	0.6–5 m	11–60 km	1:5,000–1:10,000
SPOT1-4	10–20 m	60 km	1:25,000
ALOS AVNIR-2	10 m	70 km	1:25,000
ASTER	15, 30 & 90 m	60 km	1:30,000–1:50,000 +
Landsat TM/ETM	30 m (15 m pan)	185 km	1:50,000 (1:30,000 with pan) +

VHR satellite: IRS Pan, ALOS PRISM, Ikonos, Orbview-3, Quickbird and Worldview,

Table 20.2 Present remote sensing systems and their application areas.

Dataset/Application	ENV	AGRI	FOR	GEO	MAR	RISK	PLAN	DEF	UTIL
MERIS									
Landsat TM and ETM									
ASTER									
ALOS AVNIR-2									
SAR									
Airborne radar									
SPOT 1-4									
VHR satellite									
VHR HD satellite video									

ENV, environment; AGRI, agriculture; FOR, forestry; GEO, geology and exploration; MAR, marine and coastal; RISK, hazards and risk; PLAN, cartography and urban planning; DEF, defence, security and infrastructure; and UTIL, utilities, telecoms, media and consumer.

SAR: ERS, Envisat, ALOS PALSAR, Radarsat, Sentinel-1.

VHR satellite: IRS Pan, ALOS PRISM, Ikonos, Orbview-3, Quickbird, SPOT5, RapidEye, WV-1, GeoEye, Pleaides, WV-2; HD = High Definition.

Excellent		Useful in some circumstances	
Useful		Not normally used	

Table 20.3 Recent launches of remote sensing satellite sensors.

Launch Year	Satellite sensor	Organisation
2006	EROS-B	ImageSat International
	Kompsat-2	Korea Aerospace Research Institute
2007	TerraSAR-X	TerraSAR
	WorldView-1	Digital Globe
	Radarsat-2	RadarSat International, MDA
2008	WorldView-2	Digital Globe
	GeoEye-1	GeoEye
	GOES-O and GOES-P	GOES at Boeing
2009	EROS-C	ImageSat International
	Nigeriasat-2	SSTL
	Cryosat	ESA
2010	TerraSAR-2	TerraSAR
2011	Pleaides HR1	Airbus
	Galileo 1	Airbus
2012	Galileo 3	ESA
	SPOT 5	Airbus
2013	Landsat8 OLI & TIRS	NASA
	SkySat-2	SkyBox
2014	SPOT 7	Airbus
2015	Galileo 6	Airbus

available remote sensing sensors and the corresponding application areas they are used for. Table 20.3 presents a selection of recently launched remote sensing satellites/sensors.

20.1.1.2 Georectification, image co-registration and mosaicing

It is essential to ensure that all datasets being used are geo-rectified or georeferenced. These days all digital, earth observation data are provided as georectified products, which are accurate in x and y to some specified degree. They are always supplied in one of a few standard formats, and almost always conforming to WGS84 datum and Universal Transverse Mercator (UTM) projection, since these are global standards. If, for some reason, the data are not georectified, then the rectification process will normally be the first image processing step to be carried out. Most image processing and GIS software now provide a 'projection-on-the-fly' facility, which removes the requirement that all data conform to same specific coordinate system. Provided that the coordinate systems of each input dataset are defined numerically, one dataset can be re-projected automatically to the coordinate system of another dataset and so brought into alignment, visualised together and, if necessary, assembled into an image mosaic.

The geometric accuracy of these EO datasets, as they are delivered, is generally adequate for many applications but quite often a higher degree of accuracy is needed. This need will depend on the application in question and on the spatial resolution of the dataset. For instance, Landsat data can be considered 'medium resolution' data; it enables mapping at about 1:40,000 (at best) and its geopositioning is generally accurate to about 50m on the ground. In contrast, VHR image data (e.g.

Quickbird, WorldView and SPOT5) may enable mapping at better than 1:10,000 scale on the basis of spatial resolution. When delivered as raw products these claim a positional accuracy of ca 15 m, but to realistically map at such scales, and to do so accurately, the data require not only improved georectification but ortho-rectification to correct for the image distortions imposed by terrain relief.

In such cases, a user may perform the georectification based on measured positions (ground control points or GCPs) acquired using high quality GPS. If the data require ortho-rectification, then a DEM of suitable quality is essential. The quality of the GPS data collection, the user's capability and how they document their survey, rather than the instrument capability, are of paramount importance.

The re-sampling of the image data is inevitable during the georectification process. There is an argument to leave the georectification as the final step after thematic mapping in order to minimise the errors and distortions that may be introduced by georectification process. This approach is typically based on an image processing focused mentality, rather than on a practical one, and so it rather depends on what you intend to do with the data, as to when the georectification is done. Today remote sensing application projects normally involve multi-source datasets, comprising images acquired by different sensors and on different dates. The demand for georectification to ensure all the datasets conform to a standard coordinate system thus often over-rides any concern over potential (and subtle) degradation of image information and so georectification is almost always considered the first step in the production of thematic maps from remotely sensed imagery. In this context, it is far more efficient to georectify the raw data so that all derivative images are also georectified.

Though image co-registration and mosaicing can be performed between images based on the GCPs of local matching features, as discussed in Chapter 9, this process becomes redundant once the images are all precisely georectified to the same map projection system. Georectified images of the same area are in fact co-registered while adjacent images of different areas are in a mosaic based on a frame of the map projection coordinates.

20.1.2 Image processing

We suggest the consideration of the image processing of remotely sensed data in two threads: spectral information and spatial information, as described in Table 20.4. As an

Table 20.4 Sample procedures illustrating the two component threads of image processing, spectral and spatial, within remote sensing applications.

Image Processing for Remote Sensing Applications	
Spectral Information	**Spatial Information**
General enhancement for visual observation	Data fusion to improve spatial resolution
Selective enhancement	Filtering:
Enhancement based on data structure and physical models	a. Low pass
	b. High pass:
	i. Gradient
	ii. Laplacian
Image classification and segmentation	
Spectral analysis	Spatial component extraction – textural properties, image segmentation and feature extraction

example, this procedure may not cover every aspect of image processing and every application, but it serves a useful guide to the essential image processing techniques and of a workable processing strategy. Alternatively, and depending on the nature of the application, only part of procedure shown in Table 20.4 may be required within a particular project; and this is demonstrated in the teaching case studies in Chapter 21. Again, we emphasize that a remote sensing application study should be driven by scientific goals or application objectives, and not by any particular processing procedure.

More details of the processing steps given in Table 20.4 are provided in the following sections.

Spectral information enhancement and extraction

- *General enhancement for visual observation:*
  - Optimal contrast enhancement: Piecewise linear stretch and balance contrast enhancement technique (BCET) are preferred but the specific choice and configuration of contrast enhancement techniques should be decided by observation of the image histograms.
  - False colour composition (consider proper band selection based on common spectral signatures): As shown in Fig. 2.9 in Chapter 2, for an area with considerable spectral variety, BCET automatically produces an optimal colour composite image with balanced colours. Piecewise linear stretch enables you to generate a good colour composite interactively.

○ De-correlation stretch: Direct de-correlation stretch (DDS), IHS or Principal Component Analysis (PCA) de-correlation stretch. As shown in Chapter 5, a de-correlation stretch increases the colour saturation without altering the hues, making ground objects of different spectral signatures more distinctive for visual interpretation.

• *Spectral analysis:* Spectral analysis for target identification is the ultimate goal of hyper-spectral image data processing, while for broadband multi-spectral images, the purpose of spectral analysis is to analyse the spectral differences between targets thus to produce algorithms for selective enhancement and effective target discrimination. In this case, whether the image spectral profile of a target matches its true spectral reflectance signature or not is not important. What is important is to maximise the differences between different targets. BCET is a simple and effective process for this purpose. As shown in Fig. 20.1, the ETM+ spectral profiles of several rock types derived from the original image are very similar in form, with peaks in bands 3 and 5 and a sharp trough in band 4 because they are modulated by very high digital number (DN) averages in bands 3 (142) and 5 (145) and very low DN average in band 4 (77). BCET balances each band to the same DN average (110) and therefore enhances spectral differences; the spectral profiles of the same rock types derived from the same locations in the BCET dataset of this ETM+ image are distinctively different.

• *Selective enhancement:* Based on spectral analysis (pixel profile or laboratory spectral measurements), selective enhancement algorithms can be composed to highlight or segment specific targets, such as vegetation, water, red soil and clay minerals. Ratio and difference are the simplest selective techniques. The typical approach to enhance a target is to use the band of the highest DN (reflection peak) to subtract or divide by the band of lowest DN (absorption trough). The differential and ratio indices of vegetation, iron oxide and clay minerals introduced in Chapter 3 (§ 3.5) are all based on this simple principle. One may also consider compound difference and ratio images, using the summation of bands of two peaks against the summation of two troughs, in the same way. Indeed, you may create highly complex algebraic operations, and with good mathematical logic, but please do not get lost in doing so!

• *Enhancement based on data structure and physical meaning:*
○ Atmospheric correction: Atmospheric scattering effects add a constant to multi-spectral images making them look hazy. The spectral bands of shorter wavelength, for example, blue and green bands, are more severely affected than those of longer wavelength. Removal of this constant can significantly improve image contrast, and thus visual quality, and also refine the functionality of ratio technique for topography suppression. There are many techniques for atmospheric correction. The simplest, crude correction is to shift the minimum of an image histogram to zero by

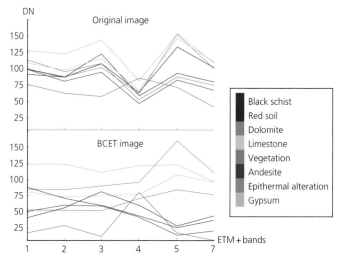

Fig. 20.1 Landsat 7 ETM+ spectral profiles before and after BCET stretch.

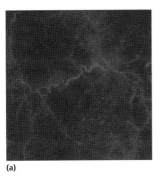
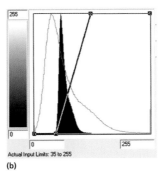
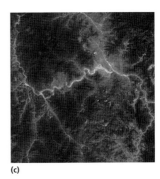

(a)   (b)   (c)

Fig. 20.2 (a) This ETM+ band 1 image is rather hazy and lacks contrast because of the added constant of atmospheric effects; (b) the atmospheric effects are shown in the solid histogram as the gap between the minimum DN, 35, and 0. Automatic 99% clipping using a piecewise linear stretch effectively removes the added constant of atmospheric effects and stretches the image histogram to fill the display range of 0–255, as shown by the line delineated histogram. (c) The resultant image shows significantly enhanced contrast with haze completely removed.

clipping or piecewise linear stretch as illustrated in Fig. 20.2. The operation is equivalent to the well-known 'dark pixel subtraction' technique, proposed by Chavez (1989), but performed more efficiently.

○ PCA and eigenvector analysis: As fully discussed in Chapter 7, PCA is based on the statistical structure of multi-spectral image data. The analysis of eigenvectors can tell us how each *PC* image is composed from the spectral bands of the original image and thus how a particular target will be highlighted in a particular PC. This comprises the so-called 'FPCS' (Feature-oriented PC Selection) technique, one of the most effective techniques for PCA-based selective enhancement.

○ Simulated reflectance: As introduced in Chapter 3 (§ 3.7), we can derive simulated reflectance and thermal emittance from a multi-spectral image with both reflective spectral bands and thermal bands, such as TM/ETM+, OLI and ASTER, based on a simplified model of solar radiation to the earth. This technique suppresses topography and enhances the spectral signature of ground objects according to their true spectral reflectance. Thus a simulated reflectance image is directly comparable to its corresponding spectral band and is easy to interpret. In contrast, other techniques, such as ratio and differencing, enhance targets' spectral signatures indirectly in a combination of two or more bands. While the simulated reflectance technique is for direct enhancement of spectral reflectance of individual image bands, ratio and differencing techniques achieve selective enhancement of a particular target on the basis of its spectral properties among several spectral band images.

○ HRGB: This technique introduced in Chapter 5 (§ 5.4) is the most effective method of suppressing topographic shadowing and of condensing the spectral information from up to nine spectral bands into a three-band colour composite. The HRGB image does not facilitate easy visual interpretation without reference to simple colour composites because the spectral properties of ground objects are indirectly presented. It is, however, very good for classification but caution must be taken: since the hue value is in the 2π range of a colour wheel, those colour vectors with hues around 0 and 2π are numerically very different but are in close proximity in the RGB colour cube, as detail discussed in § 5.4.

• *Image classification and segmentation*: Multi-spectral images, as well as multi-source datasets, may be treated as one multi-variable dataset, and so statistical classification algorithms can be applied to produce classification maps, automatically or semi-automatically, from them. Many image features relating to geology and environment can be easily picked up 'by eye' on the basis of our knowledge-based understanding of spatial patterns and spectral properties, while many of the tasks that appear to be easy actually turn out to be 'mission impossible' for classification. Therefore, image classification cannot replace visual interpretation and, quite often, a classification image still need visual interpretation. For those ground objects that can be discriminated or identified purely by their spectral properties, classification is the most effective way of mapping them.

○ Thresholding: This simple technique is very effective for highlight particular targets. For instance, a thresholded TM5/TM7 ratio image can reveal those pixels representing hydrated alteration minerals that may indicate the presence of mineral deposits.

○ Statistical classification: Reiterating what we learned in Chapter 8, supervised classification is based on image training and is often guided by user knowledge that may be biased or incorrect. Unsupervised classification, whilst sounding like and often appearing like an automatic technique in many image processing software packages, is not one. An unsupervised classification image requires interpretation and can be significantly affected by the choice of initial parameters (most software packages provide only default values!) even though the algorithms have self-optimization functionalities. For both classification approaches, the statistical decision rules dictate the classification accuracy.

○ Spectral angle classification: The greatest advantage of this algorithm is that classification is independent of any illumination variation or, in other words, it is not affected by topographic shading. Though performed on multi-spectral imagery, that is, raster data, it is essentially a vector classifier that treats each pixel of an N band image as an N dimensional vector in N dimensional feature space.

Spatial enhancement

• *Image fusion*: Fusion of a lower spatial resolution colour composite with a higher spatial resolution panchromatic image can generate an apparently higher resolution colour composite that combines both the high spatial resolution of the panchromatic image and the higher spectral resolution of the colour composite. However, we must realise that the image fusion, no matter which technique is used, does not improve the spatial resolution of spectral information in the original colour composite. The processing is for visual observation and interpretation but not for quantitative analysis. In Chapter 6, we introduced following three fusion techniques.

○ SFIM: The smoothing filter-based intensity modulation is a spectral preservation fusion technique; it maintains the fidelity of the spectral information from the original colour composite. In other words, it does not introduce colour distortion. The technique is, however, very sensitive to image co-registration accuracy. Any co-registration errors may produce subtle edge blurring effects that degrade the image sharpness. The pixel-wise co-registration introduced in Chapter 11 (§ 11.3.4) resolved the problem. SFIM is widely used to produce standard pan-sharpen imagery products from the panchromatic band and multi-spectral bands of the same sensor system.

○ IHS fusion is achieved by intensity replacement (with a higher spatial resolution image) in an intensity, hue and saturation coordinate system. It is insensitive to miss-registration and thus produces sharply enhanced fusion images even if the refined textures actually mismatch with the spectral edges. This technique inevitably introduces distortion of colour and albedo. This distortion will be severe if the spectral range of the three bands forming the colour composite is very different from that of the higher resolution image used for intensity replacement.

○ Brovey transform: This is based on direct intensity modulation. It has the same merits and drawbacks as the IHS fusion technique, in terms of fusion quality, but is more efficient in processing as it does not require forward and inverse RGB-IHS transformations.

• *Filtering*: As a neighbourhood processing technique, filtering may not 'honour' the original intensity information of an image. Instead, it brings out the spatial relationships between a pixel and its neighbours. On the other hand, all images acquired through an optical system are, to a degree, filtered images.

○ Low pass filtering: The main objective of this is to remove noise at the cost of spatial resolution but there are many edge preserve low pass filters that reduce noise with minimal spatial information loss. For smoothing a classification image, only those filters without numerical calculations should be used.

○ Gradient filters: As a first-derivative-based filter, it performs directional enhancement. We can configure a gradient filter to enhance the linear features in a particular direction. Important advice in this context is that if linear features in a direction are already very obvious in the unfiltered image, there is no need to apply a gradient filter to enhance this direction at all because it will also enhance subtle features and thus dilute the already clear lineaments.

○ Laplacian filters: As a second-derivative-based filter, it is non-directional. It enhances image textural edges in all direction. A Laplacian filter is often the first step for texture extraction. One variant of the Laplacian filter is the 'sharpen' filter, which is equivalent to adding a Laplacian filtered image (textures) to the original image. The result is an edge-sharpened image. Such processing is for visualisation and can aid visual interpretation.

- *Spatial component extraction*: Based on neighbourhood processing, many spatial components of a raster dataset may be derived and extracted, such as local contrast, local variance and edge intensity. With DEM data, many of the extracted spatial components have specific physical significance, for instance slope angles, slope aspects, slope curvature and flow grids. These properties relating to surfaces have already been described in Chapter 17 and will be further discussed in the next section.

20.1.3 Image interpretation and map composition

When we interpret images to map particular thematic information, either manually or using software annotation tools, we are effectively working on a vector layer. The primary advice here is to start from the easiest and most obvious and to work towards the most difficult and complex.

Progressing from images to thematic maps is where image processing and GIS merges. As a complete processing cycle, we briefly describe image interpretation and map composition here as the final stage of image processing. We will revisit some contents in this part in the next section in greater detail. Thematic mapping for various application areas may be different but all follow a generic route. The list below describes a general procedure for geological and environmental mapping but it is generically applicable to the mapping of other thematic information.

- *Map format:* Map page set up, geographic coordinates, scale bar, north arrow, title, legend and other relevant general annotation form a standard template for image interpretation and map composition. These are common and standard tools in image processing and GIS software suites, and whilst they can be applied and modified at any stage of the work, issues such as map scale may greatly affect the detail at which the interpretation can and should be performed, and so should always be considered at the start. Bear in mind that it is considered good practice to capture data at the greatest detail possible, right from start since it is far easier to reduce detail than to add or re-capture it later on.
- *Basic geographic (cultural) information*: Well-enhanced colour composites can provide more than adequate information for the interpretation of:
  - ○ Man made features – cities, towns, roads, rail and cultivated areas;
  - ○ major drainage systems – rivers, coast and water bodies.

- *General interpretation:*
  - ○ Separation of land and sea: Thresholding using infrared bands often enables the masking of sea pixels from an image, thus enabling more effective enhancement of particular land objects. Be careful, however, to be very critical when choosing the threshold and be aware that dark shadows may cause you problems in this respect.
  - ○ Vegetation: Standard false colour composites and vegetation indices enable interpretation of vegetated cover, whether natural or otherwise, and so are important sources of landuse and/or land cover information.
  - ○ Identify major land cover categories: Agriculture, forest and urban areas (industrial and residential).
- *Geological and environmental interpretation:* This stage forms a dynamic process involving image processing and interpretation. Specific information can be extracted from images produced using purposely designed techniques, as described in § 20.1.2. From start to finish, the interpretation becomes progressively enriched.
  - ○ Separate the bedrock (solid geology) from any superficial deposits (drift geology): Use both spectral and textual information to do this, such as simple colour composites or edge-sharpened colour composites, or colour composites involving fusion of high spatial resolution. Examine different band combinations and look for targets that may be spectrally similar but texturally different, or vice versa.
  - ○ Interpret major rock types: Again, use colour composites of differing band combinations. Use PCA to identify areas that are spectrally distinctive, and then try to establish why. Derive simulated reflectance to enable comparison with laboratory data. Produce spectrally complex images, such as via HRGB, to try to distil the spectrally significant information.
  - ○ Highlight specific targets: Use differences and ratios, PCA and FPCS to highlight particular target materials. Be careful here to remember that if you have success in highlighting a particular target that you think may be significant, you must also be able to explain why the particular technique has been successful. For instance, discrimination of hydrated minerals and red soils using TM/ETM+ difference images of bands 5-7 and 3-1 must also be related back to the known spectral signatures of these materials, in order to understand how they work.
  - ○ Structural features: Interpretation of lineaments, faults, fractures and fold axes can, in many cases, be

achieved without filtering, so that filters should be applied only when such features are not clear or obvious. In some cases, faults and folds are revealed not by their textural characteristics alone, but by the combination and spatial context of spectral variations and topographic shading. In fact, the routine extraction of image 'lineaments' should be avoided since these often have no sub-surface geological basis.

○ Classification: This can be quite effective for lithological mapping based on spectral signatures of rocks and minerals, when these are clearly visible at the surface, such as where exposure is continuous. Classification 'falls down' where there are irregular spatial patterns, such as those caused by intense faulting/fracturing or where anthropogenic features (cultivated land patterning) are present. One could argue that in such cases where rock exposure is continuous, good 'old fashioned' interpretation of geology is more reliable. It may prove useful when working with very large areas and/or where unknown spectral variations of unknown significance exist, which might not be in the interpreter's knowledge base and so might otherwise be overlooked. It is our experience that most experienced geologists prefer to avoid classification altogether when interpreting images.

Completion of the composition of a thematic map in this way may be the final stage of a remote sensing application project but more often it forms the beginning of the GIS 'modelling' or spatial analysis part of a project, which may have a much wider scope, may involve data from widely different sources, and may include other remotely sensed images acquired for very different purposes. In other words the image-based geological interpretations we have described here may form a tiny part of a much larger and broader project remit.

20.2 Remote sensing-based GIS projects: From images to thematic mapping

• Important preparatory considerations: The regional context and setting of the area being studied. Understanding of the wider context helps to anticipate the variety and types of targets that will require interpretation. This also includes the climate of the area. Tropical and temperate regions will suffer from cloud cover and data acquisition may be problematic. These areas are also likely to be densely vegetated and, whilst this is not a problem if landuse or agriculture are the applications of interest, it will limit the depth and detail of any geological interpretation. Images acquired in tropical areas, even if cloud free, will suffer from haze in the visible bands, which will require correction, and in some cases may render the first three bands unusable for image processing and interpretation. Arid and semi-arid areas make interpretation of ground objects relatively easy since the spectral signatures of the rocks are less likely to be obscured by those of vegetation and thick soils. In some desert environments, wind-blown deposits may also obscure the bedrock and hinder the interpretation of the solid geology beneath.

• A suggested generic procedure: This will always follow the same scheme of three broad phases. The first phase will begin with problem/objective definition, data acquisition/collection, followed by data integration, image processing and analysis, then by interpretation and end with map production and output. This phase should then be followed by fieldwork to verify the results (phase two). Phase three should then involve a refinement of the processing and interpretation, in response to the additional knowledge gained during fieldwork, to arrive at a more complete and realistic interpretation map.

• Mapping using thematic layers derived from remote sensing: Data integration and visualisation.

• GIS 'modelling' based on multi-source data: This demands the integration of data as described above. The point of this exercise is to incorporate different and complementary datasets, in an attempt to describe or model some potentially complex phenomenon.

○ Multi-source data integration and spatial analysis: Involving other data, such as geological maps, geophysical data and geochemical data and DEM, which have themselves been processed, interpreted or classified in a particular way, leads to the identification of some complex objective, such as a hazard assessment or a site selection exercise.

20.3 An example of thematic mapping based on optimal visualisation and interpretation of multi-spectral satellite imagery

A real remote sensing application project does not necessary involve all the image processing and GIS modelling techniques as described in the two previous sections. It is always important to remember that for any application, the project should be problem-driven rather than processing-driven. In this section, we present a simple example to demonstrate the use of basic image processing techniques and GIS mapping functionality to produce a digital geological interpretation map. Although the theme is geological mapping, the general approach is applicable for visualisation and interpretation of images for other themes too. This case study has been set up as coursework for an introductory course in remote sensing and GIS as part of our undergraduate teaching schedule.

20.3.1 Background information
20.3.1.1 Study area
The study area lies in Almeria province in south-east Spain and is illustrated in Fig. 20.3. The environment is characterised by a semi-arid climate (it is Europe's only semi-desert), sparse natural scrub vegetation, localised intense (covered and irrigated) horticulture, economic extraction of gypsum (and other materials); the bedrock geology is very well exposed over large areas, with little soil development and almost no vegetation. The area is well-known for the teaching of field geology, geomorphology and geography.

Despite the limited geographical extent of the area, its geology is varied, which is one of the reasons for its popularity in field teaching. Superficial deposits consist of Quaternary palaeo-alluvial fans and red soils, and Holocene alluvial fans and gravels in ephemeral river channels. The solid geology consists of Palaeozoic graphitic and garnet-mica schists, later Permo-Triassic phyllites and dolomites (in thrust sheets). These form a series of basement massifs between which are sedimentary basins filled by Messinian sediments, including reef limestones, marls and gypsum. To the south-east, these basin and range terrains are separated from the Cabo de Gata, a volcanic terrain of the same age as the basin sediments but which is exotic, i.e. has been transported to its current location by plate tectonics. This volcanic terrain is typified by acid-intermediate, island-arc volcanics (andesites and dacites), some of which have been subject to late-stage epithermal alteration and mineralization. The area has quite a long history of gold exploration and mining operations near the town of Rodalquilar (now closed).

Tectonically, the area is still active and is structurally more complex than it may at first appear. There are major faults, such as the Carboneras fault (a conservative plate boundary, marked by a major transcurrent fault), numerous oblique-slip faults synthetic and antithetic to this, in addition to a great many minor neotectonic faults.

20.3.1.2 Data
A Landsat 7 ETM+ subscene acquired in June 1999 has been used. The data are georectified to map projection UTM (zone 30N) and datum WGS84, and could be said to provide a 'GIS-ready' mapping base-layer. From this dataset, a series of enhanced colour composites are produced, as information sources for the interpretation of geological features. In addition, a regional geological map, at a scale of 1:200,000 and a shaded relief image derived from SRTM 30 m resolution DEM data, are also provided for reference.

Both the latter two datasets provide information at scales that are much coarser than the images being used, and this sometimes causes a little initial confusion on the part of the students. They are only beginning to learn about mapping scales and acceptable/workable levels of detail for (a) discrimination and (b) feature extraction, and do not at this stage have the conviction (or experience) to know that what they will achieve is an interpretation map that should be far more detailed, more up to date, and positionally more correct than the published regional map. Neither do they realise that the reason for performing this exercise, in reality, may be because geological map information does not exist at the required scales for a particular area, so that mapping from remotely sensed data sources may be the only way. Either way, in a real case, you would always collect as much background information (publications and maps) as you possibly could, to equip yourself, but when it comes to processing the images, you should try very hard not to allow your interpretation to be biased by that information; a great deal of the value of a remotely sensed image interpretation lies in its independence and objectivity.

20.3.2 Image enhancement for visual observation
For this project, the objective of the image processing is quite simple, to produce a few good colour composites from the given Landsat 7 ETM+ image data.

Colour composites of Landsat ETM+ image bands: 321, 432 and 531 (RGB) are recommended for general-purpose visual interpretation of all ground target types, natural and anthropogenic. The bands 321 RGB true

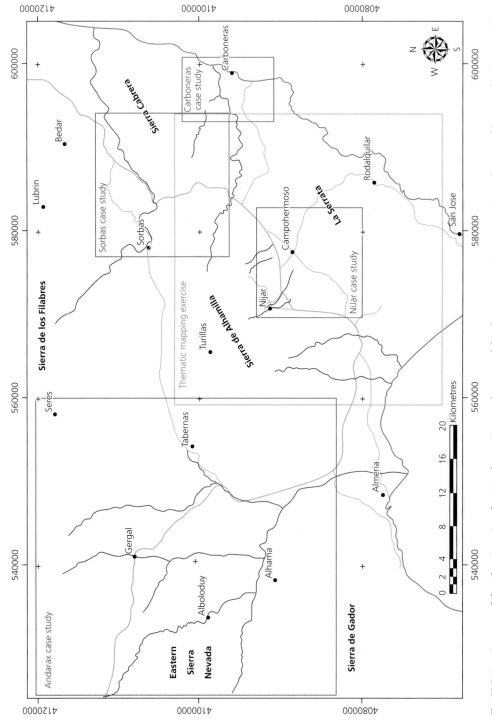

Fig. 20.3 Location map of the Almeria region showing the positions and extents of this thematic mapping exercise (green box) in addition to the four Spanish case studies described in § 21.1, Sorbas; § 21.2, Carboneras; § 21.3, Nijar; and § 21.4 Andarax (red boxes). Major towns, rivers and coast (fine and bold blue lines, respectively), main roads and motorways (fine and bold brown lines, respectively).

colour composite shows the land surface, similar to the way we see it with the naked eye in field. The 432 RGB standard false colour composite effectively highlights vegetation in red tones but also major ground objects such as water, soils and gross lithological variations. In semi-arid terrain, where the level of exposure is high and soil and vegetation cover are low, the 531 RGB colour composite is almost always the best image for discriminating lithological variations.

We recommend a simple processing procedure: contrast enhancement using BCET (or linear contrast enhancement with appropriate clipping) to balance the colours followed by DDS to increase the colour saturation.

The images shown in Fig. 20.4 illustrate the effects of this procedure on the image dataset for this project. The three colour composites are shown in columns a, b and c, with row 1 representing the raw image composites before contrast enhancement. All three are subject to colour

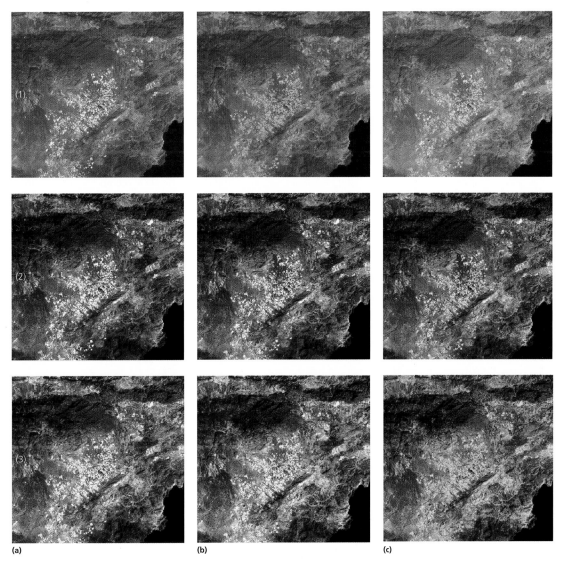

(a)   (b)   (c)

Fig. 20.4 Landsat 7 ETM+ colour composites of bands 321 RGB in column (a), 432 RGB in column (b); and 531 RGB in column (c); and colour composites of the original bands in row (1); after BCET enhancement in row (2); and with DDS after the BCET in row (3).

bias to a certain degree and a lack of contrast. After the BCET enhancement, the colour bias in each composite is removed by the contrast enhancement and colour enriched, as presented in the second row of Fig. 20.4.

Following this, the DDS with the achromatic parameter $k=0.5$ is applied to further enhance the colour saturation of the three colour composites. These are shown in the third row of Fig. 20.4 and present vegetation, red soil, water and drainage patterns, alluvial fans, and various rock types vividly in bright and distinctive colours. These three images are then used for visual interpretation in the next stage of the project to generate a digital geological map using a GIS system.

20.3.3 Data capture and image interpretation

You should begin by considering the location of the area being mapped. The most obvious considerations in this case are its regional geological setting and its climate. The former will help you to anticipate the tectonics and lithological characteristics. In this instance, the area lies in the Betic Cordillera of south-east Spain, the rocks range in age from lower Palaeozoic to recent and have undergone two phases of orogenic deformation (Variscan and Alpine) and the region is still tectonically active today. The latter will point to the nature of the terrain surface: whether it is vegetated, weathered to any great depth, is subject to persistent cloud and so forth. In this case, the area is classified as Europe's only semi-desert (with ca 300 mm rainfall pa), there is very little soil development, the atmosphere is hazy from time to time, and there is very little vegetative cover (what vegetation there is related closely to ephemeral drainage and irrigation). As a consequence, what is recorded in remotely sensed imagery represents an almost complete record of surface geological exposure across the region, which makes geological interpretation relatively straightforward. The arid climate makes vegetation a very useful indicator of the presence of ground and surface water; the appearance of localised patches of healthy vegetation usually reveal small rural settlements supported by springs, which are themselves controlled by lithology and structure. Even large scale agriculture may reveal similar geological control of regional water supplies since this is always more cost effective than piping in water from elsewhere. The aspect of mountain areas will also affect the distribution of areas that can support natural vegetation and woodland, that is, north-facing slopes; its

presence will need to be considered in interpretation of spectral properties of ground targets. Understanding landuse can also therefore be an important factor in interpreting the geology.

Interpretation of different themes in multiple layers:

- Structure of the map project file: The data will likely be organised slightly differently from case to case, because of differences in the specific GIS software used. Essentially though it is sensible to keep solid and drift geological features in separate layers. At this simple level of data capture, it is desirable to capture lithological areas as polygons. This makes for a rather more rapidly constructed map than the more correct method of capturing arcs and later building topology to construct polygons. This choice of strategy rather depends on the time available to complete the task and the software tools available to you. Doing this the 'quick', non-topologically correct way means that there are certain limitations on the complexity of information that be captured and conveyed; slivers and gaps, and island polygons will have to be avoided. This method is perfectly acceptable if the final product is required only as a single map product for reference and if no further spatial analysis will be required of the geological polygons.

- Other features such as quarries can also be easily stored as simple polygons. Faults on the other hand, by their inherent nature, are stored as linear features in a polyline file. Other *cultural* data can also be captured/imported and stored but should be stored separately from the interpreted features, but could be grouped together for convenience. Such features could include towns (points), roads (polylines) and drainage (polylines). In addition to the images that are the source data for the interpreted features, there are other raster images in the database, namely the DEM and a regional geological map. Again these raster data layers are, by their nature, stored differently and separately from the vector features but could usefully be grouped together as reference layers or in two groups, for example, satellite images and regional data.

- Use of an interpretation guidance table: During the practical work, students are advised to use a table, such as the one illustrated in Table 20.5, to help familiarise themselves with and note down the appearance of various ground features, as they appear in each colour composite (321, 432 and 531) and to use the suggested legend symbols (or ones of their own making) to annotate their interpretation. This forms an important

Table 20.5 A sample image interpretation guidance table—an aide memoire.

Ground Objects	Description of the Features in Landsat Images			Legend Symbols
	321RGB	432RGB	531RGB	
Natural vegetation				
Horticulture				
Urban areas				
Quarries				
Alluvial fans				
River debris				
Schists				
Dolomites				
Limestones				
Marls				
Gypsum				
Andesite				
Dacite				
Solid and drift geological boundaries				
Inferred geological boundaries				
Faults (major)				
Faults (minor)				

step in understanding the way the displayed spectral bands determine the colour of features in each image. The connection is made between relative reflectivity in particular wavebands (Landsat bands 1, 2, 3, 4 and 5 in this instance) and image brightness in particular colour layers (red, green and blue). For instance, iron-oxide rich red soils appear bright red in the 321(RGB) image but are greenish in both 432(RGB) and 531(RGB) images since the relative reflectance of iron-oxides is high in Landsat band 3 and low in bands 2 and even lower in 1.

- The items listed in the table also provide as a hint towards the lithologies the students should expect to see in this area. In this example, the students have visited many localities during fieldwork in the previous year and so are familiar with its geology and geomorphology; they simply need reminding of what they saw and learned. They also need encouragement to make the link between the appearance of rocks in the field and hand specimen, with their appearance in the image, and to treat the work rather like a complicated puzzle to which there may be no definitive answer.

- Procedure: We always recommend spending some time just looking at the images, familiarising yourself with the database and software and examining how different targets appear in each image band combination. During this process, you will probably begin conceptually sub-dividing the study area into geologically significant zones (or terrains), before actually capturing any new data. This will help you to understand which are the most important features to convey in the final map. During this stage you should also establish the optimum scale at which to capture features, according to the spatial resolution of the images, probably around 1:35,000 in this case, where you can see maximum surface detail but not individual pixels.

- You should soon begin to feel confident about identifying boundaries between objects that are spectrally and texturally different – these will probably be the most obvious and largest, eye-catching features. When you are ready to capture some features, start by capturing those most obvious lithological outcrops, that is, their boundaries, and when you have created a feature, remember also to enter an identifying, descriptive

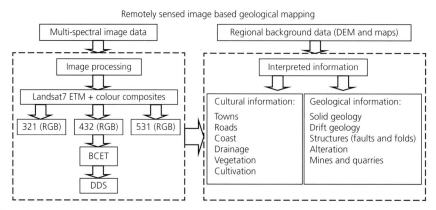

Fig. 20.5 Workflow chart, summarising the processing and interpretation procedures of the mapping project described in this section.

name for the feature in its attribute table. Doing this for each feature as you create it will save time later (when you may not remember quite so clearly what you were thinking at the time). At this stage you will almost certainly not be able to give the outcrop a specific geological identity, that is, you may have little idea about the lithology, but that does not matter at this stage. You will build up a series of units identified perhaps as 'sedimentary_1, _2, _3' or 'volcanic_1, _2' etc. As you proceed in this way, moving around the image, you will find outcrops that look spectrally and perhaps texturally similar to some that you have already captured, so you should soon find that you have several polygons with the same identifying code or label. You can of course amend these descriptions as you proceed.

- After identifying all the obvious features, you will then begin to find boundaries between lithologies that are only subtly different and perhaps spectrally complex in themselves. For instance, they may be spectrally similar but texturally different, in which case you may surmise that they may represent chemically similar lithologies that are not of the same stratigraphy or facies. Alternatively, they may be spectrally different but texturally contiguous, in which case you could conclude that they represent local variations of lithology that have common structure, such as suites of metamorphic rocks in mountain ranges that have been subject to regional deformation. As you progress around the image, capturing lithological information, many questions will arise and as you attempt to answer those questions, calling on your own geological knowledge and experience, you will get closer to giving more

precise geological names to the outcrops. After some time, it is advisable to stop and do something else, returning later with a fresh and critical eye, to go over what you have done and refine it. You may repeat this process a number of times. During these times, it is always a good idea to zoom back out from your detailed observations, to a more regional scale since it is very easy to get 'bogged down' in the detail and to spend more time on one small area than a) you can afford and b) is necessary for the scale of the map you are creating. We also provide an image of very high spatial resolution (a orthorectified digital aerial photograph in this case) to provide almost outcrop-level spatial detail. This allows the visualisation of image textural information relating to bedding planes in sedimentary rocks, or to jointing and fracturing on a level that facilitates more meaningful interpretation.

- Geological structure: A considerable amount of common sense and logic is required to be successful and this comes with experience, confidence and clear thought. You will also, no doubt, find objects and features that you cannot identify or understand at all, and these are the ones that should be recorded in your notebook as features that require verification during fieldwork.

20.3.4 Map composition

The finished product will include the interpreted geological information plus sufficient cultural information to make the map navigable, along with items normally found on any map, such as a coordinate grid, scale bar, north arrow, annotation, title and map legend to explain colours and symbols used on the map.

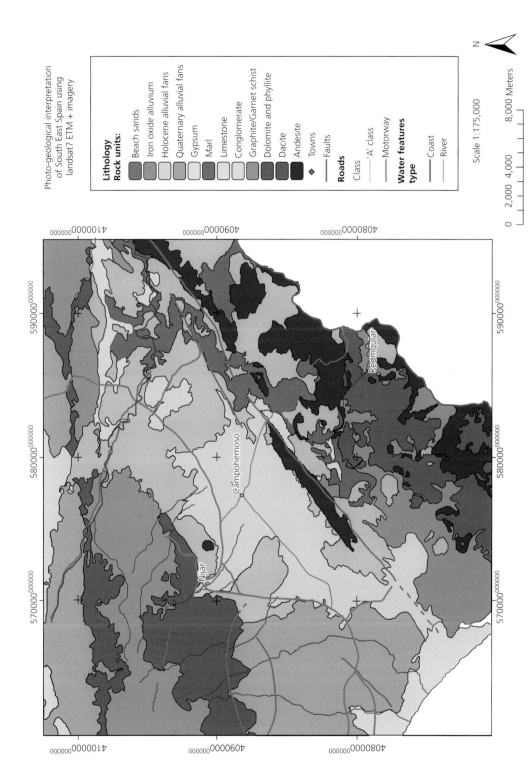

Photo-geological interpretation
of South East Spain using
landsat7 ETM + imagery

Lithology
Rock units:

Beach sands
Iron oxide alluvium
Holocene alluvial fans
Quaternary alluvial fans
Gypsum
Marl
Limestone
Conglomerate
Graphite/Garnet schist
Dolomite and phyllite
Dacite
Andesite
◆ Towns
━━ Faults

Roads
Class
━━ 'A' class
━━ Motorway

**Water features
type**
━━ Coast
━━ River

Scale 1:175,000

0   2,000 4,000   8,000 Meters

N

Fig. 20.6 An example of the finished interpretation map.

Care needs to be taken in the selection of colours for the interpreted polygons in particular. Bearing in mind that in the world of digital mapping using highly functional software, a dazzling array of colours and symbols is available, it is important to take a step back and re-consider the objectives considered at the start of the exercise (in § 20.1). Remember what the application theme and overall objectives are; what aspect of the map should be the most obvious one to the eye of the intended end user of the map? Ensure the map scale for the final map is a sensible number (preferably a whole number rather than some obscure fraction) and that all the information you want to show appears sensibly laid out on the page. Make sure that the map is not over-crowded with either cultural information or labels. Will the symbols/labels be discernable/legible in the final scaled version? It is often worth visualising at 1:1 scale and then making one or more test prints of the map to establish this. Remember to consider who might be using the map – consider if it conveys sufficient background/ ancillary information to sufficiently explain what the map shows, how it was produced, what it represents and so forth to the untrained eye; assume that a layperson will need to make use of it and then assess whether he or she will actually be able to understand it.

Given that the database contains height data, in the form of a DEM, both the images and the finished map can then be visualised in pseudo-3D.

Theoretically, the finished map should be of a potentially publishable standard. In this class exercise, the results will always fall short of that standard but will be an impressive achievement for each student nonetheless and will look something like the example shown in Fig. 20.6.

When the final map is complete, it is then output to one of a number of standard formats, such as pdf, tiff or jpg, or to some format that supports zooming and some query functions (such as ArcPublisher), ready for supply to its final destination, where ever that may be. Extremely useful at this stage is the ability to output the coordinate reference information with the publishable map, using a *world file*, either a .tfw or .jfw (as described in Chapter 13), so that it can then be displaced in any other GIS system as a map database product.

20.4 Summary

This is a very important chapter since it sets out a kind of recipe for a logical way to approach a typical project using remotely sensed data to achieve an application objective. In this case, we use basic geological mapping as an example, but the topic could easily be landuse, environment, agriculture or water resources. The important thing common to all such projects is the strategy. We have thus tried to provide some valuable and important rules of thumb, which we know from the experience of doing this type of work. There can surely be no better way to learn anything than by simply doing it; what we have done here is to accelerate the learning process by steering you away from the many known potential pitfalls that lie along the way. From defining the project goals through to extracting the elusive thing at the very end – that is, the real image information – this chapter forms a simple and generic formula for doing so.

CHAPTER 21

Thematic teaching case studies in SE Spain

In this chapter, we discuss several teaching case studies on specific themes, using image data of SE Spain, to demonstrate remote sensing applications in earth and environmental sciences. Each case emphasizes different parts of the general strategy (described in the last chapter), but all follow the same route from image processing to information extraction and finally to thematic mapping.

21.1 Thematic information extraction (1): Gypsum natural outcrop mapping and quarry change assessment

The Sorbas area, in Almeria province of south-eastern Spain, contains one of the largest and most significant gypsum deposits in the world. The large-scale economic extraction and environmental conservation of the natural gypsum karst landscape are in direct conflict. In this case study, multi-temporal TM/ETM+ images are used to map the distribution of natural gypsum outcrops and to chart the temporal changes in the extents and location of gypsum quarrying, to provide objective information relating to the impact of the extraction industry on the regional environment. The main objectives of the study are:

- Identify and map the natural outcrops of gypsum;
- extract gypsum quarries and investigate the changes of gypsum quarries in 16 years.

The study comprises three parts:

1 Multi-spectral image enhancement for gypsum mapping;
2 gypsum quarry extraction;
3 multi-temporal comparison for quantitative assessment of the change of gypsum quarries.

21.1.1 Data preparation and general visualisation

Three TM/ETM+ images with 8-year intervals, acquired in 1984, 1992 and 2000 (Table 21.1), are used in this study. The Landsat 7 ETM+ image acquired in 2000 was downloaded from the Global Land Cover Facility (glcf.umiacs.umd.edu/index.shtml), which has been accurately rectified to Universal Transverse Mercator (UTM) N30 based on WGS84 datum. The other two images were co-registered to the ETM+ image to conform to the same map projection. As the three images were acquired by the same type of sensor system and from similar orbits, although onboard different satellites, the geometric deformations between them are mainly linear translation and rotation. The simplest linear polynomial transform therefore produces the best co-registration accuracy.

As explained in Chapter 20 (§ 20.1.2), to optimise the spectral analysis and visualisation, balance contrast enhancement technique (BCET) has been applied to produce BCET datasets corresponding to each of the three images.

Remote sensing can reduce the workload of field investigation significantly but cannot replace it. Field knowledge of the gypsum outcrops and existing geological maps are of great assistance in sampling to produce image spectral profiles and to assess the results of this study. Where field knowledge and existing maps are unavailable, the understanding of the spectral properties of major ground objects and target minerals is essential, while general image visualisation is the starting point to gain this knowledge. Figure 21.1 is a colour composite of the ETM+ bands 541 in RGB with BCET and direct de-correlation stretch (DDS) enhancement. The image

Image Processing and GIS for Remote Sensing: Techniques and Applications, Second Edition. Jian Guo Liu and Philippa J. Mason.
© 2016 John Wiley & Sons, Ltd. Published 2016 by John Wiley & Sons, Ltd.

displays vegetation in green and various rock types in a variety of different colours. The very bright patches in this image are produced by the gypsum quarries. Spectral profiles of gypsum can be sampled in these quarries and nearby areas of outcrop.

Table 21.1 The TM/ETM+ images used in this case study.

Satellites	Sensors	Path/Row	Acquisition Date
Landsat 4	TM	199/034	19 July 1984
Landsat 5	TM	199/034	25 July 1992
Landsat 7	ETM+	199/034	8 August 2000

21.1.2 Gypsum enhancement and extraction based on spectral analysis

As a hydrated mineral, gypsum is spectrally similar to clay minerals relating to alteration characterised by high reflectance in TM/ETM+ band 5, a broad short-wave infrared (SWIR) spectral band centred at 1.65 μm, and strong absorption in TM/ETM+ band 7, a broad SWIR spectral band centred at 2.2 μm. A simple TM/ETM+ bands 5 and 7 differencing or ratio can therefore selectively enhance both targets but cannot achieve the separation between them (Fig. 21.2). Figure 21.3(b) shows the ETM+ image spectral profiles of a known gypsum quarry and natural outcrop, an epithermal

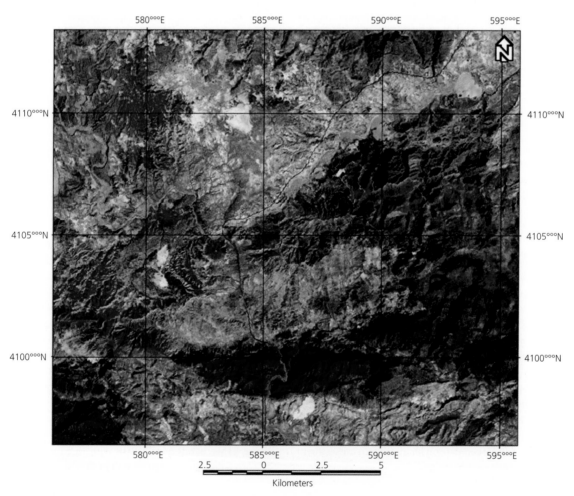

Fig. 21.1 Landsat 7 ETM+ bands 541 RGB colour composite with BCET and DDS enhancement to show vegetation, lithology and quarries of the study area. The scale bar in this image serves as a reference to all the images, which cover exactly the same area, in this case study.

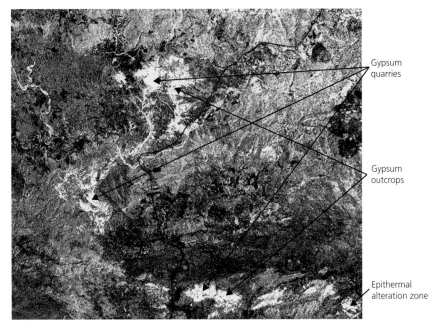

Fig. 21.2 Difference image of ETM+ band 5 minus band 7. Most white patches are either gypsum quarries or gypsum outcrops except the one in the bottom right corner that is an epithermal alteration zone.

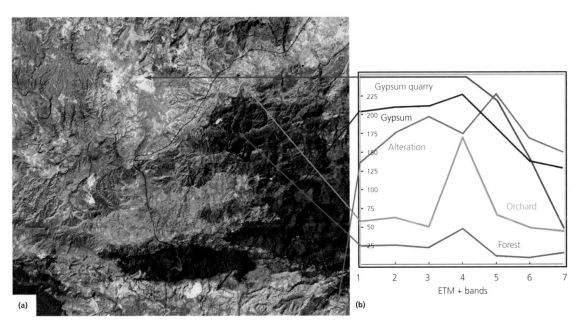

Fig. 21.3 (a) Colour composite of ETM+ bands 457 RGB with BCET and DDS enhancement. Gypsum outcrops and gypsum quarries are uniquely highlighted in yellow colours while vegetation is in red and reddish orange, and alteration clay minerals are the same as several rock types in cyan. (b) ETM+ image spectral signatures of gypsum quarry, outcrop of gypsum, alteration zone, orange orchard and pine tree forest. The arrows point to the spectral sample position in the image.

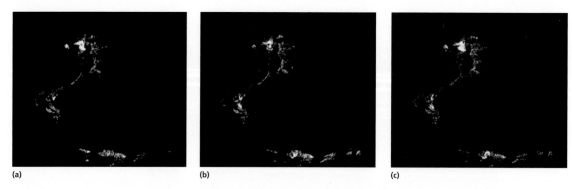

Fig. 21.4 Images of gypsum extraction and mapping: (a) 1984; (b) 1992; and (c) 2000.

alteration zone and vegetation. The unique spectral feature of gypsum that makes it different from clay minerals is that it has slightly higher reflectance in band 4 (nearer infrared) than in band 5, where alteration clay minerals have a strong absorption trough in band 4. Thus, a simple BCET DDS colour composite of TM/ETM+ bands 4, 5 and 7 in RGB highlights gypsum uniquely in yellow separating from the alteration clay minerals in cyan, as shown in Fig. 21.3(a). The spectral sample points are denoted in this image as well.

The spectral signature of gypsum bears some similarity to that of vegetation in TM/ETM+ bands 4, 5 and 7 as shown in Fig. 21.3, but the so-called 'red edge' feature of strong absorption in red band (TM/ETM+ band 3) in contrast to very high reflectance in near infrared (NIR) band (TM/ETM+ band 4) is unique to vegetation. With these observations of spectral signatures in mind, a simple technique is designed to extract gypsum with the following algebraic and logical operations:

$$If \ \frac{TM4 - TM3}{TM4 + TM3} > 0.1 \ then \ 0 \ else \ if \ TM4 - TM5 \geq 0$$
$$then \ TM5 - TM7 \ else \ 0 \qquad (21.1)$$

In the above formula, the first condition is the Normalised Difference Vegetation Index (NDVI) to eliminate vegetation; the second condition $TM4 - TM5 \geq 0$ is to exclude clay minerals. Thus gypsum (both outcrops and quarries) is extracted in a single image with a threshold $TM5 - TM7 \geq 10$ as shown in Fig. 21.4.

The key difference between the gypsum quarries and natural outcrops of gypsum is the very high albedo of the smooth quarry floor in visible spectral range

[Fig. 21.3(b)]. Thus a slight modification of eqn. 21.1 to add the red band in the final operation will extract the gypsum quarries only:

$$If \ \frac{TM4 - TM3}{TM4 + TM3} > 0.1 \ then \ 0 \ else \ if \ TM4 - TM5 \geq 0$$
$$then \ TM5 - TM7 + TM3 \ else \ 0 \qquad (21.2)$$

The extracted gypsum quarries in 1984, 1992 and 2000 are presented in Fig. 21.5. As shown in each of the corresponding histograms, a threshold is set in the bottom of the trough between the main peak on the left and a small hump on the right that represent the quarry pixels. The threshold sets the digital numbers (DNs) of quarry pixels to 255 and others 0.

21.1.3 Gypsum quarry changes during 1984–2000

We can display the gypsum quarry extraction images of year 2000, 1992 and 1984 in red, green and blue to formulate an RGB colour composite as shown in Fig. 21.6. The colours of the extracted quarries in this colour composite indicate the temporal change and development of gypsum quarrying in the region as interpreted in Table 21.2.

The quarry 1 is the largest in the image. The patch is mostly white, indicating the quarry was already in great scale in 1984 and in operation throughout the following years to 2000. The surrounding yellow belt along the east and north margin is the quarry expansion during 1984–1992, while the red belt surrounding the south half of the quarry indicates the quarry expansion during 1992–2000. The green patches on the north edge of the quarry are interesting; they denote the areas quarried

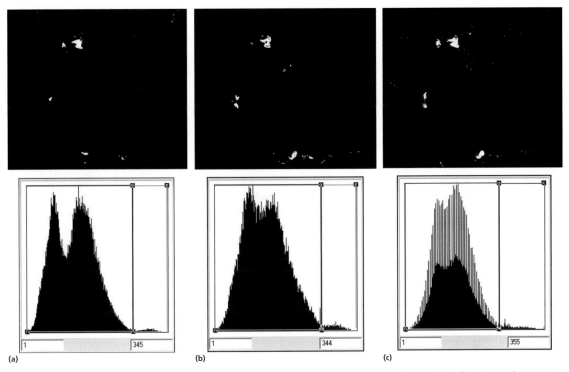

Fig. 21.5 Gypsum quarry extraction images and their corresponding histograms and thresholds: (a) 1984; (b) 1992; and (c) 2000.

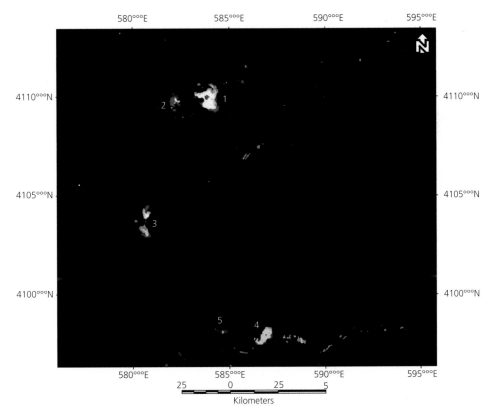

Fig. 21.6 Colour composite of gypsum quarry extraction images of year 2000 in red, 1992 in green and 1984 in blue. The colour interpretation is detailed in Table 21.2.

during 1984–1992, and then the ground was probably restored. The small pink patch in the centre of the white area marks the lowest bottom of the quarry, and it can easily become a water pond after a heavy rain as shown in the field photo in Fig. 21.7. The pink colour of this

little patch implies that it was filled with water on the date of the 1992 image and was dry on the dates of the 1984 and 2000 images.

The quarry 2 was started before 1984, as indicated by a small white patch in the east of the quarry. Significant development occurred after 1992 to the west, as illustrated in red. The quarry 3 was started in the north part before 1984 as well and then saw considerable expansion during 1984–1992 shown in yellow and green and 1992–2000 in red.

The quarry 4 is in a major belt of gypsum natural outcrops as highlighted in Fig. 21.4. Shown in yellow colour, this second largest quarry in the image was started after 1984 and quickly reached the scope recorded in the 1992 image. The expansion after 1992 till 2000 was limited. There are some blue, green and cyan patches nearby along the gypsum outcrop belt. These patches are the abandoned quarry trials in the years before 2000. In particular, the quarry 5 in blue was sizable before 1984, but the quarrying operation was ended before 1992.

Table 21.2 Interpretation of colours of Fig. 21.6.

Colour	Interpretation
White	Gypsum quarries since before 1984 (in 1984, 1992 and 2000 images)
Yellow	Gypsum quarries since after 1984 (in 1992 and 2000 images)
Red	Gypsum quarries after 1992 (only in 2000 image)
Green	Gypsum quarries after 1984 and before 2000 (only in 1992 image)
Blue	Gypsum quarries before 1992 (only in 1984 image)
Cyan	Gypsum quarries before 2000 (in 1984 and 1992 images)
Magenta	Gypsum quarries before 1984 and after 1992 (in 1984 and 2000 images)

Fig. 21.7 Field photo of the gypsum quarry 1 in Fig. 21.6 taken in 2003. The lowest part of the quarry has become a water pond with vegetation.

Besides the major quarries described above, there are some scattered isolated dots in Fig. 21.6, that are not likely gypsum quarries. These could be some casual diggings of gypsum as well as incorrectly extracted pixels.

21.1.4 Summary of the case study

In this case study, we demonstrated how to design simple and effective image processing techniques to map gypsum outcrops and extract gypsum quarries based on image spectral profile analysis. Though the image spectral profiles of six TM/ETM+ reflective multi-spectral bands and a thermal band are fairly crude in comparison with laboratory spectral profiles, the diagnostic spectral property of gypsum can be enhanced to achieve effective discrimination for accurate extraction of gypsum outcrops and quarries. Usually, broadband multi-spectral image data are only adequate for ground object discrimination but not identification; however, the identification can be easily achieved through minimal field investigation guided by these images of target-oriented enhancement and thematic extraction. A field investigation indeed forms an essential part of this type of case study project. All the quarries and natural outcrops extracted in the resultant images of this project had been verified in our field trips for remote sensing master's students.

21.1.5 Questions

1 In order to separate gypsum quarries from gypsum natural outcrops, TM3 is added in to the last operation of eqn. 21.2. Can TM1 or TM2 be used instead of TM3, and why?
2 Following the logic of eqns. 21.1 and 21.2, try to design a ratio-based image processing procedure to enhance gypsum and extract gypsum quarries.

21.2 Thematic information extraction (2): Spectral enhancement and mineral mapping of epithermal gold alteration and iron-ore deposits in ferroan dolomite

The Carboneras area lies south-east of the town of Sorbas, on the eastern coast of Almeria province, Spain (refer back to Fig. 20.3). A regional NE-SW oriented tectonic system, known as the Carboneras fault zone, cuts through the area into complicated jumbled slices of Palaeozoic and Mesozoic basement schists, phyllites and dolomites, together with pockets of Tertiary volcanic rocks.

Intense epithermal alteration has resulted in the enrichment of economic minerals within a small gold deposit in the study area. Some exploration was carried out but later abandoned because of the low grade and limited size of the deposit. However, the extensive alteration zone serves as a good test site to demonstrate the application of multi-spectral remote sensing for mineral exploration. Another mineral of economic interest in this area, which has been actively mined elsewhere, is iron ore found here within Triassic dolomite. Although closely associated geographically, the ferroan dolomite deposits were accumulated through a quite different process from the iron oxides associated with the epithermal gold deposits; the highly fractured dolomites have become enriched via weathering and leaching (i.e. they are gossans type deposits).

With two distinctively different mineralization systems within a small vicinity, we use this case study to demonstrate effectiveness of simple multi-spectral enhancement techniques for mineral exploration with two objectives:

• Locate argillic-siliceous alteration zone, the Carboneras gold prospect;
• locate Triassic ferroan dolomite and iron minerals (limonite).

Using 11-band Airborne Thematic Mapper (ATM) and Terra-1 ASTER 14-band images, the study comprises three steps:

1 Image datasets preparation;
2 ASTER image processing and analysis for regional prospection;
3 ATM image processing and analysis for target extraction.

21.2.1 Image datasets and data preparation

Two images, an ATM image taken in 1991 (NERC UK Airborne Remote Sensing Facility) and a Terra-1 ASTER image taken in 2002, are used in this study. The details of ATM and ASTER sensors in comparison with Landsat TM/ETM+ are listed in Table 21.3. It is important to notice that the ASTER three spectral groups of visible light and nearer infrared (VNIR), SWIR and thermal

Table 21.3 Comparison of spectral bands and spatial resolution of the ASTER and ATM images used in this case study with the Landsat TM/ETM+.

Sensor Systems	Terra-1 ASTER			Landsat 3-7 TM/ETM+			ATM (7.5 m)	
Spectral region	Band	Spectral Range (μm)	Spatial Resolution (m)	Band	Spectral Range (μm)	Spatial Resolution (m)	Band	Spectral Range (μm)
VNIR						30	1	0.42–0.45
				1	0.45–0.53		2	0.45–0.52
	1	0.52–0.60	15	2	0.52–0.60		3	0.52–0.60
	2	0.63–0.69		3	0.63–0.69		4	0.605–0.625
							5	0.63–0.69
							6	0.695–0.75
	3N	0.78–0.86		4	0.76–0.90		7	0.76–0.90
	3B	0.78–0.86		Pan	0.52–0.90	15	8	0.91–1.05
SWIR	4	1.60–1.70	30	5	1.55–1.75	30	9	1.55–1.75
	5	2.145–2.185		7	2.08–2.35		10	2.08–2.35
	6	2.185–2.225						
	7	2.235–2.285						
	8	2.295–2.365						
	9	2.360–2.430						
TIR	10	8.125–8.475	90	6	10.4–12.5	TM 120 ETM+ 60	11	8.5–13
	11	8.475–8.825						
	12	8.925–9.275						
	13	10.25–10.95						
	14	10.95–11.65						

infrared (TIR) are not only in different spatial resolution but also different radiometric quantization range. The VNIR and SWIR bands are in 8 bits value range while the thermal bands are in 12 bits.

It is typical for airborne image data that the ATM image is subject to various localised geometric distortion caused by unstable imaging status from an aircraft. As the first step for data preparation, the ATM image was rectified to the ASTER image that complies with UTM N30 based on ED50 datum. The warping transformation was a cubic polynomial fitting derived from 25 ground control points (GCPs) and the bilinear re-sampling was applied to produce the rectified output image from the input image DNs. Because of the significant irregular distortion of the ATM image, the RSM of GCPs ranges from 1 to 22 pixels even though these GCPs were quite carefully selected. We therefore do not expect an accurate co-registration between the two images, and the two images can only be processed and analysed separately for comparison. The rectified ATM image is in a curved irregular shape, indicating nonlinear geometric distortion of the image to the reference map projection.

Figure 21.8 is a merged display of standard false colour composites of ASTER (bands 321 in RGB) and ATM (bands 753 in RGB) images. It shows that the rectified ATM image in the middle fairly well matches to the ASTER image; however, a close look reveals considerable discrepancy between the two. Efforts for integrated processing and image fusion-based analysis of the two images will introduce more errors rather than benefits. In this case study, we use low spatial resolution ASTER image for regional prospective study and high spatial resolution ATM image to focus on the area of interest for mineral targets extraction.

21.2.2 ASTER image processing and analysis for regional prospectivity

This case study has been chosen because we already know the area quite well through image study and earlier field investigations. However, if presuming that we knew little about the area but recognised it might have the potential of mineralization based on the regional geological setting, then the first step to study

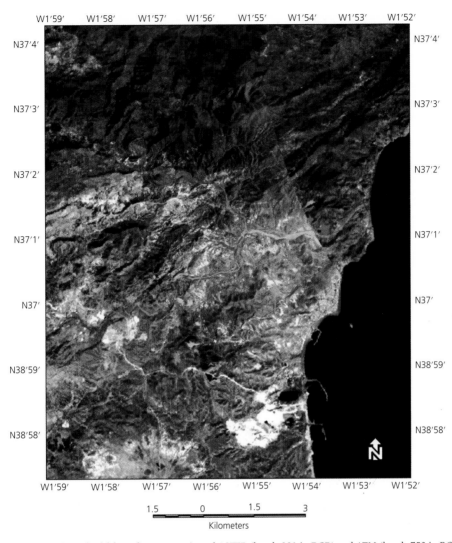

Fig. 21.8 Merged display of standard false colour composites of ASTER (bands 321 in RGB) and ATM (bands 753 in RGB) images.

the area would be to conduct a regional prospection using a satellite image with adequate spatial resolution and large coverage. The ASTER image with three 15 m spatial resolution VNIR bands, six 30 m spatial resolution SWIR bands and five 90 m spatial resolution thermal (TIR) bands serves the purpose well. The particular advantage of the ASTER image is its very high spectral resolution in SWIR and TIR bands.

Without knowing the mineral targets, well-established techniques should be tried first, from visualisation to general spectral enhancement. First, colour composites of ASTER bands 421 in RGB with BCET and DDS are generated, as shown in Fig. 21.9. These colour composites

present rich information of general lithology and geological structure (reference the geological map in Fig. 20.6) as well as rivers, quarries and manmade structures, but they do not show obvious features indicating minerals.

As we studied before, simple standard differencing and ratio techniques can effectively locate clay and hydrated minerals. ASTER imagery has six SWIR bands with band 4 equivalent to TM band 5, and bands 5–9 are high spectral resolution bands within a narrow range of 2.145–2.43 μm that largely overlap the spectral range of TM band 7. Most alteration-related clay minerals and hydrated minerals have a deep absorption trough in the

(a) **(b)**

Fig. 21.9 ASTER bands 421 RGB colour composites: (a) BCET; and (b) DDS.

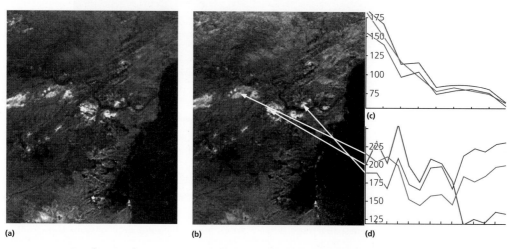

(a) **(b)** **(c)** **(d)**

Fig. 21.10 (a) ASTER *Band*4 − *Band*6 differencing image; (b) ASTER *Band*4/*Band*6 ratio image; (c) spectral profiles of original ASTER bands 1–9 (VNIR and SWIR); and (d) spectral profiles of the BCET stretched ASTER bands 1–14 (VNIR, SWIR and TIR).

narrow spectral rang depicted by ASTER band 6, thus we expect that ratio or differencing images between ASTER bands 4 and 6 can highlight potential targets of such. As shown in Fig. 21.10, both *Band*4 − *Band*6 and *Band*4/*Band*6 highlight an obvious bright belt in the middle of the image. The two techniques do not show much difference in the results. Spectral profiles of selected bright pixels in several patches from the original image data of VNIR and SWIR bands 1–9 appear to have similar shapes with the diagnostic absorption features,

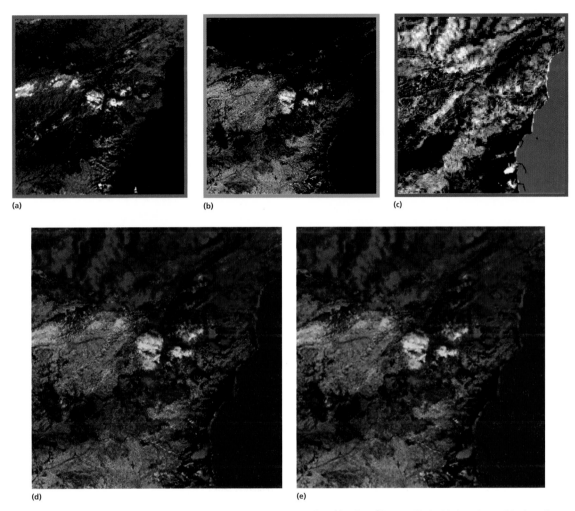

Fig. 21.11 Generation of compound differencing colour composite: (a) $Band4 - Band6$ generally highlights clay and hydrated minerals; (b) $2 \times Band4 - Band6 - Band3$ eliminates gypsum and highlights clay minerals of the alteration zone only; (c) $Band4 - Band6 + Band11 - Band3$ highlights quartz; (d) compound colour composite of (a) in red, (b) in green and (c) in blue; and (e) compound colour composite smoothed with a 3×3 smoothing filter.

implying the same type of alteration or hydrated minerals [Fig. 21.10(c)]. However, the spectral profiles of the same points from the BCET data of bands 1–14 (the thermal bands were re-scaled into 8 bits value range) in Fig. 21.10(d) are clearly in three distinctive groups:

- Argillic alteration: high reflectance in band 4, strong absorption in both band 6 and band 3, and low reflectance in thermal bands 10–14.
- Siliceous alteration: similar spectral signature to the argillic alteration in VNIR and SWIR bands, but the high emission in thermal bands 10–14 is diagnostic to underpin quartz and silica-rich minerals.

- Gypsum: similar to the above two groups in SWIR bands, but there is a reflectance peak rather than absorption trough in NIR band 3. Recall the last case study; this spectral profile is similar to the ETM+ spectral profile of gypsum illustrated in Fig. 21.3.

Based on the above spectral signatures, the following compound differencing colour composite can enhance the three different minerals in distinctive colours:

- Red: $Band4 - Band6$. Generally highlight all the alteration clay minerals and hydrated minerals, Fig. 21.11(a).
- Green: $2 \times Band4 - Band6 - Band3$. Because of the high reflectance of gypsum and strong absorption of

alteration clay minerals in band 3, this difference eliminates gypsum while further enhancing alteration clay minerals, Fig. 21.11(b).

• Blue: *Band4 – Band6 + Band11 – Band3*. The difference between band 11 and band 3 eliminates argillic alteration for its low thermal emission in band 11, and suppresses gypsum for its high reflectance in band 3, leaving silica being further enhanced as the brightest pixels, Fig. 21.11(c).

Therefore, in this compound difference colour composite [Fig. 21.11(d)], argillic alteration is in bright yellowish, as it is bright in both red and green layers; quartz siliceous alteration is bluish white, as it is bright in all the RGB layers but brighter in the blue layer; and gypsum is in bright reddish/orange because it is bright in the red layer and dark in the blue layer. There are many noise-like blocky edge effects in this image as the result of different spatial resolutions of the VNIR, SWIR and TIR band groups. This artefact can be effectively suppressed by a 3×3 smoothing filter [Fig. 21.11(e)].

The regional prospection using the ASTER image successfully located an alteration zone and separated two different types of alterations. However, for more detailed study on these detected small alteration targets, the relatively low spatial resolution of the ASTER image is not adequate. Thus high spatial resolution airborne remote sensing study focused on this alteration zone is undertaken.

21.2.3 ATM image processing and analysis for target extraction

The Airborne Thematic Mapper is an airborne version of Landsat TM/ETM+ across-track scanners, but it has finer spectral resolution with total of 11 spectral bands (10 reflective spectral bands in VNIR and SWIR spectral regions and 1 broad thermal band), as shown in Table 21.3. The spatial resolution of ATM is dictated by the flight altitude, which is 7.5 m for this dataset.

With the regional prospection results using the ASTER image, the obvious starting point of the detailed study using the ATM image is to repeat the technique used to produce a higher resolution version of the equivalent results generated from the ASTER image. One of the advantages of the ATM image is that its thermal band has the same spatial resolution as the reflective spectral bands, and therefore it is possible to generate high quality colour composites of simulated reflectance, as introduced in § 3.7.

We start from simple colour composites and simulated reflectance colour composites. Figure 21.12 shows colour composites of (a) ATM bands 10-5-2 in RGB with DDS enhancement and (b) simulated reflectance of the same bands. This band combination is equivalent to TM bands 731 in RGB displaying the clay mineral absorption SWIR band in red, red band in green and blue band in blue. The DDS and the simulated reflectance colour

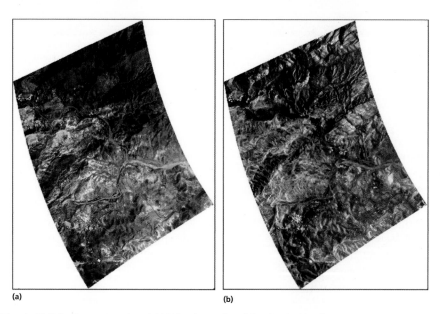

(a)   (b)

Fig. 21.12 ATM bands 10-5-2 colour composites: (a) DDS enhanced; and (b) simulated reflectance.

composite are spectrally similar, but the topographic shadows in the simulated reflectance colour composite are subdued with spectral variation further enhanced. Comparing these images with the ASTER compound difference colour composite in Fig. 21.11, the cyan patches on the left side of the image are gypsum outcrops; the brown coloured patch left to the river junction where two channels merge into one is the epithermal alteration zone. The much higher spatial resolution of the ATM data indeed brings out a lot of details of these mineral targets. Among others, the most eye-catching features in these two images are the red patches in the top half of the images. These are iron-ore deposits in ferroan dolomites. Showing in red, these iron deposits are characterised by high reflectance in ATM band 10, where both alteration clay minerals and gypsum have strong absorption, and are therefore not depicted by the ASTER compound differencing colour composite in Fig. 21.11.

The simulated reflectance colour composite of ATM bands 972 in RGB [Fig. 21.13(a)] displays similar phenomena as the bands 10-5-2 RGB composites except

vegetation in distinctive bright green, but the iron deposits are less red and the epithermal alteration zone is more reddish making the two less distinguishable. The spectral profiles extracted from the BCET processed ATM image data [Fig. 21.13(b)] indicate that the ferroan dolomite has high reflectance in both ATM bands 9 and 10 but the reflection peak is at band 10, which is a diagnostic feature distinguishing ferroan dolomite from all the other mineral targets in the study area. The ATM spectral profiles of argillic alteration, siliceous alteration and gypsum are similar to those obtained from the ASTER image shown in Fig. 21.10(d). As a trial, the compound difference colour composite using ATM band combination equivalent to the ASTER compound differencing colour composite in Fig. 21.11(d) was produced as follows:

- Red: $ATM9 - ATM10$ highlights both the argillic alteration and gypsum.
- Green: $2 \times ATM9 - ATM10 - ATM8$ highlights argillic alteration only.
- Blue: $ATM9 - ATM10 + ATM11 - ATM7$ highlights siliceous alteration.

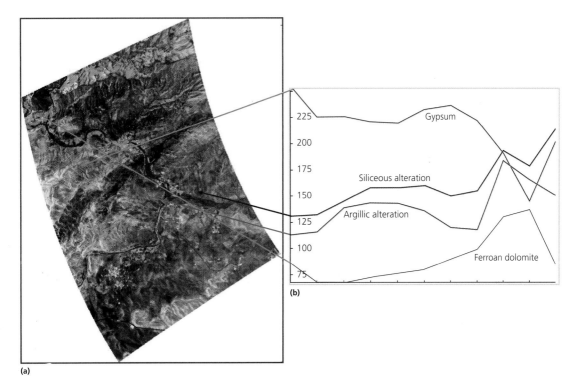

(a)

(b)

Fig. 21.13 (a) Colour composites of simulated reflectance of ATM bands 9-7-2 in RGB; and (b) Spectral profiles of gypsum, siliceous alteration, argillic alteration and ferroan dolomite derived from BCET stretched ATM image data.

Fig. 21.14 The compound difference colour composite of $ATM9 - ATM10$ in red, $2 \times ATM9 - ATM10 - ATM8$ in green and $ATM4-ATM6+ATM11-ATM3$ in blue.

Following this same principle, the image in Fig. 21.14 presents argillic alteration in bright yellow to green, gypsum in bright orange red and siliceous alteration in bluish white. However, as the ATM band 10 is a rather broad SWIR band in comparison with the ASTER band 6, which targets the deepest absorption of most clay and hydrated minerals. The same is true for the thermal band, which depicts the diagnostic thermal emission feature of quartz, whereas the ATM compound difference colour composite does not enhance these minerals as distinctively as the ASTER. It is not surprising that the image does not enhance the iron deposits in ferroan dolomites either, since it is not designed for the purpose.

As mentioned before, the key spectral feature making the iron-ore deposits in ferroan dolomites different from the argillic alteration zone and gypsum is its higher reflectance in ATM band 10 than in band 9, but this feature is shared by many other rock types and is thus not diagnostic. We would like to produce a colour composite that enhances ferroan dolomite, argillic alteration zone and gypsum only. Since this is not easily achieved using arithmetic operations, we consider a combined approach using both differencing and the feature-oriented PC selection method (FPCS, see § 7.2). Principal components analysis was applied to bands 2 to 10; the band 1 of this ATM dataset is very noisy and therefore discarded. Table 21.4 presents the matrix of eigenvectors of covariance matrix of the 9 bands. The *PC5* is dominated by the difference of ATM band 10 minus band 9 depicting the key spectral feature of ferroan dolomites. The difference between band 7 and band 5 enhances vegetation as well in the *PC5*. As shown in Fig. 21.15(a), ferroan dolomites are bright while both argillic alteration and gypsum are very dark in the *PC5* image. The *PC3* is largely a weighted summation of all the VNIR bands subtracting the summation of SWIR bands 9 and 10. As implied in the ATM spectral profiles in Fig. 21.13(b), the operations for the *PC3* will produce high values for gypsum and very low values for argillic alteration zones and the ferroan dolomites, and thus the negative *PC3* highlights argillic alteration zones and ferroan dolomites while suppressing gypsum, as shown in Fig. 21.15(b). Again we use the difference image of $2 \times ATM9 - ATM10 - ATM8$ to highlight argillic alteration only, as shown in Fig. 21.15(c). Finally a colour composite is generated as below:

- Red: *PC5* highlights ferroan dolomites and suppress argillic alteration and gypsum.
- Green: negative *PC3* highlights both ferroan dolomites and argillic alteration and suppress gypsum.
- Blue: $2 \times ATM9 - ATM10 - ATM8$ highlights argillic alteration only.

The resulting colour composite in Fig. 21.15(d) shows ferroan dolomites in distinctive yellow because they are bright in both red and green layers, and argillic alteration in bright cyan because of its high values in green and blue layers. It is interesting to notice that the image depicts gypsum in rather characteristic deep blue while none of the three images forming this colour composite is aiming to highlight gypsum. As shown in Fig. 21.15, gypsum is suppressed as very dark features (this is distinctive as well!) in both *PC5* and negative *PC3*, while in the compound difference image it is in intermediate tones. Consequently, gypsum is clearly enhanced as deep blue in the resulting colour

Table 21.4 Eigenvectors of covariance matrix of nine ATM image bands.

Covariance Eigenvectors	PC1	PC2	PC3	PC4	PC5	PC6	PC7	PC8	PC9
Band2	0.320	0.395	0.292	−0.679	0.196	−0.046	0.354	−0.160	0.065
Band3	0.348	0.325	0.187	−0.046	−0.019	0.054	−0.631	0.450	−0.363
Band4	0.343	0.286	0.091	0.320	−0.215	−0.162	−0.184	−0.117	0.755
Band5	0.339	0.199	0.071	0.430	−0.297	−0.192	0.298	−0.395	−0.535
Band6	0.347	−0.208	0.234	0.262	0.039	0.481	0.476	0.495	0.089
Band7	0.331	−0.490	0.248	−0.038	0.223	0.342	−0.353	−0.544	−0.012
Band8	0.312	−0.558	0.078	−0.154	−0.064	−0.704	0.035	0.247	0.002
Band9	0.329	−0.101	−0.628	−0.347	−0.539	0.276	−0.008	−0.013	0.013
Band10	0.329	0.111	−0.592	0.188	0.693	−0.106	0.036	0.013	−0.014

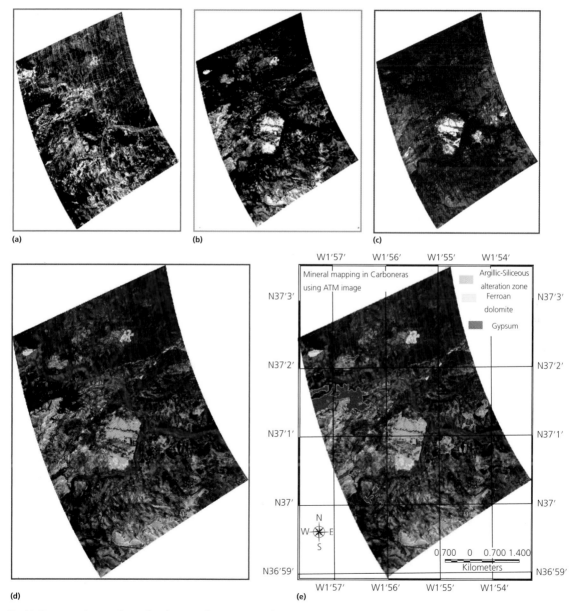

Fig. 21.15 (a) *PC5* image of ATM bands 2–10; (b) *PC3* image of ATM bands 2–10; (c) compound difference image of $2 \times ATM9 - ATM10 - ATM8$; (d) colour composite of (a), (b) and (c) in RGB; and (e) interpretation of the three major mineralization targets in the area: argillic-siliceous alteration zone, ferroan dolomite and gypsum.

composite [Fig. 21.15(d)]. This reminds us that feature enhancement can be achieved by suppression as well as by highlighting!

Finally, as shown in Fig. 21.15(e), a simple interpretation map of argillic alteration zone, iron ore deposits in ferroan dolomites and gypsum outcrops was produced from the PC and compound difference colour composite image.

21.2.4 Summary of the case study

In this case study, we used ASTER and ATM images to demonstrate the application of multi-spectral and multi-resolution remote sensing for mineral exploration via image processing. First, the ASTER image with lower spatial resolution but large coverage was processed and analysed for regional prospection. This enabled us to focus on a much smaller area with several different types of minerals, using a higher spatial resolution ATM image for detailed study. For both datasets, image processing began from optimal visualisation of the data followed by well-established standard enhancement techniques for the presumed targets. Then image spectral profiles of pixels representing possible mineral targets located by standard techniques were carefully analysed to design further processing strategy and specific enhancement operations addressing diagnostic spectral features of the target minerals. Using a colour composite of compound difference images of ASTER, regional distribution of argillic-siliceous alteration and gypsum are highlighted, while the details of argillic-siliceous alteration, iron deposits in ferroan dolomites and gypsum are mapped by a colour composite of FPCS PCs and compound difference of the ATM image. One interesting lesson to learn from this ATM colour composite in its enhancement of gypsum is that enhancement does not always mean highlighting the target in bright pixels; suppressing the target features as very dark pixels is enhancement as well!

21.2.5 Questions

1 The correspondent ASTER and ATM formulae of the compound differencing colour composite in this case study are slightly different. Try to find the difference between these two groups of formulae and explain why these formulae have been so designed based on the image spectral profiles of argillic alteration, siliceous alteration and gypsum and the spectral bands of the two datasets.

2 Based on the ATM spectral profiles, explain why the *PC3* produces high values for gypsum and very low values for argillic alteration zones and the ferroan dolomites.

3 Explain how gypsum is enhanced in deep blue in Fig. 21.15. Comment on the lesson that we can learn from this particular scenario.

21.3 Remote sensing and GIS: Evaluating vegetation and landuse change in the Nijar Basin, SE Spain

21.3.1 Introduction

This case involves the use of multi-temporal satellite image datasets acquired during the period between 1984 and 2014, to demonstrate the nature, distribution and rate of change to landuse patterns in the Nijar Basin, in Almeria province of south-eastern Spain. It is a fairly simple case that demands accurate data co-registration (georeferencing) and the processing of multi-spectral imagery to reveal features characteristic of landuse in this area, in order to then identify where and how much change has occurred. The GIS here serves to manage and display the processed results and to enable some spatial statistical analysis.

The sustainable economy of Almeria province has for some considerable time been based on agriculture, and this is still the case today. Throughout the 1980s the style of agriculture changed radically from one of open growth of grapes, olives, nuts and other vegetables to the highly intensive production of tomatoes, melons, cucumbers, strawberries and other soft fruits and vegetables under plastic – that is, in greenhouses. Flat ground, plentiful sunshine and EU subsidies have together enabled the rapid development of this style of agriculture. The plastic covering prevents excessive evaporation of water, helps keep pests out and promotes a year-round growing season. This growth has been accompanied by a huge increase in demand for water, which has traditionally been supplied by a local aquifer located in the Messinnian sediments below. Unregulated pumping on an unprecedented scale eventually caused a depression in the local groundwater table and the incursion of saline water from the Mediterranean sea to the south. These events have been the subject of some attention and many publications exist on the prpblems it caused. Irrigation styles have now changed to drip-feed methods, which use water much more effectively, and using water supplied from small upland reservoirs under gravity. A visit by aeroplane to Almeria these days greets the tourist with a view of the 'sea of plastic' which now covers the much of the open, flat ground around Almeria. As you come into land, it is a shocking and spectacular sight. The construction of plastic greenhouses today involves the very latest technology, in vast installations; it is a very hot political topic since the benefit to the economy is undeniable yet all agree that the greenhouses are a 'blot on the landscape'.

The location of the Nijar Basin with respect to the other previously described teaching case study areas is shown in Fig. 20.3. We have already described the climatic setting of this area, in § 20.3.3, as being semi-arid (semi-desert). The 200–300 mm incident rainfall predominantly falls on the Sierras, in the months between October and April, and of that rainfall perhaps 40–50% is lost through evaporation and 5–10% lost as run-off so that perhaps 40% infiltrates and becomes groundwater and ultimately enters the aquifer.

The Nijar Basin lies between the Sierra de Alhamilla and the range of low hills, known as La Serrata, which represents the topographic expression of the Carboneras fault. The latter forms a structural and topographic trap into which terrestrial sediments have been deposited. The resulting sequence of marls, gypsum, limestones and sandstones deposited in the basin now forms the main aquifer supplying water for agriculture in the Nijar area. The geological setting is illustrated by the interpretation map shown in Fig. 20.6 and the local geography is shown in Fig. 21.16.

The objectives of this study are broadly to demonstrate the use of multi-temporal imagery in revealing landuse change in the Nijar Basin. In doing so, we hope to comment on the rate and nature of the changes, and to illustrate these graphically and quantitatively.

More specifically, the objectives are to:

- Highlight the distribution of both vegetation (natural and agricultural) and plasticulture for each year represented in our database;
- Locate and quantify areas of changing land cover.

21.3.2 Data preparation

In any study of temporal change, a set of image data acquired over a time period is required. In this particular case, the area in question has undergone quite a radical change over a period of 20 years and so the database consists of subsets of four Landsat scenes, acquired in 1984, 1989, 1992, 2000 2010 and 2014, and digital airphotographs acquired in 2004, as summarised in Table 21.5.

Generally speaking these days, all EO datasets are delivered as georeferenced products. In the 1980s and early 1990s however, this was not the case, so the first step in this case study is to georeference, or at least co-register, the older image datasets used, and to ensure that this is done as accurately as possible. The Landsat7 ETM+ dataset already conforms to WGS84 and UTM (zone 30), so the most logical step is to co-register the raw Landsat5 images to the Landsat7 ETM+ scene. As explained earlier, given that all the images were acquired by similar sensors and have similar geometry, this co-registration is best achieved using a simple linear translation and rotation.

21.3.3 Highlighting vegetation

Here we begin with standard false colour composite images, of bands 432 RGB, for general visualisation of land cover types (vegetated and non-vegetated) for each of the years of observation. For all six Landsat observations, these are shown in Fig. 21.18.

We can see clearly from the extent and form of the dense patch of red in 1984, that the vast majority of land under agriculture was not devoted to plasticulture but was grown under open skies [Fig. 21.18(a)]. We can also see that the plasticulture made its first impact in 1989 and expanded steadily to 2000. From the appearance of Fig. 21.18(e) it seems at first glance that open vegetation has increased once more but this is not the case. The red colour in evidence here represents a snapshot in time when the crops inside are particularly vigorous and we are seeing their photosynthetic response through the plastic roofs. The older style plastic greenhouses had roofs that were painted to make them opaque, but these have been replaced over time by larger, modern, rigid contructions with clear plastic roofs and automated ventilation systems.

Now that we have identified the vegetated and plasticulture areas we need to establish both the decrease

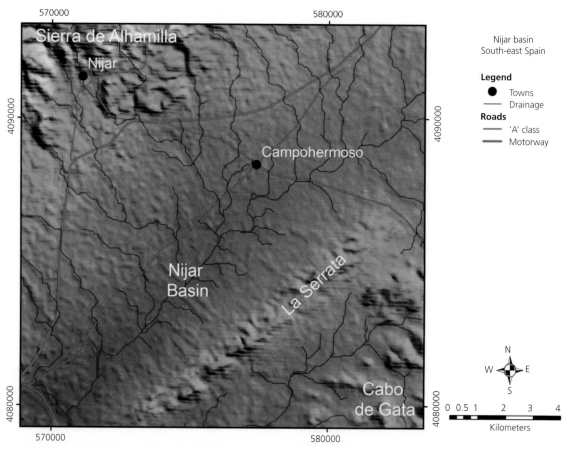

Fig. 21.16 Map of the Nijar Basin area. The background colour image is a pseudo-colour shaded relief display of the Aster GDEM dataset of this area. The field of view and scale bar here relate to all the images shown in this case.

(a)   (b)

Fig. 21.17 (a) View looking south-eastwards, from the lush, spring-fed vegetation of the upland village of Huebro, over the greenhouses of the Nijar Basin in the middle distance; and b) inside one of the older plastic-covered greenhouses constructed from wooden posts and covered with plastic-coated fabric mesh, and filled with growing tomatoes.

Table 21.5 Multi-temporal datasets used in the case study.

Dataset	Scene-Identifying Numbers	Acquisition Date(s)
Landsat 5 TM	Subset of path 199/ row 034	1984, 1989, 1992
Landsat 7 ETM+	Subset of path 199/ row 034	2000
Air photography	Ortho-quads (Junta de Andalucia air survey)	2004
Landsat 5 TM	Subset of path 199/ row 034	2010
Landsat 8 OLI	Subset of path 199/ row 034	2014

in open agriculture and the expansion of plasticulture (since they do not necessarily mirror one another). First we need identify and extract the open vegetation so we will use a simple Normalised Difference Vegetation Index or NDVI. The resultant NDVI images for each year are shown in Fig. 21.19. These allow us to visualise the extent and decline of open vegetation from 1984, when almost no plasticulture existed [Fig. 21.19(a)] and that we take as the baseline for our estimations, to 2000, when very little open agriculture remained. In 2000, the few patches of healthy vegetation represent gardens and trees growing around the town of Nijar, which is supplied with plentiful spring water, and a few isolated fields and old greenhouses that are in disrepair.

One large patch of open agriculture is noticeable (centre right) in the NDVI of every year; this represents a farm whose owner stubbornly refuses to adopt plasticulture. Sadly, his plot is now in a state of disrepair. Surrounding the central patch of cultivation in each image, the background appears in a mid grey tone, this represents the natural scrub vegetation characteristic of this semi-arid part of Spain; it has fairly similar appearance in every year.

The 1992 image [Fig. 21.19(c)] contains a linear feature running from bottom left to top right that has anomalously low NDVI values. This represents the main Malaga-Murcia motorway, which was constructed around 1991–1992. The motorway as shown in 2000 is represented by a much narrower dark line than in 1992. This is probably a result of the re-establishment of natural vegetation on the verges of the motorway, as opposed to the broad swath of ground that is stripped of

vegetation during construction in 1992. Conversely, the area around Nijar town is relatively bright in each year, since the vegetation here is found in parks and gardens inside and around the town.

21.3.4 Highlighting plastic greenhouses

To highlight and extract the plastic-covered greenhouses we must first understand their spectral properties. The roofs are what we see in these images and, on occasion, we may see something of the vegetation growing within. Their roofs are generally highly reflective and often cause saturation in the visible bands. The older greenhouse roofs are sometimes painted dark (black or grey) in winter to make them less reflective and thus absorb more radiation. In some cases where the plastic sheeting is relatively new and/ or less opaque than older greenhouses, or where the sheeting has not yet been painted, chlorophyll in the growing vegetation inside makes a contribution to the overall reflectance of the greenhouse and it appears pale pink in the standard false colour images shown in Fig. 21.18. In other cases, the plastic roofs appear less reflective because they are curved rather than flat (generally the more modern constructions). The plastic itself though does not appear to have any particularly diagnostic spectral features and so the signature of the plastic resembles that of other highly reflective targets, such as bare flat, cleared fields or the smooth level floor of a nearby gypsum quarry. Clearly, the plastic-covered targets in the images cannot simply be extracted on the basis of visible brightness alone. As with any other target, we need to look carefully at their spectral signatures to discover some diagnostic features and so a way to separate them from the other image features. Using one of the images in the data base for illustrative purposes (Landsat ETM+ 2000 in this case), the spectral profiles of some of the main ground targets are shown in Fig. 21.20.

With these observations of spectral profiles in mind, a simple formula is designed to extract pixels representing plastic covered greenhouses using the following algebraic and logical operations:

$$If\left((b4-b3)/(b4+b3)\right) < 0.1 \; and \left(b5/b7\right) < 1.1 \; then$$
$$\left(if\left((b1+b2+b3)/3\right) > 220 \; then \; 1 \; else \; 0\right) else \; 0$$

Referring also to the profiles in Fig. 21.20, the first part $((b4–b3)/(b4+b3))<0.)$ effectively removes any surface

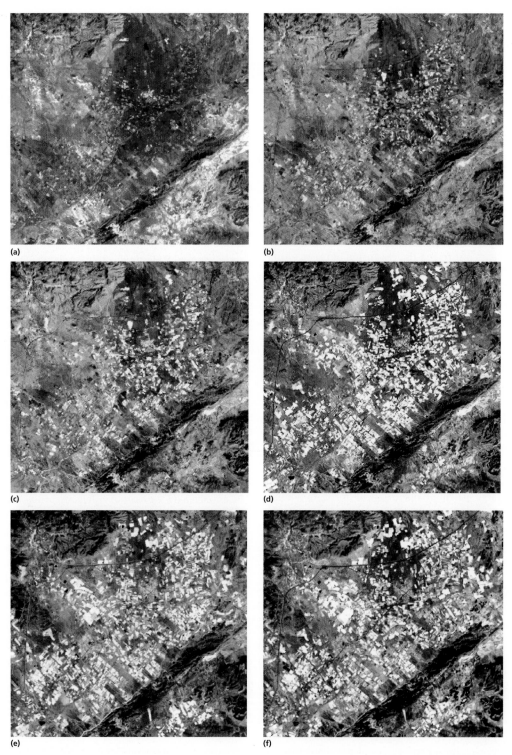

Fig. 21.18 Standard false colour composite images, bands 432 RGB (Landsat 5 and 7), and 543 RGB (Landsat 8) illustrating clearly the extent of open vegetation (in bright red tones) and plastic greenhouses (in white, bright cyan and pink tones) as observed in (a) 1984; (b) 1989; (c) 1992; (d) 2000; (e) 2010; and (f) 2014 (Landsat 8). Rocks and soils, exposed beyond the cultivated and urban areas, appear in a variety of grey, bluish-green, brown and greenish tones.

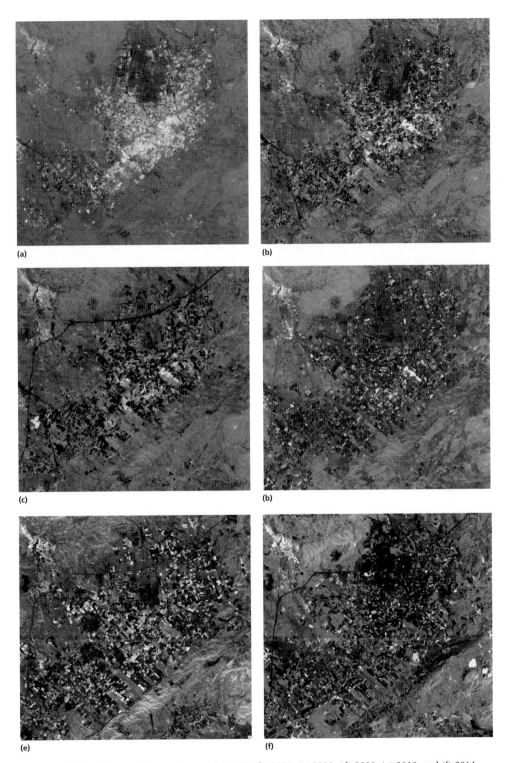

Fig. 21.19 NDVI as calculated from each image dataset: (a) 1984; (b) 1989; (c) 1992; (d) 2000; (e) 2010; and (f) 2014.

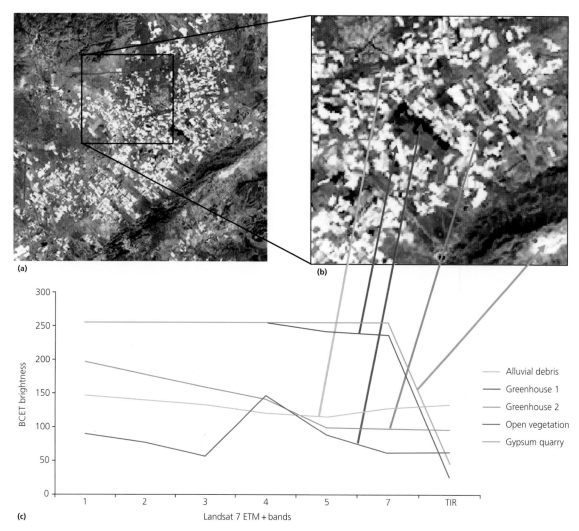

Fig. 21.20 (a) True colour composite (321RGB, 2000) of the Nijar Basin area; (b) detail of the image in (a); and (c) Landsat ETM+ BCET spectral profiles of some of the main ground targets.

that is vegetated (cultivated or natural), the second part ((b5/b7)<1.1) enhances and thresholds hydrated minerals (including gypsum), and the last part (((b1+b2+b3)/3)>220) masks on the basis of average visible brightness. All excluded pixels are then coded zero and all retained pixels coded with a value of one, thus producing the binary images shown in Fig. 21.19.

The differing illumination conditions on each acquisition date mean that the thresholds in Fig. 21.19 need to be adjusted to compensate for slight changes in relative brightness in each image. Even after such minor adjustments, we find that in Fig. 21.21(c), we have gained some pixels from the road leading to a small gypsum quarry and have partially lost one large, almost triangular shaped greenhouse from the upper centre of the area. Our algorithm excluded the latter because its roof has been painted some time before 2000 and is much darker in 2000 than it was in either 1989 or 1992.

Our dataset also contains ortho-rectified digital air-photographs (as a three-band true colour image only, i.e.

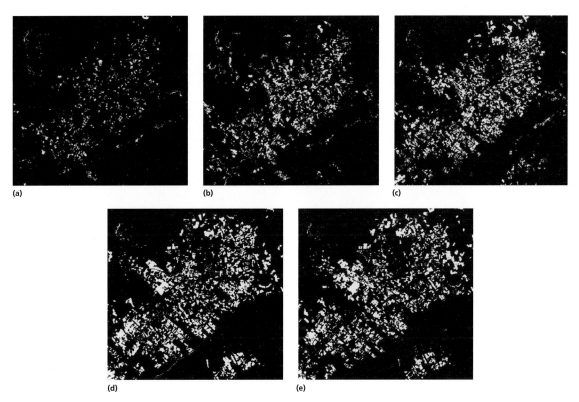

Fig. 21.21 Binary images with pixels encoded to show the extent of plasticulture for years (a) 1989; (b) 1992; (c) 2000; (d) 2010; and (e) 2014 (we have assumed that there were no greenhouses in 1984).

without near infrared) acquired in 2004. These images have 1 m spatial resolution and have been mosaiced to produce the image shown in Fig. 21.22(a). They provide much spatial detail for the interpretation of ground features, including plastic covered greenhouses. If we perform a similar procedure to classify the greenhouses from this image, though this time on the basis of relative visible brightness since we have no infrared bands, we produce the binary image in Fig. 21.22(b). Once again we have picked up the few pixels that make up the gypsum quarry and the town of Nijar but we shall ignore these.

21.3.5 Identifying change between different dates of observation

Since the images are now accurately co-registered we can use image subtraction to identify areas of change from one image to another. We may also display

different dates in different colour guns. First we need to consider two of the primary objectives: highlight areas of vegetative cover and of plastic greenhouses. With these in mind, we can use the NDVI as one measure from which identify change and the thresholded plasticulture image (produced in 21.22) as the other. We can then compare these indices for each year to estimate the proportion of land devoted to open cultivation.

The difference images in Fig. 21.23 indicate that the greatest reduction in the extent of open vegetation occurred between 1984 and 1989. The white line that appears in Fig. 21.23(b) represents the sudden localised loss of natural vegetation caused by motorway construction in 1991–1992; the same motorway path in 21.23(c) shows the opposite change in this time period and so appears black and represents the re-establishment of vegetation adjacent to the motorway. The difference image Fig. 21.23(c) also shows a bright patch around the

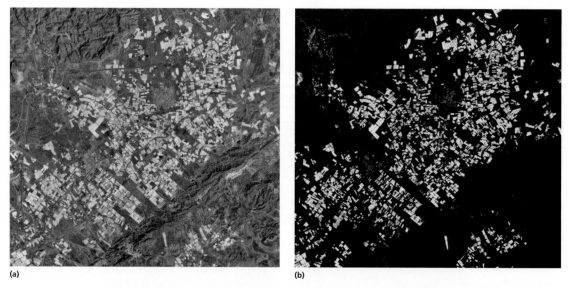

Fig. 21.22 (a) Ortho-photo true-colour mosaic (2004) of the study area; and (b) the binary plasticulture image derived from it.

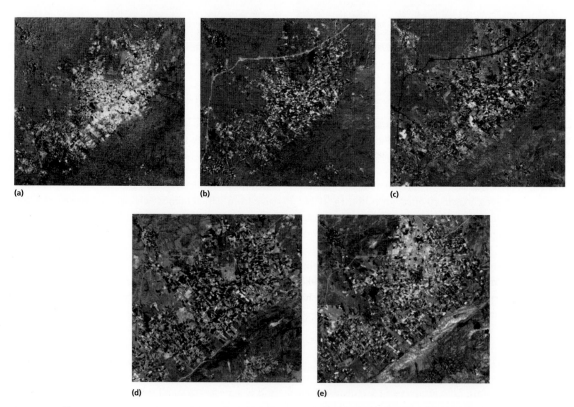

Fig. 21.23 Difference intervals between NDVI images: (a) 1984–1989; (b) 1989–1992; (c) 1992–2000; (d) 2000–2010; and (e) 2010–2014. In general, high values (bright pixels) indicate a loss of vegetation in the time period.

Table 21.6 Statistics of change: land devoted to open cultivation.

Year	1984	1989	1992	2000	2010	2014
Open cultivation (sq km)	11.7	3.2	3.5	1.3	<1	<1
Non-cultivated nor plasticulture (sq km)	155	164	164	166	109	59
Proportion of the area cultivated	8%	2%	2%	~1%	<1%	<1%

Table 21.7 Statistics of change: land devoted to plasticulture.

Year	1984	1989	1992	2000	2010	2014
Plasticulture area (sq km)	<1	2.8	5.1	15.6	23.4	21.8
Non-plasticulture area (sq km)	>166	164.2	161.9	151.4	139.9	141.5
Proportion of the area under plastic	<2%	2%	3%	10%	17%	15%

town of Nijar, representing a decrease in open vegetation between 1992 and 2000; the reasons for this are unclear and would require field investigation to explain.

Thresholding of the NDVI to selectively retain the highest values representing healthy cultivated vegetation, reclassifying them to a value of 1, and excluding the remaining values representing natural scrub vegetation and un-vegetated areas (classified to a value of zero), allows the production of a binary image representing open cultivation. If we perform this classification for each year, we may calculate the proportion of the area devoted to open cultivation and therefore an indication of the reduction over the time period, using any GIS statistical package. Table 21.6 shows the results of such calculations; we find that land devoted to open cultivation has decreased from a maximum of 8% in 1984 to a minimum of less than 1% in 2000.

Using the thresholded binary images shown in Fig. 21.21 in which we have classified plasticulture to a value of 1 against all other pixels with a value of zero, we can again calculate the areas and proportions of land occupied by plasticulture. This reveals that from our baseline in 1984, when we observe only open cultivation, plasticulture has commenced and increased to the point where it occupies ca 15% of the total study area, as summarised in Table 21.7. In other areas, such as south-west of Almeria city, this percentage is far higher. Making the same calculation of area from the binary image in Fig. 21.21(b) (2004), we find that the proportion of land devoted to plasticulture has increased again to just over 13%. The apparent drop in the proportion of plasticulture in 2014 (15%) is a result of misclassifications where modern greenhouse roofs

prove to be rather more transparent than the older style and so appear darker, hence the classification algorithm fails in a these cases. The current proportion of plasticulture is probably close to 20%.

These figures seem rather low in visual comparison with the images, but they do not account for the averaging that human eye/brain tends to perform, which ignores the areas in between the greenhouses.

21.3.6 Summary of the case study

This case study has allowed an estimate to be made of the extent of vegetation and plasticulture in this one small area, as a representation of change that is mirrored in other parts of the region. We have explained one method of doing so and some of the difficulties along the way. Clearly some assumptions and inaccuracies must be accepted, such as those introduced by the subjective application of thresholds to produce binary classifications, and these must always be recognised and acknowledged even if they cannot be eradicated.

Comparison of the extent of plasticulture in 2014 with that of open cultivation in 1984 reveals that plasticulture has more than merely replaced the traditional agriculture since it has expanded far beyond, even to some areas that might not at first seem suitable, such as the top of La Serrata ridge.

It is tempting to try to classify the greenhouses themselves and thereby the type of vegetation growing inside them, but this has, time and again, proved to be a waste of time, for a variety of reasons. In many cases, there is indeed a contribution from the photosynthesising plants to the overall reflectance of the greenhouse roof material, but the amount of that contribution is dependent on many things, not least the age and type of material making up the roof of the greenhouse. Only the newest of the roofs are of clear plastic or glass, many have been made more or less opaque though painting, some are made of mesh and some have several layers of plastic or mesh, added over the years. Since greenhouses are costly to maintain, the only safe conclusion is that those with intact roofs are generally filled by some growing cash crop and that crops are grown all year round.

The availability of, and impact on, water resources in this region is a further avenue for investigation, but one that remote sensing can contribute to in only a limited degree since the source and storage of water is underground. What we can surmise is that in this period (1984–2000) the changing style, extent and intensity of cultivation will have meant that the demand for water has increased enormously. Unregulated pumping in the past from wells in this area has caused depletion of the local aquifer that is known to have resulted in the incursion of saline water from the Mediterranean to the south, thus contaminating the aquifer and increasing salinity in soils. New irrigation techniques (such as drip-feed) and regulation of water extraction have lessened these effects, but water and soil quality remain important issues in this area and require constant monitoring.

21.3.7 Questions

1 Very high resolution (VHR) imagery provides unprecedented spatial detail of the land and of the greenhouses in this case. Do we really need this detail to carry out the kind of temporal analysis that we have done here?
2 Why might the increase in plasticulture and the decrease in open vegetation not be reciprocals of one another? What else might be happening?
3 Without resorting to the ground-based surveying of every individual greenhouse, how else might we improve on this work to more accurately estimate greenhouse coverage?

21.3.8 References

Cantliffe, D.J., Shaw, N., Jovicich, E., Rodriquez, J.C., Secker, I. & Karchi, Z. (2000) Passive ventilated high-roof greenhouse production of vegetables in a humid, mild winter climate. *International Symposium on Protected Cultivation in Mild Winter Climates: Current trends for sustainable technologies*. 7–11 March, Cartagena-Almeria, Spain.

Millington, A.C. & Alexander, R.W. (2000) Vegetation mapping in the last three decades of the twentieth century. In: R.W. Alexander and A.C. Millington (eds), *Vegetation Mapping: From Patch to Planet*, pp 321–31. Wiley, Chichester.
Zukowskyj, P.M., Alexander, R.W., Teeuw, R. & Faulkner, H.F. (2004) Changing vegetation density in Almeria province, south east Spain. *NERC ARSF Session Remote Sensing and Photogrammetry Annual Conference*. September, Aberdeen.

21.4 Applied remote sensing and GIS: A combined interpretive tool for regional tectonics, drainage and water resources in the Andarax basin

21.4.1 Introduction

This case involves the use of multi-spectral imagery to improve our understanding of the regional geology, tectonics and hydrology in the Tabernas-Andarax basin of Almeria province, Spain. The focus is on the applied use of remote sensing and GIS in the interpretation and exploration of a region's water resources. Whilst the topic seems to concern mainly geological concepts, the reason for concentrating on these phenomena is to reveal the controls on the sustainable rural economy in this semi-arid area. The geography and location of this case study area are illustrated in the map in Fig. 21.24 and in Fig. 20.3.

21.4.2 Geological and hydrological setting

Unlike the Nijar Basin to the east, horticulture in the Andarax basin is predominantly devoted to open agriculture with the use of greenhouses only recently increasing. Water has been plentiful enough here to support the prolific growth of citrus fruits, not possible in the Nijar area, but only through systematic irrigation. The area under active irrigation is reported to have increased to some 14000 ha in the last 20 years (Pulido-Bosch 1994; Callego *et al.* 2007). The region is also said to have been in rural economic 'boom' since the 1960s. The inadequate surface water resources and low and irregular rainfall has meant that the demand from Almeria can only be satisfied by groundwater abstraction.

The Andarax basin comprises some 250 sq km of alluvial valley infill and 2000 sq km of mountain area, which consists of four mountain ranges: Sierra de Alhamilla, Sierra de los Filabres, Sierra de Gador and Sierra Nevada. The hydrological network consists of the

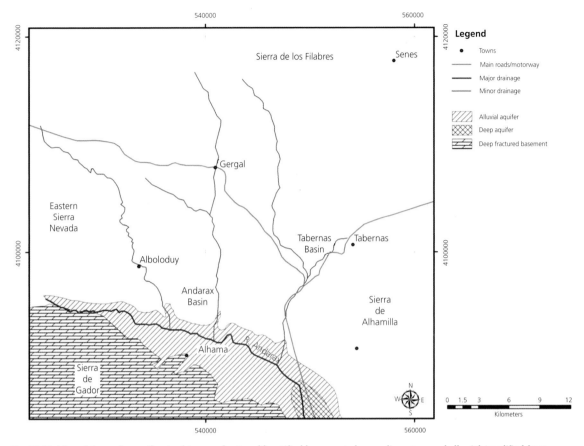

Fig. 21.24 Map of the main aquifers in this area: fractured karstified basement, deep sedimentary and alluvial (modified from Pulido-Bosch 1994).

catchments of the Andarax, Nacimento and the ramblas of Tabernas and Gergal.

Most rivers are ephemeral streams and are dry for much of the year but the gauge at El Chono is very close to the Sierra de Gador and there is permenant runoff measured here. Runoff is between a minimum of < 0.5 mm/month^{-1} and a periodic max of 4 mm/month^{-1} in 'flash-flood' type events during autumn and winter months.

Water resources are recycled continuously, resulting in a concentration of solutes and a gradual degradation of water quality. This degradation has been severe enough, in the last few years, to cause a marked decrease in productivity and a decline in local economic terms. By contrast, the areas of the Andarax Valley nearer to Almeria city have experienced these effects far less since their economic stability is lifted by the growth of Almeria city, which it owes partly to the successes of plasticulture in

neighbouring areas (Pulido-Bosch *et al.* 2005). Research by Pulido-Bosch *et al.* (1992) involved analysis of piezo-metric data (collected by the IGME) in boreholes between the towns of Almeria and Gador along the Andarax. All boreholes show a significant decrease in water levels between 1973 and 1987. Their work also showed that the salinity of groundwater increases south-eastwards, towards Almeria. Discharge from the basement aquifer occurs mainly through springs and below ground level directly into the river Andarax. Recharge is through winter rain and snowfall on the Sierra de Gador, and to an extent on the Sierra Nevada to the north-west.

The characteristic lithologies of the area range in age from the Permo-Triassic to Quaternary. Those of the Sierra Gador basement massif comprise Alpujarride nappes of dolomites, schists and marbles. The Andarax basin is filled by a sequence of sedimentary units of Neogene (Messinian and Pliocene) and Quaternary age. To the north, the

Fig. 21.25 The Andarax river valley looking west, with the Sierra de Gador in the left background. The barren higher slopes appear in great contrast to the lush growth in the valley floor.

Sierra de los Filabres is dominated by mica schists (in this area). There are three main aquifers that provide groundwater to the area. These are deep, fractured and karstified basement carbonates of the Sierrra de Gador, unconsolidated Messinian sediments in the Andarax Valley, and above these, shallow Pliocene and Quaternary alluvial and deltaic sediments that are exposed along the length of the valley (Pulido-Bosch 1994).

The entire region is tectonically active today. The most recently recorded deformation in this area is related to movement along the Carboneras fault (Bousquet & Phillip 1976), and has produced faults, oriented approximately NNW-SSE, with vertical displacements of the order of 10 m. These can be seen in the Quaternary fan deposits at many localities of this region. The Sierras de Gador, de los Filabres and the Andarax basin are separated by much older major basin-bounding faults, whose displacement history is complex, many of which have been reactivated in recent times and many now provide pathways for water.

21.4.3 Case study objectives

The overall objective of this study is to demonstrate the uses of multi-temporal, multi-spectral imagery in improving the understanding of geology and water resources, in a semi-arid area that is supported, dominantly, by a rural economy. The value of land here is relatively low here, and solar energy is plentiful so the real limiting factor on sustainable economy has always been the availability of water. The questions are therefore, what controls the presence of water and what can we glean from the data we have about its presence.

With remote sensing we 'see' only the surface of the ground and therefore only the exposed rocks, soils and vegetation; from these we must extract (or infer) information relating to water resources at depth. The study area is quite large (25 × 25 km) so we chose medium resolution imagery for the task (in this case Landsat). Our approach begins by enhancing surface features that then enable us to interpret sub-surface phenomena: vegetation type and distribution; geology and potential aquifers; and

Table 21.8 Multi-temporal datasets used in the case study.

Dataset	Scene-Identifying Numbers	Acquisition Date(s)
Landsat 5 TM	Subset of path 199/ row 034	1984, 1989, 1992
Landsat 7 ETM+	Subset of path 199/ row 034	2000
Air photography	Ortho-quads (Junta de Andalucia air survey)	2004
SRTM 3 DEM	30 m spatial resolution	2001/2014
ASTER DEM	Subset area of Landsat path 199/row 034, 30 m spatial resolution	2001

the surface expressions of structural features, that is, faults and joints (fractures in general) that act as pathways for water once it penetrates the surface.

The main objectives are therefore to locate natural and cultivated vegetation, distinguish the main litho-tectonic units, enhance and identify the main structural elements, and to examine how these pieces of evidence reveal and explain the connections between landuse, water and geology.

These goals will be achieved by

1 Preparing a multi-temporal, multi-spectral, medium resolution dataset;

2 Simple directed processing of multi-spectral imagery;

3 Interpretation of land cover, geology and geomorphology

Our database for this study consists of multi-temporal Landsat images, digital air-photographs images and digital elevation models (DEMs; both ASTER and SRTM) as summarised in Table 21.8. As in the previous example all data conform to WGS84 datum and UTM zone 30.

21.4.4 Landuse and vegetation

To get a first glimpse of the importance of water is in this environment, one need only look at the distribution of vegetation since it cannot survive without water. Both the cultivated vegetation and natural plant cover are important here. The transportation of water is expensive and might be prohibitive to economic agriculture, if it were necessary, so the existence of extensive areas of cultivated land suggests that plentiful, natural water supplies are in close proximity. The density and intensity of cultivation in and along the Andarax river valley suggests that water is indeed plentiful here. In contrast the distribution of

natural vegetation on adjacent hillsides reveals several things including the whereabouts of the main recharge areas (where rainfall is incident), areas where the rocks and soils hold water and areas where evapo-transpiration is relatively low (largely on north-facing slopes).

Standard false colour composites of bands 432 (RGB) reveal significant patterns in the distribution of vegetation (Fig. 21.26). The most obvious one is the marked absence of vegetation in the centre of the area, in the 'badland' terrain occupied by Messinian and Pliocene sediments of the Tabernas-Andarax basins. What little vegetation exists here follows closely the stream networks that drain southward into the Andarax. In contrast, the Andarax river valley, which runs from west to east along the northern margin of the Sierra de Gador before turning southward towards Almeria and the Mediterranean contains the most noticeable area of dense vegetation. Along this valley cultivation is intense; fruits and vegetables of every kind, especially citrus varieties, have been grown here for hundreds of years. Like the Nijar Basin, cultivation here has only become systematic and intense in recent decades and the impact on water quality is well documented.

The Sierras are characterised by an even covering of natural upland scrub vegetation, which gives them a pale reddish tinge in the 432(RGB) image. The effect of topography on this natural vegetation distribution can be clearly seen across the watershed of the Sierra de los Filabres in the northern part of the study area. On the south-facing slopes of the Sierra (south of the watershed) any kind of vegetation seems extremely sparse, whereas the north-facing slopes are covered with healthy vegetation and in many places this local climatic effect of topography allows the successful and growth of managed coniferous forest. The geology and hydrology of the Sierra de Los Filabres is rather different from that of the Sierra de Gador to the south. Here the dominance of relatively impermeable rocks and south-facing slopes means that evapo-transpiration is high whereas porosity and rainfall infiltration are low here. The slopes are largely barren and devoid of lush vegetation in these south-facing areas. On closer inspection, the sparsity of vegetation on the southern side can be seen to be punctuated by several tiny flushes of lush vegetation growth in valleys, high on the flanks of the mountains. At these locations several villages can be seen, as small white patches in Fig. 21.27, on these slopes, high up (ca 1000 m above sea level) and in a linear arrangement. Below each village there is noticeably more healthy vegetation, than above. Given the altitude, climate

Fig. 21.26 Colour composite of bands 432 RGB DDS (2000) revealing the extent and distribution of vegetation across the Andarax-Tabernas basin, in red tones. The darker, more contiguous areas of red represent managed upland forest. The remainder of the upland areas are covered by scrub vegetation that appears as a pale reddish tinge.

Fig. 21.27 Landsat 432 RGB subset of the image shown in Fig. 21.26 (north-eastern corner). The Sierra de los Filabres villages of Olula de Castro, Castro de Filabres, Velefique and Senes (from south-west to north-east) appear in white on the largely barren south-facing slopes, high up and in a linear arrangement. There is noticeably more healthy vegetation below each village than immediately above, suggesting a structurally controlled water pathway. Notice also the dense natural vegetation (and forestry) on the northern side of the range's watershed.

and lack of aquifer here, we conclude that the villages only survive here because of a perennial supply of water. We also notice that the villages are aligned and therefore interpret that they are located where the valley floors are intersected by fault(s) or fracture(s) (black dashed line in Fig. 21.27) that bring water to the ground surface at natural springs.

The Sierra de Gador is composed dominantly of dolomite, with some phyllites and schists. The dolomite is fractured and karstified and presents considerable secondary porosity; as mentioned earlier, it forms the major aquifer here. The river flows from west to east and south, along the northern margin of the Sierra de Gador, and into the Mediterranean at Almeria. A very noticeable strip of healthy vegetation can be seen along its length and on the lower reaches of some of its tributaries. Again we must consider what water source is great enough to support such prolific cultivation here. The reason is that at this margin, water held in the

dolomitic basement aquifer, is forced to the surface at numerous fault springs and then feed into the Andarax. The simple model, for the regional geology of this area, is of a classic half-graben, with the southern margins of all the basins being faulted and the northern margins being gently sloping. This is of course an over simplification but it is a useful model that fits quite well here.

We also notice that no such cultivated river valley exists on the gently sloping southern margin of the Sierra de los Filabres. The reasons for this are complex and rely at least partly on the fact that the lithologies here are dominantly impermeable, so that no large volume aquifer exists on that side of the basin. So the plentiful water here could be considered an accident of geological evolution.

Looking at the standard false colour composites of the Andarax Valley in detail, it can be seen that there has been little change in the extent of cultivation along the valley over the last 20 years or so. The image of 2000 [Fig. 21.28(d)] reveals the appearance of some plastic

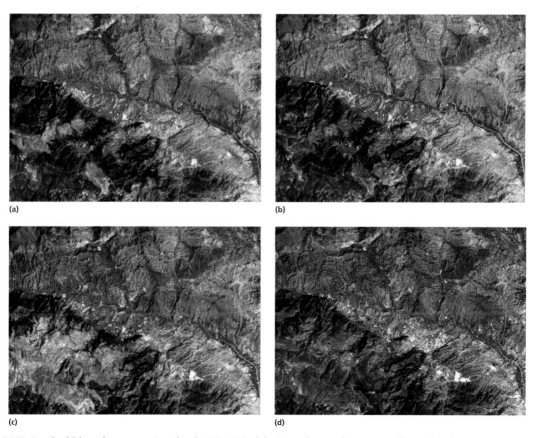

(a)   (b)

(c)   (d)

Fig. 21.28 Standard false colour composites (bands 432 RGB) of the immediate Andarax river valley and the lower reaches of its main tributaries, showing the extent and distribution of vegetation between 1984 and 2000: there has been very little change in this time. (a) 1984; (b) 1989; (c) 1992; and (d) 2000.

Fig. 21.29 Grey-scale image representing the NDVI, as calculated from the muilti-spectral data of the Tabernas-Andarax basin. The highest values (white) appear where there is cultivated vegetation and forested areas, followed by north-facing slopes of the Sierras, which are populated by healthy natural vegetation (pale grey), with the remaining low-lying basin areas being populated by a very thin, chlorophyll-poor covering of scrub vegetation (darkish grey). The recently completed motorway link to Granada and the main channel of the Andarax river valley show the lowest values (black).

greenhouses replacing open cultivation but no substantial change in geographic extent of the cultivated area. The town of Alhama de Almeria (Alhama meaning 'spring') seems also to have expanded in recent years, perhaps because the success of cultivation has meant increased prosperity, attracting growth and development. Some of the land in and around the town, which was clearly cultivated in 1992 now appears to be urban or devoted to greenhouses.

Closer inspection of the central part of the Tabernas-Andarax basin, using a calculated NDVI (Fig. 21.29), reveals a thin covering of vegetation that appears to be relatively less dense or less photosynthetic, or both. This represents the natural scrub that is characteristic of this and most semi-arid areas. The covering is patchy and its distribution is affected by small scale topographic and lithological (porosity) variations and, in areas of unconsolidated materials, by potential vulnerability to surface erosion. The highest values (white) appear where there is cultivated vegetation and forested areas, followed by north-facing slopes of the Sierras that are populated by healthy natural vegetation (pale grey), with the remaining low-lying basin areas being populated by a very thin, chlorophyll-poor covering of scrub vegetation (darkish grey). The recently completed motorway link to Granada and the main channel of the Andarax river valley show the lowest values (black).

21.4.5 Lithological enhancement and discrimination

Now we draw our attention specifically to the geology, although as we can see from our observations of vegetation above, it is difficult to avoid the geology since it has a controlling influence on many natural phenomena in this area.

If we begin once again with simple colour composites and the general enhancement of rocks and soils, we may start with a simulated true colour image (Fig. 21.30), to visualise the main targets of interest in

the way we would see them with the naked eye. We notice that the Sierras appear in dark greys, browns and blueish tones, with the sediments and soils of the basins appearing in paler greys, dull cyan, buff brown and greenish tones. We can make some simple but confident divisions of these broad classes from this image alone, as shown in Fig. 21.30. Urban areas, buildings and greenhouses appear near white and vegetated areas appear in very dark grey. These make up the majority of the area. Noticeable against this background are a few areas of reddish-brown, the largest of which lies to the west of

Fig. 21.30 Landsat 7 ETM+ simulated true colour image (321 RGB, DDS) of the Tabernas-Andarax basin, showing exposed metamorphic basement lithologies in brown (dolomites) and bluish tones (mica schist), with unconsolidated Neogene and Quaternary sediments in cyan-grey, buff brown, pale brown and greenish tones. The boundaries between these appear quite clearly and have been indicated by dashed white lines. Vegetation appears dark green and urban areas appear near white, as do areas that have been cleared for cultivation or greenhouse construction.

Fig. 21.31 Colour composite of bands 531 RGB DDS for general geological discrimination. In this image, the dolomites and mica schists of the basement massifs appear in pinkish/reddish-brown and darkish blue tones respectively. Neogene and Quaternary sedimentary rocks appear as a variety of pinkish, buff-brown, greenish, pale blueish-grey and yellow-brown tones (interpreted bedding traces have been indicated in dashed white lines). Vegetation appears in very dark red and Quaternary red soils in bright olive-green tones.

the town of Gergal. The reddish colour in this image indicates higher reflectance in band 3 and low reflectance in bands 2 and 1, and represents something that looks red to the naked eye.

If we also look at a colour composite of bands 531 RGB DDS (Fig. 21.31) which as we know is often the best combination for geological discrimination in semi-arid areas, we are presented with very useful image revealing many of the significant lithologies very clearly in a variety of vivid colours. Looking first at the Neogene and Quaternary basin sediments we see that they produce quite a complex pattern of colours and textures in the centre of the image. The Sierras appear are clearly distinguishable on the basis of both tone and texture: the dolomites appear in reddish-brown tones while mica schists

appear in bluish-purple tones. The Andarax-Tabernas basin is filled with Neogene and Quaternary sediments and these appear in a variety of green, grey, blue, pinkish, pale brown and yellow-brown tones. Some division of these lithologies and a hint of folding are interpretable from this area (as shown in Fig. 21.31). In this band combination, vegetation appears in very dark reddish tones because it too has relatively low reflectance in bands 3 and 1 and high reflectance in band 5, but less high than the rocks and soils that have their reflectance maxima at these wavelengths. We notice that in this image, the patch of soils to the west of Gergal (appearing red in the previous Fig. 21.30) appears in bright olive green tones, indicating high reflectance in band 3 with lower reflectance in band 5 and very low reflectance in band 1.

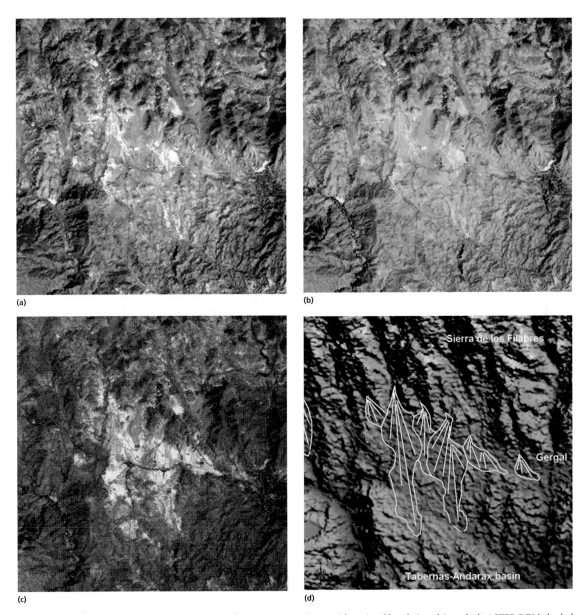

(a)

(b)

(c)

(d)

Fig. 21.32 Images of the Gergal area: (a) 432 RGB; (b) 531 RGB; (c) iron-oxide ratio of bands 3 and 1; and (d) ASTER DEM shaded relief image overlain by a simple interpretation of the fan systems (dashed red lines indicate faults, solid white lines indicate alluvial fan systems, the motorway is shown in brown for reference).

Looking at this area in more detail, and referring back to our 432 standard false colour composite [Fig. 21.27(a)] we see that what appears red in the 321 true colour image, also appears in greenish tones, again indicating high reflectance in band 3 (green colour gun here) but this time low in both bands 4 and 2. This reflectance pattern is characteristic of iron oxide and hydroxide minerals (such as haematite and goethite), which also have a typically reddish appearance to the naked eye. The area also has a noticeably smooth texture and appears relatively flat compared to the surrounding rocky hillsides, and we may conclude that it represents an 'pocket' of accumulated sediment or soil. Generation of an iron-oxide ratio image using bands 3/1 [Fig. 21.32(c)], indicates several connected patches with high iron-oxide content, relative to the surrounding

Fig. 21.33 Red coloured alluvial fans near Gergal, looking south towards the flanks of eastern Sierra Nevada (rising to the right or west) and the Sierra de Gador (almost invisible in the far distance, centre left).

rocks and soils. These patches of ground suddenly become rather interesting: why do we have a series of isolated areas of reddish soils at this location (rich in iron oxides) and not elsewhere.

Examination of the ASTER 30 m DEM of this area, as shown in Fig. 21.32(d) (as a shaded relief image) reveals two breaks in topography, one to the north of the soil patches and a more subtle one to the south. The soil patches appear aligned in a north-west south-east direction and each is elongate in a north-south direction. Careful geological interpretation made from these images and the DEM suggests a series of alluvial fans that drain to the south from the Sierra de los Filabres and are trapped by the topography (see also the photograph in Fig. 21.33). They lie along a series of sub-parallel NW-SE oriented faults, these are related to the main basin-forming fault systems. The northern of these faults has uplifted the land to the north, and the southern one has uplifted to the south, producing a small graben between the two. Comparison of the colour composite of bands 531 with the DEM

[Fig. 21.32(b) and 21.32(d), respectively], reveals that the basement lithologies in purple and brown tones correspond with the topographic highs. This suggests a rather classical model of uplift induced erosion and transport of debris from a mountain front and into a fault controlled topographic trap, producing isolated pockets of sediment. The next question is the reddish colour. The source area of the debris in these pockets lies immediately to the north and consists largely of mica schists, which contain a lot of biotite, and this breaks down very readily and may release its iron to the soils. The unconsolidated nature of the transported debris would then facilitate rapid oxidisation of the iron to give the red colour; this may explain the presence of the iron oxides in these soils.

The next question relates to the boundaries between the lithologies of the Sierras and basins – can we see these? The answer is yes, but the contacts are not always crisp, linear or well defined. We have already mentioned two of these (and that they are mainly structural) – the faulted contact to the south (between the Andarax

Valley and the Sierra de Gador), and the gentle dip slope rising northwards to the Sierra de los Filabres. The other forms the northern boundary of the Andarax Valley that lies near Alboloduy and this is described in the next section

21.4.6 Structural enhancement and interpretation

One rather commonly adopted method of enhancing structural features in remotely sensed images comprises spatial filtering, using one or more kernels of varying form and dimension to enhance or suppress features of varying orientation. Choosing which band of a multispectral dataset presents a further choice. We might prefer to spatially enhance the band or dataset that has the highest spatial resolution. In the case of the Landsat 7 ETM+ or Landsat 8 OLI datasets, this would be the 15 m panchromatic band. In the case of the Landsat 5 however, with no 15 m band we must choose one of the VNIR bands; band 2 or 3 would constitute a good choice since these are less affected by haze than band 1. The reason for spatial filtering in medium resolution images is that we often cannot see those structural features of interest and so need a little textural help. We can use filters to enhance information of different frequencies, high or low. For structural features we may use a high pass filter to enhance systematic changes in image tone and contrast in the hope of detecting the surface expressions of faults and fractures. At the opposite extreme, filtering of a VHR image, such as the 2004 air-photography, would seem rather pointless in this sense, since it already gives us unprecedented detail of ground surface features and we can interpret even quite small scale faults and fractures directly (refer back to the table of image mapping scales in § 20.1). In fact a VHR image may overwhelm us with so much spatial detail that we can no longer see the really significant regional scale structures; clearly there is a balance to be struck and we need to think carefully about what we want to achieve when we (a) choose the data and (b) decide how to process it.

The structural trends in the study area are complex and the dominant features lie largely on a E-W orientation. These structures comprise the major basin-forming normal faults that were opening and lifting the Sierras out of the seas some 15 million years ago. Several other older compressional structures (thrusts) exist within the

basement complexes of the Sierras and these also have an approximately E-W orientation. There are several other structural trends in the basin sediments, produced largely by Quaternary and post-Quaternary faults on approximately N-S and NW-SE orientations. These are largely normal or oblique-slip faults where the dominant slip component is vertical.

Highlighting the older basement structures using directional filters is problematic since they are commonly thrusts, which are low angle structures and produce distinctly non-linear, rather sinuous surface expressions. There tend to be highlighted more easily by outcrop (lithological) variations and relationships and so best interpreted visually. This is certainly the case along the south-eastern margin of the eastern Sierra Nevada, near the town of Aboloduoy [Fig. 21.34(c)]. Here the surface trace of several thrust faults is highlighted by the presence of relatively highly reflective Neogene sediments (marls containing gypsum). These have been thrust up, as slices between metamorphic rocks (phyllites), from the basin onto the flanks of the Sierra. They form an eye-catching bright east-west oriented stripe of ground, the uppermost edge of which is characteristically sinuous, hinting at its low-angled relationship with the hillside. The aerial photographic subset in Fig. 21.34(c) provides great detail, at visible wavelengths, of ground tones and textures but only very subtle tonal differences between the Neogene sediments and the phyllites. In such situations, both spectral detail form Landsat and spatial detail from the digital photograph are vital to the understanding of structural context.

So we must concentrate on the basin-bounding normal faults, and the Quaternary normal faults. Given that there are two dominant trends, E-W and NW-SE, we could use Sobel filters or variants thereof, to selectively pick out surface topographic features that may indicate structural control. Alternatively, we may chose a Laplacian filter, to highlight any high frequency textural information and yield a generally sharpened image or gradient filters to enhance features of a particular direction that we know to exist. The results of such experiments are illustrated in Figs. 21.35 and 21.36. The use of a 3×3 Laplacian filter yields the image shown in Fig. 21.35, this image is effective at highlighting drainage patterns, watersheds and landuse changes. The image is texturally complex and subtle textural changes can be seen across some of the larger structures but otherwise the result proves unhelpful in showing features that we do not

Fig. 21.34 Thrusted contact near the town of Alboloduy: (a) Colour composite of bands 531RGB showing the slice of thrusted basin sediments (marls and gypsum) in bright yellow and orange tones surrounded by mica schists (blues) to the north and other basin sediments (cyan, red and greenish tones) to the south (black box indicates the coverage of the aerial photograph in (c); (b) photograph of basement metamorphic rocks and basin sediments thrusted to the north (left), looking eastwards along the line of the thrusted contact; and (c) 2004 aerial photograph showing VHR detail of the thrust belt near the town of Alboloduy [the red spot shows the location at which the photograph in (b) was taken].

already know to exist. Directional Sobel filters yield slightly more promising results as the images in Fig. 21.26 Show. These images represents the use of a simple 3×3 Sobel filters (of the form described in § 4) to highlight features oriented north-south [Fig. 21.36(a)] and east-west [Fig. 21.36(b)]. In these images, the main basin bounding faults can be discerned, thought only because we already know they are there. The most conspicuous features are the bed of the Andarax and its main tributaries, and the watershed of the Sierra de los Filabres.

Fig. 21.35 Landsat ETM+ band 2 enhanced using a Laplacian filter. This filtered image is most ineffective for the detection of faults and fractures. Only the most systematic and extensive features are detectable, such as the WSW-ENE trending ridge through the Tabernas basin, which is clearly visible. The drainage, vegetation and urban areas near Alhama are highlighted, as is the sharp boundary between the Sierra de Gador and the Andarax basin sediments.

By far the most effective method of interpreting and so extracting structural features is using a DEM. Since the features we are looking for are sub-surface phenomena, only some of which may intersect the surface, the main way we can interpret them is by looking at the physical surface to identify systematic topographic expressions that may have structural significance, and hopefully correlating some of them with image spectral variations to make an interpretation. Although the image spectral variations along the boundaries between these terrain units seem subtle and complex, when we focus on the DEM we are able to see several important pieces of information. The Andarax-Tabernas basins appear clearly in the low lying areas (darker blues in the DEM shaded relief image) while the high ground of the Sierras is visible in bright oranges. The abrupt change in slope between the Sierra Gador can be seen in contrast with the gradual northward slope up to the Sierra de los Filabres. We also begin to see the subtle topographic changes of the alluvial fan systems at Gergal (and Tabernas) in addition to their location with respect to the regional structure, as shown in Fig. 21.37. These are evidence of relatively recent tectonic activity.

If we then calculate the slope angle (in degrees in this case) from the DEM surface, we can exaggerate these

(a)

(b)

Fig. 21.36 Landsat ETM+ band 2 enhanced using a Sobel filter to enhance linear features on (a) N-S and (b) E-W orientations, using the filters described in the text. Many of the main basin bounding faults are highlighted in these images.

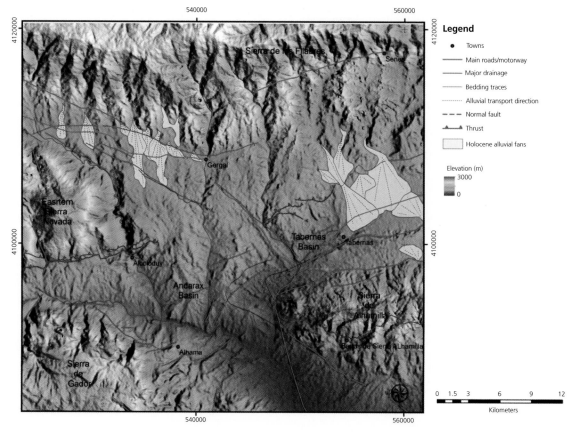

Fig. 21.37 SRTM DEM shaded relief image of the Andarax-Tabernas basin, with interpreted structures and alluvial fan systems overlain.

expressions to see them rather more easily (as shown in Fig. 21.38). The main basin bounding faults and many others are revealed by systematic and relatively abrupt changes in slope angle. In contrast, the areas occupied by recent alluvial fans are characterised by slope angles of less than 10 degrees, and in the central part of the fan systems, less than 3 degrees.

If we then combine all the fragments of geological knowledge we have gained so far, we can produce a simple regional litho-tectonic interpretation map. This presents the main lithological groups, the major structural elements, recent depositional features (tectonically triggered alluvial fans systems) and the regional aquifers together, thus providing us with a tool to better understand the connections between topography, geomorphology, geology, water and agriculture/land cover (landuse). The result of this compilation is shown in Fig. 21.39.

21.4.7 Summary of the case study

This case study is a rather good example of one where we let the images speak for themselves in guiding us towards features of potential interest, rather than approaching the work with a very fixed agenda. When faced with an unknown area and a remit of exploration and understanding, it is often sensible to begin by looking at the most obvious, eye-catching features and then proceeding gradually to the more complex issues. It has also become very clear that water and geology are intricately linked in this area, so that one cannot understand the former without considering and discovering the latter. It is a complex area, and so we have attempted to cover only a few of the more interesting features here.

We find that convolution filtering is not always useful or effective in identifying faults from imagery – you will enhance many surface edges and textural boundaries that bare no connection to sub-surface structures; you

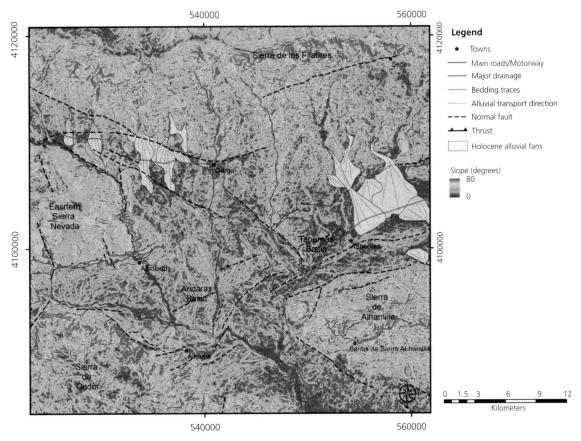

Fig. 21.38 Slope angle as calculated from the SRTM DEM, in degrees, and displayed using a red-blue colour lookup table. Interpreted structures and the alluvial fans interpreted earlier are shown.

may be lucky enough to find some faults. The point to be learned is that filtering should be used when you have a very clear objective and one that is related to purely surface targets. In fact, finding faults in images is more about using your eyes and applying your geological knowledge than about processing images. Quite often it is a matter of what you do not see that may point to the existence of sub-surface control. For instance, the presence of the three villages, aligned and high up in the mountains without obvious means of water supply, could point to a leaky pipe but in this environment is more likely to indicate the presence of a spring line. Your understanding of its geological or geomorphological context will then be required to determine whether that spring line is controlled by underlying stratigraphic or structural control. The observation of topographic expression along that line may not be

sufficient indication in itself since the expression could still be produced by either form of control. Clearly quite a bit of detective work is required. Only when you have spectral, topographic and stratigraphic agreement can you be reasonably sure that you have identified a fault; however, here again your knowledge and experience will come into play. In any case, you will certainly want to make a field visit to the location to satisfy yourself and to better understand the geometries of the features on the ground.

The farmers here become victims of their own success since increased demand for their products lead to massive expansion, as in the Nijar area, and increasing pressure on existing resources. This leads to many questions about sustainable development and the effective management of renewal resources. In this case the cost is not only the reduction of available water but also its

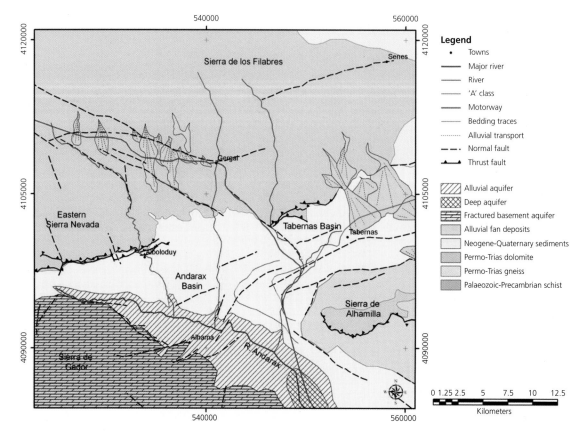

Fig. 21.39 Summary regional geological interpretation, bringing together lithological and structural information interpreted from the image data and DEM, to produce a regional litho-tectonic map; this can then be used for the better interpretation of natural water resources. The aquifer types shown here are those described earlier in the text (modified from Bosch *et al.* 1992).

quality. Here again, we study a problem retrospectively, leading one to question the benefit of doing so, given that the real problem has been known about for some time. In doing so, however, we learn many things and begin to understand the necessary approach and methods, and we can then potentially establish a methodology to be applied in other areas.

21.4.8 Questions

1 Which other surface parameters could be usefully applied here?
2 How would you go about estimating how much rainfall enters the groundwater (and the Gador aquifer) here?
3 What limitations are imposed by the data (Landsat and SRTM DEM) on the interpretation in this case? What data would you advise?

21.4.9 References

Gallego, M.C., Garcia, J.A., Vaquero, J.M. & Mateos, V.L. (2006) Changes in frequency and intensity of daily precipitation over the Iberian Peninsula. *Journal of Geophysical Research*, 111, D24105.

Pulido-Bosch, A., Sanchez Martos, F., Martinez Vidal, J.L. & Navarrete, F. (1992) Groundwater problems in a semiarid area (low Andarax river, Almeria, Spain), 1992. *Environmental Water Geological Sciences*, 20 (no. 3), 195–204.

Pulido-Bosch, A., Sanchez Martos, F., Navarrete, F. & Martinez Vidal, J.L. (1994) Agricultural practices and groundwater contamination in the lower Andarax basin (Almeria, Spain). In: *Water Down Under 94: Groundwater Papers*, national conference publication, no. 94/14, pp. 445–9. Institution of Engineers, Australia.

Van Cauwenbergh, N., Pinte, D., Tilmant, A., Frances, I., Pulido-Bosch, A. & Vanclooster, M. (2008) Multi-objective, multiple participant decision support for water management in the Andarax catchment, Almeria. *Environmental Geology*, 54, 479–89. doi:10.1007/s0025-007-0847-y.

CHAPTER 22

Research case studies

This chapter is based on the authors' published research papers. The intention of this chapter is not to cover every aspect of remote sensing applications but instead, using several case studies, to share our experiences with you on the following:

- How to think through and formulate an application research project;
- How to design and develop the most effective image processing techniques and strategy for extracting the required thematic information from images;
- How to establish the most representative and powerful GIS model to serve the objectives of the project;
- How to approach the data analysis and the presentation and critical assessment of results.

22.1 Vegetation change in the Three Parallel Rivers region, Yunnan Province, China

22.1.1 Introduction

In this case study, multi-temporal Landsat 5 TM and Landsat 7 ETM+ image data were used to assess the vegetation coverage change. With a simple and effective methodology based on the Normalised Difference Vegetation Index (NDVI), the study aims to identify areas subject to rapid vegetation destruction, as well as to detect any possible signs of vegetation revival, in the Three Parallel Rivers region in south-western China (Liu & Meng 2005).

The area lies within an N-S orogenic belt where the edge of the Eurasian plate is being compressed from the west by the underlying eastward subducting Indian plate. This continental scale tectonic movement has squeezed and uplifted the terrain dramatically to form the north-south oriented Hengduan Mountains, which lie contrary to the dominant east-west trend of the major mountains farther to the north. In this intensely sheared N-S tectonic zone, three great sub-parallel rivers flow in deeply cut valleys separated by high mountains (World Heritage Nomination – IUCN Technical Evaluation 2003). These three rivers from west to east are: Nujiang River (Salween in Burma), Lancang River (Meigong in Vietnam) and Jinsha River (the upper reach of the Yangtze). The Three Parallel Rivers region was awarded the prestigious status of 'World Heritage' site by the United Nations Educational, Scientific and Cultural Organization (UNESCO) in 2003 (World Heritage 27 COM 8 C.4 2003) for its great diversity of landscape, vegetation, animal species and human culture (Natural site datasheet from WCMC). With this new status, the conflict between economic development and environmental protection has intensified. The balance between the two will decide the fate of this rare natural beauty. With abundant water resources and tremendous potential for hydroelectric power, vegetation is a key factor for maintaining a healthy ecological system. Once it is destroyed, severe erosion will occur and the damage will not be restricted to the local environment but will extend farther downstream to Burma and Vietnam.

22.1.2 The study area and data

The study area is in the north-west corner of Yunnan Province, China, and adjacent to Burma in the west (Fig. 22.1). It is within a TM/ETM+ scene of path-raw 132-041 extending 28°6′52″ to 26°45′48″N; 98°23′13″

Image Processing and GIS for Remote Sensing: Techniques and Applications, Second Edition. Jian Guo Liu and Philippa J. Mason.
© 2016 John Wiley & Sons, Ltd. Published 2016 by John Wiley & Sons, Ltd.

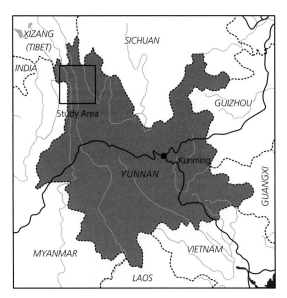

Fig. 22.1 Location map of the study area in the Three Parallel Rivers region, Yunnan Province, China.

Table 22.1 Image data of the study area.

Image 132-041	Dates (y-m-d)	Temporal separation	Seasonal difference
TM	1994-11-15	6 years 1 month	40 days
ETM+	2000-12-25	and 10 days	

to 99°54'53"E, covering much of the Three Parallel Rivers region. The three great rivers are almost parallel to one another in this area; at their closest, they are no more than ca 64 km apart.

The data were acquired from the Global Land Cover Facility, UMIACS (University of Maryland Institute for Advanced Computer Studies), USA. The TM image used in this study was taken on 15 November 1994 and the ETM+ image 25 December 2000; the temporal separation between the two is 6 years, 1 month and 10 days, while the seasonal difference is 40 days (Table 22.1). Image co-registration quality is crucial for multitemporal image comparison. Both images have been ortho-rectified to WGS84 NUTM47 to a high accuracy at source and, as a result, the two images are precisely coregistered without visually observable mismatches even when viewed at pixel level.

22.1.3 NDVI Difference Red, Green and Intensity (NDVI-D-RGI) composite

The NDVI is a well-established and robust technique for mapping vegetation based on its diagnostic absorption feature in red (R) spectrum and very high reflectance in nearer infrared (NIR) spectrum (Gausman 1974; Lilesand & Kiefer 2000). These two spectral ranges are denoted as bands 3 and 4 in TM and ETM+ image data. One of the advantages of NDVI is that it is normalised to a standard value range from -1 to 1 and thus the NDVIs derived from different images are comparable in the same value range. The technique has been widely used for assessment of changes of vegetation status, landuse patterns and ecological parameters (Cihlar *et al.* 1991; Lambin & Ehrlich 1997; Mantovani & Setzer 1997; Wang *et al.* 2001; Li *et al.* 2002).

The main purpose of the study is not simply mapping of vegetation but of the changes in vegetation coverage in the region. To this end, we have composed a simple and effective method to highlight the areas subject to significant change using multi-temporal NDVIs incorporating threshold criteria as described below.

NDVI-D-RGI composite:

$$
\begin{aligned}
\text{Red}: \quad & \text{If } NDVI1 > C1 \text{ AND } (NDVI1 - NDVI2) \\
& > T1 \text{ then } NDVI1 \text{ else } NULL \\
\text{Green}: \quad & \text{If } NDVI2 > C2 \text{ AND } (NDVI2 - NDVI1) \quad (22.1) \\
& > T2 \text{ then } NDVI2 \text{ else } NULL \\
\text{Intensity}: \quad & \text{Band } 4 \text{ of ETM}+
\end{aligned}
$$

where *C* and *T* are *vegetation criterion* and *vegetation difference threshold*. The value range for both parameters is [0, 1]. The numbers 1 and 2 denote the time sequence of the two images in comparison.

The NDVI-D-RGI composite defined by eqn. 22.1 produces a vegetation change image. The red layer highlights vegetation destruction (areas covered with healthy vegetation on the imaging date 1 but no longer on date 2) in red; while all the unchanged areas, either with or without vegetation on both dates, are output as null. Similarly, the green layer highlights the areas of vegetation revival (no vegetation on date 1 but with vegetation on date 2) in green and leaving all the unchanged areas as null. Overlaying these red and green layers on the ETM+ band 4 as an intensity layer presents vegetation destruction in red, revival in green and unchanged areas as achromatic imagery background. ETM+ band 4 was chosen as the intensity layer for its high intensity from vegetation and white appearance of snow. Snow appears in black in band 5 and 7 for its absorption in short-wave infrared (SWIR) spectral range.

For vegetation comparison, it is vital to acquire the multi-temporal images taken in the same month/season – or better, on the same date/week. Unfortunately, this is not often possible. Small seasonal differences in image acquisition date may produce non-negligible effects, preventing a fair comparison for vegetation change assessment. If the image of *NDVI*1 is taken in a much warmer (or greener) season than that of *NDVI*2, a direct comparison between the two images may falsely indicate deterioration of vegetation even if the actual vegetation coverage and conditions are not really changed. Conversely, if the image of *NDVI*1 is taken in a much colder season than that of *NDVI*2, an incorrect conclusion of vegetation revival may be reached. The *vegetation criteria*, $C1$ and $C2$, and the *vegetation difference thresholds*, $T1$ and $T2$, in eqn. 22.1 allow adjustment to reduce the seasonal bias for vegetation change assessment and control the significance level of the vegetation change to be detected.

The *vegetation criteria*, $C1$ and $C2$, decide if an NDVI value is acceptable as vegetation or not, thus eliminating non-vegetation pixels. For TM and ETM+ images, the DNs of vegetation in the NIR (band 4) should be significantly higher than those in red (band 3), therefore NDVI of vegetation should be always positive, and thus $C1 > 0$ and $C2 > 0$ ensure positive values of NDVI to remove obvious non-vegetation areas. Higher thresholds of vegetation criteria, $C1$ and $C2$, set harsher conditions to reject more pixels from being recognised as vegetation.

The *vegetation difference thresholds* $T1$ and $T2$ set the significance levels of vegetation changes between the two images. For the red layer in eqn. 22.1, the vegetation pixels in the *NDVI*1 image are displayed in red only when their values are greater than their corresponding pixels in *NDVI*2 image by a difference of no less than $T1$. In this way, pixels showing no significant vegetation change will be eliminated as null. Similarly, $T2$ will eliminate pixels showing no significant vegetation change in the green layer. Relatively high $T1$ and $T2$ thresholds ensure a critical assessment of significant vegetation change (either destruction or revival) while low values of $T1$ and $T2$ make an NDVI-D-RGI composite sensitive to changes to both vegetation conditions and coverage. The C parameters are partially controlled by the corresponding T parameters. For a given T, the NDVI difference defined in eqn. 22.1 is not sensitive to the

variation of C when $C < T$. For instance, for a given $T1$, any pixel of $C1 < NDVI1 < T1$ will be eliminated unless the corresponding *NDVI*2 has a negative value that makes up the difference of $T1 - C1$. The C parameters only take strong effects to the NDVI difference when $C > T$.

In general, a higher *vegetation criterion* and a higher *vegetation difference threshold* should be set for the NDVI image taken in a warmer (or greener) season so as to compensate for the vigorous effect of vegetation. The value of $C1$ or $C2$ should be set proportional to the seasonal greenness, that is, the greener the vegetation on the imaging date is, the higher its *vegetation criterion* should be, but this simple principle is not applicable in areas of high relief. NDVI cannot effectively suppress topography and may yield much lower values for vegetation in dark shadows than on illuminated slopes. A high *vegetation criterion* ($C1$ or $C2$) removes too many vegetation pixels in areas of dark shadows, but a low criterion makes this parameter nearly redundant if $C < T$. The *vegetation criterion* is therefore effective only for seasonal compensation in low relief and flat areas.

A more effective compensation can be achieved by setting different values for $T1$ and $T2$ depending on the seasonal greenness difference between the two images in eqn. 22.1. A higher *vegetation difference threshold* should be set to the NDVI image of a warmer (greener) season. For instance, if *NDVI*1 is taken in a greener season than *NDVI*2 in eqn. 22.1, then we should set $T1 > T2$. The difference between $T1$ and $T2$ decides the strength of the compensation to seasonal greenness bias.

The specific parameter setting and its effects can be adjusted and judged empirically, with reference of NDVI image statistics. Comparison between the standard false colour composites of the two dates, in conjunction with the NDVI-D-GRI composite, can help to make effective parameter settings and to ensure accurate detection of evident vegetation changes.

22.1.4 Data processing

As shown in Table 22.1, the temporal separation between image 1 (TM) and image 2 (ETM+) is 6 years, 1 month and 10 days, while the seasonal difference between them is 40 days. The imaging date of the TM image is in a warmer month, 15 November, than that of the ETM+, 25 December. Without any actual vegetation change, the TM image would appear 'greener' or higher average NDVI than ETM+ image. To ensure that the weak signal of vegetation in deep shadows is not eliminated, the *vegetation criteria* for both images were set to a low positive value $C1 = C2 = 0.1$. As shown in Table 22.2, the mean and median of the TM NDVI image are significantly higher than those of the ETM+ NDVI image. This is likely caused by both the seasonal bias and the significant reduction of vegetation coverage over 6 years. To compensate for the seasonal effects, the general setting for the *vegetation difference thresholds* is $T1 > T2$. Three different

sets of $T1$ and $T2$ with increasing thresholds for vegetation change were applied for comparison (Table 22.2). All these settings are slightly favourable to vegetation revival (green pixels) in the resulting NDVI-D-RGI composite to avoid exaggeration of vegetation destruction.

Linking NDVI-D-RGI composites with standard colour composites of TM and ETM+, we can observe how different settings affect the vegetation change detection, as illustrated in Fig. 22.2. The number of pixels identified

Table 22.2 TM and ETM+ NDVI statistics and settings for parameters C and T.

Colour	Images: NDVI	NDVI statistics			C1 C2	T1 T2		
		Mean	Median	Std. Dev				
Red	TM: *NDVI*1	0.299	0.327	0.225	0.1	0.20	0.25	0.30
Green	ETM+: *NDVI*2	0.204	0.205	0.195	0.1	0.15	0.15	0.20

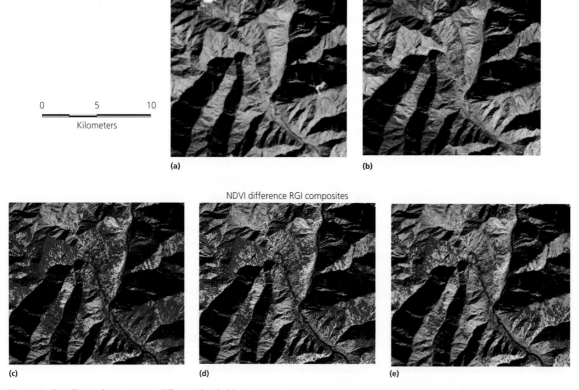

Fig. 22.2 The effects of T (*vegetation difference threshold*) parameter setting on NDVI-D-RGI composites: (a) the 1994 TM 432 RGB image; (b) the 2000 ETM+ 432 RGB image; (c) the NDVI-D-RGI composite derived from $T1 = 0.20$, $T2 = 0.15$; (d) the NDVI-D-RGI composite derived from $T1 = 0.25$, $T2 = 0.15$; and (e) the NDVI-D-RGI composite derived from $T1 = 0.30$, $T2 = 0.20$.

as vegetation destruction (red pixels) decreases considerably with the increasing $T1$ while the number of pixels representing vegetation revival (green pixels) decreases with the increasing $T2$. The first set of parameters ($T1 = 0.20$, $T2 = 0.15$) with $T1 - T2 = 0.05$ is equivalent to boosting the mean of $NDVI2$ by 0.05 and the resulting image in Fig. 22.2(c) can be interpreted as change in both vegetation condition as well as coverage. The second set of parameters ($T1 = 0.25$, $T2 = 0.15$) with $T1 - T2 = 0.1$ boosts the mean of $NDVI2$ to a level slightly higher than that of the $NDVI1$. With the increased $T1$ and $T1 - T2$, the resulting image in Fig. 22.2(d) more critically targets the severe vegetation destruction that caused the decrease in vegetation coverage [reference Fig. 22.2(a) and (b)]. The image derived from the third set of parameters ($T1 = 0.3$, $T2 = 0.2$) in Fig. 22.2(e) may well represent too harsh an assessment; many pixels showing obvious changes in vegetation coverage in Fig. 22.2(d) were eliminated by the high *vegetation difference thresholds*.

Apart from the dominant image features relating to changing vegetation coverage, there are several sources of error that produce odd features in the NDVI-D-RGI composites. The snow coverage in the two images varies according to the season and the weather conditions. A vegetated area with snow cover in the TM image but without snow cover in the ETM+ image will appear in green in the NDVI-D-RGI composite, meaning incorrect indication of vegetation revival, because the snow is recognised as no vegetation in the TM image. The opposite results in a red patch in the NDVI-D-RGI composite and indicates a similar false alarm of vegetation destruction. The vegetation coverage change detected along the edge of the permafrost zone in the high mountains must therefore be verified carefully. Clouds introduce the same type of errors in the same logic.

Both snow and clouds have strongest reflectance in the blue spectrum, recorded in TM/ETM+ band 1, and thus can be effectively eliminated using a blue band threshold in NDVI generation as below:

$$NDVI1(\text{TM}): \quad \text{If } B \text{ and } 1 < 70 \text{ then } (Band4 - Band3)$$
$$/(Band4 + Band3) \text{ else } NULL$$
$$NDVI2(\text{ETM}+): \text{If } B \text{ and } 1 < 80 \text{ then } (Band4 - Band3)$$
$$/(Band4 + Band3) \text{ else } NULL$$

The thresholds for the two images are set slightly different as the average DN level of the ETM+ band 1 is slightly higher than that of the TM band 1.

As shown in Fig. 22.3, the high mountain was covered with both snow and clouds in the TM image [Fig. 22.4(a)] but covered with snow only in the ETM+ image [Fig. 22.3(b)]. In the NDVI-D-RGI composite without the blue band thresholding [Fig. 22.3(c)], clouds and some scattered pixels of snow are wrongly recognised as vegetation revival in green. The green patches and scattered green pixels corresponding to clouds and snow are effectively removed in the image Fig. 22.3(d) by the blue band thresholding, but there are still some residual green patches that are caused by the cloud shadows in the TM image, and these cannot be easily removed.

The NDVI-D-RGI composites used for interpretation of regional vegetation changes in the following section are all with the blue band thresholding.

22.1.5 Interpretation of regional vegetation changes

Visual observation indicates that the NDVI-D-RGI composite with the second set of parameters (Fig. 22.4) provides a well-balanced estimation of the vegetation change. In this image, areas subject to significant vegetation destruction have been effectively identified, and there is no obvious evidence of the exaggeration of subdued vegetation features (possible seasonal effects) in the ETM+ image as destruction of vegetation. On the other hand, the image is 'kind' even to subtle vegetation revival, and this is ensured by the large $T1 - T2$ difference. This image is used as the principal image for interpretation. The statistics of vegetation changes derived from the NDVI-D-RGI composites, of the three different parameter settings, are summarised in Table 22.3. Again, the statistics derived from the image of the second parameter setting are used as the basis for discussion, while the statistics of the other two images serve as lower and upper limits.

The TM/ETM+ NDVI-D-RGI composite in Fig. 22.4 illustrates changes to vegetation coverage during the 6-year interval. In general, the region is largely covered by natural vegetation (forests, bushes and grass), particularly in mountainous areas. The limited areas devoted to agriculture usually occur in relatively flat areas or wide valley bottoms, but they are not cultivated in winter. Therefore the vegetation change detected using the winter TM and ETM+ images in this study is little affected by the change of cultivation in these crop fields.

In reference to the whole study area, the widely spread red patches in the NDVI-D-RGI composite indicate that vegetation coverage has decreased rapidly. As

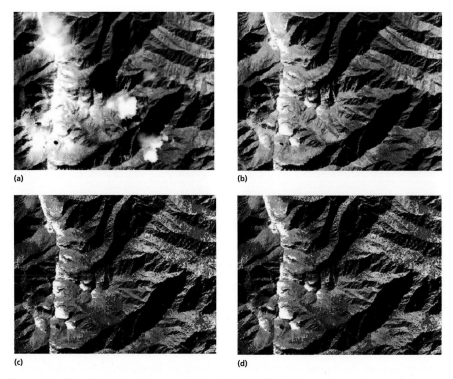

(a)   (b)

(c)   (d)

Fig. 22.3 The effects of cloud, cloud shadow and snow. (a) The 1994 TM 432 RGB image where the mountain was covered by both snow and clouds. (b) The 2000 ETM+ 432 RGB image where the mountain was covered by less snow than 1994 and without clouds. (c) The large green patches in the NDVI-D-RGI composite are not true vegetation revival; these are clouds and cloud shadows in the TM image while the scattered green pixels along the ridge of the snow mountain are caused by snow retreat in the ETM+ image. (d) After applying the blue band threshold, green pixels resulted from snow retreat and clouds are effectively removed, while there are still some residual green patches produced by cloud shadows that cannot be easily eliminated.

Table 22.3 Statistics of vegetation changes in the Three Parallel Rivers region (the numbers in bold are used in the text).

Area name Pixel Hectare	Parameter		Destruction			Revival			Net
	T1	T2	Pixel	Hectare	%	Pixel	Hectare	%	%
Whole study area	0.2	0.15	4442614	360851	16.0	477969	38823	1.7	14.3
27765180	**0.25**	**0.15**	**2585403**	**209999**	**9.3**	**477969**	**38823**	**1.7**	**7.6**
2255227	0.3	0.2	1448928	117689	5.2	288444	23429	1.0	4.2
Nujiang River Region	0.2	0.15	1124343	91317	22.0	98760	8022	1.9	20.1
5109595	**0.25**	**0.15**	**727621**	**59101**	**14.2**	**98760**	**8022**	**1.9**	**12.3**
415029	0.3	0.2	443325	36009	8.7	67492	5482	1.3	7.4
Lancang River Region	0.2	0.15	1092695	88754	18.7	85836	6972	1.5	17.2
5837570	**0.25**	**0.15**	**626844**	**50915**	**10.7**	**85836**	**6972**	**1.5**	**9.2**
474156	0.3	0.2	334762	27191	5.7	48762	3961	0.8	4.9
Jinsha River Region	0.2	0.15	1069892	86902	12.4	153304	12452	1.8	10.6
8663020	**0.25**	**0.15**	**560907**	**45560**	**6.5**	**153304**	**12452**	**1.8**	**4.7**
703654	0.3	0.2	288746	23453	3.3	84170	6837	1.0	2.3

shown in Table 22.3, in the 2,255,227 hectare area, 209,999 hectare of vegetated land in 1994 became barren in 2000; the reduction is 9.3%, according to the second set of parameters. Visual interpretation indicates that most of the noticeable green patches are caused by cloud shadows in the TM image. Even if we accept these green patches as vegetation revival, the net vegetation destruction would still be 7.6%.

The most stunning features in Fig. 22.4 are the concentrated red patches along the Nujiang River indicating an alarming rate of vegetation destruction, particularly in the northern section of the river, as shown in the NDVI-D-RGI composite [Fig. 22.5(a)]. The standard false colour composites of TM and ETM+ images of the region illustrate many areas where healthy vegetation existed in 1994 [Fig. 22.5(b)] but no

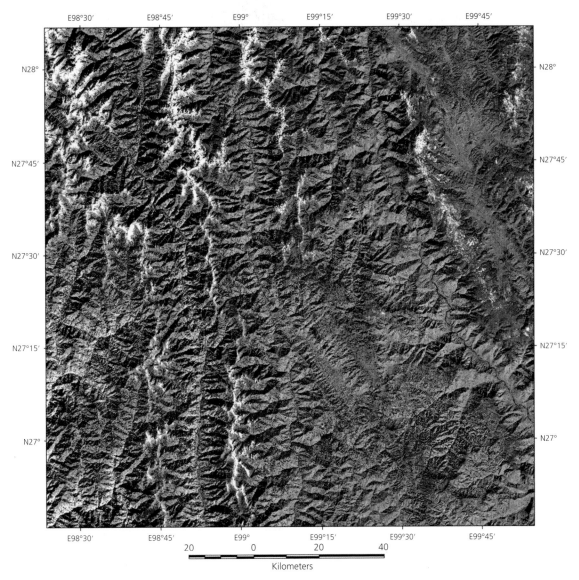

Fig. 22.4 The NDVI-D-RGI composite derived from 1994 TM and 2000 ETM+ images. The image shows the change in vegetation coverage over the 6 years between the two imaging dates; red indicates vegetation destruction during the period, green indicates areas of vegetation revival, whilst the grey-scale background presents the areas that are unchanged.

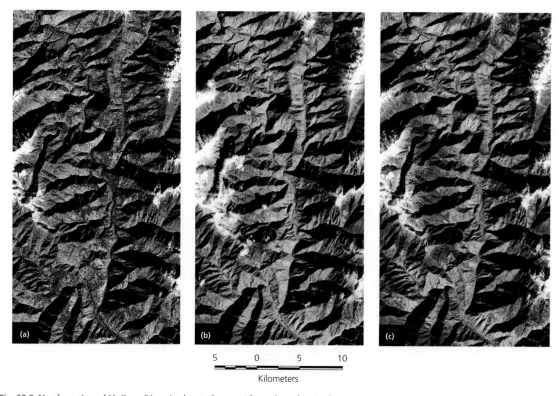

Fig. 22.5 North section of Nujiang River in the study area. The red patches in the NDVI-D-RGI composite (a) indicate devastating destruction of vegetation, which can be clearly seen by comparing the standard false colour composites (b) 1994 TM 432 RGB image with (c) 2000 ETM+ 432 RGB image. The green patches in (a) are largely caused by cloud shadows rather than vegetation revival.

longer in 2000 [Fig. 22.8(c)]. The reduction of vegetation coverage in the 405,129 hectare area along the Nujiang River is about 59,101 hectare, ~14.2% based on the second set of parameters (Table 22.3). The area also shows a slightly higher revival rate (1.9%) than the regional average (1.7%), but this information is not reliable. Scattered clouds in the 1994 TM image appear mainly in this area, where it is nearly completely cloud free in the 2000 ETM+ image. As a result, these limited recognisable green patches are nearly all produced by cloud shadows in the TM image (see Fig. 22.3). There is no clear evidence of vegetation revival in the catchments. The severe destruction of vegetation is mainly along the river, forming a belt. This may explain the dramatic recent increase in flood and mudflow hazards in the areas farther downstream. In comparison, Nujiang River catchment was much better covered by vegetation than the catchments of the other two rivers, but better vegetation cover means greater potential for destruction.

Parallel to and to the east of the Nujiang River is the Lancang River. The vegetation destruction in the Lancang River catchments was the worst among the three rivers, and its status was already poor in 1994 when the TM image was taken, leaving lower potential for further deterioration. Consequently, the decrease in vegetation coverage in the Lancang River catchments appears not as significant as that in the Nujiang River catchments, but the degradation of vegetation was still severe, particularly on mountain slopes along the west bank of the river, shown as scattered red patches in the NDVI-D-RGI composite (Fig. 22.6). The vegetation coverage reduction rate calculated from the second set of parameters for the 474,156 hectare area along the Lancang River is about 50,915 hectare (10.7%), as shown in Table 22.3. The scattered green spots around snow mountain peaks in this area are, in general, caused by snow cover that has retreated slightly in the 2000 ETM+ image, but a few green patches noticeably indicate new plantations (Fig. 22.7). Accounting for all the green pixels as vegetation revival, the net vegetation destruction is 9.2%.

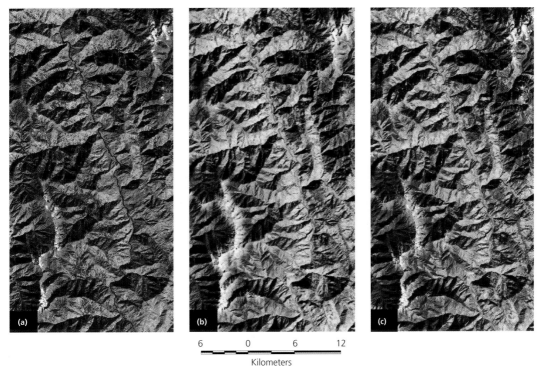

Fig. 22.6 Middle section of Lancang River in the study area. The NDVI-D-RGI composite (a) illustrates severe destruction of vegetation along the west side of the river where the already poor coverage of vegetation has further deteriorated, as shown in (b) 1994 TM 432 RGB image; and (c) 2000 ETM+ 432 RGB image in comparison.

The Jinsha River lies farther to the east of the Lancang River. Jinsha River catchments in the scene show relatively good vegetation coverage and the least change during 6 years as compared with the other two rivers (Fig. 22.4). The general trend of vegetation in the Jinsha River catchments was still in the direction of degradation, as indicated by widely spread red spots in the NDVI-D-RGI composite, as well as a few isolated red patches of obvious vegetation destruction (Fig. 22.8). Most scattered green spots in this part of the image indicate weak revival of vegetation on some slopes (Fig. 22.8). Taking all the green pixels as vegetation revival, the net reduction of vegetation coverage in the 703,654 hectares of the Jinsha River catchments in the study area is 4.7%, according to the second set of parameters, which is significantly lower than the areas along the other two rivers (Table 22.3).

22.1.6 Summary

The data derived from this study only represent the changes of vegetation coverage between 1994 and 2000.

During a field investigation in 2004 and 2006, we noticed that the local government and local population have put great effort into tree plantation and the protection of natural vegetation, but their efforts may be cancelled out by the massive developments in road building and other engineering work in an attempt to fulfil the demands of rapidly growing tourism since the Three Parallel Rivers region was granted a World Heritage site. The picture in Fig. 22.9 was taken in Nujiang valley in 2006; the massive vegetation destruction and soil loss are obvious and the damage caused by road cutting was devastating.

As the assessment is sensitive to subjective choices of parameters C and T, the accuracy of the statistics of vegetation destruction derived from this study is subject to detailed verification, but the evidence of severe destruction of vegetation and rapid reduction of vegetation coverage shown by this study is unequivocal. The three sets of parameters are all favourable to vegetation revival rather than to destruction. The vegetation destruction in shadowed areas can only be significantly underestimated because of weak signals. We therefore believe that the evidence of severe vegetation destruction based

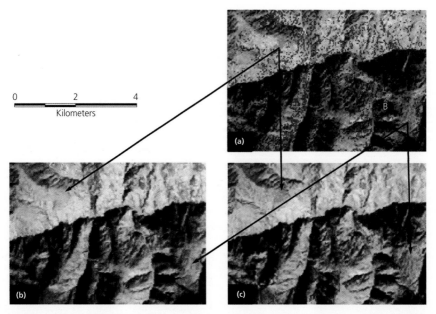

Fig. 22.7 A new plantation is detected in the NDVI-D-RGI composite (a). The green patches denoted by A and B in (a) were barren in 1994, as shown in the TM 432 RGB image (b), but then covered by healthy vegetation in 2000 shown in red in the ETM+ 432 RGB image (c).

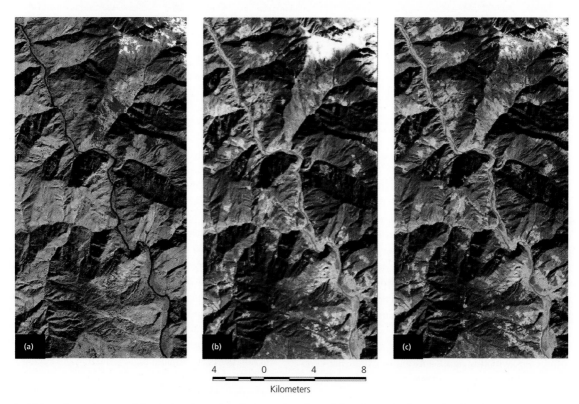

Fig. 22.8 In the catchments of the Jinsha River, a few obvious red patches in a largely grey background in the NDVI-D-RGI composite (a) reveal limited areas of severe vegetation destruction in an otherwise well-preserved region. In the obvious red patches highlighted by (a), healthy vegetation shown in (b) 1994 TM 432 RGB image has been completely stripped, as confirmed by (c) 2004 ETM+ 432 RGB image. The scattered subtle green spots on the slope north of the obvious red patches in (a) imply weak revival of vegetation; the phenomena are reflected by increased redness at the corresponding locations in (c) in comparison with (b).

Fig. 22.9 Field photo taken in April 2006 on the road along the Nujiang Valley. The picture shows the destruction of vegetation coverage by road cutting and excessive cultivation.

on the second set of parameters is reasonably close to the true situation and it is more likely an underestimation rather than an exaggeration.

On the other hand, we realise that the vegetation destruction is much easier to detect than revival. Most features of severe vegetation destruction are the result of human activities (such as logging, burning and engineering), which can be produced in a short period and with clear boundaries. In contrast, vegetation revival is a long and gradual process, and its features are subtle, scattered and more sensitive to seasonal effects. With these factors in mind, vegetation revival could be underestimated.

22.1.7 References

Cihlar, J., St-Laurent, L. & Dyer, J.A. (1991) Relation between the Normalized Difference Vegetation Index and ecological variables. *Remote Sensing of Environment*, 35 (no. 2), 257–77.
Gausman, H.W. (1974) Leaf reflectance of near-infrared. *Photogrammetric Engineering and Remote Sensing*, 10, 183–91.
Lambin, E.F. & Ehrlich, D. (1997) Land cover changes in Sub-Saharan Africa (1982–1991): Application of a change index based on remotely sensed surface temperature and vegetation indices at a continental scale. *Remote Sensing of Environment* 61 (no. 2), 181–200.
Li, B., Tao, S. & Dawson, R.W. (2002) Relations between AVHRR NDVI and ecoclimatic parameters in China. *International Journal of Remote Sensing*, 23, 989–99.
Lilesand, T.M. & Kiefer, R.W. (2000) *Remote Sensing and Image Processing*, fourth edn. John Wiley & Sons, New York.

Liu, J.G. & Meng, M. (2005) Destruction of vegetation in the catchments of Nujiang river, 'Three Parallel Rivers' region, China. In: *Proceedings of the 2005 IEEE International Geoscience and Remote Sensing Symposium (IGARSS 2005)*. 25–29 July 2005, Seoul, Korea.
Mantovani, A.C.D. & Setzer, A.W. (1997) Deforestation detection in the Amazon with an AVHRR-based system. *International Journal of Remote Sensing*, 18, 273–86.
Natural site datasheet from WCMC. (n.d.) www.unep-wcmc.org/sites/wh/Three_Parallel.html.
Wang, J., Price, K.P. & Rich, R.M. (2001) Spatial patterns of NDVI in response to precipitation and temperature in the central Great Plains. *International Journal of Remote Sensing*, 22, 3827–44.
World Heritage Nomination – IUCN Technical Evaluation, Three Parallel Rivers of Yunnan Protected Areas (China) ID N° 1083. (2003) http://whc.unesco.org/archive/advisory_body_evaluation/1083.pdf.
World Heritage 27 COM. (2003) WHC-03/27.COM/24, Paris, 10 December 2003, 91–92.

22.2 GIS modelling of earthquake damage zones using satellite imagery and digital elevation model (DEM) data

22.2.1 Introduction

The Ms 8.0 (Mw 7.9) earthquake that occurred in Wenchuan County, Sichuan Province, China, on 12 May 2008 caused widespread damage and devastation to rural communities and the economy. The terrain of the entire region has been weakened and is now highly susceptible to long-term slope instability that will trouble this region for many years to come.

The epicentre of the Wenchuan earthquake was near the Yingxiu town on the Yingxiu-Beichuan Fault, which is in the central part of the Longmenshan complex fault system along the eastern edge of the Tibetan Plateau separating the mountain range from Chengdu Basin in the east (Fig. 22.10). The main shock was followed by thousands of aftershocks along the seismic fault zone, for a period of more than a month, resulting in tremendous multiple and cascading hazards and loss of human life. The widespread destruction of infrastructure accounts for a total economic loss of over 110 billion USD (Xie *et al.* 2009). These effects were induced both by direct earthquake shaking and by a great number of large-scale earthquake-induced geohazards. The most

common forms of these geohazards in the region are landslides and mudflows (Chigira *et al.* 2010; Dai *et al.* 2011), which account for 15% of deaths (Wang *et al.* 2009a). The Beichuan county town is a typical example. It is situated in the main rupture zone along the Yingxiu-Beichuan Fault (Cui *et al.* 2011), in a deeply incised valley surrounded by very steep mountainous slopes. The town was completely destroyed by the strong earthquake and the ensuing massive landslides, which involved entire mountain slope collapse (Fig. 22.11).

Satellite remote sensing imagery data were widely used to identify and map the distribution of earthquake-induced geohazards (Chigira *et al.* 2010; Godard *et al.* 2010; Ouimet 2010), in addition to providing geomorphologic and landscape categorisation using DEMs (Tang *et al.* 2009; Qi *et al.* 2010; Dai *et al.* 2011). Several GIS models for landslide and mudflow susceptibility after the earthquake have been proposed using similar

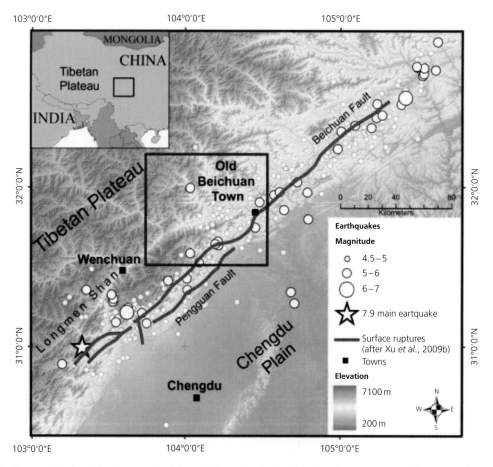

Fig. 22.10 The location of the Wenchuan earthquake and the major aftershocks along the Longmenshan fault zone. The study area is indicted by the blue box. (Adapted from Ouimet 2010 and Xu *et al.* 2009b.)

Fig. 22.11 Earthquake-destroyed Beichuan county town. Standard false colour composite images of ALOS AVNIR-2 bands 2, 3, and 4 in red, green and blue taken on (a) 31 March 2007 before and (b) 4 June 2008 after the earthquake. Massive landslides triggered by the earthquake can be seen clearly in the post-earthquake image; these have caused the burial of collapsed buildings and the blockage of rivers. (c) A field photograph of the destroyed Beichuan town (looking northwards).

sets of geological and topographical factors in the form of multiple variables in weighted linear combination (Su *et al.* 2010; Wang *et al.* 2010; Tang et al. 2011a). However, none of these models fully integrate the earthquake deformation as a controlling factor for susceptibility mapping of landslides and mudflows in this region that was seriously damaged by the Wenchuan earthquake event. We believe that modelling the earthquake-induced hazards and damage is quite different from conventional geohazards susceptibility modelling for general cases; it should be considered in line with seismic intensity zonation.

Seismic intensity is widely used to assess damage levels induced by an earthquake. Unlike the earthquake magnitude scales, which express the seismic energy released by an earthquake, seismic intensity denotes how severely an earthquake affects a specific place. The China Seismic Intensity Scale (CSIS) is a national standard and is called *Liedu* in Chinese. It was formally established by the China Earthquake Administration (CEA) in 1980 but has been in use since the 1960s. Liedu is similar to the European Macroseismic Scale 1998 (Grünthal, 1998) as well as to the Modified Mercalli intensity scale (Wood & Neumann, 1931) that categories seismic impacts into 12 degrees of intensity. Table 22.4 shows the Liedu scale of the most recent update in 2004 from the National Standard GB/T 1772-1999 (Chen *et al.* 1999). Before the earthquake, the maximum intensity expected in the Wenchuan region was VI~VIII Liedu, but the actual intensity caused by the earthquake ranged between VIII and XI; this matches field observations in Beichuan county town. The particular devastation in Beichuan town was, however, a combined result of destructive shaking and massive landslides triggered by the earthquake because the town was in a deeply incised valley surrounded by the high, steep mountain slopes (Fig. 22.11): it is extremely prone to widespread and large-scale slope failures. Our field investigations have shown that many areas along the Yingxiu-Beichuan fault were subject to the same level of earthquake shock but experienced far less significant damage than Beichuan town because their locations were not independently prone to geohazards. For effective hazard management and prevention of earthquake-induced damage, the lesson to be learnt is that the seismic intensity zonation system should be enhanced by the characterisation of localised variation of damage levels that an earthquake may produce.

In this case study, using the Beichuan area in the Wenchuan earthquake zone as an example (Fig. 22.10), a GIS-based approach to earthquake damage zone modelling using satellite remote sensing and DEM data is presented (Liu *et al.* 2012). The novelty is to take into account the co-seismic ground deformation as an important modulating factor in modelling the susceptibility of earthquake-related geohazards, together with conventional multi-criteria factors that draw on geological and topographical variables derived from broadband multispectral satellite imagery and DEM data, such as rock competence, slope, proximity to drainage and fracture density. While the differential synthetic aperture radar (SAR) interferograms derived from cross-event ALOS (Advanced Land Observation Satellite) PALSAR (Phased Array type L-band Synthetic Aperture Radar) image pairs provide broad-scale co-seismic deformation zones relating to seismic intensity, far more detailed damage levels within each deformation zone can be mapped based on local geo-environmental factors. The modulation effect of earthquake deformation greatly enhances the susceptibility in the areas where the majority of the ensuing landslides and debris-flows actually took place. When this susceptibility model is further integrated with the mapped surface destruction caused by the earthquake, the final output no longer represents, though directly links to, the conventional measure of seismic intensity, and we call it *earthquake damage*. The earthquake damage zone map represents both the current damage status as well as the future damage (hazard) potential.

22.2.2 The models

Many studies, as well as our own field investigations, indicate that slope failures (mainly landslides and debris-flows) are the dominant but not exclusive forms of geohazard triggered by the Wenchuan earthquake. In this case study, we use the word 'geohazard' as a general term to represent all the forms of catastrophic slope failure, to emphasize the hazardous aspect of this natural process of surface evolution.

Our basic principles for geohazard GIS modelling are that geohazard susceptibility cannot be quantitative at regional scale, that it has to be subjective and arbitrary at certain levels and it can never be entirely objective. Whilst a simple model seems rather crude and arbitrary, from our own experience we suspect that a complex and apparently quantitative model may never really work at all. There is thus a fine balance for the modelling: too crude and the model will fail to effectively categorise the theme, whilst too much detail may make the model too narrow and localised, thereby restricting its applicability. In addressing this balance, we have adopted a simple numerical yet qualitative model.

The GIS-based modelling of susceptibility to geohazards and the modelling of current damage levels are two separate but closely linked issues. Given various geo-environmental conditions relating to geohazards, an earthquake will intensify the overall susceptibility to those geohazards. We therefore consider the GIS

Table 22.4 Table of China Seismic Intensity Scale (CSIS) – Liedu.

Liedu (Seismic intensity)	Sensed by people on the ground	Damage degree of buildings		Other damage phenomena	Horizontal ground motion	
		Damage phenomena	Mean damage index		Peak acceleration m/s²	Peak speed m/s
I	Insensible					
II	Sensible by very few still indoor people					
III	Sensible by a few still indoor people	Slight rattle of doors and windows		Slight swing of suspended objects		
IV	Sensible by most people indoors, a few outdoors; a few people wake up from sleep	Rattle of doors and windows		Obvious swing of suspended objects; vessels rattle		
V	Commonly sensible by people indoors, sensible by most people outdoors; most wake up from sleep	Noisy vibration of doors, windows and building frames; falling of dust, small cracks in plasters, falling of some roof tiles, bricks falling from a few rooftop chimneys		Rocking or flipping of unstable objects	0.31 (0.22–0.44)	0.03 (0.02–0.04)
VI	Most unable to stand steady, a few scared to running outdoors	Damage occurs: cracks in the walls; falling of roof tiles; some rooftop chimneys crack or fall apart	0–0.10	Cracks in riverbanks and soft soil; occasional burst of sand and water from saturated sand layers; light fractures on some of brick chimneys	0.63 (0.45–0.89)	0.06 (0.05–0.09)
VII	Majority scared to running outdoors, sensible by people cycling or in moving motor vehicles	Light destruction induced: localized destruction, cracking, may continue to be used with minor repairs or without repair	0.11–0.30	Riverbank collapse; frequent burst of sand and water from saturated sand layers; many cracks in soft soils; moderate destruction of most brick chimneys	1.25 (0.90–1.77)	0.13 (0.10–0.18)
VIII	Most swing about, difficult to walk	Moderate destruction: structural destruction requiring repair to recover to usable status	0.31–0.50	Cracks appear in hard dry soils; severe destruction of most brick chimneys; trees break on canopy tops; building destruction result in casualty of people and domestic animals	2.50 (1.78–3.53)	0.25 (0.19–0.35)
IX	Moving people fall	Severe destruction: severe structural destruction with localized collapse, often beyond of repair	0.51–0.70	Many cracks in hard dry soils; fractures and dislocations may appear in bedrocks; landslides and collapses are common; brick chimneys collapse	5.00 (3.54–7.07)	0.50 (0.36–0.71)
X	Bicycle riders may fall; people in unstable state may fall away; sense of being thrown up	Majority of buildings collapse	0.71–0.90	Mountains collapse and earthquake faults appear; destruction of bridge arches founded on bedrock; brick chimneys are destroyed from foundation	10.00 (7.08–14.14)	1.00 (0.72–1.41)
XI		Nearly all the buildings collapse	0.91–1.00	Extended earthquake faults; massive landslides and mountain avalanches		
XII				Intense deformation of the land surface and drastic landscape change.		

Notes: The mean damage index is a special parameter in the Chinese Liedu system. It ranges from 0 to 1 to categorise the building damage levels from Liedu scale VI to XI. It is used for building regulation in areas subject to high seismic intensity.

Notes about qualifiers: very few, < 10%; few or some, 10– 50%; most, 50–70%; majority, 70–90%; commonly, > 90%.

Source: http://www.csi.ac.cn/ymd/flfg/fd007.htm (last access on 17 June 2011).

modelling of susceptibility to earthquake-related geohazards as a conventional geohazard susceptibility model
modulated by seismicity, hence:

$$E_{gh} = G \times S \qquad (22.2)$$

Here E_{gh} represents the susceptibility of earthquake-
related geohazard, while G and S are the geohazard susceptibility model and seismicity, respectively. The
modulation relationship between G and S represents the
fact that the two variables co-exist all the time and
intensify each other.

The *earthquake damage* zonation model should consider both geohazards already induced by the earthquake and any future geohazard potential, defined by
eqn. 22.2, and is therefore given as

$$E_{dz} = H \times E_{gh} = H \times G \times S \qquad (22.3)$$

Here E_{dz} is the earthquake damage zonation and H
mapped geohazard induced specifically by the earthquake. As shown later, the mapped geohazard variable
H is assigned values of 1 for 'none' and 2 for 'geohazard'. As such it acts like a switch via the multiplication
operation, that is, doubling the values of E_{gh} only for
existing earthquake-induced geohazards and with
values remaining unchanged for other areas where no
actual geohazards have yet occurred. A map produced
using eqn. 22.3 not only reflects the current status of
earthquake damage but also indicates the future potential of hazard damages.

We conducted extensive field investigations on the
earthquake-induced geohazards, guided by satellite
images, shortly after the earthquake (Liu & Kusky 2008;
Tang *et al.* 2009) and then in the summers of 2008 and
2009. The choice of geological/topographical variables
relating to slope instability geohazards and their relative
importance are decided on the basis of these first-hand
observations and in reference to our previous work in
the Three Gorges region (Liu *et al.* 2004; Fourniadis *et al.*
2007a, 2007b). Accordingly, the most important variables chosen are rock competence (Rc), slope (Sp),
drainage (Dr) and faults/fractures (Ff). These variables
are in common with those in all the published geohazard
susceptibility models mentioned before, but we do not
use elevation as a viable variable as others did. There
is no definite relationship between mass wasting and elevation; it all depends on slope. Indeed, high elevation
tends to allow steep slopes to develop but not necessarily,
such as on high plateau plains of young geomorphology.

The relationship between elevation and slope instability
is largely biased to the dominant elevation range of an
area, and therefore a model including elevation can only
be a local model. We thus propose a geohazard susceptibility model based on four variables, as below:

$$G = Rc \times Sp + Dr + Ff \qquad (22.4)$$

This formula involves both multiplication (intra-variable
modulation) and summation (linear combination). The
multiplication is used to modulate the key variables, slope
angle and rock competence. The relationship between
these two is a modulation because any slope must also be
composed of a lithology. The modulation between rock
competence and slope ensures a representative lithology/
slope classification but does not duplicate its effects. For
instance, the same slope angle is regarded significantly less
stable in soft rock with a competence rank of 3, than in a
hard rock with a competence rank of 1, in case of debris-
flows. Second, drainage and faults/fractures often overlap,
especially when river channels follow weak deformation
zones (along faults and joints) but they do not necessarily
co-exist or interact with one another. Their contributions
to geohazard susceptibility can be quite different. Whilst
faults/fractures provide planar discontinuities on which
rock-slides may develop, drainage channels often form
the confining pathways for debris-flows. We therefore
consider Dr and Ff as individual 'add-on' variables in
linear combination with the rock competence modulated
slopes. However, to minimise the duplication effect of Dr
and Ff when they do co-exist and to reflect both the relationships between rock competence and fault/fracture
and drainage channels, Ff and Dr are assigned quite different ranking values for rock-slides and debris-flows, as
shown later in § 22.2.4 and § 22.2.5, to characterise the
different effects when these two variables interact with
rock competence.

Replacing eqn. 22.4 in eqn. 22.2 and eqn. 22.3, we
have

$$E_{gh} = S \times \left(Rc \times Sp + Dr + Ff \right) \qquad (22.5)$$

$$E_{dz} = S \times \left(Rc \times Sp + Dr + Ff \right) \times H \qquad (22.6)$$

There are no weights specified for any of the additive
items in the above equations. As a numerical qualitative
(not quantitative) modelling process, each variable is
assigned a sequence of ranks, the range of which determines the relative contribution of the variable to the
model and so effectively acts as a weighting scheme. For
instance, the rank range of drainage (Dr) is 0-2 and that

of fault/fracture (*Ff*) is 0-4 for rock-slides, indicating that *Ff* has a higher contribution than *Dr* for mapping the susceptibility to this type of geohazard. As the relative importance of different variables is implemented via their ranking scale, it is unnecessary to further complicate the models with an extra hierarchy of weights. Deriving quantitative weights of various variables is difficult. A complex statistical approach may appear to be quantitative, but it could be misleading without consideration of basic principles and might not stand up to simple reality checks. The fate of the final result of such an approach is sealed from the start. If a variable that is in fact not directly relevant is put into the system, a statistical procedure will produce a weight for it anyway, and the weight could be significant.

The major types of geohazard, rock-slides and debris-flows, have different mechanical characteristics, and so the chosen variables will influence the model differently for each type. For instance, debris-flows often occur in rocks of low competence, and/or in unconsolidated sediments, whilst rock-slides are more likely to happen in competent lithologies that are well bedded and fractured (especially if the fractures were caused or re-activated by earthquake shock). The two types of geohazards are therefore modelled separately, using the same input variables, but with different ranking schemes. Since the relationship between the two types is one of co-existence but not overlap and is accumulative, the final earthquake-related geohazard susceptibility map is a weighted linear combination of the likelihoods of rock-slide and debris-flow occurrence. Field evidence strongly indicates that rock-slides are the more frequent geohazards triggered by the earthquake; they are of massive scale and extent and constitute a more significant threat to the inhabitant population than do debris-flows in this region. Debris-flows tend to be of a much smaller extent than rock-slides, but they are very widely spread (Tang *et al.* 2009; Chigira *et al.* 2010; Dai *et al.* 2011; Tang et al. 2011b). To reflect this basic observation, the weighting for rock-slides (E_{ghr}) is empirically assigned twice that of debris-flows (E_{ghd}):

$$E_{gh} = 2E_{ghr} + E_{ghd} \qquad (22.7)$$

The model described here cannot include all conceivable factors influencing geohazard susceptibility. Instead our idea is to keep the model simple, and therefore appropriate for regional assessment, using the most significant factors that are derivable from satellite remote sensing data without dependency on detailed field investigation, which is often logistically difficult and/or restricted. As a result, important variables that are heavily dependent on field work, such as the relationship between slope azimuth (aspect) and the dip-direction of sedimentary rock formations, have to be omitted in this case. We believe it is more reliable to avoid using this variable than to misrepresent it through a paucity of data.

22.2.3 Derivation of input variables
22.2.3.1 Imagery and DEM data
Source data for this study are from JAXA's (Japan Aerospace Exploration Agency) ALOS, Landsat TM and ASTER GDEM (Global Digital Elevation Model), as listed in Table 22.5. For the multi-criteria GIS modelling,

Table 22.5 Datasets used in this study, the dates of acquisition and sources (courtesy to all data sources).

Satellites	Sensor/Data types	Imaging dates		Data source
		Before	**After**	
ALOS	AVNIR-2: Four VNIR spectral bands (blue, green, red and nearer infrared), 10 m spatial resolution.	31/03/07	04/06/08	JAXA provided data free for emergency natural hazard response
	PALSAR: L-band (23 cm wavelength). SLC data with approximately 5 m pixel size for differential InSAR	05/03/08 17/02/08	05/06/08 19/05/08	
Landsat 5	TM: 6 VNIR and SWIR bands with 30 m spatial resolution and 1 thermal band with 120 m spatial resolution	p129r3801/05/1988 p130r3826/06/1994		GLCF, NASA / University of Maryland
Terra-1	30 m resolution Global DEM derived from ASTER stereo images (ASTER GDEM)	Multiple		NASA/JPL

all the chosen variables are derived from satellite imagery data and DEM. This ensures objective and full data coverage to the study area, not biased by accessibility to field investigation and local data.

22.2.3.2 Rock competence map

Lithology has a background control on susceptibility in that certain rock types are less resistant or are unlithified and are therefore likely to be more erodable and more easily dissected, transported and re-deposited. The mechanical properties of rocks (such as the degree of fracturing and weathering) are far more influential here than any specific rock type. We therefore consider the

physical competence of rock formations as a fundamental factor in influencing geohazard occurrence.

Searching the satellite data archives, to identify suitable images for regional lithological mapping, reveals that the best cloud-free multi-spectral images of the study area are two Landsat TM scenes (see Table 22.5). Figure 22.12 shows a colour composite of Landsat TM image bands 5, 3 and 1, displayed in red, green and blue enhanced using the direct de-correlation stretch (DDS) (§ 5.3), of the study area within the Mianyang metropolitan district. The band combination in this image is generally effective in differentiating major rock types by enhancing the rich spectral variation and sharp textual features of lithological

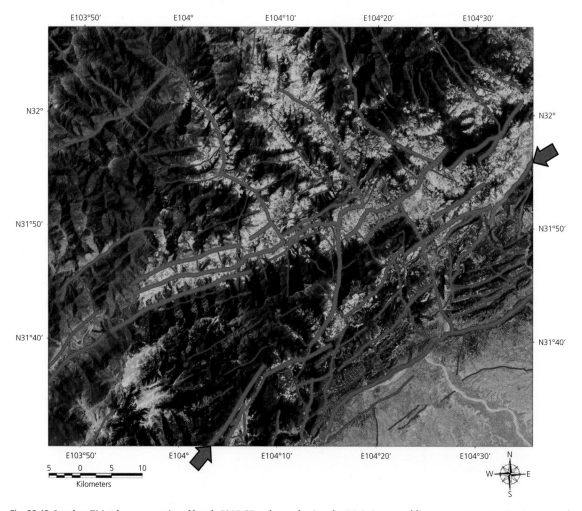

Fig. 22.12 Landsat TM colour composite of bands 531RGB enhanced using the DDS. Structural lineaments representing interpreted faults and fractures are shown as red lines; the two blue arrows outside the image frame indicate the approximate position and orientation of the Yingxiu-Beichuan fault zone.

units and their boundaries. This image was therefore used for the majority of geological interpretation. Rock competence can be quite reliably mapped from this colour composite via visual interpretation of colour, texture and landforms, in combination with field investigation carried out in the summers of 2008 and 2009.

According to field investigation and publications (e.g. Xu *et al.* 2009a), medium to thickly bedded limestone and sandstone formations, and metamorphic rocks, are dominant in the mountainous region of the study area. These rocks are usually highly competent but may become less competent when heavily fractured and weathered within and near fault zones, such as those that form valleys and residual hills around the flanks of large mountains along the Yingxiu-Beichuan fault zone. We have classified these rock units into two broad rock competence groups, hard (high competence) and relatively soft rocks (medium-low competence), rather than differentiate on the basis of lithological type and stratigraphic units. Toward the south-east of the mountain range, near the Chengdu Plain, there is a belt of narrow thrust folds characterised by inter-bedded limestone and sandstone. This belt is classified as high-medium competence. Further to the south-east, in the lower right corner of the study area, there lies the western margin of Chengdu Plain which comprises low-lying Tertiary sandstones and unconsolidated Quaternary deposits of low competence that form alluvial terraces and platforms. In summary, the study area can thus be mapped into four classes of rock competence (Fig. 22.24):

- High competence: thick layers of limestone, sandstone and metamorphic rocks;
- high-medium competence: folded, inter-bedded limestone and sandstone;
- low-medium competence: highly fractured and weathered rocks;
- low competence: Tertiary sandstone and Quaternary unconsolidated sediments

The rock competence map (Fig. 22.13) that we produced resembles the simplified geological map of the area (e.g. Tang *et al.* 2011b) in major boundaries but our map categories correspond to lithology instead of stratigraphic units.

Rock competence has varying influence on different types of geohazards and causes rocks to respond differently to earthquake shock. In general, rock-slides tend to

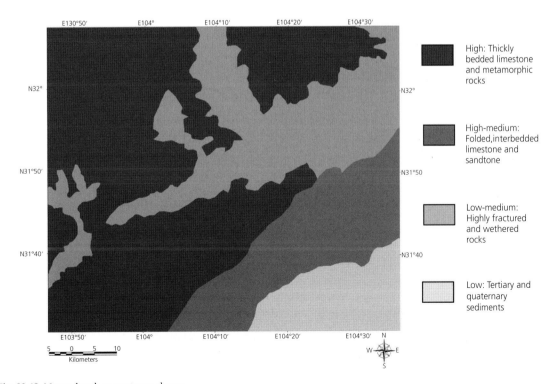

Fig. 22.13 Mapped rock competence classes.

Fig. 22.14 The massive Daguangbao rock avalanche (looking NNW direction). The head scarp (back wall in the view) is about 600 m high. The sliding body (partially shown at lower right) had a volume of about 7.5×10^8 m³ (Huang *et al.* 2011) and moved eastward, obliquely to the bedding, which dips to NW.

be more common in lithologies of medium competency and above, where failures occur along bedding planes and fractures. During periods of intense earthquake shock, massive failures (rock-slides and rock avalanches) may occur in highly competent rock formations, as in the case of the Daguangbao rock avalanche (Fig. 22.14). Conversely, slopes composed of low competence materials are more prone to debris-flows. Rock competence is therefore ranked differently for these two geohazard types, as shown in Table 22.6.

22.2.3.3 Slope map

Slope is the single most important variable controlling a mass wasting process, such as rock-slides and debris-flows. As the gradient of elevation, a slope layer can be easily derived from a DEM, using a 3×3 slope angle kernel, based on the following formula:

$$\theta = \tan^{-1} \sqrt{\left(\frac{\partial z}{\partial x}\right)^2 + \left(\frac{\partial z}{\partial y}\right)^2} \qquad (22.8)$$

Here z represents elevation of a pixel at position *(x,y)* in the DEM data.

Table 22.6 Rank values for rock-slides and debris-flows in each rock competence class.

Rock competence	Ranking	
	Rock-slide	**Debris-flow**
Low	0	3
Low-medium	1	2
High-medium	3	1
High	2	0

An image of slope angles of the study area was generated from the ASTER GDEM. It was then divided into 4 classes by thresholding: gentle [0°–10°], intermediate (10°–30°], steep (30°–40°] and very steep >40° (Fig. 22.15, Table 22.7). To reflect its relative importance in comparison with the other variables, slope is given a rank range of 0–4.5 with an increment of 1.5 based on empirical assessment of the modelling results with different boosting factors. The modulation of slope with rock competence, $Rc \times Sp$, produces values in the range 0–13.5. The maximum value for rock-slides is likely to

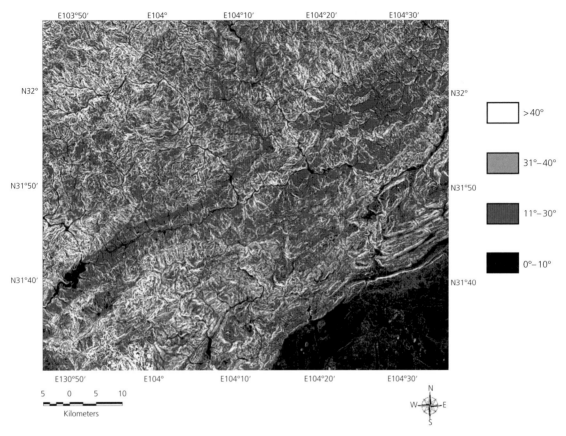

E103°50'   E104°   E104°10'   E104°20'   E104°30'

N32°   N32°

N31°50'   N31°50

N31°40'   N31°40

> 40°

31°–40°

11°–30°

0°–10°

E130°50'   E104°   E104°10'   E104°20'   E104°30'

5   0   5   10

Kilometers

N
W —+— E
S

Fig. 22.15 Map of slope classes.

Table 22.7 Rank values for slope angle classes.

Slope angle (deg)	Ranking
0–10	0
11–30	1.5
31–40	3
>40	4.5

be incurred on steep slopes composed of rocks of high to medium competency, while the effective value range for debris-flows should be much narrower since steep slopes tend not to form on soft lithologies.

22.2.3.4 Proximity to drainage

Proximity to drainage has an indirect influence on slope instability since undercutting by rivers at the toes of slopes is a common trigger of debris-flows, and the channels themselves tend to form the preferential

pathways for the debris-flows. Here the area affected by fluvial erosion is decided by the magnitude of the river channels and water flow capacity, which is in turn controlled by rainfall intensity. For instance, a small stream can swell dramatically to carry a significant amount of water and sediments during a heavy storm. The average annual precipitation in the region is about 800–900 mm and the highest recorded rainfall, in a 24-hour period, is ca 334.7 mm, and this occurred during 2008. In consideration of this, we generated a drainage network buffer map (Fig. 22.16) from the ASTER GDEM with four stream-order levels that are given different buffer widths and rank values, as listed in Table 22.8. We give higher order streams wider buffer zones and higher ranks because large river channels affect larger areas and are more likely to trigger bank collapse and landslides. According to the failure mechanism and materials involved, fluvial erosion has a more significant influence on debris-flows than rock-slides. We therefore

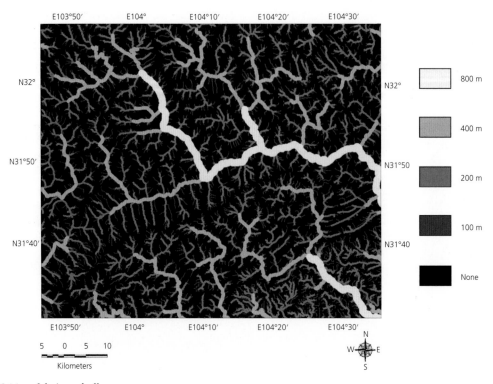

Fig. 22.16 Map of drainage buffer zones.

Table 22.8 Rank values for rock-slides and debris-flows in each drainage buffer zone.

Drainage order	Buffer width (m)	Ranking	
		Rock-slide	Debris-flow
None	–	0	0
3	100	0.5	1
4–5	200	1	2
6–7	400	1.5	3
8–9	800	2	4

assigned the debris-flow buffer zone twice the rank of that for the rock-slides (Table 22.8). As the relationship between proximity to drainage and the other variables is a linear combination, the areas outside drainage buffer zones are assigned a zero value.

Several publications specified either uniform river buffer zones (e.g. 1000 m, Chen *et al.* 2009) or buffer zones of variable width (e.g. 500–1500 m, Tang *et al.* 2011a), which are significantly wider than the buffer zones that we specified in Table 22.8 (100–800 m). In this mountainous area, river channels are largely developed in relatively competent rock formation forming deep and narrow valleys, and their direct influence to slope stability is often constrained within a narrow belt along the river. As shown in Fig. 22.11(b), in the most hazardous area around Beichuan town, the farthest edges of massive landslides are still within 1000 m from where the river flows through. The extent of a large landslide along a river should not be used as a direct measure for river buffer zone definition, because it is the combined result of several factors and in particular the slope. The direct effect of river undercutting and swelling should extend no greater than the lower half of the landslide. The varying width buffer zones that we specified for this region are therefore justified.

22.2.3.5 Fracture density
Proximity to faults and fractures (mapped as structural lineaments) also favour the occurrence of geohazards since these zones of brittle deformation provide the passage of water and facilitate weathering and weakening of the rocks. In the context of seismically triggered geohazards, the proximity of a fault is particularly

relevant if it ruptured during the earthquake. Field evidence indicates that faults and fractures have direct impact on surface damage. A location close to the epicentre of an earthquake could escape from destructive rupture if it is not in the immediate vicinity of an active fault or fracture, whilst one situated at a distance from an epicentre could experience very severe damage and destruction if it lies on an active fault.

The same TM colour composite image (Fig. 22.11) was used as the principal image for visual interpretation of structural lineaments, with maps in relevant publications (Chen & Wilson, 1996; Lin *et al.* 2009; Wang *et al.* 2009b; Xu *et al.* 2009b; Jia *et al.* 2010) used as reference to produce a lineament map (Fig. 22.12). It is important to keep the drainage network in mind during the structural lineament mapping, since many river channels will either follow the weak zones along faults or will appear as lineaments and be included in both the drainage buffer map and structural lineament map. Such areas may receive double emphasis in the earthquake-related geohazard susceptibility model, although in many cases these are indeed the most vulnerable localities to intense damage. For our image-based structural interpretation, which is a highly subjective process, we paid great attention to the characteristic features of faults and fractures rather than drawing any straight lines between vaguely aligned geomorphologic features. We have verified the major fault systems through field investigation (Kusky *et al.* 2010). Whilst having made many in-situ observations, which are often subjective as well, we believe that satellite images provide more accurate and reliable spatial information about zones representing the mechanical breakup of rocks by faults, on both regional and local scales.

Considerable uncertainty and subjectivity are unavoidable in mapping faults and fractures either from images or from field observation. Coupled with the concept that faults and fractures may deform and destabilize a much wider region than the immediate area of fault displacement, and since there will always be more faults and fractures actually present than can ever be represented at a particular map scale, it was deemed more relevant to consider the density of fracturing rather than the positions of each fracture. A fracture density map was then generated in several steps. First the lineament map was encoded such that structural lineaments of different magnitude were assigned a series of weights (Table 22.9); for instance, a major, regional fault was given a weight of 4, and

Table 22.9 Structural lineament weights.

Lineament magnitude	Weights
None	0
Local	1
Sub-major	2
Major	3
Regional	4

Table 22.10 Fracture density ranking.

Fracture density	Ranking	
	Rock-slide	Debris-flow
None	0	0
Low	1	0.5
Medium	2	1
High	3	1.5
Very high	4	2

a minor, local fault a weight of 1. Density (*D*) is calculated as a value representing the total lineament length of each rank, encountered within a circular search radius of 3 km, multiplied by the weight value for each rank and divided by the area of the search radius, as follows:

$$D = \frac{\left(\sum_n wL \right)}{A_{SR}} \qquad (22.9)$$

where *L* represents feature length, *w* indicates the weight, and A_{SR} is the area of the search radius. The resultant density value, in the range of 0–3.2 km/km², is then assigned to the pixel DN (digital number). Finally, the fracture density map was simplified into four broad density levels based on its histogram (density distribution), as shown in Table 22.10 and Fig. 22.17.

Since they often occur in competent rock formations, faults and fractures have a far greater significance in leading to the occurrence of rock-slides than of debris-flows. Many massive rock-slides were triggered by the earthquake along existing faults and fractures. The same fracture density class is therefore assigned twice the ranking value for rock-slides as for debris-flows (Table 22.9). In comparison with the drainage buffer map (Table 22.8), the ranking for the two types of geohazards is complementary, and therefore the duplicated impact of drainage and fractures is reduced.

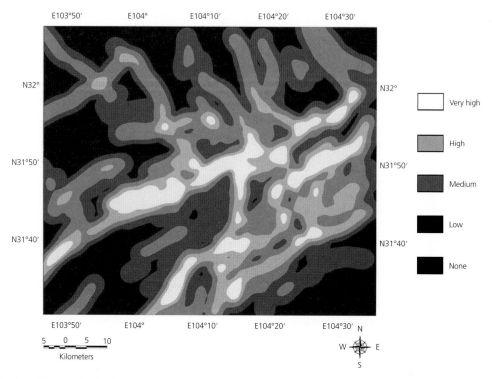

Fig. 22.17 Map of fracture density classes.

22.2.3.6 Co-seismic deformation zone

One of the most significant impacts of the Ms 8.0 Wenchuan earthquake on this region is long-term geohazard susceptibility; the strong tremors having reduced shear strengths to a minimum, opened pathways for water percolation and thus weakened the entire body of the mountain belt along the Longmenshan fault zone. New faults and fractures, as well as those re-activated by the earthquake, have destabilised many otherwise stable slopes, making the whole region more vulnerable to geohazards than ever before, and for many years to come. Geohazards become more frequent and on a greater scale; there has been frequent media coverage of massive landslides in the region since the earthquake. For instance, the Chediguan bridge on the Dujiangyan-Wenchuan motorway was destroyed by a massive rockslide and rock fall on 14 July 2009 after moderate rain.

The 2008 Wenchuan earthquake has intensified and exacerbated the problems caused by geohazards for the entire region; its impact upon our model of the susceptibility to earthquake-related geohazards can be considered directly proportional to the co-seismic deformation as a modulation of the geohazard susceptibility without

the earthquake, as indicated before. The JAXA's ALOS provided speedy coverage of PALSAR data in a short period after the main shock and then released cross-event interferometric PALSAR data to support the international research effort to study the seismic fault zone using the DInSAR technique. The PALSAR data, in combination with ASTER GDEM data, have enabled us to generate differential SAR interferograms of the region that are similar, if not identical, to the products from other research groups. The DInSAR data have been largely used for investigation of co-seismic deformation along the Longmenshan fault zone (e.g. Hashimoto *et al.* 2010) but rather little for studying earthquake-induced geohazards. As shown in Fig. 22.18(a), the co-seismic deformation is mapped by 2π wrapped interferometric fringe patterns. Each fringe represents ground displacement of a half-wavelength of the L-band PALSAR (118 mm) in the SAR look direction. In the central part of the seismic fault zone within the near vicinity of earthquake raptures, the deformation is so great and chaotic that the cross-event SAR signals have lost coherence, leaving no clear fringe patterns to quantify the deformation. The unwrapped interferogram in Fig. 22.18(b) presents

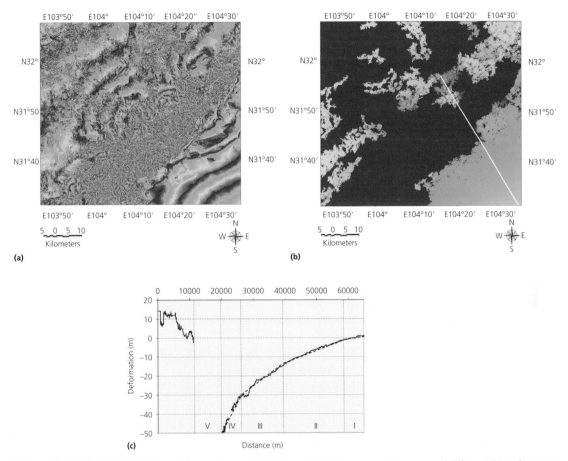

Fig. 22.18 ALOS PALSAR DInSAR data of the study area. (a) Differential interferogram; (b) unwrapped differential interferogram; and (c) deformation profiles plotted from the unwrapped interferogram along the white line in (b). In (c), the vertical red dot lines define different co-seismic deformation zones I–V from low to very high on the foot wall of the Yingxiu-Beichuan fault zone, and the oblique black dot lines indicate the average deformation slope in each zone except zone V, where deformation is not measurable because of de-coherence.

the magnitude of the deformation in the LOS direction, but the de-coherence zone is masked off because unwrapping is impossible in this zone.

In the interferogram, the density of fringes corresponds to the gradient of the unwrapped interferogram or the deformation gradient. The deformation gradient is considered a more effective measurement of seismic intensity than the deformation magnitude, although one is often associated with the other. Since the unwrapped interferogram is very noisy and contains many 'holes' caused by de-coherence and localised sudden changes, calculation of gradient on a pixel-wise basis is deemed unreliable and is often unrepresentative of regional trends. For the earthquake damage

modelling presented here, simply quantified zones of different deformation levels are adequate, while the detailed and precise deformation gradient at each data point (image pixel) is not really necessary. This zonal definition is achieved by plotting deformation profiles from the unwrapped interferogram, perpendicular to the fault zone, and then locating sections of different deformation slopes. This is illustrated in Fig. 22.18(c), where the vertical red dashed lines define different co-seismic deformation zones I–V, from low to very high, on the foot-wall (to the SE to the fault trace) of the Yingxiu-Beichuan fault zone, and the oblique black dashed lines indicate the average deformation slope in each zone, except zone V, where deformation is not measurable because of de-coherence. Obviously, the deformation slope increases and becomes progressively more significant from low to high deformation zones.

Using the deformation slopes of the deformation profiles as control and in combination with the observed fringe densities in the interferogram [Fig. 22.18(a)], a co-seismic deformation map with five deformation levels was generated (Fig. 22.19). The highest deformation level is assigned to the de-coherence zone, although the actual deformation here is unknown. Careful observation of the interferogram reveals faintly detectable fringes within the de-coherence zone, and these are significantly denser than the sparse patterns outside the de-coherence zone. Ranking of this highest deformation class is therefore assigned two levels higher than the class immediately below it (Table 22.11). The ranks given here are locally relative; even the 'Low' class represents considerable co-seismic deformation.

The SE boundary between deformation zone V (very high) and IV (high) in Fig. 22.19 roughly follows the Yingxiu-Beichuan fault, which is a right lateral thrust dipping NW. Thus all areas of very high deformation and most areas of high deformation are on the hanging

wall of the fault. By incorporating co-seismic deformation, our models can better reflect the fact that the earthquake-induced geohazards in this region are concentrated in the hanging wall of Yingxiu-Beichuan fault where the ground shaking was particularly strong (Chigira *et al.* 2010; Ouimet 2010; Dai *et al.* 2011).

22.2.3.7 Mapping the earthquake-induced geohazards

In this heavily vegetated region, the earthquake-induced geohazards are typically characterised by broken ground that is stripped of vegetation in the AVNIR-2 images taken shortly after the earthquake, as illustrated in

Table 22.11 Rank values for classes of co-seismic deformation.

Co-seismic deformation class	Ranking
Low	1
Low-medium	2
High-medium	3
High	4
Very high	6

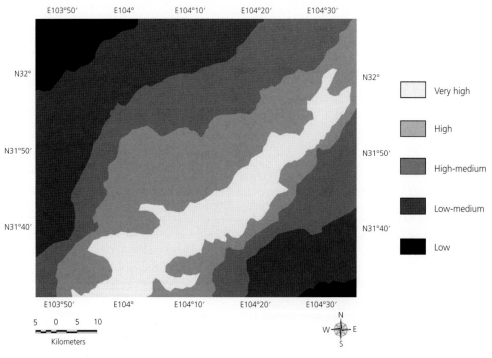

Fig. 22.19 Co-seismic deformation zones.

Fig. 22.11. Such areas can be quite accurately mapped using NDVI (Di *et al.* 2010; Zhang *et al.* 2010). We used an inverse NDVI (INDVI) with a threshold (*T*):

$$INDVI = \frac{R - NIR}{R + NIR} > T \qquad (22.10)$$

$$\text{For ALOS AVNIR} - 2 \text{ image}: \quad INDVI_{ALOS} = \frac{B3 - B4}{B3 + B4} > T \qquad (22.11)$$

where *R* represents the image of the red spectral band and *NIR* the near-infrared; *B3* and *B4* represent ALOS AVNIR-2 band 3 and band 4.

Because of the cloudy and rainy weather conditions during the period immediately after the Wenchuan earthquake, the AVNIR-2 image that we obtained is degraded by haze and partially covered by clouds. The red and NIR bands are, however, less affected by haze, and the signatures of geohazards can often penetrate the haze and thin clouds to be picked up by the INDVI. Thick clouds completely block out any ground information and can be wrongly identified as geohazards using the INDVI because of similar spectral shape, with DNs in band 4 of the NIR spectral range being lower than those in band 3 of the red spectral range (Fig. 22.20). The shadows of thick clouds can also be misidentified as geohazards for the same reason (Fig. 22.20). A unique spectral property of cloud is its high reflectance in all visible bands; in particular, the DNs of cloud in band 1 (Blue) are very high and significantly higher than those of other ground objects. A threshold of 220 for band 1 is chosen empirically to

eliminate pixels with DN > 220 as clouds. Broken ground caused by geohazards has much higher reflectance than cloud shadow in all the AVNIR-2 bands, whilst vegetation in topographic shade has very similar low reflectance to that of cloud shadow areas in the visible bands but much higher reflectance in band 4 (Fig. 22.20). A threshold of 25 for band 4 is therefore chosen empirically to eliminate pixels with DN < 25 as cloud shadows while retaining vegetated slopes that are in topographic shade. The whole processing procedure is then as follows:

$$\text{If } B1 > 220 \text{ then } NULL \text{ else}$$
$$\text{If } B4 < 25 \text{ then } NULL \text{ else}$$
$$\text{If } (B3 - B4)/(B3 + B4) > 0.2 \text{ then } (B3 - B4)/(B3 + B4) \text{ else } NULL$$

$$(22.12)$$

With the threshold of *INDVI* > 0.2, this simple approach effectively extracts nearly all broken ground caused by geohazards (the red layer of Fig. 22.21); it also largely prevents barren fields from being mistaken as geohazards (Fig. 22.20). Showing similar spectral properties to local soils and rocks, the urban areas in this region can be mis-identified as geohazards. It so happens, however, that in much of this study area, urban areas were indeed severely damaged by the earthquake, and so the classification as geohazard is valid in some areas but not in the vast Chengdu Plain. After processing, there are still some thin clouds, at the edges of thick cloud, that cannot be completely removed and are incorrectly mapped as geohazards. Figure 22.21 is a merged illustration of the mapped

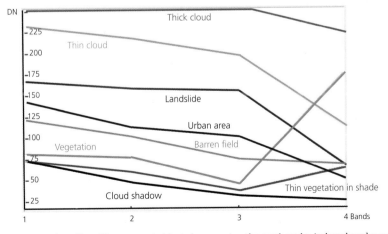

Fig. 22.20 AVNIR-2 image spectral profiles of key ground objects for mapping the earthquake-induced geohazards.

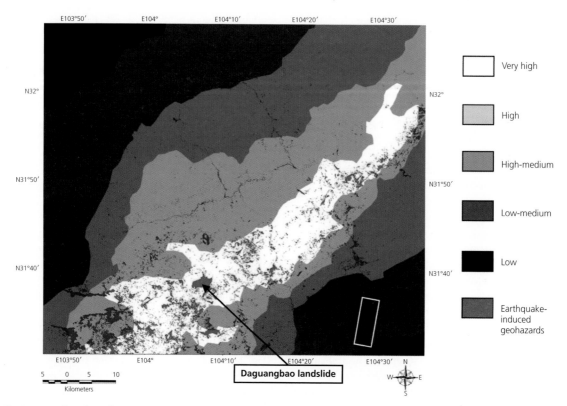

Fig. 22.21 Earthquake-induced geohazards (in red) overlain on the co-seismic deformation zones. The white box in the bottom right indicates misidentification of a patch of thin clouds as geohazards. The largest patch of mapped geohazards is Daguangbao landslide, indicated by the arrow.

geohazards overlain on the co-seismic deformation zone map. The box in the bottom right corner of the image in Fig. 22.21 indicates the largest patch representing such a misclassification. It becomes clear that, in general, large-scale landslides are concentrated in the very high deformation zone on the hanging wall of the fault regardless of the other variables.

The mapped geohazards are assigned a value of 2, with 1 representing the background when applied to the earthquake damage zone model defined by eqn. 22.6. As a modulation factor, the value 2 doubles the susceptibility scores of the pixels representing geohazards, whilst the value 1 keeps the scores of non-geohazard pixels unchanged in the final map of earthquake damage zones. In this way, the correctly identified pixels of geohazards in Fig. 22.21 will be enhanced in the model defined by eqn. 22.6, while the pixels that are misidentified as geohazards will be suppressed and degraded because of their low scores of other variables of geohazard susceptibility.

Cloudy and highly hazy weather conditions are not favourable for multi-spectral remote sensing, but such conditions are common in many regions of the world, and especially in the Wenchuan earthquake zone of Sichuan Province, China. For rapid response to an earthquake disaster, the most valuable image data acquisition period is just shortly after the event whilst the weather condition is often poor. The simple image processing approach that we present here enables the effective extraction of crucial information of earthquake-induced geohazards from images degraded by haze and clouds and therefore is widely applicable in the rapid response to natural disasters.

22.2.4 Earthquake damage zone modelling and assessment

With all the input variables prepared, the earthquake damage zone modelling comprises two steps as defined in eqns. 22.5 and 22.6: susceptibility mapping of earthquake-related geohazards forms the first step, using all the variables except the layer of mapped earthquake-induced geohazards that is applied in the second step to produce the final map of earthquake damage zones.

22.2.4.1 Susceptibility mapping of earthquake-related geohazards

Applying different variable rankings for rock-slides and debris-flows, as described earlier, to eqn. 22.5, the corresponding susceptibility maps were derived as shown in Fig. 22.22. The initial value range for rock-slide susceptibility is [0–117] and that for debris-flow [0–90]. The slightly wider value range of rock-slide susceptibility is caused by the fact that rocks of high and high-medium competence can form very steep slopes (see Tables 22.6 and 22.7) and therefore the score of high susceptibility for rock-slides should be higher than that for debris-flows. These value ranges were then subdivided into four susceptibility levels via multi-level thresholds based on the natural break points in the image histograms: corresponding to classes of low, medium, high and very high susceptibility to geohazards. From these two maps, we can observe:

- The spatial distribution of the classes of both geohazard types is strongly controlled by co-seismic deformation in the first order.
- The high to very high susceptibility classes have quite different distributions in the two maps, with strong

control exerted by the lithology-slope modulation. In the case of rock-slides, the high to very high susceptibility classes occur mainly in rock formations of high-medium and high competency, whilst for debris-flows, the high to very high susceptibility classes occur dominantly in lithologies of low-medium competency, and in highly fractured and weathered rocks.

- The differences in distribution of the medium to low susceptibility classes between the two types of geohazards is controlled mainly by fracture density in the case of rock-slides and by drainage buffer zones for debris-flows.

The two layers were then combined based on eqn. 22.7. The modelling process results in a value range of [0–297]; this value range was classified into four susceptibility levels in the same way, and the final susceptibility map of the earthquake-related geohazards was then produced [Fig. 22.23(a)]. This map shows that the very high susceptibility class forms a belt along the Yingxiu-Beichuan fault zone and is coincident with the very high co-seismic deformation zone on the hanging wall of the fault. For comparison, the susceptibility map of geohazards without the modulation by co-seismic deformation using the model eqn. 22.4 is shown in Fig. 22.23(b), and this implies that the mountain areas of medium competency rocks and the banks of major rivers would be subject to the highest geohazard susceptibility even

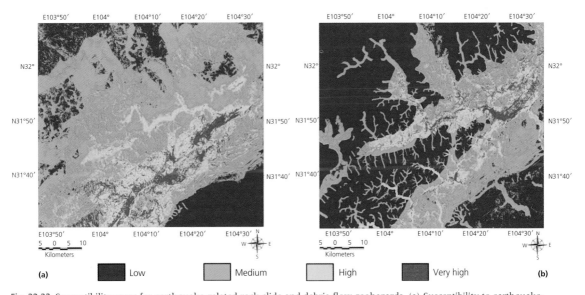

Fig. 22.22 Susceptibility maps for earthquake-related rock-slide and debris-flow geohazards. (a) Susceptibility to earthquake-related rock-slide geohazards; and (b) susceptibility to earthquake-related debris-flow geohazards.

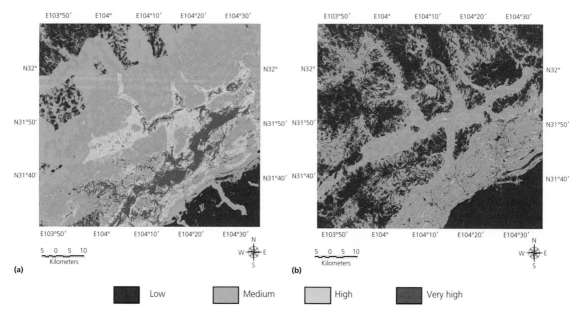

Fig. 22.23 Final susceptibility map of earthquake-related geohazards in comparison with a map derived from the geohazard susceptibility model without modulation of co-seismic deformation. (a) Susceptibility map of earthquake-related geohazards, including both rock-slides and debris-flows, produced using eqn. 22.7; and (b) susceptibility map of geohazards produced using eqn. 22.4, without earthquake impact.

without the occurrence of an earthquake. In the model defined by eqn. 22.5, the earthquake event imposes a strong impact on the result. High mountains characterised by competent rock formations, which would not otherwise be particularly susceptible to geohazards, become highly vulnerable during and after the earthquake. This agrees with reality, as discussed in the following section.

22.2.4.2 Mapping the earthquake damage zones

An earthquake damage zone map should present both the current damage caused by the earthquake (the seismic intensity) and the future potential damage from geohazards under the changed geo-environmental conditions caused by a major earthquake event. For instance, the massive Wenchuan earthquake and its aftershocks fractured otherwise stable mountain formations rendering this high relief region susceptible to frequent major rock-slides and debris-flows for many years to come. We therefore proposed to model earthquake damage zones by modulating the geohazard susceptibility with the mapped earthquake-induced geohazards as defined in eqn. 22.6. The final map of earthquake

damage zones is shown in Fig. 22.24. Considering that this *earthquake damage* zone map is different from the earthquake-related geohazard susceptibility map [Fig. 22.23(a)] only in areas of existing, mapped geohazards where the score is doubled, the same thresholds were applied to classify the earthquake damage zone into four levels: low, medium, high and very high. This simple approach assures that the mapped geohazards in the areas of highest susceptibility level are saturated at the same level, whereas those at lower susceptibility levels are raised to higher levels.

The two maps [Fig. 22.23(a) and Fig. 22.24] are very similar because most damage caused by the earthquake does indeed occur in the areas of very high susceptibility to earthquake-related geohazards. The map of earthquake damage zones in Fig. 22.24 clearly illustrates that the very high damage zones form a belt along the Yingxiu-Beichuan fault, which is subject to very high co-seismic deformation, and it is characterised by rock formations of high-medium and high competency. The destroyed Beichuan county town and most of the large landslides triggered by the Wenchuan earthquake are concentrated in this belt, and rock-sliding is the dominant failure mechanism. The probability of geohazard

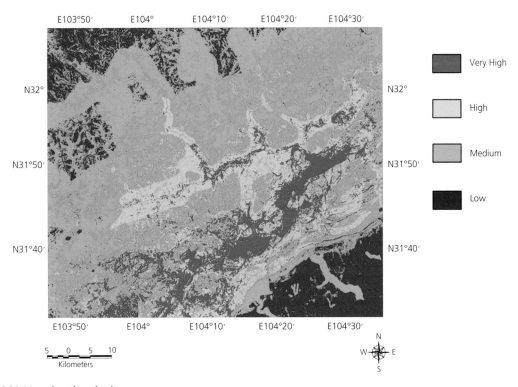

Fig. 22.24 Map of earthquake damage zones.

occurrence across the region has been significantly raised by the Wenchuan earthquake event.

The China Earthquake Administration produced a new Liedu map of the region shortly after the earthquake, as shown in Fig. 22.25. From Chengdu to the most intensively damaged zone, along the Longmenshan fault, the Liedu is ranked from VII to XI. The chosen study area of Beichuan region lies in zones of Liedu VIII to XI. With the western margin of Chengdu Plain as a relatively low damage class and Beichuan county town as a very high damage class in Fig. 22.24, our four mapped classes roughly correspond to the Liedu values VIII, IX, X and XI. The Liedu map in Fig. 22.25 is significantly lacking in detail, although it is effective in guiding regional engineering standards for earthquake resiliency of buildings and infrastructure. In comparison, the earthquake damage zone map that we produced, in Fig. 22.24, presents far more detail of earthquake damage levels, which can be used to support regional and local management for recovery from the earthquake disaster and to rebuild the region with enhanced measures to protect from future geohazards.

22.2.4.3 Assessment

For verification, the mapped geohazards in Fig. 22.21 were coded with the colours of their corresponding susceptibility levels, as defined in Fig. 22.23(a), using simple image visualization techniques. The resulting image in Fig. 22.26 demonstrates that most mapped geohazards occur in areas of very high and high susceptibility, as shown in red and yellow colours. The main discrepancies lie in the lower left and the lower right corners, where mapped geohazards are of medium and low susceptibility, shown in blue and cyan. These are caused by the incorrect identification of thin clouds as being geohazards, as indicated before. These discrepancies have little impact on the final map of earthquake damage zones (Fig. 22.24) because the model is robust in its description of the overall zonation of the earthquake damage based on all the variables, and thus the incorrectly mapped geohazards in areas of low susceptibility cannot be raised to high levels. For instance, the area of thin clouds that was misidentified as geohazards, illustrated by a box in Fig. 22.21, is finally classified as low damage level in Fig. 22.24. This indeed

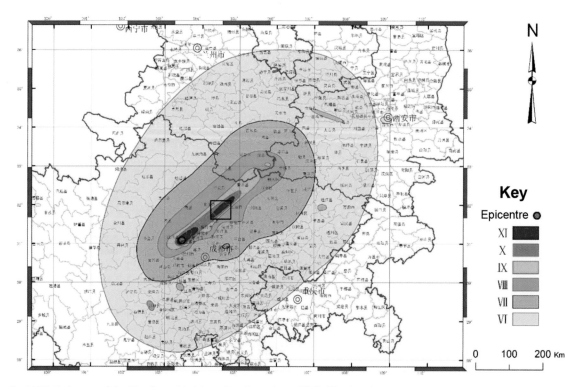

Fig. 22.25 Liedu map of the Wenchuan Ms 8.0 earthquake region published by the China Earthquake Administration in 2008 after the earthquake. The inserted box indicates the study area (http://www.cea.gov.cn/manage/html/8a8587881632fa5c0116674a0183 00cf/_content/08_09/01/1220238314350.html).

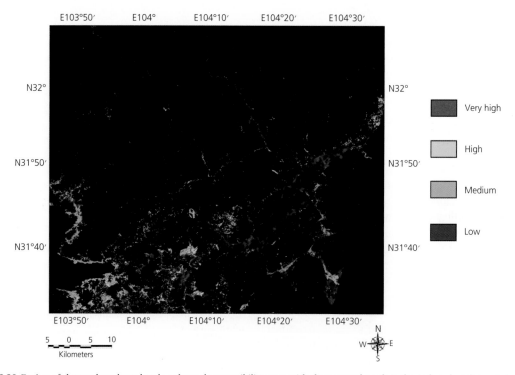

Fig. 22.26 Fusion of the earthquake-related geohazard susceptibility map with the mapped earthquake-induced geohazards for verification.

Table 22.12 Mapped earthquake-induced geohazard pixels at different levels of susceptibility of earthquake-related geohazards.

Total number of pixels	Number of geohazard pixels	Geohazard (Ghzrd) pixels per susceptibility level			
		Very High (VH)	High (H)	Medium (M)	Low (L)
55598057	4027523	1062761	1228957	1198205	537600
Percentage	7.24%	26.39%	30.51%	29.75%	13.35%
	Ghzrd/Total	VH/Ghzrd	H/Ghzrd	M/Ghzrd	L/Ghzrd

demonstrates the advantages of multi-variable modelling, showing it to be statistically correct and reliable.

Whilst a quantitative accuracy assessment of the earthquake damage zone map is impossible, calculation of the percentages of pixels of earthquake-induced geohazards in different damage levels, from the image in Fig. 22.26, presents a useful evaluation of the performance of the model. As shown in Table 22.12, 7.24% of the pixels of the study area are mapped as earthquake-induced geohazards, among which ~57% are of very high and high susceptibility levels, ~30% of medium and 13.35% of low levels. The mapped geohazards pixels of low susceptibility are in dark blue in Fig. 22.26, and these are mainly caused by incorrect identification of thin clouds as earthquake-induced geohazards, although a few pixels of this type (in the Chengdu Plain) do represent geohazards. In this context, we can estimate that the mapping inaccuracy of earthquake-induced geohazards is no worse than 13.35%. While the earthquake damage model eqn. 22.6 effectively eliminates error pixels of the mapped earthquake-induced geohazards, certainly much more than 87% (could be up to 95%) of correctly mapped geohazard pixels lie in medium to very high levels of susceptibility of earthquake-related geohazards, indicating that the susceptibility model of earthquake-related geohazards eqn. 22.5 is highly effective.

It is worth noting that the well-known and largest slope failure (a rock avalanche), the Daguangbao landslide (see Figs. 22.14 and 22.21), lies in medium to high susceptibility levels (cyan and yellow in Fig. 22.26). This suggests that this location was not the most susceptible to geohazards, even during a strong earthquake. However, when the mapped earthquake-induced geohazards layer in Fig. 22.21 is applied to modulate the susceptibility of earthquake-related geohazards model eqn. 22.5 in the earthquake damage model eqn. 22.6, areas of this type are boosted to higher damage levels in

the final map of earthquake damage zones as the scores are doubled, while on the other hand, the areas of incorrect classification of earthquake-induced geohazards cannot be promoted because of their very low scores in the model eqn. 22.5, as shown in Fig. 22.24. Analysis of Figs. 22.23(a), 22.24 and 22.26 indeed reveals the areas of massive landslides that are not in the highest susceptibility level. These massive landslides are most likely to link to deep-seated processes according to the study so far; for instance, pre-existing deep fractures may have played an important role in the genesis of the Daguangbao rock avalanche (Xu et al. 2009a; Chigira et al. 2010; Dai et al. 2011; Huang et al. 2011). This implies that while our susceptibility model for earthquake-related geohazards is effective for mass wasting of shallow processes, it is inadequate to predict the deep-seated massive slope failures even though the final earthquake damage model elevated these areas to the highest damage level. Specifying geological factors that characterise the deep-seated slope failures and incorporating these factors into the current models will be focal points for further research.

22.2.5 Summary
In this application example, we have presented a multi-variable modelling procedure for the mapping of susceptibility to earthquake-related geohazards and of earthquake damage zones, based on the case study of the Beichuan area where extreme devastation and destruction were triggered by the Ms 8.0 Wenchuan earthquake event.

The proposed GIS model for earthquake damage zone mapping is characterised by a combination of geohazard susceptibility factors with co-seismic deformation and mapped earthquake-induced geohazards. In developing this model, we step away from the conventional descriptive approach of seismic intensity, by incorporating measurable variables of co-seismic deformation and geological/topographical conditions to characterise not only the destruction immediately after an earthquake but also the future potential for damage as a consequence of the earthquake. The result from this model no longer strictly refers to seismic intensity, though it is closely related, and we call it *earthquake damage*.

The study demonstrates that under the circumstance of a strong earthquake, co-seismic deformation takes the first-order regional control in causing earthquake-induced geohazards and susceptibility to future geohazards, whilst other variables can be used effectively to categorise local hazard levels. The final map of earthquake damage zones fairly accurately reflects the ground reality and matches the geohazards inventory. The importance of the map is not only to show what has already happened but also to indicate where geohazards are most likely to happen in future, as a long-term impact of the Wenchuan earthquake in the Beichuan region. It can therefore be used as guidance for engineering measures in the re-development of infrastructure and for regional, long-term, post-disaster management.

22.2.6 References

Chen, D., Shi, Z., Xu, Z., Gao, G., Nian, J., Xiao, C. & Feng, Y. (2008) China Seismic Intensity Scale (in Chinese). http://www.csi.ac.cn/ymd/flfg/fd007.htm [accessed on 10 June 2011].

Chen, J. Li, J., Qin, X., Dong, Q. & Sun, Y. (2009) RS and GIS-based statistical analysis of secondary geological disasters after the 2008 Wenchuan earthquake. *Acta Geologica Sinica*, 83, 776–85.

Chen, S.F. & Wilson, C.J.L. (1996) Emplacement of the Longmen Shan Thrust-Nappe belt along the eastern margin of the Tibetan plateau. *Journal of Structural Geology*, 18, 413–30.

Chigira, M., Wu, X. Inokuchi, T. & Wang, G. (2010) Landslides induced by the 2008 Wenchuan earthquake, Sichuan, China. *Geomorphology*, 118, 225–238.

Cui, P., Chen, X., Zhu, Y., Su, F., Wei, F., Han, Y., Liu, H. & Zhuang, J. (2011) The Wenchuan earthquake (May 12, 2008), Sichuan Province, China, and resulting geohazards. *Natural Hazards*, 56, 19–36.

Dai, F.C., Xu, C., Yao, X., Xu, L., Tu, X.B. & Gong, Q.M. (2011) Spatial distribution of landslides triggered by the 2008 Ms 8.0 Wenchuan earthquake, China. *Journal of Asian Earth Sciences*, 40, 883–95.

Di, B., Zeng, H., Zhang, M., Ustin, S.L., Tang, Y., Wang, Z., Chen, N. & Zhang, B. (2010) Quantifying the spatial distribution of soil mass wasting processes after the 2008 earthquake in Wenchuan, China. A case study of the Longmenshan area. *Remote Sensing of Environment*, 114, 761–71.

Fourniadis, I.G., Liu, J.G. & Mason, P.J. (2007a) Landslide hazard assessment in the Three Gorges area, China, using ASTER imagery: Wushan-Badong. *Geomorphology*, 84, 126–44.

Fourniadis, I.G., Liu, J.G. & Mason, P.J. (2007b) Regional assessment of landslide impact in the Three Gorges area, China, using ASTER data: Wushan-Zigui. *Landslides*, 4, 267–78.

Godard, V., Lavé, J., Carcaillet, J., Cattin, R., Bourlès, D. & Zhu, J. (2010) Spatial distribution of denudation in Eastern Tibet and regressive erosion of plateau margins. *Tectonophysics*, 491, 253–74.

Grünthal, G., ed. (1998) European Macroseismic Scale 1998 (EMS-98). Cahiers du Centre Européen de Géodynamique et de Séismologie 15, Centre Européen de Géodynamique et de Séismologie, Luxembourg.

Hashimoto, M., Enomoto, M. & Fukushima, Y. (2010) Coseismic deformation from the 2008 Wenchuan, China, earthquake derived from ALOS/PALSAR images. *Tectonophysics*, 491, 59–71.

Huang, R., Pei, X., Fan, X., Zhang, W., Li, S. & Li, B. (2011) The characteristics and failure mechanism of the largest landslide triggered by the Wenchuan earthquake, May 12, 2008, China. *Landslides*, 9, 131–42.

Jia, D., Li, Y., Lin, A. & Wang, M. (2010) Structural model of 2008 Mw 7.9 Wenchuan earthquake in the rejuvenated Longmenshan thrust belt, China. *Tectonophysics*, 491, 174–84.

Kusky, M.T., Ghulam, A., Wang, L., Liu, J.G., Li, Z. & Chen, X. (2010) Focusing seismic energy along faults through time-variable rupture modes: Wenchuan earthquake, China. *Journal of Earth Science*, 21, 910–22.

Lin, A.M., Ren, Z.K., Jia, D. & Wu, X.J. (2009) Co-seismic thrusting rupture and slip distribution produced by the 2008 M-w 7.9 Wenchuan earthquake, China. *Tectonophysics*, 471, 203–15.

Liu, J.G. & Kusky, T.M. (2008) After the quake: A first hand report on an international field excursion to investigate the aftermath of China earthquake. *Earth Magazine*, October, 18–21.

Liu, J.G., Mason, P.J., Clerici, N., Chen, S., Davis, A., Miao, F., Deng, H. & Liang, L. (2004) Landslide hazard assessment in the Three Gorges area of the Yangtze River using ASTER imagery: Zigui-Badong. *Geomorphology*, 61, 171–87.

Liu, J.G., Mason, P.J., Yu, E., Wu, M.C., Tang, C., Huang, R. & Liu, H. (2012) GIS modelling of earthquake damage zones using satellite remote sensing and DEM data. *Geomorphology*, 139–40, 518–35.

Ouimet, W.B. (2010) Landslides associated with the May 12, 2008 Wenchuan earthquake: Implications for the erosion and tectonic evolution of the Longmen Shan. *Tectonophysics*, 491, 244–52.

Qi, S., Xu, Q., Lan, H., Zhang, B. & Liu, J. (2010) Spatial distribution analysis of landslides triggered by 2008.5.12 Wenchuan Earthquake, China. *Engineering Geology*, 116, 95–108.

Su, F., Cui, P., Zhang, J. & Xiang, L. (2010) Susceptibility assessment of landslides caused by the Wenchuan earthquake using a logistic regression model. *Journal of Mountain Science*, 7, 234–45.

Tang, C., Zhu, J. & Liang, J. (2009) Emergency assessment of seismic landslide susceptibility: A case study of the 2008 Wenchuan earthquake affected area. *Earthquake Engineering and Engineering Vibration*, 8, 207–17.

Tang, C., Zhu, J. & Qi, X. (2011a) Landslide hazard assessment of the 2008 Wenchuan earthquake: A case study in Beichuan area. *Canadian Geotechnical Journal*, 48, 128–45.

Tang, C., Zhu, J., Qi, X. & Ding, J. (2011b) Landslides induced by the Wenchuan earthquake and the subsequent strong rainfall event: A case study in the Beichuan area of China. *Engineering Geology*, 122, 22–33.

Wang, F.W., Cheng, Q.G., Highland, L., Miyajima, M., Wang, H.B. & Yan, C.G. (2009a) Preliminary investigation of some large landslides triggered by the 2008 Wenchuan earthquake, Sichuan Province, China. *Landslides*, 6, 47–54.

Wang, M., Qiao, J. & He, S. (2010) GIS-based earthquake-triggered landslide hazard zoning using contributing weight model. *Journal of Mountain Science*, 7, 339–52.

Wang, Q.C., Chen, Z.L. & Zheng, S.H. (2009b) Spatial segmentation characteristic of focal mechanism of aftershock sequence of Wenchuan earthquake. *Chinese Science Bulletin*, 54, 2263–70.

Wood, H.O. & Neumann, F. (1931) Modified Mercalli intensity scale of 1931. *Bulletin of the Seismological Society of America*, 21, 277–83.

Xie, F.R., Wang, Z.M., Du, Y. & Zhang, X.L. (2009) Preliminary observations of the faulting and damage pattern of Ms 8.0 Wenchuan, China, earthquake. *Professional Geologist*, 46, 3–6.

Xu, Q., Pei, X. & Huang, R., eds. (2009a) *Research on Mega Scale Landslides Triggered by Wenchuan Earthquake*. Publication House of Science, Beijing (in Chinese).

Xu, X., Wen, X., Yu, G., Chen, G., Klinger, Y., Hubbard J. & Shaw, J. (2009b) Coseismic reverse- and oblique-slip surface faulting generated by the 2008 Mw 7.9 Wenchuan earthquake, China. *Geology*, 37, 515–8.

Zhang, W., Lin, J., Peng, J. & Lu, Q. (2010) Estimating Wenchuan earthquake induced landslides based on remote sensing. *International Journal of Remote Sensing*, 31, 3495–508.

22.3 Predicting landslides using fuzzy geohazard mapping: An example from Piemonte, north-west Italy

22.3.1 Introduction

This section of the chapter describes the use of multi-form, digital image data within a GIS-based multi-criteria analysis model, for regional assessment of risk concerning slope instability (Mason *et al.* 1996; Mason *et al.* 2000; Mason & Rosenbaum 2002). Landslides are documented throughout the Piemonte region of NW Italy, but they are a significant problem in the area known as the 'Langhe', a range of hills southeast of Turin (Fig. 22.27), where slope instabilities have been experienced over a prolonged period. Exceptionally severe storm events pass across southern Europe quite regularly. One in November 1994 produced extensive flooding and widespread mass movements, leading to many fatalities and consequential damage to property.

Response to all natural disasters demands an assessment of the hazard and some prediction of the likelihood of future such events. Recent research into the understanding and quantification of the problem in Piemonte has also included the creation of a persistent scatterer system for monitoring slope movements using InSAR (PSInSAR) (Meisina *et al.* 2008a and 2008b). The regional agency for environmental protection (*Agenzia Regionale per la Protezione Ambientale*, or ARPA Piemonte) presented the latest state of the Piemonte landslide inventory (*SIstema Informativo FRAne in Piemonte*, or SIFRAP) at the World Landslide Forum in 2011. SIFRAP forms a region-wide framework for the monitoring and management of hydrogeological instabilities. Its production began in the 1970s through the work of the Geological Survey of Piemonte and forms a branch of the Italian landslides inventory (*Inventario dei Fenomeni Franosi in Italia*, or IFFI). SIFRAP was first released in 2005, and it concerns flooding, landslides and geomechanics. In 2005 the number of single phenomena in Italy was placed at 490,000, of which some 35,000 are in Piemonte. SIFRAP incorporates air-photo interpretations and a GIS-based geological database and since 2009 also includes PSInSAR results (from ERS and ENVISAT data) as part of its ongoing monitoring system. A total of 35 new landslides have been identified this way; a further 426 'anomalous' areas have since been interpreted as landslides. About 300 landslides are now monitored by ARPA Piemonte using inclinometers, piezometers and extensometers, many of which were installed in response to the 1994 event. More complex instruments have also been installed at a few sites with threatened towns or important structures over the last 10 years or so.

In cases where landslides occur across a sizeable region, remote sensing may be the only readily and rapidly available source of information concerning the terrain, particularly in the aftermath of a devastating event or where erosion could rapidly remove the evidence. Remote sensing gives us information about the surface morphology (arcuate scarp, hummocky ground, tension cracks and disrupted drainage), changes in vegetation as a result of increased water content, and soils that contain a lot of water (i.e. are poorly drained). The remotely sensed

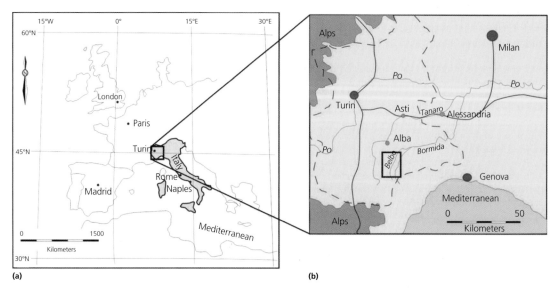

Fig. 22.27 (a) Map of western Europe, showing the location of Piemonte; and (b) map of the Langhe area surrounding the Belbo and Bormida valleys in Piemonte, NW Italy (the dashed line indicates the Piemonte regional administrative boundary, and the small bold box indicates the study area).

information is supported by field measurements using reflectance spectroscopy and X-ray diffraction (XRD), which provides direct information concerning the local soil mineralogy. Oxidised iron is an indicator of intense leaching and weathering of iron-bearing rocks and gives a very distinctive signature in remotely sensed imagery as well as colour to soils. Intensely weathered and fractured rocks are zones of inherent weakness that may indicate locations where mass movement is likely to be initiated. Clays and iron oxides have therefore been employed as the two main soil mineralogical targets within this investigation, to establish their association with landslide occurrence with a view to establishing their utility as geohazard indicators for mass movement on a regional scale.

An important aspect of this GIS-based study was to identify the temporal and spatial distribution of areas liable to movement, including the location of potential slip surfaces. This study considered the geomorphological and mineralogical expressions of mass movements, in addition to some engineering considerations, with a view to producing the geohazard assessment.

22.3.2 The study area

The Langhe Hills of Piemonte (Fig. 22.27) lie on the flank of the southern-most arc of the western Alps, on the margins of the plain of the River Po, and comprise a series of gently dipping (7°–12°) Miocene sediments of

Oligocene (Aquitanian) age (ca 26 Ma). The gently north-westerly dipping strata produce a series of NE-SW trending asymmetric valleys with south-east facing, gently dipping slopes and north-west facing steep scarp slopes.

Fine-grained, smectite-bearing argillaceous rocks, such as claystone, mudstone, siltstone and shale, dominate this region and usually occur as alternating sequences of porous sandstones and impermeable mudrocks. Stratigraphy of this nature is particularly prone to differential weathering, swelling and erosion.

The area has also been isostatically active since glacial times, and the geomorphology of the upper Langhe river basins suggests that the area has undergone significant Quaternary uplift and rotation (Biancotti 1981; Embleton 1984). This has caused a marked change in drainage characteristics, including river capture away from the Cuneo plain north-east into the Alessandria plain (which is several hundred metres lower).

22.3.2.1 History of slope instability

Major slope movements in the area have been documented over the last hundred years or so, by Sacco (1903), Boni (1941), Cortemilia and Terranova (1969), Govi (1974) and Tropeano (1989). Govi and Sorzana (1982) were the first to draw attention to the similarities between the various landslides in the region, noting that a close relationship existed between the timing of the landslides and the period of antecedent rainfall. They also observed that some failures occurred on slopes that had been affected by similar failures in the past. They also inferred that human action, such as the construction of road cuts, terraces, alteration of natural drainage systems and dumping of waste into fissures, can be significant factors for initiating slope instability.

One interesting aspect to this case is that the Langhe experiences very heavy rain and some flooding each winter, yet the literature suggests a periodicity to major landslide events (with past events in 1941, September 1948, November 1951, October 1956, November 1968, February 1972, March 1974, November 1994 and March 2011). The map shown in Fig. 22.28 illustrates the distribution of landslides produced during major landslide events. This map suggests significant clustering of landslide events.

Between 4 and 6 November 1994, during a severe cyclonic weather event, several hundred millimetres of rain fell on Piemonte. The average rainfall during each day of the 1994 storm was 33 mm, contrasting with the average monthly rainfall of around 140 mm for November in Piemonte (Polloni *et al.* 1996). In fact between 200 mm and 300 mm of rain fell between 2 and 6 November, with 90% of this falling on 5 November. On 6 November, the region received the greatest recorded rainfall in 80 years (up to 25 mm/hour). The 2011 storm event saw similar volumes of rainfall between 12 and 16 March but only 50–100 mm rainfall fell in this specific area.

In 1994 groundwater storage capacities of the river basins were exceeded and the water table reached surface levels; subsequently rainfall could only escape by flowing overland, causing widespread flooding. In total, 70 people were killed, several thousand people were rendered homeless, 200 settlements (towns and villages) were affected and over 100 bridges were damaged or destroyed. The total damage was estimated at approximately 10 billion USD, within an area comprising ca 30% of the region (Polloni *et al.* 1996). In March 2011 the flooding was widespread across Piemonte but less so than in 1994; many landslides were re-activated, but only a few in this area.

22.3.2.2 Slope movement types and geotechnical considerations

Principally, two types of failure are observed in the region: superficial debris-flows and block-slides (single-surface and compound, rotational and translational, and both first-time and re-activated). These are illustrated in Figs. 22.29 and 22.30 and their characteristics described in Table 22.13.

Previous research (Bandis *et al.* 1996; Forlati *et al.* 1996; Polloni *et al.* 1996) indicated a number of conditioning factors and situations; from their results the following conditions have been identified as being significant in triggering slope failures in this case:

- The rocks and soils are in a severely weakened state at the time of failure;
- Antecedent rainfall is critical in the initiation of debris-flows;
- The most frequent slope gradients where failures occurred are between 30° and 40° for debris-flows and between 10° and 15° for block-slides;
- The position of the groundwater level (relative to ground level) and rainfall intensity are critical to slope stability – if these are both high, then slopes may become unstable even at very low slope gradients;
- The slope failure planes (in block-slides) are pervasive, between 100 and 200 m in length, and occur at clay-rich layers;
- Failure planes contain high contents of the swelling clay, smectite (montmorillonite).

Work by Bandis *et al.* (1996) also yielded valuable geotechnical parameters for the rocks and soils in this region that were used in preparing the data for the hazard assessment.

22.3.3 A holistic GIS-based approach to landslide hazard assessment

22.3.3.1 The source data

A digital image database was compiled for this work from a number of sources. It included a DEM, created photogrammetrically from ASTER imagery, from which slope angle (degrees) and slope aspect (degrees, as a bearing from north) were calculated. Geological boundaries, drainage and infrastructure were digitised from 1:25,000 and 1:50,000 scale paper maps. Multi-temporal SPOT panchromatic and Landsat Thematic Mapper (TM) image data were used to locate known landslides produced by the 1994 storm event and to derive rock, soil and landuse information.

22.3.3.2 Data selection and preparation

A number of significant criteria were identified on the basis of direct field observation and published work:

1 Slope morphology – surface parameters, slope gradient and aspect, were calculated from the DEM, on a

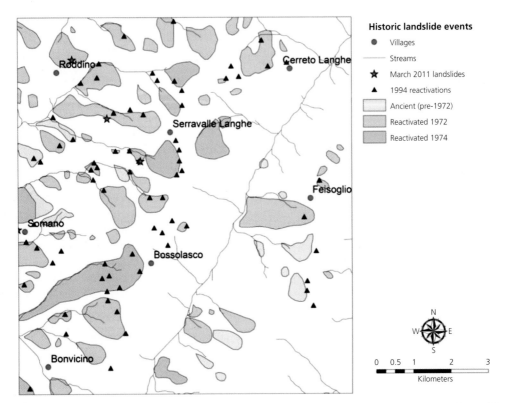

Fig. 22.28 Locations of historical landslides in the Langhe region occuring between 1972 and 2011. The size and shape of the coloured polygons indicate that entire hillslopes are often involved, either as large single but complex and composite landslides, or as clusters of smaller landslides.

pixel-by-pixel basis. Fuzzy functions were used to normalise these criteria to a common scale, using control points derived from field evidence.

2 Field evidence suggests that block-slides occur frequently in close proximity to roads and that debris-flows tend to be channelled by first- and second-order stream-valley morphology. Two criteria images were generated to represent Euclidian distance from roads and from drainage features, using information extracted from satellite images and published maps.

3 Geotechnical measures – both block-slides and debris-flows in the Langhe can be considered as planar failures and as such can be treated as 'infinite slopes' at such a regional scale of assessment (Taylor 1948; Skempton & DeLory 1957; Brass et al. 1991). The lack of pore pressure and shear strength information, combined with observations that these slope failures include both rock and soil, permit such a simplified approach rather than attempting to apply a more rigorous method. A version of this

model was used to produce a 'factor of safety' image layer as follows:

$$F = \frac{shear\ strength}{shear\ stress} = \frac{c' + (\gamma - m\gamma_w)z\cos^2\alpha\tan\phi'}{\gamma\,z\sin\alpha\,\cos\alpha}$$

(22.14)

where c' (effective cohesion) = 0.005 kN/m² (c'_{res} = 0 kN/m²); γ (bulk unit weight) = 24 kN/m³; m (ratio of water table depth to failure surface depth) = 1.0; γ_w (unit weight of water) = 10 kN/m³; z = depth to failure surface; α = slope angle; and ϕ' = effective friction angle.

The infinite slope equation was applied directly, on a pixel-by-pixel basis, to the slope gradient and aspect images. The other parameters were interpolated from results presented by Bandis et al. (1996). Based on field evidence, clearly defined ranges in slope aspect could be defined for the block-slides (240°–020°) and debris-flows (020°–240°). Again on field evidence, maximum block thickness (z) was taken as 10 m for block-slides

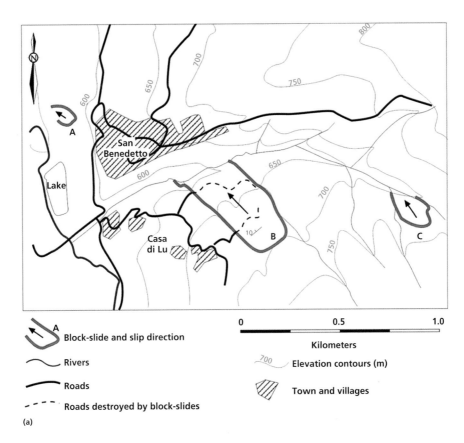

Legend:

- Block-slide and slip direction
- ~~~ Rivers
- ▬▬ Roads
- - - - Roads destroyed by block-slides

0   0.5   1.0
Kilometers

700 Elevation contours (m)

▨ Town and villages

(a)

(b)

Fig. 22.29 (a) Map of the San Benedetto area, indicating locations and slip directions of rotational/translational (complex) block-slides (A, B and C); and (b) view of block-slide B (looking southeastwards from San Benedetto town).

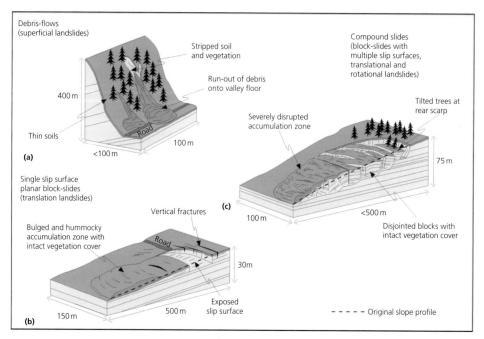

Fig. 22.30 Slope movement types observed in the area.

Table 22.13 Characteristics of slope movement types.

Type	Debris-flows	Block-slides
Movement mechanism	Superficial sheet flows (and some slides)	Translational (and rotational) simple and compound block slides
Slopes	20–53° (generally 20–40°)	5–15°
Attitude	At high angles to bedding	At low angles to bedding
Width/length aspect ratio	0.05–0.3	0.3–0.5
Depth	Ca < 1.5 m	1–10 m (simple); 20–30 m (compound)
Material involved	Top soil/regolith and vegetation	Rock, soil and vegetation
Other characteristics	Commonly related to slope concavities, drainage gullies and hollows; commonly in wooded areas	Incipient phase marked by ground swelling; open fractures and tension gashes above the crown prior to failure.
Timing and rates	Occur after major rainfall events. Rapid movement (a few metres/second)	Variable movement rates (between 10 m/hour to 100 m/hour)
Nature	Highly destructive	Large area of the ground unbroken (simple slides); Considerable disruption to the ground surface (compound slides)

and 3 m for debris-flows. For the materials occurring on low angle dip slopes, friction angle was assigned a residual value of 10° (characteristic of marls) and for materials on scarp slopes the angle was assigned to be 25° (characteristic of sandstones). Cohesion was taken as 0.004 kN/m², an average based on laboratory test

results for the marl and mudstones on dip slopes, and 0.005 kN/m² for sandstones on scarp slopes (Bandis *et al.* 1996); these are small values, as weakened materials are assumed. Eyewitness accounts indicate that a state of steady seepage occurred at surface level for some time after failure, so the 'ratio of water table depth to failure surface' or *m* can be assumed to equal 1.0. 'Map algebra' was then used to calculate the infinite slope equation to generate a factor of safety (*f*) map, using the following 'map algebra' expression, as constructed in ERMapper's formula editor:

$$\text{if } i1 > 0 \, and \, i1 < 20 \, or \, i1 > 240 \, and \, i1 < 360 \, then$$
$$(0.004 + (24 - 1 * 10) * (10 * (cos(i1 * (pi/180)) * cos(i1 * (pi/180)))) * tan(10 * (pi/180))))/$$
$$(24 * 10) * sin(i2 * (pi/180)) * cos(i2 * (pi/180))) \, else \, if \, i2 > 18 \, then$$
$$(0.005 + (24 - 1 * 10) * (3 * (cos(i2 * (pi/180)) * cos(i2 * (pi/180)))) * tan(25 * (pi/180))))$$
$$/(24 * 10 * sin(i2 * (pi/180)) * cos(i2 * (pi/180))) \, else \, null \qquad (22.15)$$

where *i1* = pixel values in the slope aspect image; *i2* = pixel values in the slope angle image; and *null* = pixels representing areas not at risk to failure and therefore excluded from processing algorithm (no value).

4 Clay content and leached zones – image ratios were derived from Landsat TM data to create two indices revealing selected ground characteristics: (1) iron-oxide content, from TM bands 3/1 (wavelengths 0.63 to 0.69 / 0.45 to 0.52); and (2) hydrated mineral (including clay) content, from TM bands 5/7 (wavelengths 1.55 to 1.75 / 2.08 to 2.35). The distribution of iron-oxide rich areas of soil is included in the analysis as an indirect indication of the presence of highly fractured zones (where iron is preferentially leached from the rocks and soils around) and therefore of potential instability. The Tasselled Cap transform was used to produce a soil wetness index. This transform, used to derive indices such as 'brightness', 'greenness, and 'wetness' from remotely sensed images, was developed by Crist and Cicone (1984a and b). These three image-derived indices provide information about leaching and fracturing of the ground (iron-oxide), the water retentive properties of the soils (hydrated mineral and clay content) and soil moisture (wetness), which could then be used as evidence in the GIS geohazard assessment of conditions leading to instability.

22.3.3.3 Multi-criteria evaluation

The model used here is based on the analytical hierarchy process, where there are both *factors* and *constraints* input to the model. The factors have been prepared using a variety of fuzzy membership functions, and their significance was evaluated using a pairwise comparison matrix, as shown in Table 22.14. A selection of factors is illustrated in Fig. 22.31.

The parameters controlling the fuzzy membership thresholds were selected on the basis of field observation and published work, as described previously. These control points and function types and forms used here are summarised in Tables 22.14 and 22.15.

22.3.3.4 Hazard and risk maps

The probability of occurrence of both spatial and temporal events needs to be determined with respect to the mass movement hazard. Varnes (1984) defines a hazard as being the probability of occurrence of a potentially damaging phenomenon within a given time

Table 22.14 Pairwise comparison matrix for factors influencing block-slide hazard.

Factors	Slope	Aspect	FS	Rddist	Drdist	Wet	Fe	Clay
Slope	1	1	1	1/3	1/5	1/2	1	1/4
Aspect	1	1	1	1/3	1/9	1	1/3	1/5
FS	1	1	1	1/3	1/7	1/3	1/3	1/8
Rddist	3	3	3	1	1	3	3	1
Drdist	5	9	7	1	1	3	3	3
Wet	2	1	3	1/3	1/3	1	1/2	1/6
Fe	1	3	3	1/3	1/3	2	1	1/3
Clay	4	5	8	1	1/3	6	3	1
Factor weights	0.052	0.045	0.039	0.187	0.292	0.070	0.089	0.226

Where Slope, gradient; FS, factor of safety; Rddist, distance from roads; Drdist, distance from drainage; Wet, wetness index; Fe, iron-oxides index and Clay, hydrated mineral index .

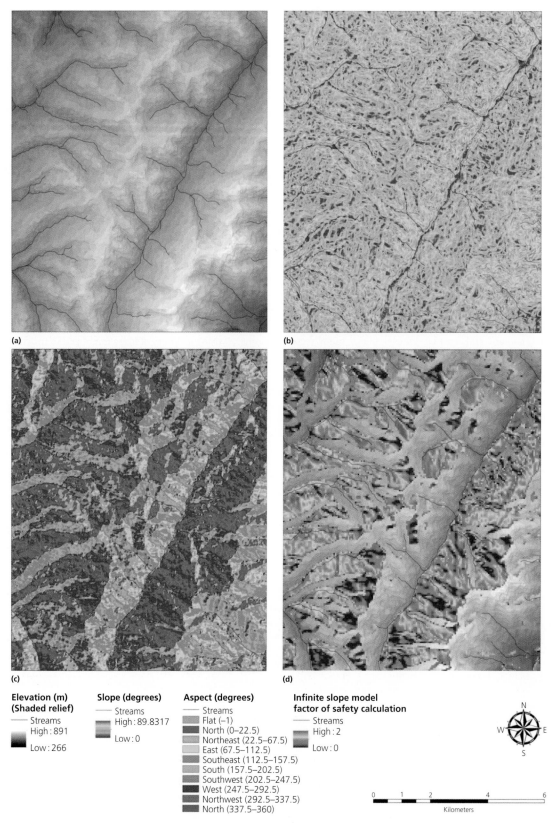

Elevation (m)
(Shaded relief)
— Streams
High : 891
Low : 266

Slope (degrees)
— Streams
High : 89.8317
Low : 0

Aspect (degrees)
— Streams
Flat (–1)
North (0–22.5)
Northeast (22.5–67.5)
East (67.5–112.5)
Southeast (112.5–157.5)
South (157.5–202.5)
Southwest (202.5–247.5)
West (247.5–292.5)
Northwest (292.5–337.5)
North (337.5–360)

Infinite slope model
factor of safety calculation
— Streams
High : 2
Low : 0

0 1 2 4 6
Kilometers

Fig. 22.31 (a) Shaded relief map; (b) slope; (c) aspect, derived from the digital elevation model; and (d) the factor of safety calculation using the slope and aspect images and the infinite slope model.

Table 22.15 Fuzzy membership functions used to prepare input criteria.

Factors	Fuzzy membership function and control points	
	Block-slides	**Debris-flows**
Slope gradient	Sigmoidal, positive symmetric	Sigmoidal, increasing monotonic
Slope aspect	Sigmoidal, symmetric	Sigmoidal, symmetric
Factor of safety	Linear, monotonic decreasing	Linear, monotonic decreasing
Distance from roads		
Distance from drainage	Sigmoidal, monotonic decreasing	Sigmoidal, monotonic decreasing
Wetness		
Iron oxides	Sigmoidal, monotonic increasing	Sigmoidal, monotonic increasing
Hydrated minerals		

and in a given area. The relationship between hazard, risk and vulnerability can be expressed as:

$$RISK(r) = HAZARD(h) * VULNERABILITY(v)$$

$$(22.16)$$

Decisions concerning the hazard being considered can be computed within the GIS by employing rules based on logic. Where data values have been measured directly, 'hard' decision rules can be formulated. This is difficult to achieve in reality, and generally 'soft' decisions have to be established on the basis of experience, prior knowledge and judgement, that is, 'belief' in the possible outcomes.

Uncertainty in this case concerns both the natural variability of the data and the lack of evidence about the significance of the data. This can be extended to consider whether a slope could become unstable as a result of an adverse combination of parameters.

The hazard map for block-slides is shown in Fig. 22.32. Much of the area has probability values exceeding 0.05 in Fig. 22.32(a), and the NW-facing slopes generally show values greater than 0.10. Comparison between distributions of older block-slides (1972 & 1974) reveals coincidence with the areas of highest hazard. Comparison with the current SIFRAP landslide classification and inventory [Fig. 22.32(b)] shows considerable agreement between areas with high values and those of known and active landslides.

A map of relative vulnerability can be derived from a generalised landuse dataset and by assigning values to the classes, between 0 and 1, as measures of their relative cost value (with 1 representing the highest vulnerability). The hazard maps for block-slides and debris-flows can then be multiplied by the vulnerability map to produce landslide risk maps.

22.3.4 Summary

The study of slope stability and geohazard assessment has attracted a great deal of attention as concern has grown for the safety of urban development encroaching on upland areas. The results show that planar failures are more likely to occur on the NW-facing dip slopes, but that if the soil/rock interface is taken as a potential discontinuity, planar failures may also occur on scarp slopes.

Comparison of the hazard maps generated by GIS with the current distribution of landslides in the SIFRAP database reveals the general applicability of the methodology. It is acknowledged, however, that the dataset used in this work is incomplete and contains errors, and that work in this area has continued (Luino 1999; Guzzetii 2000; Godio & Bottino 2001; Canuti et al. 2004; Meisina et al. 2008a, 2008b). Furthermore, the planar, infinite slope model is known to be a simplification of the actual failure mechanisms operating, but the geohazard map computed in this manner seems to reasonably reflect the observed occurrences of landslides in the Langhe region.

Image information relating to landslides is complex and contains two important components: texture and spectral detail. Remote sensing has been widely and successfully used to detect landslides in the past (Murphy & Vita-Finzi, 1991; Murphy & Bulmer 1994). It also provides a very convenient source of time-dependent information.

Geomorphological studies indicate that the Langhe region is still dynamic in terms of post-Alpine, post-glacial crustal uplift and that the mass movements are an ongoing natural, slope-dynamic consequence of this uplift. Recent research also points to a close link between slope instability in the Alps and periodicity in the Holocene climate

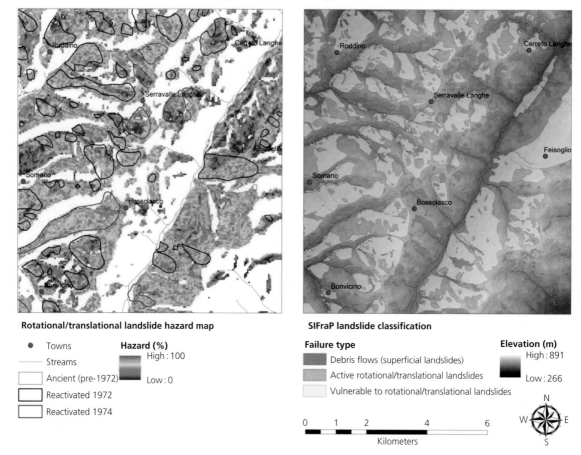

Rotational/translational landslide hazard map

- Towns
- Streams
- Ancient (pre-1972)
- Reactivated 1972
- Reactivated 1974

Hazard (%)
High : 100
Low : 0

SIFraP landslide classification

Failure type
- Debris flows (superficial landslides)
- Active rotational/translational landslides
- Vulnerable to rotational/translational landslides

Elevation (m)
High : 891
Low : 266

0 1 2 4 6
Kilometers

N
W E
S

Fig. 22.32 (a) Block-slide (rotational and translational landslides) hazard map, representing probability of occurrence, as a worst case scenario. Debris-flows are excluded from this particular hazard map and they are confined to the white areas. (b) Extract from the ARPE-Piemonte SIFRAP landslide classification inventory (2011) showing close similarity with the hazard prediction for rotational and translational landslides.

(Canuti *et al.* 2004). This implies that landsliding in the region is a long-standing phenomenon and is likely to remain so. Such situations are not uncommon, so the continuing development and exploitation of new technologies to help understand, monitor and mitigate the effects of such geohazards are vitally important.

The hypothesis that incorporation of digital information within geohazard assessment utilising GIS can significantly improve risk management in areas such as the Langhe has been considered for:

- Compilation of thematic information from remote sensing, geomorphology (elevation and its derivatives), and land usage;
- Stability analysis of selected slope profiles and of the whole study area;

- Multi-criteria hazard assessment (using probability and decision support tools) to compute geohazard maps.

It can be concluded that for image-based studies utilising satellite data, the most significant information is morphological since it is these features of shape and pattern that are detectable using remotely sensed imagery. There are certain important features that are needed for the correct identification of mass movements, for example, arcuate scarp, tension cracks and hummocky displaced ground. Detection of such features in imagery is, however, helped considerably by prior knowledge of the likely mechanisms and the prevailing state of activity.

GIS provides a flexible and effective tool for slope stability assessment and the production of thematic maps. The multi-criteria GIS approach provides a practical means for aggregating significant attributes (factors) influencing slope instability, a flexible means of combining individual factors reflecting their relative influence on the system controlling the outcome, and of course a convenient way of displaying the information.

22.3.5 Questions

1 Is the WLC method the most appropriate to use in a case like this?
2 What other datasets could be considered?
3 Why do the extents of the older landslides not match exactly the areas predicted in the hazard assessment and risk maps?
4 What can we say about the temporal constraints on a hazard prediction of this kind?

22.3.6 References

Bandis, S.C., Delmonaco, G., Margottini, C., Serafini, S., Trocciola, A., Dutto, F. & Mortara, G. (1996) Landslide phenomena during the extreme meteorological event of 4-6 November 1994 in Piemonte Region in N. Italy. In: K. Senneset (ed.), *Landslides*, pp. 623–8. Balkema, Rotterdam.

Biancotti, A. (1981) Geomorphologia dell'Alta Langa (Piemonte Meridionale). *Memori Societe Italiano Scienze Naturale*, Milano, 22 (no. 3), 59–104.

Boccardo, P. (1995) Integrazione di dati multispectrali acquisiti da differenti sensori montati a bordo di piattaforme aeree e satellitari: Il caso dell'alluvionale del novembre 1994 in Piemonte. *Telerilevamento, GIS e Cartografia al servizio dell'Informazione Teritoriale*, VII Convegno Nazionale AIT (Associazione Italiana di Telerilevamento) Chieri (TO), 17–20 ottobre, CSEA-Bonafus, 315–20.

Boccardo, P., Lingua. A. & Rinaudo, F. (1995) Integrazione di dati planoaltimetrici con immagini acquisite da piattaforma satellitare: Il caso delle aree esondate durante l'evento alluvione del novembre '94 in Piemonte. *Telerilevamento, GIS e Cartografia al servizio dell'Informazione Teritoriale*, VII Convegno Nazionale AIT (Associazione Italiana di Telerilevamento) Chieri (TO), 17–20 ottobre, CSEA-Bonafus, 223–8.

Boni, A. (1941) Distacco e scivolamento di masse a Cissone, frazione di Serravalle delle Langhe. *Geofisica Pura e Applicata*, 3 (no. 3), 1–19.

Brass, A., Wadge, G. & Reading, A.J. (1991) Designing a geographical information system for the prediction of landsliding potential in the West Indies. In: M. Jones & J. Cosgrove (eds), *Neotectonics and Resources*. Belhaven Press, London.

Campus, S. & Forlati, F. (2000) Preliminary study for landslide hazard assessment: GIS techniques and a multi-variate statistical approach. In: E. Bromhead, N. Dixon & M. Ibsen (eds), *Landslides in Research, Theory and Practice. Proceedings of 8th International Symposium on Landslides*, pp. 215–20. 26–30 June, Cardiff, UK. Thomas Telford, London.

Canuti, P., Casagli, N., Ermini, L., Fanti, R. & Farina, P. (2004) Landslide activity as a geoindicator in Italy: Significance and new perspectives from remote sensing. *Environmental Geology*, 45 (no. 7), 907–19.

Chiappone, A. & Scavia, C. (2000) Study of planar sliding instability phenomena in the Langhe area (Piedmont Region, Italy). In: E. Bromhead, N. Dixon & M. Ibsen (eds), *Landslides in Research, Theory and Practice. Proceedings of 8th International Symposium on Landslides* pp. 254–60. 26–30 June, Cardiff, UK. Thomas Telford, London.

Cortemiglia, G.C. & Terranova, G. (1969) La frana di Cigliè nelle Langhe. *Memorie Societa Geologica Italiana*, 8, 145–53.

Crist, E.P. & Cicone, R.C. (1984a) A physically based transformation of Thematic Mapper data – the TM Tasselled Cap. *IEEE Transactions on Geoscience and Remote Sensing*, GE-22, 256–63.

Crist, E.P. & Cicone, R.C. (1984b) Application of the Tasselled Cap concept to simulated Thematic Mapper data. *Photogrammetric Engineering and Remote Sensing*, 50 (no. 3), 343–52.

Delmonaco, G., Dutto, F. & Mortara, G. (1995) Landslides and precipitation. *In:* Casale, R. & Margottini, C. (eds), *Meteorological Events and Natural Disasters*, pp. 69–73. ENEA CR Casaccia, Rome.

Di Maio, C. & Onorati, R. (2000) Swelling behaviour of active clays: The case of an oversonsolidated, marine origin clay. In: E. Bromhead, N. Dixon & M. Ibsen (eds), *Landslides in Research, Theory and Practice. Proceedings of 8th International Symposium on Landslides*, pp. 468–74. 26–30 June, Cardiff, UK. Thomas Telford, London.

Embleton, C., ed. (1984) *Geomorphology of Europe*. Macmillan Press, London.

Forlati, F., Lancellotta, R., Osella, A., Scavia, C. & Veniale, F. (1996) The role of swelling marl in planar slides in the Langhe region. In: K. Senneset, K. (ed), *Landslides*, pp. 721–5. Balkema, Rotterdam.

Godio, A. & Bottino, G. (2001) Electrical and electromagnetic investigation for landslide characterization. *Physics and Chemistry of the Earth, Part C: Solar, Terrestrial & Planetary Science*, 26 (no. 9), 705–10 .

Govi, M. (1974) La frana di Somano (Langhe Cuneesi). *Studi Trentati di Scienze Naturale*, 51, 153–65.

Govi, M. & Sorzana, P.F. (1982) Frana di scivolamento nelle Langhe Cuneesi Febbraio-Marzo 1972, Febbraio 1974. *Bolletina della Associazione Mineraria Subalpina*, Anno XIX (1–2), 231–63.

Guzzetti, F. (2000) Landslide fatalities and the evaluation of landslide risk in Italy. *Engineering Geology*, 58 (no. 2), 89–107.

Hutchinson, J.N. (1988) Morphological and geotechnical parameters of landslides in relation to geology and hydrogeology. In: C. Bonnard (ed), *Landslides: glissements de terrain. Proceedings of 5th International Symposium on Landslides*, pp. 3–35. 10–15 July, Lausanne, Switzerland.

Luino, F. (1999) The flood and landslide event of November 4–6, 1994 in Piedmont Region (Northwestern Italy): Causes and related effects in Tanaro Valley. *Physics and Chemistry of the Earth, Part A*, 24 (no. 2), 123–9.

Luino, F., Ramasco, M. & Susella, G. (1993) Atlanti dei centri abitati instabili piemontese. *CNR-IRPI Torino-REGIONE PIEMONTE* S.P.R.G.M.S. GNDCI, Publication Number 964, 16–30.

Mason, P.J. & Rosenbaum, M.S. (2002) Geohazard mapping for predicting landslides: The Langhe Hills in Piemonte, NW Italy. *Quarterly Journal of Engineering Geology & Hydrology*, 35, 317–26.

Mason, P.J., Palladino, A.F. & Moore, J.McM. (1996) Evaluation of radar and panchromatic imagery for the study of flood and landslide events in Piemonte, Italy, in November 1994. *Proceedings of the European School on Floods and Landslides: Integrated Risk*, 19–26 May, Orvieto, Italy.

Mason, P.J., Rosenbaum, M.S. & Moore, J.McM. (2000) Predicting future landslides with remotely sensed imagery. In: E. Bromhead, N. Dixon & M. Ibsen (eds), *Landslides in Research, Theory and Practice. Proceedings of 8th International Symposium on Landslides*, pp. 1029–34. 26–30 June, Cardiff, UK. Thomas Telford, London.

Meisina, C., Zucca, F., Notti, D., Colombo, A., Cucchi, A., Bianchi, M., Colombo, D. & Giannico, C. (2008a) Potential and limitation of PSInSAR technique for landslide studies in the Piemonte Region (Northern Italy). *Geophysical Research Abstracts*, 10, EGU2008-A-09800.

Meisina, C., Zucca, F., Notti, D., Colombo, A., Cucchi, A., Savio, G., Giannico, C. & Bianchi, M. (2008b) Geological interpretation of PSInSAR data at regional scale. *Sensors*, 8 (no. 11), 7469–92, doi:10.3390/s8117469.

Murphy, W. & Bulmer, H.K. (1994) Evidence of pre-historic seismicity in the Wairarapa Valley, New Zealand, as indicated by remote sensing. *Proceedings of the Tenth Thematic Conference on Geologic Remote Sensing*, 9–12 May, San Antonio, Texas, 341–51.

Murphy, W. & Vita-Finzi, C. (1991) Landslides and seismicity: An application of remote sensing. *Proceedings of the Eighth Thematic Conference on Geologic Remote Sensing*, April 29–May 2, Denver, Colorado, 771–84.

Polloni, G., Aleotti, P., Baldelli, P., Nosetto, A. & Casavecchia, K. (1996) Heavy rain triggered landslides in the Alba area during November 1994 flooding event in the Piemonte Region (Italy). In: K. Senneset (ed.), *Landslides*, pp. 1955–60. Balkema, Rotterdam.

Sacco, F. (1903) La frana di Sant'Antonio in territorio di Cherasco. *Annali - Reale Accademia di Agricoltura di Torino*, 46, 3–8.

Skempton, A.W. & DeLory, F.A. (1957) Stability analysis of natural slopes in London clay. *Proceedings of the 4th International Conference on Soil Mechanics and Foundation Engineering*, 2, 378–81.

Taylor, D.W. (1948) *Fundamentals of Soil Mechanics*. Wiley, New York.

Tropeano, D. (1989) An historical analysis of flood and landslide events, as a tool for risk assessment in Bormida valley. *Suolosottosuolo*, Congresso Internazionale di Geoingegneria, 27–30 September, Turin, Italy, 145–51.

Tropeano, D. (1995) Evento alluvionale del novembre 1994 in Piemonte, interventi di studio effettuati dall'IRPI-CNR di Torino: sintesi delle osservazione. *Geoingegneria Ambientale Mineraria*, 32 (nos. 2–3), 135–45.

Varnes, D. (1984) Landslide hazard zonation: A review of principles and practice. *Commission on Landslides of the International Association of Engineering Geology*, United Nations Educational Social and Cultural Organisation (UNESCO), Natural Hazards, No. 3.

22.4 Land surface change detection in a desert area in Algeria using multi-temporal ERS SAR coherence images

As indicated in Chapter 10, multi-temporal SAR interferometric coherence imagery is a useful information source for the detection of random changes of the land surface. In this case study, three coherence images derived from three ERS-1 SAR images of an arid area of the Sahara desert in Algeria revealed some interesting phenomena, including distribution of mobile sand, erosion along river channels, variation of ephemeral lakes and seismic survey lines (Liu *et al.* 2001).

22.4.1 The study area

The area chosen for study is in eastern Algeria near the border with Libya in North Africa, 100 × 100 km, at approximately 27°–28°N and 8°–9°E. The Atlas Mountains separate the warm and temperate region along the coast of the Mediterranean from the vast hot arid desert: the Sahara. With very low humidity levels from 5% to 25%, the rainfall is rare, the solar radiation is intensive and the diurnal variation of temperature is large in the region (Ahrens 1994).

The very low precipitation and excessive evaporation make the desert hyperdry, barren and almost completely devoid of surface vegetative cover. This absence of a binding agent allows the loose sand or topsoil to migrate according to the prevailing wind patterns. It has been observed that the desert in this region is expanding northwards, with the vegetation of marginal lands being stripped for firewood or animal fodder, further exposing fragile soils to erosion.

As shown in a colour composite of a Landsat TM image (Fig. 22.33), the main geographic features of the study area are large expanses of flat bare rock or gravel plains broken up by escarpments, gully networks and ephemeral drainage channels, some of which flow into lakes or depressions. Large parts of

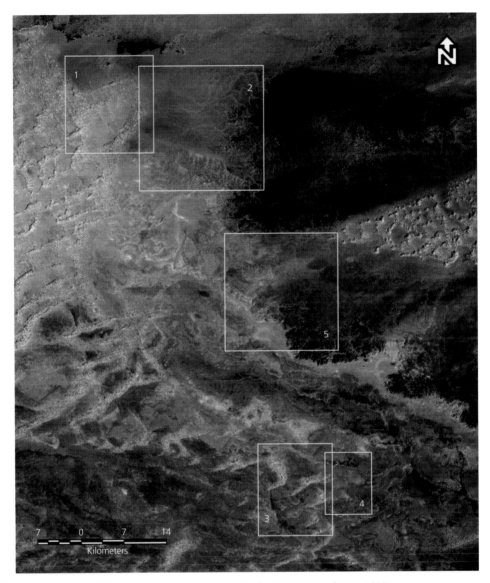

Fig. 22.33 Study area: Landsat TM colour composite of band 4, 2 and 1 in RGB (10 February 1987).

the region are covered with seas of sand, with linear, barchanoid and star dune types present, as well as thin sand sheets.

22.4.2 Coherence image processing and evaluation

Three scenes of ERS-1 SAR raw data of the study area acquired on 8 September 1992, 13 October 1992 and 28 September 1993 were processed by a SAR processor to produce single look complex (SLC) images. We name the three scenes in time sequence as *Alg1*, *Alg2* and *Alg3*. Among the three SLC images, *Alg2* was used as the reference scene, while *Alg1* and *Alg3* were used as target scenes to be co-registered to the reference. Three coherence images with 35, 350 and 385 days temporal separation were thus produced using eqn. 10.23 and named as *Coh12*, *Coh23* and *Coh13*.

There are several de-correlation factors that cause the loss of coherence in multi-temporal coherence imagery

(Zebker & Villasenor 1992; Gens & van Genderen 1996). Besides the temporal change of the land surface, which is the objective of the study, the major factors reducing coherence level are the baseline distance and the local slope as discussed in § 10.5.

Decorrelation caused by baseline separation is an inherent factor of the multi-pass, multi-temporal interferometric SAR system. The component of baseline perpendicular to radar look direction ($B_\perp$) represents the difference in view angles for the same ground object between the two observations. The phase of a radar return signal is decided by the vector summation of all the scatterers within a ground resolution cell. If $B_\perp$ is significant, the radar beam will illuminate the same ground target at considerably different angle and the collective effects of the relevant scatterers will result in a certain degree of random variation of phase. Thus the coherence decreases with the increase of $B_\perp$ as characterised in eqn. 10.24 in Chapter 10 and thus a short $B_\perp$ is generally preferred for coherence-based random change detection.

We can calculate the theoretical coherence values of the three coherence images from eqn. 10.24 using the nominal parameters of the ERS-1 SAR system. These data together with the actual average coherence values of the whole scene, a high coherence flat area and a gully-dissected area are shown in Table 22.16. The theoretical coherence value declines steadily with the increase of $B_\perp$. The average coherence for the whole scene is much lower than the theoretical value for all the three coherence images because of very low coherence resulting from temporal de-correlation in the large areas covered by mobile sand. It is interesting to notice that the average coherence over the gully-dissected area for *Coh23* is higher than that for the full scene and it is significantly higher than those for *Coh12* and *Coh13*, which are lower than their correspondent full scene averages.

For further analysis, the ratios between the actual and the theoretical coherence values were calculated (Table 22.16). The ratio data give an evaluation of the relationships among $B_\perp$, local slope, temporal changes and coherence level. For the full scene, $\rho_{actual}/\rho_{theory}$ ratio declines with the increase of temporal separation because the random changes of land surface accumulate with time. In the gully area, however, $\rho_{actual}/\rho_{theory}$ value increases significantly between *Coh12* and *Coh23* when $B_\perp$ decreases from 263 m to 105 m. This is due to the spatial de-correlation effect on directly radar-facing slopes (Lee & Liu 1999), which becomes more severe with the increase of $B_\perp$. For the same reason, $\rho_{actual}/\rho_{theory}$ ratio decreases gently with the increase of $B_\perp$ in the flat stable area.

From Table 22.16, it is obvious that *Coh13* has the poorest quality of coherence because of the largest $B_\perp$ among the three coherence images, and it covers the repeated temporal range of *Coh12* and *Coh23*. Therefore, *Coh13* is only used when necessary in the following interpretation for change detection.

22.4.3 Image visualisation and interpretation for change detection
22.4.3.1 Principles of interpretation
The study area has a very stable environment. The possible factors causing random changes of land surface are sand movement, erosion and deposition caused by wind or occasional flash flooding and limited human activities mainly relating to oil exploration. These changes will cause the decrease and loss of coherence and form dark features remarkably obvious against the high coherence background of a stable barren land surface. With three images taken with 35 and 350 and 385 days temporal separation, simple logical analysis is effective for interpreting the nature of the changes. Typically, there are

Table 22.16 Coherence data of the three coherence images of the study area.

		Coh12		*Coh23*		*Coh13*	
Time separation (days)		35		350		385	
$B_\perp$ (metre)		263	$\dfrac{\rho_{actual}}{\rho_{theory}}$	105	$\dfrac{\rho_{actual}}{\rho_{theory}}$	368	$\dfrac{\rho_{actual}}{\rho_{theory}}$
Theoretical ρ_{theory}		0.7958		0.9185		0.7143	
Actual ρ_{actual}	Full scene	0.5142	0.6461	0.5569	0.6063	0.4279	0.5990
	High coherence area	0.6677	0.8390	0.7939	0.8643	0.5460	0.7644
	Gully area	0.3900	0.4901	0.6104	0.6646	0.3496	0.4894

Table 22.17 Coherence scenarios and logical interpretations.

Scenario	Coherence level		Interpretation
	Coh12 (35 days) 8/9-13/10/92	*Coh23* (350 days) 13/10/92-28/9/93	
1	High	High	Stable, no change.
2	High	Low	Stable, then substantial change after 13 October 1992.
3	Low	High	Sudden change within the first 35 days then stable in the following 350 days.
4	Low	Low	Continually substantial change over whole period. OR Sudden change in 35 days followed by substantial change in 350 days.
5	Medium	High	Slight change until 13 October 1992 and then stable.
6	Medium	Low	Slow and progressive change over whole period.

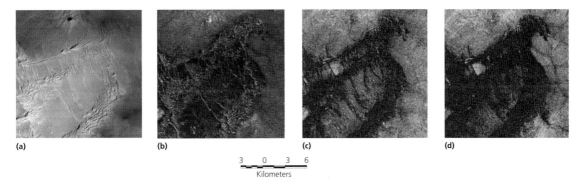

(a) (b) (c) (d)

3 0 3 6
Kilometers

Fig. 22.34 Dune boundary definition and mobile sand detection: (a) colour composite of Landsat TM band 421 in RGB (Box 1 in Fig. 22.33); (b) multi-look amplitude image of *Alg1* scene; (c) *Coh12*, the coherence image of 35 days separation; and (d) *Coh23*, the coherence image of 350 days separation.

six possible scenarios between *Coh12* and *Coh23*, as shown in Table 22.17

22.4.3.2 Sand movement (boxes 1 and 3 in Fig. 22.33)

Several types of sand dunes are present in the area, including transverse barchan and linear types and star dune networks. These are generally evident on TM imagery [Fig. 22.34(a)], which shows the morphology and structure of individual dune features. However, to define the boundaries of a dune or dune-field and to identify thin sheets of mobile sand are not always possible using TM or other types of optical imagery, particularly when the spectral properties of sand are very similar to the solid basement, as shown in Fig. 22.34(a). SAR amplitude imagery is even less adequate for the task, as shown in Fig. 22.34(b), because the tone

variation of the image is relevant to surface roughness rather than spectral or dynamic properties.

In contrast, based on quite different principles, coherence imagery is very effective for dune boundary delineation and mobile sand sheet identification, thus enabling a critical assessment of dune movement and sand encroachment. The loose sand grains on dune surfaces or thin sand sheets on a solid basement plain are subject to continuous movement under the wind even though the dune is static as whole. The sand movement causes random changes of the micro-geometry of scatterers on the sand-covered land surfaces and thus results in loss of coherence over a very short period, as characterised by scenario 4 in Table 22.17. The very dark decoherence features of mobile sand over a bright high coherence background are not only direct evidence of sand mobility but also effectively delineate the outlines of active dunes and optically indiscernible thin sand sheets. These data are not easily obtainable over a large region using other earth-observation techniques.

As illustrated in Fig. 22.34(c), complex boundaries of three chains of dunes (barchan and linear types) in the region are sharply defined in the *Coh12* image as de-coherence patches over a high coherence background. The boundaries are distinctive and definite. With 350-day temporal separation, the *Coh23* [Fig. 22.34(d)] reveals a thin sheet of mobile sand spreading into the inter-dune areas making the whole dune field a nearly continuous de-coherence patch. The central part of the dune-field is typically characterised by scenario 6 in Table 22.17 as medium coherence in *Coh12* and low coherence in *Coh23*, indicating continuous transport of the sand sheet as it is swept over the barren land surface.

The dune positions are defined effectively in coherence images of 35, 350 and 385 days temporal separation.

A colour composite of the three coherence images may reveal possible dune migration, which occurred during the 385-day period. For a colour composite of *Coh12* in red, *Coh23* in green and *Coh13* in blue, a quickly migrating barchan dune would be presented as a dark de-coherence feature with narrow trailing edge in red and windward edge in green (Liu *et al.* 1997a). As illustrated in Fig. 22.35, this diagnostic pattern is not evident, a discovery not unexpected for following reasons:

- The large formations of approximately 1 km wide are static as a whole. These large formations consist of small barchanoid ridges 50 m wide, which themselves are likely to be the migrating features, but the migration cannot be detected in the largely de-coherent background of the large sand formations.

- Even dunes migrating rapidly at 20 m per year would not produce a substantial signal on images of 35- and 350-day intervals, at a coarse pixel resolution of around 30 m.

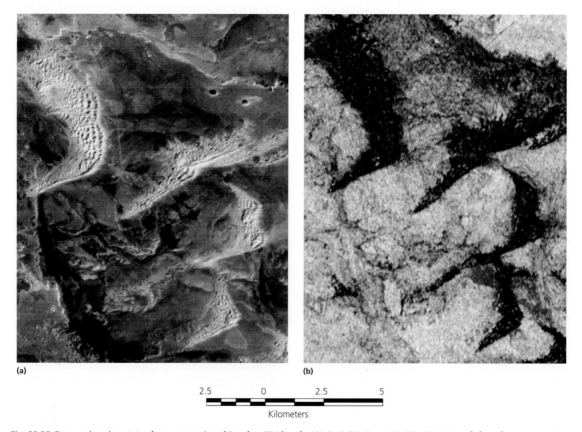

(a)

(b)

2.5   0   2.5   5
Kilometers

Fig. 22.35 Dune migration: (a) colour composite of Landsat TM bands 421 in RGB (Box 3 in Fig. 22.33): and (b) colour composite of *Coh12*, *Coh23* and *Coh13* in RGB; a migrating barchan dune would be presented as a dark de-coherence feature with narrow trailing edge in red and windward edge in green. This diagnostic feature does not appear in the image.

In order to make a serious attempt to identify dune migration, coherence imagery with a much longer temporal separation is required.

22.4.3.3 Ephemeral lakes and water bodies (box 4 in Fig. 22.33)

The RGB colour composite of coherence images is an effective aid for logical analysis of various events of land surface changes. The area defined by a box in the coherence colour composite Fig. 22.36 (a) presents an obvious red patch. It appears to correspond strongly with a

bright cyan feature on the TM 421 colour composite [Fig. 22.36(d)], which is defined as a shallow ephemeral lake in a reference map of the area (DMAAC 1981). The analysis of TM multi-spectral information indicates that the lake was nearly dry when the TM image was taken on 10 February 1987 (there was no precipitation in January and February 1987 according to the data from the GPCC website). As shown in Fig. 22.36(e) and (f), the lake patch is not particularly dark in near infrared band TM4 and very bright in the thermal band TM6. This characteristic is contradictory to the typical water

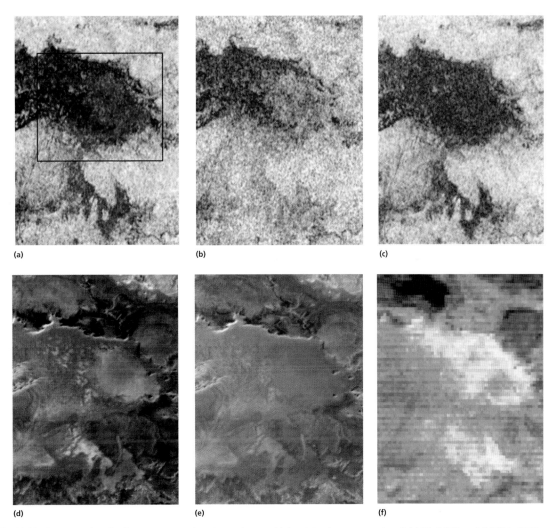

(a)   (b)   (c)

(d)   (e)   (f)

Fig. 22.36 The spectral and coherence properties of an ephemeral lake: (a) colour composite of *Coh12*, *Coh23* and *Coh13* in RGB. The rectangular box indicates an area with progressive decreasing of coherence that is better shown by comparison between (b) *Coh12* image and (c) *Coh23* image, where the patch is becoming darker; (d) colour composite of Landsat TM bands 421 in RGB (Box 4 in Fig. 22.33) indicates that the feature is in cyan colour and likely a water body; however, lack of water absorption in (e) TM band 4 image and strong thermal emission in (f) TM band 6 image indicate there was no water when the image was taken.

spectral signature: strong absorption in TM4. The area in fact presents an unusual spectral property: high albedo and high thermal emission. In general cases, high albedo objects would have low thermal emission (Liu *et al.* 1997b). The exceptional cases may occur for crystallised transparent material with strong internal scattering such as snow, gypsum and salt. It is reasonable to presume that this dried saline lake basin is covered with salt deposits.

Red pixels in Fig. 22.36(a) are those coherent in *Coh12* [Fig. 22.36(b)], but not in *Coh23* [Fig. 22.36(c)], logically implying a slow changing environment that appears relatively stable in the short-term (35 days), but the accumulated progressive change is substantial over a much longer period (350 days). It is therefore suggested that the lake basin was dry during the initial 35 days with a relatively stable surface. This condition allows medium to high coherence in *Coh12*. Then in the following 350 days, the lake possibly experienced re-charges of floodwater, temperature variation over a considerable range and repeated salt mineral crystallisation due to the water level change. Any of these processes can produce random changes significant enough to result in decoherence in *Coh23*. This explanation is supported by monthly average precipitation data (GPCC 1992–93; Rudolf *et al.* 1994) of the area during the period as shown in Fig. 22.37. There was 6–10 mm precipitation in winter 1992 and 5–6 mm in autumn 1993, which is adequate to cause seasonal re-charge to the lake.

Numerous similar patchy features can be identified in this region using the same logic and methodology, which correspond well with ephemeral lakes in the TM421 RGB colour composite [Fig. 22.36(d)]. Obviously, a confident identification of these desert lakes cannot be achieved without the TM colour composite. The extra contribution of the SAR multi-temporal coherence image is the detection of the dynamic activities of these lakes.

22.4.3.4 Drainage pattern and erosion (box 2 in Fig. 22.33)

Though the dominant agent of erosion in the Sahara desert is the prevailing wind, occasional and isolated intense rainstorms can cause local flooding and rapid fluvial erosion/deposition. Multi-temporal coherence imagery can provide direct evidence of this process. As there is an acute lack of information on the spatial location and temporal frequency of such erosion/deposition events, coherence imagery represents a valuable potential source of such data.

As shown in Fig. 22.38(a), the coherence image *Coh12* illustrates an area with high coherence over the initial 35-day period. There are no obvious water channel features except for a small section of channel approximately 30 m wide in the bottom right corner of the image with low coherence. The subsequent *Coh23* image [Fig. 22.38(b)], on the other hand, exhibits two separate major channels as obvious de-coherence features in a bright background of high coherence. These features are very eminent from the east and gradually become less pronounced further downstream towards the west. This characteristic reflects localised flooding from isolated storms, coupled with high transmission losses and evaporation causing surface flow to diminish downstream. As shown in Fig. 22.37, there was no precipitation during September and October 1992. We can therefore assume that the channels were dry and stable in the initial 35 days. In the subsequent 350 days, there were 6–10 mm precipitation in winter 1992 and 5–6 mm in autumn 1993. These rainfall events could have

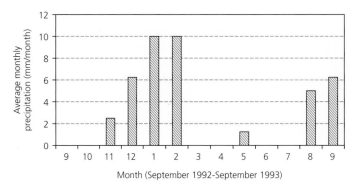

Fig. 22.37 Average monthly precipitation from September 1992 to September 1993 for the study area in Sahara (E7°–9°, N27°–28°) (Global Precipitation Climatology Centre homepage: www.dwd.de/research/gpcc).

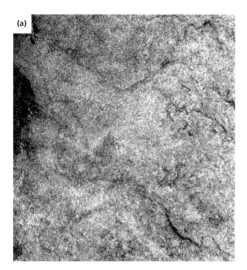 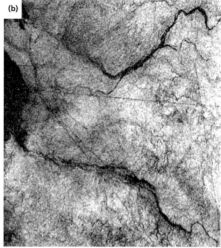

Fig. 22.38 Drainage channels: two major rivers in the region are not visible in (a) the 35-day separation coherence image *Coh 12*, but clearly shown as de-coherence features in (b) the 350-day separation coherence image *Coh 23*.

caused seasonal flash flooding in the rivers and resulted in active erosion/deposition, resulting in loss of coherence in the *Coh23* image.

22.4.3.5 Geophysical survey lines (box 5 in Fig. 22.33)

We have alluded to this unexpected finding in Chapter 10. As presented in Fig. 10.9, the coherence image *Coh23* observed between 13 October 1992 and 28 September 1993 reveals a mesh of straight lines that are not present in the relevant SAR multi-look images. These lines are clearly visible over this approximately 40×40 km area, with single lines up to 40 km long at a spacing of 2–3 km. Obviously, these are the results of anthropogenic disturbance over the periods between the repeated SAR image acquisitions. It is known that Sonatrach, the national oil company of Algeria, operated in the area during the period; the coherence image *Coh23* exposed considerable details of seismic survey.

Seismic survey lines will result in de-correlation between the SAR images taken before and after the survey as long as the swath is equivalent to or wider than the SAR image resolution cell and the random disturbance caused by engineering work is greater than half a wavelength of the radar beam (2.83 cm) in its slant range direction. The features are therefore detectable using the coherence image in such a largely stable environment. However, the disturbance to the ground, after land

surface recovering, is not great enough to significantly alter the average intensity of return SAR signals corresponding to each pixel of a SAR multi-look image, and the features are therefore not visible in Fig. 10.9(a).

This case demonstrates a unique function of coherence imagery as a tool for monitoring environmental impact of human activities.

22.4.4 Summary

The primary value of coherence imagery lies in its ability to record very subtle random changes on the land surface in an otherwise stable environment. If the average random change of dominant scatterers within a resolution cell exceeds a half of the radar wavelength in the direction of slant range, it will cause total de-correlation. Changes at this small scale are usually not detectable on conventional optical imagery. The change detection technique based on multi-temporal SAR coherence imagery is fundamentally different from the SAR interferogram-based measurement technique. Differential SAR interferometry is capable of measuring centimetre-level land surface deformation (a consistent block movement) but not workable with the random changes in the same scale range. On the other hand, coherence imagery can detect the centimetre-level random changes but cannot provide quantitative measurements.

In an arid environment such as the Sahara discussed here, the predominantly bare desert surface forms an extremely stable landscape that retains high coherence over very long periods (several years). The contrast between this bright background and the dark de-correlation signatures of any random changes enables detection and delineation of unstable features. It is this property that facilitates the spatial and temporal mapping of surface processes with a confidence unrivalled by other earth-observation techniques, and over areas too large or inaccessible for effective field surveys. This case study demonstrates the potential of SAR coherence imagery to detect and interpret changes in a desert environment. For example, persistent de-correlation over short time intervals is a direct evidence of sand mobility.

The lack of precipitation data in remote desert regions often hampers attempts to research the contribution of catastrophic fluvial erosion to arid landscapes. A sequence of short time scale (monthly), frequent coherence images could provide critical objective information on the temporal and spatial distribution of localised sporadic flood events. Coherence imagery also provides an effective way to detect human-induced disturbances over various time intervals.

A multi-temporal SAR coherence image presents an objective record of irregular land surface changes between two SAR image acquisitions as de-coherence features. Such low coherence phenomena can easily be distinguishable only when they are in sharp contrast to a high coherence background. The technique is most effective to detect changes in a largely stable environment, such as desert, but needs more sophisticated analysis in an unstable environment with many other de-correlation factors.

22.4.5 References

Ahrens, C.D. (1994) *Meteorology Today: An Introduction to Weather, Climate, and the Environment.* West Publishing Company, Eagan, Minnesota, 514–21.

Corr, D.G. & Whitehouse, S.W. (1996) Automatic change detection in spaceborne SAR imagery. *Proceedings of AGARD Conference – Remote Sensing: A Valuable Source of Information*, October, Toulouse, France. AGARD-CP-582, Paper No. 39. NATO, 39-1-7.

DMAAC. (1981) *Tactical Pilotage Chart of Algeria, Libya*, Series TPC, sheet H-3D, edition 1. Scale 1:500,000. Defence Mapping Agency Aerospace Center, St. Louis, Missouri.

Gens, R., & van Genderen, J.L. (1996) SAR interferometry – issues, techniques, applications. *International Journal of Remote Sensing*, 17, 1803–35.

GPCC. (1987). *Global Precipitation Climatology Centre* homepage: www.dwd.de/research/gpcc.

GPCC. (1992–93). *Global Precipitation Climatology Centre* homepage: www.dwd.de/research/gpcc.

Hagberg, J.O., Ulander, L.M.H. & Askne, J. (1995) Repeat-pass SAR-interferometry over forested terrain. *IEEE Transactions on Geoscience & Remote Sensing*, 33, 331–40.

Ichoku, C., Karnieli, A., Arkin, Y., Chorowicz, J., Fleury, T. & Rudant, J.P. (1998) Exploring the utility potential of SAR interferometric coherence images. *International Journal of Remote Sensing*, 19, 1147–60.

Lee, H. & Liu, J.G. (1999) Spatial decorrelation due to the topography in the interferometric SAR coherence image. In: *Proceedings of the International Geoscience and Remote Sensing Symposium (IGARSS'99)*, 26 June–2 July, Hamburg, Germany. IEEE, Piscataway, New Jersey, 485–87.

Liu, J.G., Capes, R., Haynes, M. & Moore, J.McM. (1997a) ERS SAR multi-temporal coherence image as a tool for sand desert study (dune movement, sand encroachment and erosion). In: *Proceedings of the 12th International Conference and Workshop on Applied Geologic Remote Sensing*, 17–19 November, Denver, Colorado. ERIM, Ann Arbor, Michigan, 478–85.

Liu, J.G., Moore, J.McM. & Haigh, J.D. (1997b) Simulated reflectance technique for ATM image enhancement. *International Journal of Remote Sensing*, 18, 243–55.

Liu, J.G., Black, A., Lee, H., Hanaizumi, H. & Moore, J.McM. (2001) Land surface change detection in a desert area in Algeria using multi-temporal ERS SAR coherence images. *International Journal of Remote Sensing*, 22, (no. 13), 2463–77.

Rudolf, B., Hauschild, H., Rueth W. & Schneider, U. (1994) Terrestrial precipitation analysis: Operational method and required density of point measuremenets. *Global Precipitations and Climate Change*, 26, 173–86.

Schwäbisch, M., Lehner, S. & Winkel, N. (1997) Coastline extraction using ERS SAR interferometry, 3rd ERS Symposium, 18–21 March, Florence, Italy. http://florence97.ers-symposium.org/12/10/florence.

Smith, L.C. & Alsdorf, D.E. (1997) Flood monitoring from tandem ERS phase coherence maps: Ob River, Siberia, 3rd ERS Symposium, 18–21 March, Florence, Italy. http://florence97.ers-symposium.org/12/10/florence.

Touzi, R., Lopes, A., Bruniquel, J. & Vachon, P.W. (1999) Coherence estimation for SAR imagery. *IEEE Transactions on Geoscience and Remote Sensing*, 37, (no. 1), 135–49.

Zebker, H.A. & Villasenor, J. (1992) Decorrelation in interferometric radar echoes. *IEEE Transactions on Geoscience and Remote Sensing*, 30, (no. 5), 950–9.

CHAPTER 23

Industrial case studies

This chapter describes two industrial case studies conducted by the authors in collaboration with other co-workers. Some of the issues surrounding these cases are highly confidential and so the material here has been confined to the aspects of the work that are directly related to remote sensing and GIS, and largely to the methodological aspects of the projects. In the first case (§ 23.1), the work has been carried out jointly with Mr Anders Lie of NunaMinerals A/S. The second case (§ 23.2) represents a small part of a wider project funded by UNICEF and carried out for Gibb Africa Ltd, by Image Africa Ltd (UK) and Aquasearch Ltd (Kenya).

23.1 Multi-criteria assessment of mineral prospectivity in SE Greenland

23.1.1 Introduction and objectives

Here we describe the data and methodology used to enable a multi-disciplinary assessment of prospectivity for a number of economic commodities, nickel, copper and PGE (platinum group elements) in previously unexplored terrains of south-east Greenland. Since this was a very large and ambitious project, we cannot do justice to its full complexity here, and so this chapter contains a summary of the data preparation and methodology aspects, rather than the results, which in detail are confidential. Some early, regional results, in the form of an example prediction map, are shown here since it is difficult to convey the concepts without visualising the results to some extent.

The results were intended mainly as a tool for attracting investors for the further development of any worthy areas identified during the project. The main drivers for conducting this work are explained in this chapter and lie in the fact that the south-east coast of Greenland comprises vast and relatively unexplored Archaean and Proterozoic terrains that had already shown potential for hosting mineral discoveries. It is ideal for a GIS-based assessment using remotely sensed and other regional geoscientific data and so is a good case study example for this book.

The project involves three independent phases: first, the testing, compilation and assessment of rock and sediment sample material; second, the systematic, multi-parameter spatial analysis of remotely sensed and all available geoscientific data; and finally, the exploration and ground validation of target areas pinpointed during the course of the project. The study area consists of two parts: the Reference (SW Greenland) and Survey (SE Greenland) areas, as illustrated in Fig. 23.1. The general approach is to use the 'fingerprints' of known mineral occurrences in the Reference area to help predict new occurrences in the Survey area by identifying significant spatial patterns in the various datasets we have at our disposal. This approach is not new to exploration, nor is it new in GIS; the use of GIS to conduct this kind of regional spatial analysis is well documented elsewhere (Bonham-Carter et al. 1988; Knox-Robinson 2000; and Chung & Keating 2002, to name but a few) but in Greenland at the time was a novel tactic.

23.1.2 Area description

The project concerns 24,500 km² of highly exposed, poorly explored Archaean and early Proterozoic shield terrain that has significant inferred potential for hosting

Image Processing and GIS for Remote Sensing: Techniques and Applications, Second Edition. Jian Guo Liu and Philippa J. Mason.
© 2016 John Wiley & Sons, Ltd. Published 2016 by John Wiley & Sons, Ltd.

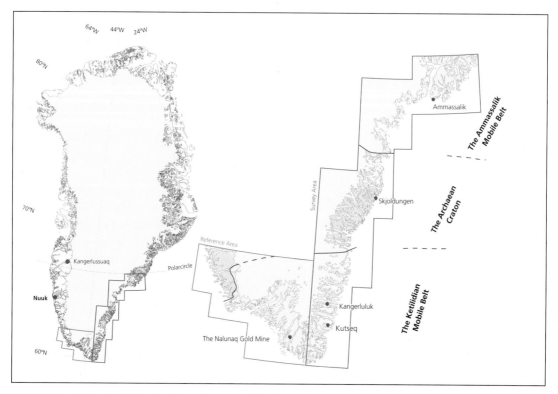

Fig. 23.1 (a) Map of Greenland showing the study area; and (b) detail showing the project Reference and Survey areas in southern and south-eastern Greenland.

mineral deposits and that stretches from Kap Farval in the south to the Ammassalik peninsula in the north. The terrain is extremely remote and only a handful of expeditions have been conducted to this part of Greenland (these are well documented). We built on that knowledge during our own fieldwork in the summers of 2006–2008, and we visited some of the same locations, plus a great many more. Logistically, fieldwork is hampered by the persistent presence of icebergs along some sections of the coast, making ship traffic hazardous. Helicopter reconnaissance is hindered by the lack of any re-fueling station, meaning that fuel must be carried onboard ship. Fieldwork can be done by ship alone but it restricts accessibility to near-shore localities, which means that far less distance can be covered. A mixture of the two is optimum.

Being in the 'rain-shadow' of the Greenland ice sheet means that the east coast terrain is generally dry and barren. Compared to the west coast, it is also steep, largely ice-covered and almost devoid of vegetation and wildlife. The altitude and steepness further necessitate

the use of a helicopter to conduct effective field reconnaissance and sampling work.

The Reference area is well studied and a great wealth of data and experience has been gleaned from it. The Survey area, between Kap Farval and Ammassalik Island has, in contrast, been mapped only at regional scales. The geological understanding of the west coast far exceeds that of the south-east coast and this may partly explain why very little commercial exploration has been conducted here. The map of known mineral occurrences in Greenland is testament to this as it can be seen that there are many more on the west than on the east coast and almost none on the south-east (Fig. 23.2); it is well known that the potential of finding new occurrences is perceived to be greater where occurrences have already been found. The discrepancy in understanding, knowledge and mineral potential between west and east coast terrains is being addressed but it will take time, so to gain parity in such a large area, effectively and relatively quickly, a novel approach was necessitated.

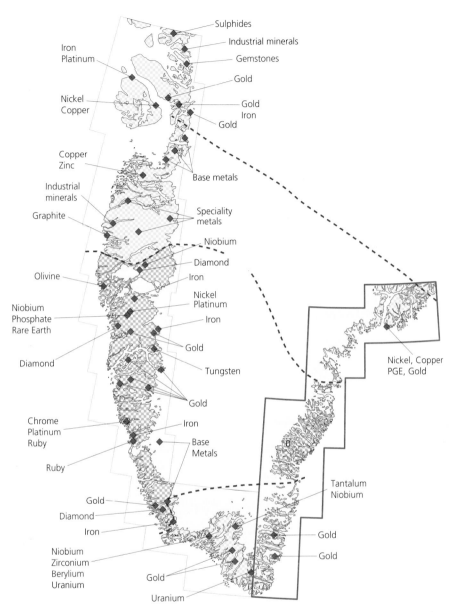

Sulphides
Industrial minerals
Gemstones
Gold
Iron
Platinum
Nickel
Copper
Gold
Gold
Iron
Gold
Copper
Zinc
Base metals
Industrial
minerals
Speciality
metals
Graphite
Niobium
Diamond
Olivine
Iron
Niobium
Phosphate
Rare Earth
Nickel
Platinum
Iron
Gold
Diamond
Tungsten
Gold
Chrome
Platinum
Ruby
Iron
Base
Metals
Ruby
Nickel, Copper
PGE, Gold
Tantalum
Niobium
Gold
Diamond
Iron
Gold
Niobium
Zirconium
Berylium
Uranium
Gold
Gold
Uranium

Fig. 23.2 Mineral occurrence map of Greenland with the project Survey area shown in red. Lithotectonic terrain boundaries are shown as blue dashed lines. For each terrain it can be seen that far more discoveries have been made on the west coast than on the south-east, despite the fact that the geological framework is the same.

23.1.3 Litho-tectonic context – why the project's concept works

The Precambrian shield of south-east Greenland comprises three distinct basement provinces: Archaean terrain reworked during the early Proterozoic (Ammassalik Mobile Belt, of the Nagssugtoqidian Orogen), Archaean terrain almost unaffected by Proterozoic or later orogenic activity (the North Atlantic Archaean Craton), and juvenile early Proterozoic terrain (the Ketilidian Mobile Belt or Orogen). This project involves assessment of all three terrains.

It is thought that the Nagssugtoqidian Orogen extends beneath the Greenland ice-cap to the east coast and that it can be closely correlated with the Torngat Orogen of

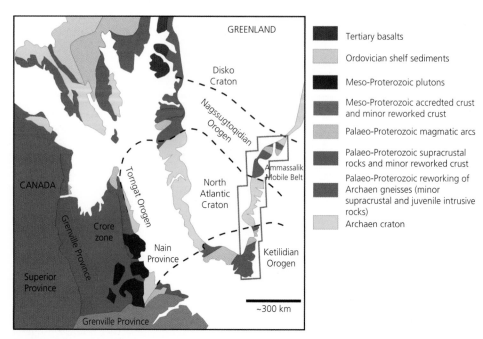

Fig. 23.3 Proterozoic reconstruction of Greenland and Canada, showing the Atlantic-Arctic litho-tectonic trend (Van Gool *et al.* 2002. Reproduced with permission from NRC Research Press). The approximate position of the Survey area is shown by the red polygon.

north-eastern Canada (van Gool *et al.* 2002), see Fig. 23.3. It is also thought that Greenland is closely related to similar, well explored terrains in Finland and Russia. We conclude that the east coast has all the same litho-tectonic suites and litho-geochemical characteristics as the west coast and as provinces in Canada, and that it should therefore have similar potential to yield mineral deposits. This is our justification for exploring for a series of mineralisation styles, which are already known to exist elsewhere.

23.1.4 Mineral deposit types evaluated

This sizeable area also comprises many different litho-structural and geochemical settings; a number of well-known mineral deposit models are therefore being evaluated within these terrains. Each type has been characterised according to economic commodity, geological setting and pathfinder minerals (primary and secondary), to aid the selection of input layers used in the generation of prediction maps. These include komatiitic hosted nickel-copper-PGE, mafic-ultramafic intrusion-hosted deposits, lode gold and calc-alkali porphyry deposits. The deposit model focused on here is the komatiitic hosted nickel-copper-PGE type. To adequately describe and

illustrate the methodology and results for all commodities and deposit models evaluated in this project would far exceed the scope of this chapter, and for the purposes of illustrating the methodology, is unnecessary; thus we focus on one of them, komatiite-hosted Ni-CU-PGE.

23.1.5 Data preparation
23.1.5.1 Published maps

Maps from the east coast are available at rather coarser scales than other parts of Greenland, with the exception of a few localities where detailed work has been undertaken by students, and these have been made use of where possible. The Reference area and the Lindenow Fjord area are covered by 1:100,000 scale maps and these were also made use of. Maps at a scale of 1:500,000 formed the backbone of the extraction of geological background knowledge and were used to target the fieldwork during summer campaigns (two map sheets cover the Reference and Survey areas). These regional maps contain considerable internal geometric distortions, making it very difficult to georectify them accurately; this emphasises the importance of the ASTER imagery in providing an accurate basemap for interpretation and data capture.

23.1.5.2 Lithology

The published maps have been used to guide the interpretation and discrimination, using ASTER imagery, of supracrustal packages, mafic, ultramafic and alkali intrusion and generally any non-gneiss/granite outcrops. The absolute positions of the identified outcrops have then been corrected using the ASTER imagery.

For simplicity, four categories have been created and coded with integer values of 1 to 4. These are ultramafic rocks (4), mafic meta-volcanic supracrustal packages (3), gabbros and other mafic igneous intrusions (2) and other alkali igneous intrusive bodies (1). Since the vast majority of exposed rocks in this part of Greenland are unmineralised crystalline lithologies such as gneisses, granites and granodiorites, we are really interested in identifying any other outcropping lithology. In the case of komatiitic hosted Cu-Ni-PGE deposits, the mineralisations occur in massive sulphide form and these are often small, dark and almost impossible to identify directly using remote sensing. We must therefore concentrate on identifying any potential hosts such as ultramafic intrusions and mafic volcanogenic supracrustal packages (which may also contain ultramafic rocks). Likewise, any mapped gabbroic intrusions should be included since they may be associated with ultramafic bodies that have not yet been mapped (given that mapping is not particularly detailed). We have not included the younger, 'Gardar' intrusive bodies since these are related to Atlantic opening events and are not significant for this deposit type. At the first stage of prediction, to avoid giving bias to one lithology type or another, we use a background value of zero to represent gneiss-granite exposures and a value of 1 for all other lithologies (see Table 23.1).

We also accept that the maps will not show all lithological outcrops of interest, and that some may have been incorrectly mapped, so we hope to detect others from the ASTER imagery.

23.1.5.3 Structure

Structural features have also been extracted from the maps and since these are at a coarse scale, the features extracted tend to represent only major faults and structural sutures; again, their positions are often inaccurate and require correcting using the ASTER imagery. The vast majority of fractures and faults in the spatial modelling database have been interpreted from the ASTER imagery, working at a scale of about 1:50,000.

For the purposes of the spatial modelling exercise, these interpreted (and mapped) structures are taken to represent zones of fracturing (potentially including faulting, shearing, jointing and other planes or zones of weakness and discontinuity). These are considered important as conduits and potential destinations for mineralised fluids, which are otherwise too small to be directly detected in the remotely sensed imagery and may not be obvious during fieldwork. Since they are represented as linear features but are taken to represent wider zones in reality, buffers are calculated around the linear features at variable distances, dictated by the relative tectonic significance of the structure. For instance, major terrain boundaries are coded with the highest rating value and minor faults the lowest; these values are then used in producing the buffers with the largest buffer distance for the most significant structures and so on. The actual distances used were chosen on the basis of field experience and published literature, and vary between 500 m and 10000 m. A fuzzy distance

Table 23.1 Numerical coding for thematic input layers to the spatial modelling of komatiitic-hosted Ni-Cu-PGE prospectivity.

Thematic layer	Class represented	Original values	Buffer distance (m)	Coded values
Lithology	Ultramafic bodies	4	–	1
	Mafic intrusives (e.g. gabbros)	3	–	1
	Mafic supracrustal (meta-igneous)	2	–	1
	Alkali intrusives	1	–	1
	Crystalline basement (gneiss & granite)	0	–	0
Structure	Terrain boundary	3	10000	1
	Major	2	500	1
	Minor	1	300	1
	'Unfractured' (massive) basement	0	0	0

rated buffer layer was then created in which buffer zones are coded from a value of one, closest to the structure, decreasing (almost linearly) to zero over the distance set by the structure's significance. The remaining, unfractured background is assigned a value of zero (see Table 23.1).

23.1.5.4 Mineral occurrences

Mineral occurrence data were derived from several sources and their function is twofold, to provide evidence data for the prospectivity predictions, and to provide data for the later cross-validation of those prediction maps (not shown in this case study).

These data are simply point positions of known mineral occurrence but with no numerical value as such, except for nominal values to identify the commodity (or commodities) found at that location. The occurrences used here have been derived from the Geological Survey of Denmark and Greenland (GEUS) south-west Greenland database, from the Ujarassiurit public geochemical database (managed by the Bureau of Minerals and Petroleum, of Greenland) and from the laboratory results of our own fieldwork.

For the komatiite-hosted Ni-CU-PGE deposit model a total of 78 known mineral occurrences (containing pathfinder elements for this deposit model) were used as evidence in the spatial modelling. The point localities were coded (as in Table 23.2), then buffered at a distance of 300 m, to allow for inaccuracies in the positioning

Table 23.2 Mineral occurrences used and their reclassified values.

Mineral occurrence	Commodity represented	Original values	Coded values in binary raster
Ni, Cu	Dominantly nickel plus copper	1	1
Cu, Au	Dominantly copper plus gold	2	1
Au	Gold	3	1
Au, Cu	Dominantly gold plus copper	4	1
PGE	Platinum group elements	5	1
Various	All other elements	6–99	0
–	Background	–	0

information, then converted to a binary raster with a value of 1 for any occurrence and zero for the background.

23.1.5.5 Remotely sensed image data

Some 60 ASTER Level 1B scenes have been used and these formed the most reliable and accurate basemap framework for all other mapping and data capture; they were also used to generate a series of spectral indices.

Ideal acquisition time is during the summer since the snow disappears in June and reappears in September, making July and August the ideal months; all the data used in this project were acquired in those months, between 2001 and 2006.

Digital elevation models (DEMs) were first generated from each ASTER scene, using the onboard ephemeris data as reference. Very limited ground control data exists (only at limited sites visited during scant field visits) and this is insufficient for any practical purposes here. The DEMs were then used to ortho-rectify each ASTER scene.

Prior to any image processing, the ASTER images were pre-processed to mask very brightest targets, ice (and snow) and the darkest targets, water, from the data. Without doing this, the enhancement of geological targets becomes very difficult, if not impossible. Generally speaking, sea pixels are not difficult to remove when the water is clear, since it's reflectance in the infrared is almost zero. The problem comes when water contains suspended rock flour, as is commonly the case in the upper reaches of fjords around Greenland. Reflectance from the suspended rock flour renders the water almost indistinguishable from land pixels. Defining a threshold at which to mask flour-loaded water becomes a very delicate operation, since some land pixels will be lost if you are not very careful and conservative. There will be cases where the removal of the last pixels, representing the most stubborn flour-laden waters has to be done manually, using a vector mask (which is extremely tedious and to be avoided if at all possible). With ASTER data acquired on very different dates, the algorithm used for remove ice and sea required customising for each set of illumination conditions. Fortunately, one long strip of data collected on a single day covered much of the central and southern part of the Survey area, making this job slightly less arduous. The logical expression constructed involved the examination of the visible and thermal bands over land, sea and ice, to identify suitable thresholds; an example is shown in Fig. 23.4(f).

23.1.5.5.1 Image processing for general visualisation

A standard false colour composite mosaic (ASTER bands 321 RGB) of the entire area was constructed and used both for general visualisation and interpretation, and also for field reconnaissance planning (shown in Fig. 23.4).

For optimum general visualisation and use in the field, image data that still contains at least the ice masses should be used since it is very difficult to navigate using an image from which the glaciers and ice-capped peaks have been removed. Individuals not so familiar in working with remotely sensed data often find it difficult to navigate using imagery when it is in its processed and rather abstract form. The best image for such purposes should be as simple as possible so that image features can be readily correlated with real objects on the ground.

23.1.5.5.2 Targeted spectral image processing

Ratio indices were derived from ASTER VNIR, SWIR and TIR data to highlight various chemical characteristics of rocks and minerals, whilst also suppressing topographic shadowing. These indices or *relative absorption band depth images*, as they are often described (Crowley 1989; Rowen *et al.* 2005) use the digital numbers (DNs) from the bands marking the shoulders of diagnostic absorption features, and the principle is illustrated in Fig. 23.15. Here they have been derived to highlight the spectral absorption features of our main targets: iron oxides and hydroxides, in weathered goassaniferous zones (b2/b1 and b4/b3), areas of high Mg-OH content ((b6+b9/(b7+b8)) and of silica paucity, that is, mafic and ultramafic rocks (b12/b13), where these are large enough or extensive enough to be detected.

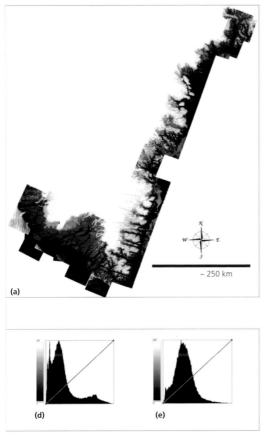

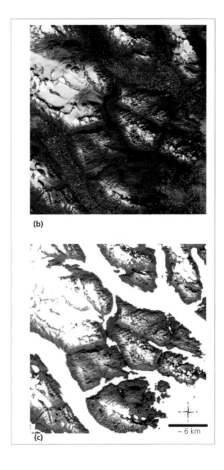

(f) if b4 > 1.6 and b10 > 5 and b1 < 100 then b3 else null

Fig. 23.4 Standard false colour composite mosaic of (a) the entire area (Reference and Survey); (b) detailed area on the east coast (indicated by the red box in (a) near Skjoldungen, (c) the same detailed area with ice, snow and sea masked out; and (d) and (e) image histograms for band 3 of the detailed area before (d) and after (e) masking of ice, snow and sea; (f) masking logical expression;

(g)

Fig. 23.4 (*Continued*) (g) Skjoldungen peninsula (looking north-westward) showing deep shadows on north-facing slopes, well illuminated south-facing slopes, summer ice and snow-fields and rock-flour-laden fjord waters with scattered icebergs.

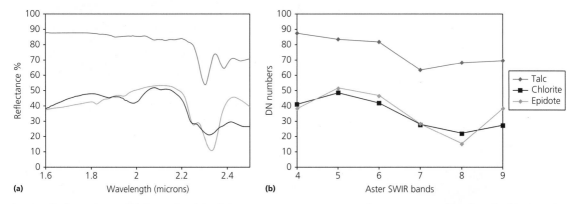

(a) Wavelength (microns)   **(b)** Aster SWIR bands

Fig. 23.5 The basis of spectral index (or band depth images) generation: (a) SWIR reflectance spectra of (mafic) talc, chlorite and epidote; and (b) the same spectra convolved to ASTER SWIR bandwidths. The sum of the band DN at the maxima either side of the feature are divided by the sum of the band minima DN, in this case, bands 6 and 9 over 7 and 8, and will highlight the broad absorption feature between 2.2 and 2.4 microns, which is characteristic of talc and especially chlorite and epidote. NB whilst the spectral range shown in (a) and (b) is the same, the ASTER band number scale in (b) is not linear and so the spectral shapes do not exactly match those of the reflectance spectra in (a).

Our primary mineralised targets are all associated with lithologies that are considerably poorer in silica content than the granites/gneisses that make up the vast majority of exposure. Many observed mineralisations have associated zones of hydrothermal alteration but in general these are of extremely limited extent and undetectable by ASTER. Spectral indices aimed at highlighting hydrated minerals (as would be normal in detecting alteration and mineralisation in other litho-tectonic settings) tends to reveal areas of weathered granite/gneiss rather than any alteration associated with mineralisation.

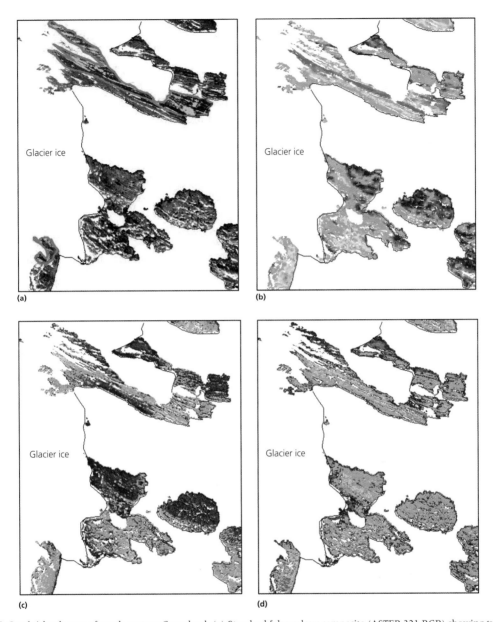

Fig. 23.6 Graah islands area of south-eastern Greenland: (a) Standard false colour composite (ASTER 321 RGB) showing two supracrustal packages as identified in published maps (green polygons); (b) mafic silicates spectral index (b12/b13) displayed as a pseudo-colour layer with increasing value from blue through to red (0–255); (c) MgOH spectral index (e.g. epidote, chlorite, talc) indicating altered mafic volcanic rocks, displayed with the same colour table as (b); and (d) ferric iron in silicates index (ferromagnesian minerals). The field of view is ca 14 km.

Some of these are illustrated in Figs. 23.6 and 23.7 for a very small part of the Survey area. The indices suggest that there are more outcrops of supracrustal packages than the ones mapped; not surprising given the mapping scale. We also see that there is variation within the mapped packages: some zones are more mafic than others [Fig. 23.6(c)] and some zones are altered [Fig. 23.6(d)]. The mafic silicates index in Fig. 23.6(c) exploits the Restrahlen feature of reflectance spectroscopy at thermal infrared wavelengths and indicates the presence of silicate minerals that are characteristic of mafic igneous rocks rather than the felsic group; here red colours indicate the mafic igneous zones within the supracrustal packages, against the felsic crystalline background of gneisses (blue). The indices shown in Fig. 23.6(d) show the intensity of absorption caused by MgOH bonds (or by carbon-oxygen bonds) in hydrated (altered) mafic igneous minerals, such as chlorite, epidote, talc, and may indicate where the supracrustal packages are intensely fractured, weathered or altered through fluid passage. Other useful indices include that highlighting oxidised iron at the surface, as an indication of gossan development, and highlighting iron in silicate phase, as an indication of the presence of mafic igneous rocks [as shown in Fig. 23.6(d)].

23.1.5.6 Geochemical sample data

The geochemical data used here is a compilation of laboratory analysis results from three sources: rock samples collected in the 1960s, rock and sediment samples collected during two targeted field reconnaissance seasons in 2006 and 2007, and data from the Ujarassiurit database. Samples from the 1960s and those collected in 2006 were also analysed using a field spectrometer (PIMA and ASD, respectively), the results of which have been very useful in characterising both the background spectral signatures and those of altered samples from known mineralised localities.

The coordinate references of some of the data points are more reliable and accurate than others. This is particularly the case for the Ujarassiorit data where, in some cases, the positions were not obtained using a GPS instrument but were approximated on a map sometime after collection.

From these data, an 'intelligent' geochemical database has been developed, principally for the purposes of this project but also for best practice in managing company assets. The data have been categorised according to the method of analysis and of field collection, the project

Fig. 23.7 Grassroots exploration fieldwork on Graah islands, east Greenland. The dark colour of the iron-rich mafic supracrustal rocks is clearly distinguishable from the paler (felsic) granitic gneisses esposed on the horizon on the far right. View looking south-eastwards along strike.

they are associated with as well as when and by whom they were collected. In this way, each result can be viewed critically, and with understanding of the likely significance and reliability of its contribution to the modelling of prospectivity.

For each particular deposit model, certain pathfinder elements were selected; nickel, copper, cobalt, chromium, gold, platinum and palladium in this case. The sample point data for each pathfinder element were then interpolated, using a minimum curvature tension (spline) polynomial function, to produce a continuously sampled representation of estimated element concentration over the entire area. For the elements used, the concentration values have been processed such that values below the detection limits (during laboratory preparation and analysis) were set to zero, and anomalously high values were clipped at an arbitrary value (of 1), and the logarithm of the remaining values were then used. This ensured that the interpolation was not biased either by extremely low values (which are of no interest) or by extremely high values (which may be produced by nugget effect, and therefore are also of low interest). The use of log values is standard practice in geochemical analysis and is appropriate here since it is the pattern (distribution) of values that are of interest rather than the absolute concentration values themselves.

The 'gridded' element data were then scaled to a 0–255 value range for input to the spatial modelling. Once again it is important that ice and sea are masked from the data before scaling to 0–255 value range and not afterwards. If scaling is done afterwards, the value range of the significant pixels (i.e. land pixels) will be suppressed (see Fig. 23.8). Remember that it is the patterns in the data (relative values) that are important here, not the absolute values themselves. We are most interested to find areas where the values are truly elevated with respect to the background values, rather than minor local perturbations in the background level. In this respect, it is also important to notice the shape and spread of the histogram. A histogram that shows a single, normally distributed population without skew (or with slightly positive skew) is highly likely to represent background levels of that element. What we would like to see is a negatively skewed histogram that has several peaks (i.e. representing multiple populations), as shown in Fig. 23.8(d), see also § 16.2.1.

A summary of the geochemical pathfinder layers used and input to the spatial prediction modelling, for this deposit model type, is shown in Table 23.3

23.1.5.7 Regional geophysical data

Regional geophysical data has also been made available to the project, in the form of gravimetric point sample records that have been corrected for free-air and Bouguer anomalies. These data were then interpolated to produce a surface representing regional gravity for the project area. It was found that although these data could be used for relative interpretations of anomalies, they could not be used within the spatial modelling without introducing a considerable regional bias, since they had not been corrected for the presence of the Greenland ice sheet that exerts a very broad effect on the data.

Since we have insufficient information to make a rigorous correction of this effect, we perform a 'make-shift' correction by subtracting a regional average, as calculated using a 31×31 average filter, from the individual values, thus removing the low frequency variation (caused by the ice) and leaving the high frequency information representing anomalies of geological origin (see Fig. 23.9). The result can then be used within the spatial modelling since the data have effectively been normalised: high values in one area have similar absolute value range, and similar meaning, to those in another location; a vital consideration with any data used in this kind of work. Once again, the interpolated, ice corrected gravity data must be masked for ice and sea before it is scaled to an 8-bit (0–255) value range, for the same reasons outlined previously.

23.1.6 Multi-criteria spatial modeling

Broadly the strategy is to conduct a semi-automated suitability mapping of the Survey area. The spectral, structural, geochemical and stratigraphic characteristics of proven mineral occurrences in the Reference area of south-western Greenland have been used as evidence from which to estimate the probability of discovering comparable mineral occurrences in the Survey area of south-eastern Greenland.

The Spatial Modelling and Prediction (SPM) software package (first developed by C.F. Chung) is being used here for the evaluation of our criteria, and to generate probability estimates (or prediction maps). Once the

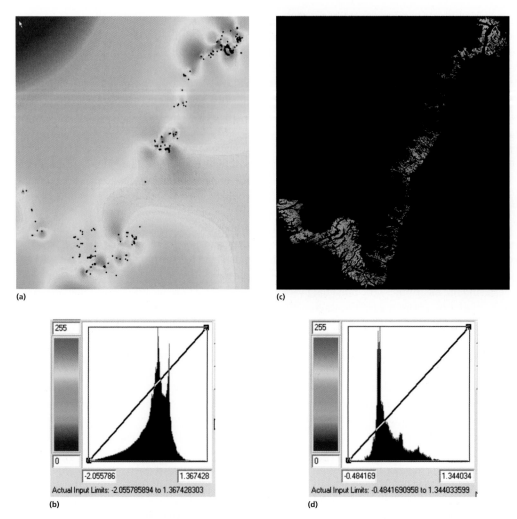

Fig. 23.8 (a) The interpolated Ni tenor grid of the entire area, with sample points (black dots), and its histogram (b); and (c) the interpolated grid of the land area and its histogram (d), after masking sea and ice. The histogram in (d) is skewed towards the low end of the value range and clearly shows several significant populations, indicated by the smaller peaks on the right. This represents the effective removal of background information and bias caused by the estimated low and high values in the ice and sea areas.

semi-automated prediction maps representing potential for identifying new targets have been produced satisfactorily, they will be weighted by market-driven factors, such as commodity price and accessibility, to identify economically favourable areas for further detailed ground validation.

23.1.6.1 Method of layer combination

All data must have been normalised to a common scale for use in the multi-criteria analysis, and in this case, a byte scale (0–255) has been used. They must also be coded so that all are positively correlated with high suitability. No input evidence must be contradictory or the results will be meaningless.

Bayesian probability discriminant analysis was chosen as the most appropriate method for combining the input layers and estimating prospectivity (as described in § 19.5.3). The main reason for doing so is that this method involves the least subjective decision making during criteria combination. Some subjectivity cannot be avoided along the way, in the capture and preparation of the data, but at the point where the input layers are

Table 23.3 Numerical coding for continuously sampled input layers to the spatial modelling of komatiitic hosted Ni-Cu-PGE prospectivity.

'Continuous' layer	Class represented	Method of processing/preparation	Coded value range
ASTER:			
Mafic content	Absence of free silica (quartz)	TIR (b13/b10)	Increasing values, on an
Iron oxides	Oxidised iron at the surface ('rusty zones')	VNIR (b2/b1)	8-bit value range (0–255)
Mg-OH	Presence of Mg-OH in mafic minerals	SWIR (b6+b9)/(b7+b8)	
Geochemical:			
Tenors (Ni, Cu, Co, Cr)	Interpolated from nickel, copper, cobalt & chromium tenors, i.e. metals in sulphide phases	Rock samples only	
Au, Pt & Pd	Interpolated gold, platinum & palladium content	Rock & sediment samples	
Geophysics:			
Gravity	Corrected for regional effect of Greenland ice sheet	Interpolated from regional Bouguer & free-air corrected ground sample point data	

combined we have to acknowledge that we do not have sufficient knowledge to apply any further rules or weighting of the input layers. In effect, we want to let the data speak for themselves and tell us what actual patterns exist. This necessity is a natural reason for choosing Bayesian probability as the method. We use known occurrences as our evidence, to gain the 'signature' (distributions) in the input layers, and use these to make predictions of the likelihood of finding new occurrences.

If known mineral occurrences are in fact associated with high values in the input layers, then the output prediction map should predict them. If so, then we should have more confidence in the map's prediction of other areas as showing potential.

One result for the deposit model described here is shown in Fig. 23.10. It reveals a series of high values areas, some more extensive and more intense than others, at various locations throughout the Survey area. Some we have confidence in, others we suspect to be exaggerated by the overwhelming presence of a few high gold values when this is used as a pathfinder element, in addition to some areas we think should appear more significant than they do here. Clearly further iterations are required. The path to achieve to a satisfactory result is not a linear or easy one, but is one where criticism is necessary, at every stage.

23.1.7 Summary

Data quality is one of the most significant issues here and probably in all cases of this type. A vital lesson is that criticism of and understanding of the data, rather than blind acceptance thereof, is absolutely crucial. Time must be spent during the data preparation stages and long before allowing it to be input to the multi-criteria spatial analysis; the old adage of 'rubbish in, rubbish out' holds very true here. There are several, very tangible reasons for conducting this project in south-east Greenland and for conducting it in the most correct way possible.

In comparison with the rest of Greenland, the depth of knowledge in the Survey area is far lower than elsewhere. The published maps are at a far coarser scale here than anywhere else. It is the only part of Greenland's coast that has never been colonized and where the logistics of travel and work have prevented all but a very few highly focused expeditions. As a consequence, this project now represents the greatest contribution to the collective knowledge base for the south-east coast of Greenland.

We have chosen the Bayesian probability method of multi-criteria evaluation since it is one of the most objective. Given the vast complexity of the area and the volume of data involved, we wanted to let the data speak for themselves and to prevent bias, as much as possible, at least in the way the data are combined. The

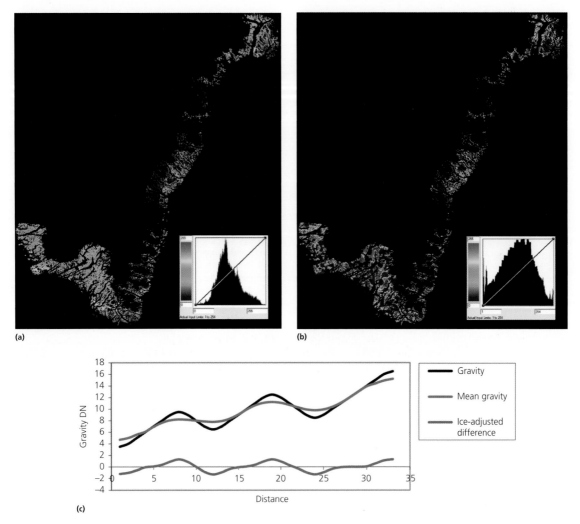

Fig. 23.9 (a) Bouguer-corrected gravimetric data with signal dominated by the effect of the Greenland ice sheet (>3 km thick); (b) ice-adjusted differential gravimetric data showing complex patterns that are caused by geological phenomena; (c) schematic representation of the ice correction applied to the gridded Bouguer-corrected gravity surface values. The black line illustrates both local variations and the regional trend imposed by the ice sheet (here sloping upwards left to right), the green line represents the mean (low frequency aspect) of this surface (as calculated using a spatial averaging filter), and the red line represents the differential surface (high frequency aspect), which reveals the significant fluctuations in gravimetric potential without the regional bias imposed by the large mass of the Greenland ice sheet.

Bayesian predictive discriminant analysis method used here identifies the signature of known occurrences in each of the input data layers and then assesses the probability of finding occurrences at other locations on the basis of the combinations of the input evidence layers (as described in Chapter 18). Some bias is unavoidable since we do not have unlimited data and the size of the

area means we must spatially sub-sample the data to make the processing manageable.

The purpose of conducting this at a very regional scale of observation and prediction is to demonstrate potential in a few target areas. These few key areas will then become the subject of further, more focused predictions. This is the logical way to proceed and the way that all

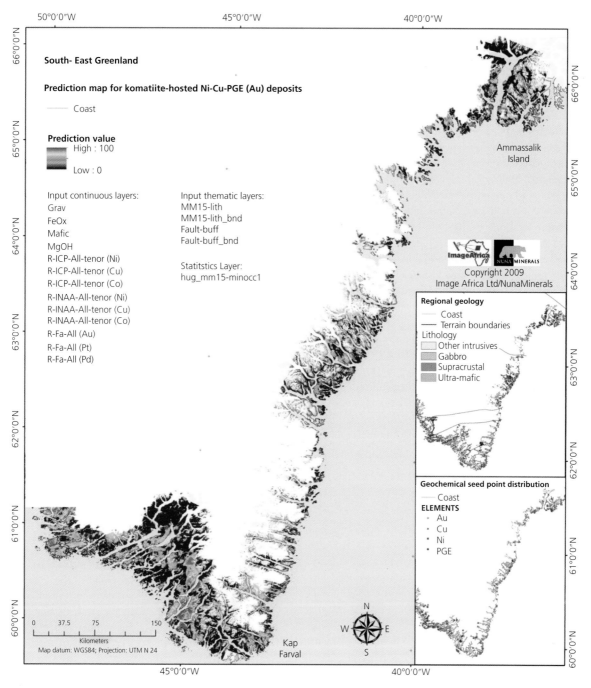

Fig. 23.10 An example prediction map representing prospectivity for komatiitic-hosted Ni-Cu-PGE deposits in the Reference and Survey areas.

studies based on remote sensing generally proceed. It would be pointless to tackle the entire area at the highest conceivable level of detail from the start. The cost of this would be astronomical and the process would be extremely inefficient since we actually expect that the vast majority of the project area will be shown to be non-prospective; just one of the drivers for this work is to prevent ourselves and others from wasting valuable time and funds in those areas.

Further work will include repeating this work for other mineral deposit types. These will necessitate the use of different spectral indices, pathfinder geochemical elements, lithological units and mineral occurrence layers, as appropriate to each model type. In each case, we have a certain expectation of where the most prospective locations are likely to be, on the basis of our own field experience. In other words, the ground 'truthing' is partly provided by our own observations, even though we have only visited a tiny fraction of the area. If our result succeeds in predicting the locations that we know to be prospective, our understanding is reinforced and we can have more confidence in the map's ability to predict previously unknown locations as being prospective. We will use later cross-validation curves to back up our own intuitive reaction in a more quantitative way. These will give an indication of the effectiveness of our prediction methodology and which of our input layers and occurrences are most significant; they may even reveal unexpected anomalies in data, occurrences or results that are significant in some previously unknown way.

We conclude that
- The size of the project area requires a novel and effective methodology to be successful.
- The paucity of data compared to other areas of Greenland means that the level of understanding here needs to be raised rapidly, to catch up.
- This part of the Greenland coast represents a geologically, tectonically, prospectively and topographically very complex area, one in which poor data quantity and quality, and little collective ground experience, means that the potential for error in this type of work is enormous. As rigorous approach as possible is therefore necessary.
- This effort represents both the most likely way to achieve some success and a significant investment for future exploration in this part of Greenland. It is, in short, the right thing to do in this context.

Despite our best efforts, there are some false positives in the results, and because of data paucity, there will almost certainly be some localities that are poorly predicted. Here our own experience and knowledge will be required to rule out the former, and a little imagination will be needed to pick out the latter. This stage of work is only the beginning. To develop this work, it may become necessary or desirable to use different methods of analysis to derive improved predictions over smaller, targeted areas. Under such circumstances, understanding of the specific geological controls involved will have developed, and therefore more rigid decision rules or more selective control over uncertainties may be applicable. We hope that several of the areas predicted in this work will form the focus for the next stage of exploration, data collection and more detailed prediction.

Since the completion of this work, detailed airborne geophysical surveys have been conducted over much of the south-eastern coast by GEUS. Stream sediments samples have also been collected and so geochemical survey data will soon have been completed. Exploration licenses have been obtained in the Ammassalik (Kitak) area and work continues.

23.1.8 Questions
1 What are the principal sources and types of uncertainty and bias in any result?
2 What is the next logical step in the refinement of this project?
3 At the more detailed stage of observation, which other criteria combination method(s) would you choose and use to critically evaluate your choice(s)?

23.1.9 Acknowledgements
This work has been funded by NunaMinerals A/S and by Image Africa Ltd, as a joint venture project. We are grateful to the GEUS for the provision of the regional geophysical data, and to Dr Bo Muller Stensgard for his valuable technical advice on the use of the spatial prediction modelling software in this context.

23.1.10 References
Agterberg, F.P. & Bonham-Carter, F. (2005) Measuring performance of mineral-potential maps. *Natural Resources Research*, 14, 1–17.
Bonham-Carter, G.F., Agterberg, F.P. & Wright, D.F. (1988) Integration of geological datasets for gold exploration in Nova

Scotia. *Photogrammetric Engineering and Remote Sensing*, 54 (no. 77), 1585–92.

Carranza, E.J.M. & Hale, M. (2002) Spatial association of mineral occurrences and curvilinear geological features. *Mathematical Geology*, 34, 203–21.

Chung, C.F. & Moon, W.M. (1990) Combination rules of spatial geoscience data for mineral exploration. *Geoinformatics*, 2 (no. 2), 159–69.

Chung, C.F. & Keating, P.B. (2002) Mineral potential evaluation based on airborne geophysical data. *Exploration Geophysics*, 33, 28–34.

Crowley, J.K., Brickey, D.B. & Rowan, L.C. (1989) Airborne imaging spectrometer data of the Ruby Mountains, Montana: Mineral discrimination using relative absorption band-depth images. *Remote Sensing of Environment*, 29, 121–34.

Knox-Robinson, C.M. (2000) Vectorial fuzzy logic: A novel technique for enhanced mineral prospectivity mapping, with reference to the orogenic gold mineralisation potential of the Kalgoorlie Terrane, Western Australia. *Australian Journal of Earth Sciences*, 47, 929–41.

Rowan, L.C., Mars, J.C. & Simpson, C.L. (2005) Lithologic mapping of the Mordor, NT, Australia ultramafic complex by using the Advanced Spaceborne Thermal Emission and Reflection Radiometer (ASTER). *Remote Sensing of Environment*, 99, 105–26.

Stendal, H. & Frei, R. (2000) Gold occurrences and lead isotopes in Ketilidian Mobile Belt, South Greenland. *Transactions of the Institution of Mining and Metallurgy*, 109, B6–13.

Stendal, H. & Schonwandt, H.K. (2000) Mineralexploration in Greenland in the 1990s. *Transactions of the Institution of Mining and Metallurgy*, 109, B1–5.

Stendal, H., Grahl-Madsen, L., Olsen, H.K., Schonwandt, H.K. & Thomassen, B. (1995) Gold exploration in the early Proterozoic Ketilidian Orogen, South Greenland. *Exploration Mining Geology*, 4 (no. 3), 307–15.

Van Gool, J.A.M., Connelly, J.N., Marker, M. & Mengel, F. (2002) The Nagssugtoqidian Orogen of west Greenland: Tectonic evolution and regional correlations from a west Greenland perspective. *Canadian Journal of Earth Sciences*, 39, 665–86.

23.2 Water resource exploration in Somalia

23.2.1 Introduction

This part of the chapter is based on work conducted as a collaborative effort between Image Africa Ltd and Aquasearch Ltd for Gibb Africa Ltd and was funded by UNICEF. The work involved the evaluation through remote sensing of the region around the city of Hargeisa, in Somalia (formerly British Somaliland), within a broader project whose remit was to identify potential bulk water resources for this city. Various criteria for such potential resources had to be satisfied to within this remit;

for instance, any suitable bulk resource should be no more than 100 km from the city, and that it should ultimately be capable of supplying a demand of 40,000 m³/d.

The general approach was to conduct a relatively detailed desk study, in two phases, followed by some targeted fieldwork. The first phase consisted of a regional-scale geological interpretation and identification of target areas that warranted more detailed investigation. Phase two comprised similar but more focused interpretation of several target areas, using two- and three-dimensional visualisation, using DEMs generated during the project, as well as spectral enhancement via image processing. The results of phase one and two image interpretation and digital mapping were carried out in parallel with, and then combined with, a detailed literature search and evaluation of borehole and other available hydrological data. The detailed nature of the two desk study phases was especially significant given the hostilities that were at the time preventing the completion of the work through field investigations, a situation that persists to this day. The work presented in this case study represents a summary of the first and second phases, as an illustration of the approach taken for a case of this type.

Specifically, the objectives of this part of the work were to provide

- Regional geological interpretation and mapping;
- Evidence of the location and extent of potential aquifers (as described by hydrogeologist M. Lane of Aquasearch Ltd);
- Recommendations on areas for further investigation.

The first objective would be satisfied largely using medium-resolution data, in the form of Landsat 7 ETM+ imagery, Shuttle Radar Topographic Mission (SRTM) DEM, published geological maps and mean annual rainfall (MAR) data. The second and third would involve the use of ASTER imagery and DEMs in addition to geological maps.

The study area occupies the some 300 sq km along the northern coast of Somalia (formerly British Somaliland) in north-east Africa and is illustrated in Fig. 23.11. It includes coastal alluvial plains in the north, volcanic escarpment, mountainous massifs, a limestone plateau and extensive desert sands to the south, and it encompasses varied climatic conditions. The average rainfall here is 380 mm p a and this falls largely on the high ground.

There are a number of potential aquifers in this region. The most significant 'extensive aquifers' with

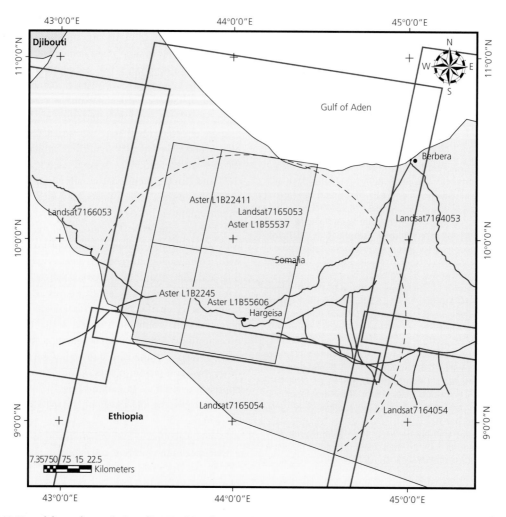

Fig. 23.11 Map of the study area in Somalia, NE Africa, showing the main cities and roads, the 100 km radius and the image scene coverage used in this work (Landsat 7: five clear labelled polygons; and ASTER: four yellow labelled polygons). The 100 km radius marker is used for scale in all regional images shown here.

potential to yield bulk water resources could be found in any or all of the solid geological units. These units are the Precambrian fractured crystalline metamorphic rocks, Jurassic limestones and sandstones (Adigrat Formation), the Nubian Cretaceous sandstone and the Eocene Auradu limestone. The second, most significant, comprise local structural basins that are dominantly filled with Jurassic limestones and younger sediments. Third are minor sedimentary aquifers, several of which were identified in previous studies in the western, central and eastern coastal plains. The alluvial and colluvial material of the coastal plains was also considered.

23.2.2 Data preparation
The datasets used in this study are listed in Table 23.4. The Landsat ETM+ images used for this work cover an area of ca 300 × 200 km, span more than one climatic zone and were acquired during April, May and September of three different years. In such circumstances, the production of a seamless image mosaic is

Table 23.4 Summary of digital data used in this study.

Data	Path/Row	Acquisition dates
Landsat 7 ETM+	164/053-054, 165/053-054, 166/053	1999, 2000 & 2001
ASTER	Four scenes	2002
ASTER DEMs (15 m)	Four scenes	2002
Maps	British Geological Survey map sheets at 1:50,000	Sheets 20-23, 32-24
Rainfall data	Mean annual rainfall (mm)	

almost impossible. Images involving combinations of bands 1, 2 or 3 also suffer from the effects of atmospheric haze, and this is most noticeable in the areas closest to the Sahara. Atmospheric clarity over the mountainous and coastal areas, by comparison, is very good. The haze effects are most noticeable therefore in the southern and south-eastern Landsat scenes, and are responsible for the poor contrast matching (and rather obvious scene boundaries) between these two and the northern and western scenes. Images involving bands 4, 5 and 7, in contrast, are affected less by haze so that the scene boundaries are less noticeable in the image shown in Fig. 23.52. In each of the image maps presented, the data have been contrast enhanced to match as closely as possible.

The MAR rainfall data were supplied in point form by Gibb Africa Ltd, and were gridded and contoured for correlation with other data. A section of the SRTM global 3 arc second DEM was used in this study since neither the 30 m ASTER Global Digital Elevation Model (GDEM) nor SRTM 1 arc second DEM (also 30 m) were available at the time. DEMs (gridded at 15 m) were also generated from each ASTER scene, using between 6 and 20 control points identified in each scene. The eastings and northings coordinate values were collected for these points, from the ortho-rectified Landsat ETM+ imagery and the corresponding height values for these points were collected from the SRTM DEM. Each of the four ASTER DEMs were generated separately and then mosaiced. The discrepancies in elevation between any of the four scene DEMs were in general less than 5 m, providing an almost seamless DEM mosaic of the target

areas, which is of sufficient accuracy to support mapping at 1:50,000 scale. Image data coverage is illustrated in Fig. 23.11.

23.2.3 Preliminary geological enhancements and target area identification

The first phase comprised a regional-scale image interpretation and desk study, to identify major areas of interest and to extract as much information on regional geology as possible, given that map information was limited. This involved the use of broadband, medium-resolution, wide swath imagery, Landsat 7 ETM+ in this case, to produce a regional geological interpretation, from which to then identify potential target areas for further investigation. In addition, the SRTM DEM was used to interpret regional structural information and to examine areas of aquifer recharge with respect to topography.

23.2.3.1 General visualisation and identification of natural vegetation and recharge areas

Once again we begin with the general visualisation of the area, using simple colour composite images (true colour and standard false colour) and a shaded relief image generated from the regional SRTM DEM.

The Landsat 432 colour composite reveals that the distribution of vegetation is not regular or even but is extremely sparse except in western and central-eastern mountains (Fig. 23.12). Calculation of a Normalised Difference Vegetation Index (NDVI) from the Landsat data and then comparison of this with the MAR data and the DEM shaded relief image (Fig. 23.13) reveals that there is close correlation between main centres of incident rainfall, persistent natural vegetation and high elevations. These areas comprise the principal areas of groundwater recharge for this region; they lie around the outer part of the 100 km zone around Hargeisa. Some small patches of vegetation can be seen in the valleys that cut the Haud plateau and at the base of the scarp slope of this plateau, to the south of Hargeisa itself.

23.2.3.2 Regional geological interpretation

Simple colour composites, band ratios and the SRTM DEM were integrated and used to produce a regional geological interpretation, which was suitable for use at ca 1:50,000 scale. Composites of bands 531 and 457

Fig. 23.12 Landsat 432 standard false colour composite image.

[Fig. 23.14(a) and 23.14(b)] formed the backbone of this work as they provided very good discrimination of metamorphic basement lithologies, carbonates, volcanics and superficial sedimentary cover. The DEM and images together allowed a very detailed structural picture to be produced [Fig. 23.14(c)], especially in the area to the south of Hargeisa, where the Nubian and Auradu sediments are faulted and dip away to the south beneath the desert sands. Structure in the metamorphic basement massifs is too complex for interpretation at this scale of detail, and so only the major terrain units were identified. A number of major sedimentary basins were identified within these massifs in the western part of the area, and these were predominantly filled with carbonate sequences.

The broad bandwidths of Landsat SWIR mean that carbonates and hydrated minerals (and vegetation where present) can be highlighted but not distinguished from one another. If we use a b5/b7 ratio we highlight both hydrated minerals (clays) and carbonates. By looking at the spectral profiles of the major units of interest, we find that the carbonates have, in general, higher DN values, in the daytime thermal infra red of Landsat (b6). If we modulate the b5/b7 ratio with the TIR, we produce an index image that highlights all the major carbonate bodies in the area, in addition to some gypsiferous deposits that are interbedded with them. The resulting image is shown in Fig. 23.15; here the main targets are all highlighted and circled to identify them.

23.2.4 Discrimination potential aquifer lithologies using ASTER spectral indices

The second phase of the investigation involved sub-regional scale-interpretation of ASTER data, to evaluate the target areas identified in phase one and to assess whether these warranted further, detailed field-based investigations.

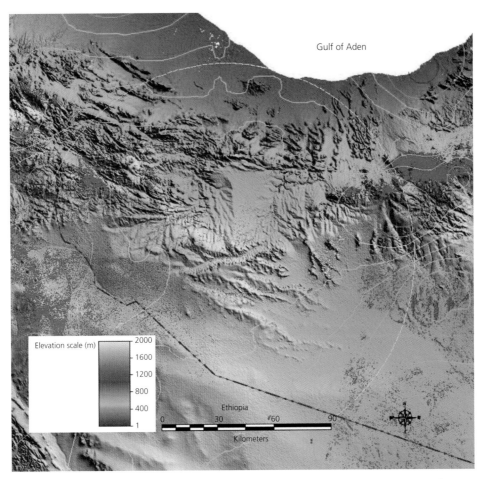

Fig. 23.13 SRTM DEM shaded relief map with thresholded NDVI (red) and MAR contours overlain, to indicate close correlation between rainfall recharge areas and natural vegetation distribution.

Having identified the lithological targets of phase one, i.e. carbonates and closely associated sandstones, ASTER is particularly suited to the task of discriminating them. Wavelength-specific variations in brightness enable the identification of carbonates (limestones, marble, calcite and dolomite), silicates (quartz, quartzites and different families of igneous rocks) and Mg-OH bearing minerals and rocks (weathered, metamorphosed and altered mafic igneous rocks), in addition to iron-aluminium silicates and iron oxides/hydroxides (both as alteration and weathering products), different families of hydrated minerals (clay alteration products, weathering breakdown products and evaporates such as gypsum and anhydrite) and, of course, vegetation.

The main target areas for investigation in phase two included the Bannaanka Dhamal, Agabar Basin and Nubian/Auradu contacts south of Hargeisa, the Waheen and Dibrawein valleys. We cannot describe all the identified target areas within this chapter so we focus on two, the: (a) Hargeisa and (b) Waheen areas.

23.2.4.1 Hargeisa area: Auradu limestone and Nubian sandstone

The objectives here are to enhance and delineate the outcrops of the main potential aquifers – the carbonates and sandstones; specifically, the Eocene Auradu limestone and the Cretaceous Nubian sandstone lying stratigraphically below it. The Nubian sandstone is the principal aquifer for the whole of eastern Africa and its

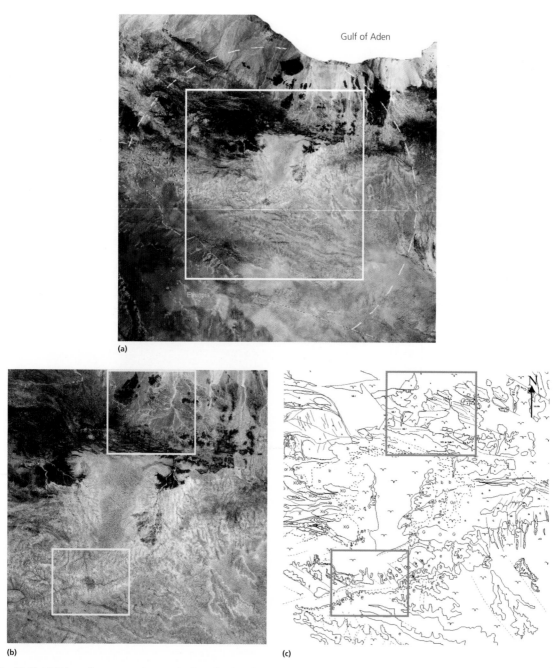

(a)

(b)   (c)

Fig. 23.14 (a) False colour composite mosaic of Landsat bands 531 RGB DDS of the entire study area for general interpretation of geological features: Sarahan desert sands appear in yellow and brown tones, with crystalline basement lithologies in dark blue and purple tones, and coastal alluvial sediments in the north in a variety of similar colours according to their provenance. The white box indicates the extent of the images shown in (b) and (c). (b) False colour composite of bands 457 RGB DDS showing carbonate rocks in greenish tones, basaltic volcanic in dark reds, desert sands in pinkish tones and crystalline basement lithologies and coastal alluvial materials in shades of blue; and (c) regional geological interpretation made from the Landsat imagery (lithological boundaries in black, structures in red). The two boxes in (b) and (c) indicate the locations of the detailed areas shown in Figs. 23.17–23.19. Field of view is 340 km. The location of Hargeisa is indicated as H (cyan).

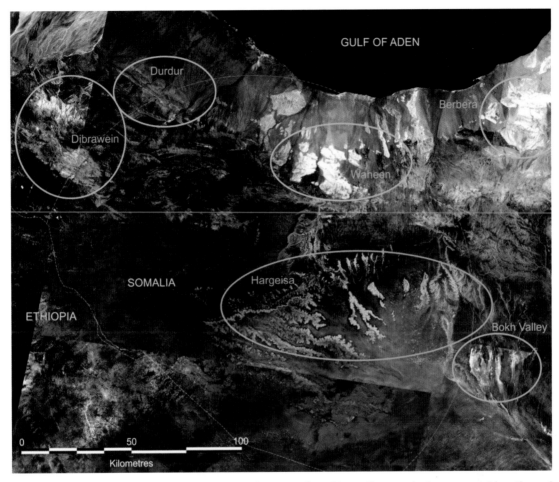

Fig. 23.15 Ratio of Landsat bands ((b5/b7)b6), revealing carbonates and possibly gypsiferous rocks that are potential aquifers and that form the target areas for detailed field-based investigation in phase two: Jurassic limestones (cyan); Eocene & Cretaceous limestones (green); transported gypsiferous debris in the Bokh Valley (pink). The 100 km circle and political borders (red and yellow dashed lines) are also shown for reference.

outcrops cover several million square kilometres – it represents the largest fossil aquifer in the world. It has been described as a hard ferruginous sandstone with several clay-rich intercalations.

Here we use relative absorption band depth images derived from ASTER VNIR, SWIR and TIR bands to highlight several target materials: iron oxides, carbonates (limestones and marbles) and silica-rich materials (sandstones, granites, gneisses), producing a composite of all three indices together, in which iron oxides/hydroxides (Fe-O and Fe-OH absorption) are enhanced and displayed in the red colour gun, silicate-rich materials (S-O absorption, by quartz sands, gneisses etc.) in

green, and carbonates (C-O absorption) or weathered and altered mafic volcanics (Mg-OH absorption) in blue. The result is a very brightly coloured image, shown in Figs. 23.16 (the entire), 23.17 (Hargeisa area) and 23.18 (Waheen area).

In Fig. 23.16, the desert sands of the Sahara in the south appear very noticeably in red tones. In the centre and to the north-west of Hargeisa, yellow colours indicate high silica content and relatively high iron oxides; here the basement is composed of crystalline gneisses and granites and the surface may have a covering of desert sand or some desert varnish. The bright greens in the northern part of the area represent almost pure

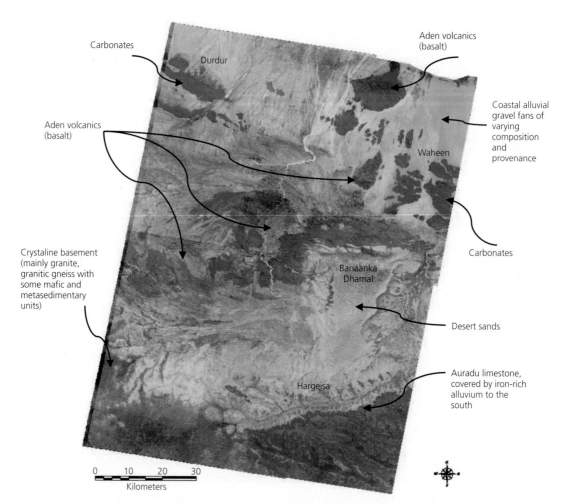

Fig. 23.16 Spectral index colour composite: iron oxides (red) highlighting red desert sands and ferromagnesian-rich lithologies, silica abundance (green) highlighting quartz sands and siliceous crystalline basement lithologies, and carbonate and Mg-OH abundance (blue), highlighting limestones and mafic volcanics; the mafic volcanics appear in magenta whereas limestones (and marbles) appear as pure blues, the coastal alluvial sediments appear in greens, oranges and reds whereas limestones covered with sands appear cyan-green and oxidised iron in sands appear yellow and orange.

quartz sands of the Waheen and coastal alluvial systems. The mafic Aden volcanics appear in magenta and bluish purple, indicating Mg-OH absorption of weathered (or altered) basalt and iron-oxide/hydroxide absorption. The remaining bright royal blue tones represent the carbonates of the Waheen (north), Dibrawein valley (west) and Auradu limestones (south).

The second objective is to attempt to identify the Nubian sandstone, specifically, if possible. The image in Fig. 23.16 suggests that silica-rich targets cover the entire northern part of this area, so that identification may not be straightforward. The 468 false colour composite in Fig. 23.17(a),

reveals the Auradu limestone in yellow tones, with the crystalline basement of the north and other areas around Hargeisa appearing in pale blue tones. There is also a narrow pink unit striking west to north-east, which in this band combination, indicates the presence of hydrated minerals. Looking at the silica index shown in Fig. 23.17(b), we see that it has two distinct silica-rich and silica-poor populations, and the boundary between them is also the boundary of the Auradu limestone (silica-poor) overlying the sand-covered granites and gneisses (silica-rich), information that we can then extract. Here we can see that the areas of greenish yellow tones in Fig. 23.17(d) have

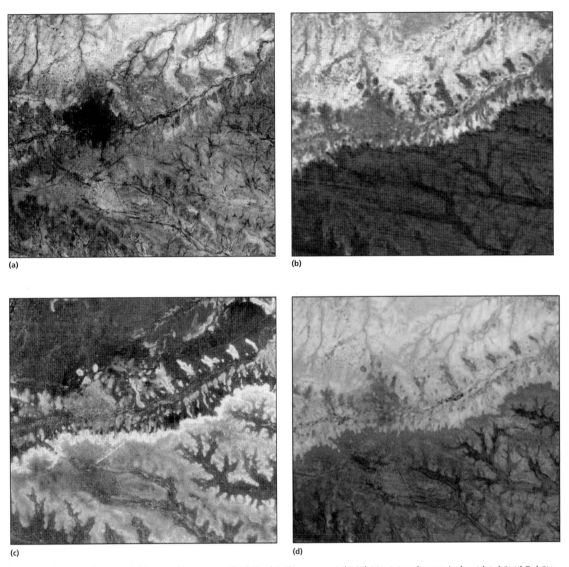

Fig. 23.17 The Hargeisa area: (a) iron-oxide content (b2/b1); (b) silica content (b13/b10); (c) carbonate index ((b6+b9)/(b7+b8)); and (d) ratio composite image produced by combining (a), (b) and (c) as RGB: desert quartz sands appearing yellow and orange, with a contribution from high silica content (green) and iron oxides (red); crystalline gneissic basement (beneath the sands) appears in yellow (high silica); the Auradu carbonates appear in bright blue tones; some patches of cyan-green appear in valleys where carbonate debris mixes with quartz sands. The upper surface of the Auradu limestone plateau, and the valleys cut into it, appear red, suggesting it has a strong iron-oxide content, perhaps produced by a surface coating or concretion. Field of view is 33 km in width.

the highest silica content. The city of Hargeisa appears as an area of low silica values. We know that the Nubian sandstone outcrops in this area but cannot easily distinguish between it and other silica-rich materials on the basis of spectral information.

If we turn to the ASTER DEM of the area we see a number of breaks in slope running across the image, to the south and the north of Hargeisa. If we calculate slope angle from the DEM [Fig. 23.18(c)], we see these breaks very clearly and notice that they coincide

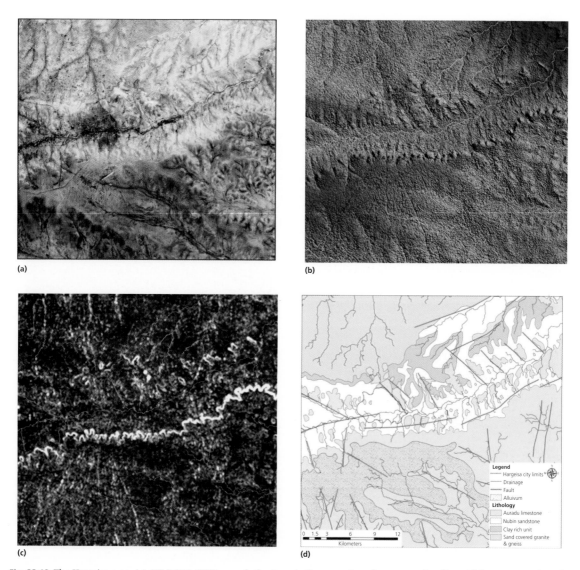

(a)

(b)

(c)

(d)

Fig. 23.18 The Hargeisa area: (a) 468 BCET (DDS) reveals the Auradu limestone in yellow tones; the silica-rich basement and sands to the north appear in pale bluish tones; a narrow pink unit is visible running west to north-east, and this colour indicates a lithology with a high clay content (the city of Hargeisa is visible as a dark blue patch in the left centre). (b) The ASTER GDEM is displayed as a shaded relief map with drainage network overlain (derived using the method described in Chapter 17). This clearly shows the breaks in slope that occur at the base of the Auradu limestone. There are very few clues to the existence of the Nubian sandstone in the topography. Between the pink layer and the yellow limestone sits the Nubian sandstone, but its spectral properties are so similar to other silica-rich lithologies it is indistinguishable from them. The field of view is as in Fig. 23.17 above. (c) ASTER DEM of the area, showing slope angle (degrees). (d) Surface geological interpretation with Nubian sandstone in shown in white.

with the scarp slopes of the dipping sedimentary units here. A subtle scarp slope can be seen in the west, that coincides with the narrow pink unit visible in Fig. 23.18(a); this can be followed eastwards but its outcrop pattern becomes more complex and it becomes more difficult to discern. The Nubian sandstone formation is reported to contain a number of clay-rich (perhaps mudstone) intercalations (Sultan *et al.* 2004); we suggest that the narrow pink band described previously represents the base of the Nubian sandstone formation. This example reveals the need for complementary datasets in conducting this type of work; the spectral information gives an honest indication of what is at the surface but the actual explanation for it may elude you until you are able introduce other data and/or to visit the site.

23.2.4.2 Alluvial gravel pockets in the Waheen

The objective here is to establish any indication of water stored in the localised alluvial basins along the coast. The images indicated that deposition (transport) direction is broadly to the north and the Gulf of Aden.

The area is characterised by a series of isolated rocky outcrops surrounded by alluvial gravels with meandering wadiis, and by the near total absence of vegetation [Fig. 23.19(a)]. The gravels appear to be silica-rich, as indicated by the green colour in Fig. 23.19(d). This implies either that there is little or no water stored in these alluvial gravels or that any water here is at a depth too great to support any plant growth. There are other similar pockets of alluvial material along the coast. Some of these show distinct signs of vegetative growth and persistence but these lie at too great a distance from Hargeisa. The standard false colour composite [Fig. 23.19(a)] suggests that no verdant vegetation exists in this area but a faint reddish colour in the south-western corner of the area suggests some plant growth. The NDVI shown in Fig. 23.19(b) suggests that some vegetation exists on the central outcrops and on the basement massif in the south-west but is absent from the alluvial plains around these outcrops.

The only vegetative growth in this area is restricted to the south-west, on the escarpment formed by the Aden volcanics [magenta tones in Fig. 23.19(d)]. The carbonates here are spectral slightly different from those of the Auradu (shown in Fig. 23.18 and Fig. 23.19), which are of Eocene age, and they more closely resemble the Jurassic limestones that outcrop further to the north-east. They contain gypsiferous units [appearing in cyan colours in the 942 composite image in Fig. 23.19(c)] and are thought to be of Jurassic age.

23.2.5 Summary

The phase one work allowed several areas to be identified as having potential for containing aquifers of one kind or another. These are Bannaanka Dhamal,

the Bokh Valley and the Dibrawein Valley. Several smaller structurally controlled basins containing recent and Jurassic sediments and alluvium were also identified, but many proved too small and at too great a distance from Hargeisa to represent any realistic potential.

The difficulty in distinguishing the Nubian sandstone formation from other quartz-rich materials illustrates the need to look at other information. Image processing can always provide at least a partial answer, but something else will be necessary to complement it and this may, again, involve fieldwork.

The phase two work allowed the following to be achieved:

- Confident discrimination of the Auradu limestone outcrops in the Hargeisa region and to the east and west of Bannaanka Dhamal, which had not previously been possible;
- Enhancement of quartz-rich solid and drift deposits in the Hargeisa and Bannaanka Dhamal regions;
- Enhancement of vegetation and soils at 1:50,000 scale;
- Confirmation of the nature and influence of faulting on the outcrop geometry around the Hargeisa region, which may affect local aquifer existence.

Sadly, phase three of this work was never completed because of the political unrest in this country. More unfortunately, this project also revealed that the main remit of finding bulk water resources to supply Hargeisa was highly likely to fail. This work suggested that although potential aquifers exist, they are either too small or too far from the city to provide viable economic supply. This project does, however, serve to illustrate a sensible strategy for other similar work.

23.2.6 Questions

1 What recommendations would you make for the development of this work, if you were given no budgetary or political restrictions?
2 Is it physically possible to distinguish between the different types of quartz (silica)-rich materials encountered here, spectrally or texturally? If so, how?
3 Having discriminated carbonates and mafic volcanic rocks (i.e. materials that exhibit CO or MgOH absorption in the SWIR), what other potential criteria could be used to differentiate between them?

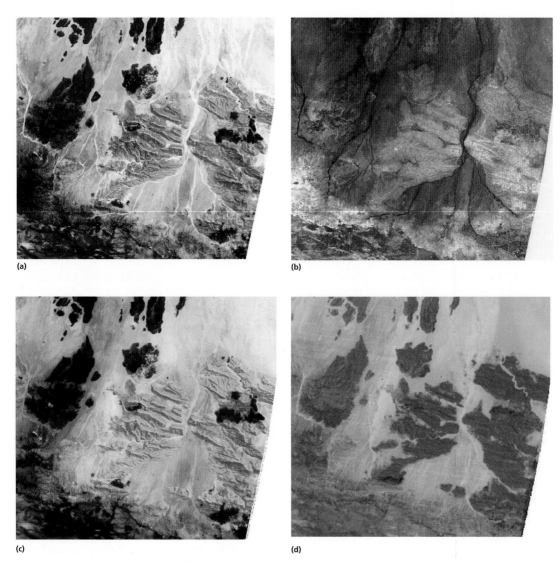

(a) (b)

(c) (d)

Fig. 23.19 Waheen region: (a) 321 DDS colour composite revealing the apparent absence of verdant vegetation; (b) NDVI indicating that some vegetation exists on the central outcrops and on the basement massif in the south-west (NDVI values range between 0 and 0.1); (c) 942 DDS false colour composite revealing central carbonates in green tones, volcanics in very dark blues and reds, gneisses in brown tones and alluvial gravels in pale pinks and yellows. Small outcrops of gypsum appear in bright cyan tones. (d) Ratio colour composite (R-iron oxides, G-silica content, B-carbonate/MgOH) as previously illustrated in Fig. 23.16.

23.2.7 References

Crowley, J.K., Brikey, D.W. & Rowan, L.C. (1989) Airborne imaging spectrometer data of the Ruby Mountains, Montana: Mineral discrimination using relative absorption band-depth images. *Remote Sensing of Environment*, 29, 121–34.

Rowan, L.C., Mars, J.C. & Simpson, C.J. (2005) Lithologic mapping of the Mordor, NT, Australia ultramafic complex by using the Advanced Spaceborne Thermal Emission and Reflection Radiometer (ASTER). *Remote Sensing of Environment*, 99, 105–26.

Sultan, M., Manocha, N., Becker, R. & Sturchio, N. (2004) Paleodrainage networks recharging the Nubian Aquifer Dakhla and Kufra Sub-Basins revealed from SIR-C and SRTM data. *Eos. Transactions AGU*, 85 (17), Joint Assembly Supplement, H31C-03.

PART IV

Summary

CHAPTER 24

Concluding remarks

Rather than simply repeating the key points from the preceding chapters, our intention here is to present more of our personal views on essential image processing and GIS techniques for remote sensing applications. We may sometimes reinforce key points that have already been made but, in doing so, we hope to convey our thoughts in a wider sense, beyond the strictest technical aspects of the book.

24.1 Image processing

1 Despite the fact that our presentation of essential image processing and GIS techniques is technique driven, the use of these techniques in their working context (within a remote sensing project) should always be application driven. In a real application project effectiveness and cost, rather than technical complexity, will dictate your data processing and we find that the simplest method is usually the best.

2 Image processing can never increase the information provided in the original image data, but the use of appropriate image processing can improve visualisation, comprehension and analysis of the image information for any particular application. There is no magic trick in image processing that will create something that does not already exist in the image, but it can be magical when image processing makes subtle things seem obvious and distinctive.

3 An image is for viewing! Always view the image before and after image processing; there are many fancy techniques to provide a so-called 'quantitative' assessment of image quality, but your eyes and brain are usually far more accurate and reliable.

4 Never blindly trust 'black box' functions in any image processing system. These functions use default settings and, for general purposes, are convenient and usually produce quite good results, but they are unlikely to produce the best results in more specific applications and are subject to unknown information loss. It is far more important to understand the principle of an image processing function than to master the operation of an image processing software package. In fact, only from a sound understanding of the principles can you fully explore and test the potential of an image processing system to its limits.

5 Colours and grey tones are used as tools for image information visualisation. Digital images can be visualized in grey tones, true colour, false colour and pseudo-colour displays. When using multi-spectral image data, beginning with simple colour composites is always recommended.

6 Our eyes are capable of recognising many more colours than grey levels, and it is therefore sensible to display a B/W image in a pseudo-colour display to get the most from it. Strictly speaking, though, a pseudo-colour image is no longer a digital image but an image of symbols; once the incremental grey levels are assigned to different colours, the sequential, numerical relationship between these grey levels is lost.

7 No matter which techniques are applied, the final step of displaying the resultant processed image will always be contrast enhancement, to stretch or compress the image to the dynamic range of the display device for optimal visualisation. The histogram provides the best guidance for all image contrast enhancement.

Image Processing and GIS for Remote Sensing: Techniques and Applications, Second Edition. Jian Guo Liu and Philippa J. Mason.
© 2016 John Wiley & Sons, Ltd. Published 2016 by John Wiley & Sons, Ltd.

8 Digital images are often provided in unsigned 8-bit integer type but, after image algebraic operations, the data type may change to real numbers. Always check the actual value range after such processing for appropriate display and correct understanding of the physical meaning of the derived images.

9 Image algebraic operations are very versatile. Four basic arithmetic operations (+, −, ×, /) enable quite a lot of image processing; more complicated algebraic operations can, of course, offer more, but it is important not to get lost in an endless 'number crunching game'. In using multi-spectral image data, spectral analysis of the intended targets is essential in guiding your image processing towards effectiveness and efficiency.

10 We usually achieve the enhancement of target features by highlighting them, but enhancement does not necessarily mean highlighting; it can also be achieved by suppression. For instance, a black spot in a white background is just as distinctive as a white spot in a black background. On the other hand, in a complicated and cluttered background, bright features are indeed more eye catching.

11 Filtering is introduced here as a branch of image processing but it is, in fact, embedded in the very beginning of image acquisition. Any imaging system is a Fourier transform (FT) filtering system. For instance, a camera is a typical optical FT system with an aperture (filter). A scene of the real world is an assembly of spatial variation in frequencies from zero to infinity. Infinitely high frequencies can only be captured by camera lenses of infinite aperture. An optical imaging system of limited physical size can only facilitate lenses with limited aperture and this will filter out the details represented by higher frequencies than its aperture allows.

12 In being a neighbourhood process, filtering enhances the relationships between the neighbouring pixels, but the relationship between a filtered image and its original image is not always apparent. An understanding of the principle and functionality of a filter is necessary for mastery of its use and for the proper interpretation of its results. For instance, both gradient and Laplacian are generally regarded as high pass filters but the functionality of these two types is quite different.

13 Convolution theorem links FT, in the frequency domain, to convolution, in the spatial domain, and thus establishes direct image filtering via a PSF (point spread function) without the computer-intensive FT and IFT. In a wider sense, convolution is just one form of local window-based neighbourhood processing tool. Many 'box' kernel filters are based on this loose concept and flexible structure but they do not necessarily conform to the strict definition of convolution.

14 Our perception of colour is based on three primaries – red, green and blue – but we do not usually describe a colour in this way, that is, red = 250, green = 200 and blue = 100. Rather, we would say that a colour is a bright, pale orange and in this way we are describing a colour based on its intensity, hue and saturation. Although colour perception is subjective and particular to an individual, we can generally establish the relationship between RGB and IHS from the RGB colour cube model, which gives us elegant mathematical solutions of RGB-IHS and IHS-RGB transforms.

15 The colour coordinate transformation based on the RGB colour cube assumes that our eyes respond to the three primary colours equally, when in fact they respond differently to each primary. For the same intensity level, we generally feel that green is much brighter than blue and red. This is because our eyes are most sensitive to the intensity variation of green despite the fact that our brains receive strong stimulation from red and that we can see red much farther away than the other two primaries. In a false colour composite of the same three bands the visual effect may, however, be quite different. Any feature displayed in red appears far more obvious and distinctive than if it is displayed in green or blue. Sometimes, simply by changing the colour assignment of a colour composite image, you may visually enhance different ground objects.

16 The saturation of a colour is generally regarded as the purity of a colour, but the actual physical meaning of colour purity is rather vague and unclear; there are several different definitions of saturation as a result. According to the RGB colour cube model, saturation is not independent of intensity; it is the portion of white light in a colour. A large portion of white light in a colour produces low saturation. The concepts and calculation formulae of saturation in this book are given following this logic. Some older definitions of colour saturation

ignore the dependency of saturation to intensity; they are not correct for the colour coordinate transformation based on the RGB colour cube.

17 We have introduced three image fusion techniques: intensity replacement (via RGB-IHS transformation), Brovey transform and smoothing filter-based intensity modulation (SFIM). All of these can be used to improve the spatial resolution of colour composites, but the RGB-IHS and Brovey transform techniques are subject to colour (spectral) distortion if the total spectral range of the low resolution colour composite is different from that of the high resolution image. SFIM is a spectral preservation data fusion technique that introduces no colour distortion, although its merits are somewhat offset by its demand for precise image co-registration, a problem that can be resolved by using the recently developed pixel-wise image co-registration technique based on advanced sub-pixel technology.

18 Principal component analysis (PCA) achieves the representation of a multi-spectral (or more generally, multi-variable) image dataset in an orthogonal coordinate system defined by the axes of its data cluster ellipsoid. Mathematically, PCA performs coordinate rotation operations from the covariance matrix of the data cluster. A PC image is a linear combination based on its corresponding eigenvector of the covariance matrix of the image dataset. Obviously, the data cluster of a multi-spectral image may not formulate a perfect ellipsoid and the PCA is therefore a statistical approximation.

19 PCA produces n independent PC images from m ($m \geq n$) correlated bands of a multi-spectral image. In most image processing software packages, the PCA function produces exactly the same number of PCs as the image bands rather than fewer. The point here is that the high rank PCs contain little information. For the six reflective spectral bands of a TM/ETM+ image, the $PC6$ image contains nearly no information but random and striping noise. The first five PCs thus effectively represent the image information from all six reflective bands and in this sense, $n < m$. On the other hand, dropping any image band will result in considerable information loss.

20 The idea of the de-correlation stretch (DS) is to reduce the inter-band correlation of three bands for colour composition by stretching the data cluster to make it spherical rather than elongated. As a result,

a colour composite with richer and more saturated colours but without hue distortion can be produced for visual analysis. DS can be achieved via either saturation stretch (IHSDS) or PC stretch (PCADS). We have proved that saturation stretch reduces inter-band correlation while PC stretch increases saturation. The direct de-correlation stretch (DDS) is the simplest and most effective technique for DS, which stretches saturation via a reduction of the achromatic component of colour without the involvement of forward and inverse transformations.

21 Instead of performing the coordinate rotation from the covariance matrix, as PCA does, the image coordinate system of the original image bands can be rotated in any direction by a matrix operation. The key point is to make a rotation that enhances the image information of specific ground object properties. The tasselled cap transformation is one such coordinate rotation operation; it produces component images representing the brightness, greenness and wetness of a land surface.

22 Classification is a very broad subject and this book only covers a small part of it: multi-variable statistical classification. Both unsupervised and supervised multi-variable statistical classifications of multi-spectral images involve the feature space partition of image data clusters on the basis of dissimilarity measures (classifiers or decision rules); these do not involve the spatial relationships among neighbouring pixels. Multi-spectral images can be spatially segmented into patches of different classes according to particular properties and spatial continuity. This type of classification is referred to as 'image segmentation', and it is not covered by this book. Hybrid multi-variable statistical classification in combination with image segmentation is also a valuable approach. 'Neural network' is another branch of classification characterised by its self-learning ability, general applicability and great computational inefficiency.

23 The ultimate assessment of the accuracy of the image classification of ground objects must be based on ground truth, even though the ground truth data themselves can never be 100% accurate. Facing such a paradox and the practical difficulties of ground truth data collection, several image data-based statistical methods have been proposed and, among them, the confusion matrix is the most

popular. The confusion matrix method produces a user's accuracy and a producer's accuracy as measures of the relative assessment of classification accuracy. These measurements are actually more applicable to the assessment of classification algorithm performance than to the indication of true classification accuracy.

24 Image co-registration and image rectification are closely related but different issues. Image rectification is used to rectify an image in compliance with a particular coordinate system. This can be done by co-registering the raw image to a map or to a geo-coded image. In the second case, the rectification is achieved via image co-registration. Image co-registration is used to make two images match one another but not necessarily any map projection or coordinate system. Image warping based on a poly-nomial deformation model derived from ground control points (GCPs) is one of the most widely used techniques for co-registration between two images or between an image and a map. The technique is versatile in dealing with images that have signifi-cantly different scale and distortion but, since it is based on a deformation framework or mesh, it does not achieve accurate co-registration at every image pixel. A different approach is the pixel-wise co-registration based on accurate measurements of displacement (disparity) between two images, at every pixel position. This can be achieved by local feature matching techniques, such as phase correla-tion. The technique ensures very high quality co-registration but its application is limited by its low tolerance to differing acquisition conditions between the two images (e.g. scale, illumination angle, spatial features and physical properties).

25 We briefly introduced interferometric synthetic aperture radar (InSAR) technology since it has become a very important tool in earth observation remote sensing. InSAR data processing is a significant subject in its own right, but it shares common ground with image processing, in optimising the visualisation of interferograms and analysing coher-ence images. For instance, ratio coherence is a typ-ical image enhancement technique for separating spatial de-coherence and temporal de-coherence.

26 Based on our ongoing research, a new chapter of phase correlation sub-pixel technology has been included in this new edition. The sub-pixel technology

extracts and exploits sub-pixel information bet-ween similar images via feature matching. This information is not directly obvious or available but enriches image capacity for wide applications through the use of precise image alignment, pixel-wise image co-registration, narrow baseline stereo matching for 3D data (e.g. digital elevation model [DEM]) generation and quantitative change detection.

24.2 Geographic information systems

1 The fundamental building blocks of all GIS are its data structures – how we encode our information right from the start can have far-reaching conse-quences when we arrive at the more in-depth ana-lytical stages of the work. The old adage of 'rubbish in means rubbish out', whilst seeming rather crude, can hold true. You must ensure that you choose the right data structure and that it is sensibly and intel-ligently structured for your needs. You should also ensure it is captured as accurately as possible from the start. At this stage, thinking about what you intend to do with the data will help when deciding how to structure it; that is, try to think several steps ahead. From the basic units of point and pixel, everything else grows.

2 For raster data we must consider the need to ade-quately represent the spatial variability of a phenomenon, bearing in mind the effect of spatial resolution and data quantisation on everything that you do beyond this step. On the other hand, we don't want to waste space by storing unnecessarily large datasets, so we try to avoid redundancy wherever possible, but this is less of a worry these days since computers are ever more powerful and increasingly efficient (almost lossless) data compression tech-niques are here to stay.

3 Consider very carefully whether you need to generate full topology in your vector dataset (remember about thinking ahead to the analytical stages). If you do, you will need to do this from the start or you face duplicating effort later on. On the other hand, you may not really need it at all, depend-ing on what you intend to do with your data, in the light of new intelligent vector data structures and fast computers.

4 You may not realise at the start but there will probably be many instances when you need to convert data between raster and vector formats, and vice versa. Clearly there are implications to the methods we choose and the kind of data involved. When you do make such conversions, it is important to pay attention to the detail of the process at pixel level, so that you are fully aware of what happens to your data.

5 It is impossible to underestimate the importance of georeferencing; it is what holds your GIS together! If you are at all concerned about the accuracy and precision of your results, then take care to understand your coordinate space (datum and projection) and the effect that it has, regardless of the application, whether it be mapping, measuring or planning. Seemingly small differences on the screen could translate to sizeable errors on the ground. Learn how to use your equipment properly, that is, your GPS, and understand the limitations of the device and what it tells you. The implications of getting it wrong should be a good incentive, but it is better that you learn first!

6 Also, bear in mind the spatial resolution of any raster data you're working and the effect this has on the precision and accuracy of the information you're recording or encoding. Also consider where in the world you are working and how this may affect your processing or calculations; for instance, consider what happens to the area represented by a square image pixel when you transform a raster dataset between polar and planar coordinate systems at high latitudes. The pixel will no longer be square and the data may need to be re-sampled, and may be re-sampled in the background without you being aware of it; the data values will be changed as a result. If the raster data have been generated by interpolation from point sample data, make sure the sample data are transformed from polar to planar coordinates before performing the interpolation, not after, to avoid this problem. Following from this, it is advisable to ensure that all raster datasets input to your model conform to the same coordinate system and spatial resolution at the start; many software packages demand this.

7 *Map algebra* operations represent the greatest area of overlap between image processing and GIS. These are our principal tools for manipulating raster and vector data. The map algebra concept was born out of raster processing but can also be applied to vector data. In fact there are many parallel operations; their behaviour may be slightly different, because of the differing data structures, but the principles and end results are conceptually the same. Sometimes the combination of two raster operations can be achieved using one vector operation; in such situations it may be better to use the vector route to achieve the result, but this will only be the case for categorical data. Once again, understanding of the principles of the process at pixel (and point) level helps you to grasp the overall meaning, relevance and reliability of the result. The fundamental link with image processing in many of these processes should always be borne in mind; if in doubt, refer back to Chapter 15.

8 The use of the Null (NoData) value is extremely important, so being aware of its uses and limitations will help to ensure that you achieve the right effect. Boolean logic is extremely useful in this and other contexts; it should be used wisely and carefully in conjunction with zeros and nulls.

9 Reclassification is probably one of the most widely used tools in spatial analysis, for the preparation of both raster and vector data. Understanding the nature of the input data value range and type, for example, via the histogram, is crucial. We stress the importance of considering the data type of both input and output layers (whether integers or real numbers, for instance, as pointed out in § 24.1 point 8), especially when using any arithmetic or statistical operations; after all, the output values should make reasonable sense.

10 We find that the geometric or attribute characteristics of one layer can be used to control operations performed on another and so can be used to influence the path of the analysis in an indirect way, and that this can be done using raster data, vector data or both. Remember also that operations do not necessarily have to be arithmetic or involve calculations to change the attributes and/or geometric properties of features, such as is the case with reclassifications and mathematical morphology.

11 *Geostatistics* is an extremely broad subject area that was born out of mineral exploration but now is applied in a great many fields. The key point of this topic is good data understanding, of both the nature of the data at the start, that is, through the use of histograms and semi-variograms, as well as how the data were collected and processed before you

received them. It is far too easy to proceed without appreciating the characteristics of the dataset, such as the possibility of hidden trends in the data, the presence of multiple statistically independent populations and/or the effects of spatial auto-correlation, clustering or randomness.

12 The next step is the choice of interpolation method, and this can be mind-boggling. The choice should be made according to the quantity, distribution and nature of your data, and on your intended use of the result, rather than on the ease of use of the tools at hand. In the end, you may not have a choice but you must understand these things first, if you are to make sense of and judge the quality and reliability of the result. Sometimes the most sophisticated method may be a waste of effort or may give incomprehensible results. On the other hand, using too simple a method may smooth out important detail or may not 'honour' the input data values. So it is important to consider your priorities and the intended use of the result in choosing your approach.

13 The exploitation of surface data is another area of strong overlap with image processing, in image convolution filtering. In image processing, we use the neighbourhood process of filtering to enhance spatial image textures, for instance, whereas here we are trying to extract quantitative measures of surface morphology, that is, measures that describe the 3D nature of that surface. Surfaces are extremely useful and powerful in a number of contexts; there's a lot more to it than merely visualising in three dimensions. Having said that, effective visualisation can be achieved in two dimensions as well as three; GIS is effective in communicating information using either of these.

14 Don't forget that surfaces can be described by both raster and vector, so once again try to think about the intended use of the surface before you decide what kind of data you need to use. Surfaces are in themselves a rich data source from which several other valuable attributes (parameters) can be extracted. We give a selection of surface parameters that are particularly useful to geoscientists of all kinds, such as slope, aspect and curvature, describing how they are calculated as well as how they might be used. We stress the very direct link between the extraction of these surface parameters and their image processing equivalents; if you need to understand

their derivation from first principles more fully, then we refer you to Chapter 4.

15 There are also many potential sources of surface data, and these are increasing in number and detail all the time. We provided a list of DEM data sources in Chapter 16, and some online resources, including the URLs of several useful sites for public data distribution, are listed in Appendix B.

16 Decision making represents another extremely broad subject area and one that extends well beyond digital mapping and the processing of spatial information; it is a kind of 'cautionary tale' that is relevant to all kinds of digital spatial analysis. The uncertainty we refer to affects everything we do in GIS and we need to be very aware of both the strengths and the limitations of our data, objectives and methods (and how they affect one another). This is the key to the successful application of the right kind of decision rules and to increasing the likelihood of making the 'good decisions' and thus the reliability of the result and to gaining the maximum reduction of 'risk' associated with a particular outcome.

17 Uncertainty creeps into our analysis in several different ways and in different forms, so we should try to understand the different situations in which these might occur, to help us handle the data correctly into the next stage of the analysis; for instance, in constructing our decision rules and then in choosing the most appropriate multi-criterion combination method. There are several ways of dealing with uncertainty in spatial analysis; you may need to use some or all of them. Being aware of the potential problem of uncertainty and doing something about it is, in general, more important than the particular method you choose to deal with it.

18 Multi-criterion analysis is a developing area, so Chapter 19 could easily have been expanded into three or four separate and weighty chapters, but we felt that this would be more detail than was appropriate in this book. We have therefore tried to give an overview of the potential methods, listed broadly in order of increasing complexity, and described their strengths and weaknesses to enable you to select the best methods for your own case, since the choice is varied. Data preparation is still the most vital step and one that will require the revision of all the preceding chapters, especially while you are

considering the structure of your 'model' and selecting input evidence layers, the values they contain and the way they are scaled, before you proceed to the combination method.

19 The most appropriate method of layer combination for your situation will depend on the nature of the case, which will then dictate the set of decision rules you employ; these depend on the quantity and quality of the data and on your understanding of the problem. In complex problems that involve natural phenomena, it is highly likely that the input evidence may not all contribute equally to the outcome. Hence you may need to rank and weight the inputs or you may need to control the influence they have within the model you use. Something that might not be apparent at the start is the possibility that all evidence may not contribute in the same way towards the problem; that is, the absence of something may be as significant as the presence of something else.

20 Such sophisticated tools as are available today mean that a highly complex and impressive-looking result is relatively easy to produce. Being able to repeat, justify and interpret that result is therefore even more important, so some method of validation becomes increasingly important too. One last important thing to remember in this context is that there is no absolutely correct method for combining multiple criteria in spatial analysis but, for any particular case, there is usually a more appropriate method; the world is indeed a fuzzy place.

24.3 Final remarks

There is clearly considerable common ground and synergy between the two technologies, image processing and GIS, and each one serves the other in some way. In our clear explanations in this book, we hope to have gone some way to de-mystifying the techniques that may previously have seemed inaccessible or a little daunting. These are all very useful tools that we have learned to use over the years and, from our collective experience and knowledge, we have also tried to inject some common sense and guidance wherever possible.

It is our intention that this book serves as a key to the door of the active and integrated application and research field of image processing and GIS within remote sensing. We hope the book is easy to read and to use, but how useful the book can be depends entirely on the reader's desire to explore.

APPENDIX A

Imaging sensor systems and remote sensing satellites

The key element in remote sensing technology is the sensor. A sensor detects and quantitatively records the electromagnetic radiation (EMR) from an object remotely; hence the term 'sensing'. For an object to be sensed, it must radiate energy either by reflecting the radiation impinging on it from an illumination source or by 'shining' by itself. If a sensor provides its own illumination source it is an *active sensor*; otherwise, if it depends on an independent illumination source, such as the sun, or the radiation from the object itself, such as the earth thermal emission, it is then a *passive sensor*. Synthetic aperture radar (SAR) is a typical active sensor system, as it sends microwave radiation pulses to illuminate the target area and receives the returned signals to produce an image. In contrast, the most commonly used panchromatic and multi-spectral optical sensors are typical passive sensors. A camera is a passive sensor in general, but it can become an active sensor when it is used with a flashlight in the dark. In this case, the camera provides its own light source to illuminate the object and meanwhile takes a picture.

A.1 Multi-spectral sensing

As a passive sensor, a multi-spectral imaging system images the earth by recording either the reflected solar radiation or the emitted radiation from the earth. The sun is the primary illumination source for the earth. For earth observation remote sensing, most passive sensor systems operate under solar illumination during the daytime; such systems range from aerial photography to satellite-borne multi-spectral scanners. These sensors detect *reflected* solar energy from the land surface to produce panchromatic and multi-spectral images. Ignoring the minor factors, we can present such an image as

$$M_r(\lambda) = \rho(\lambda)E(\lambda) \qquad (A.1)$$

where $M_r(\lambda)$ is the reflected solar radiation of spectral wavelength λ by the land surface, or an image of spectral band λ, $E(\lambda)$ is irradiance, that is, the incident solar radiation energy upon the land surface, while $\rho(\lambda)$ is the reflectance of land surface at wavelength λ.

$E(\lambda)$ is effectively the topography, as determined by the geometry of the land surface in relation to illumination. The spectral reflectance, $\rho(\lambda)$, is a physical property that quantitatively describes the reflectivity of materials on the land surface at wavelength λ. The selective absorption and reflection by a material result in variation of spectral reflectance in a spectral range, giving a unique signature for this substance. It is therefore possible to determine the land cover types or mineral compositions of the land surface based on spectral signatures using multi-spectral image data. Reflective spectral remote sensing is one of the most effective technologies for studying the earth's surface as well as that of other planets.

The US Landsat satellite family, Thematic Mapper (TM) and Enhanced Thematic Mapper Plus (ETM+), and the French SPOT satellite family HRV (High Resolution Visible) are the most successful earth observation systems, providing broadband multi-spectral and panchromatic image data of global coverage. As shown in Table A.1, this type of sensor system operates in the visible spectral range with bands equivalent to three primary colours: blue (380–440 nm), green (440–600 nm) and red (600–750 nm); as well as in the near infrared

Image Processing and GIS for Remote Sensing: Techniques and Applications, Second Edition. Jian Guo Liu and Philippa J. Mason.
© 2016 John Wiley & Sons, Ltd. Published 2016 by John Wiley & Sons, Ltd.

Table A.1 Comparison of the spectral bands of ASTER, Landsat TM/ETM+, Landsat OLI & TIRS and SPOT.

Sensor System	Terra-1 ASTER			Landsat 3–7 TM/ETM+			Landsat 8 OLI & TIRS			SPOT 1–3 HRV, SPOT 4 HRVI, SPOT 5 HRG		
Spectral Region	Band	Spectral Range (μm)	Spatial Res. (m)	Band	Spectral Range (μm)	Spatial Res. (m)	Band	Spectral Range (μm)	Spatial Res. (m)	Band	Spectral Range (μm)	Spatial Res. (m)
VNIR	1	0.52–0.60	15	1	0.45–0.53	30	1 (Coastal)	0.435–0.451	30	1	0.50–0.59	20
	2	0.63–0.69	15	2	0.52–0.60		2	0.452–0.512	30	2	0.61–0.68	
	3N	0.78–0.86		3	0.63–0.69		3	0.533–0.590		3	0.79–0.89	
	3B	0.78–0.86		4	0.76–0.90		4	0.636–0.673		Pan	SPOT1–3: 0.51–0.73	10
				ETM+ Pan	0.52–0.90	15	5	0.851–0.879			SPOT4: 0.61–0.68	10
							8 (Pan)	0.503–0.676	15		SPOT5: 0.48–0.71	2.5–5
SWIR	4	1.60–1.70	30	5	1.55–1.75	30	9 (Cirrus)	1.363–1.384	30	4	1.58–1.75	20
	5	2.145–2.185		7	2.08–2.35		6	1.566–1.651		Band 4 only on SPOT 4 HRVI (High Resolution Visible Infrared) and SPOT 5 HRG (High Resolution Geometric)		
	6	2.185–2.225					7	2.107–2.294				
	7	2.235–2.285										
	8	2.295–2.365										
	9	2.360–2.430										
TIR	10	8.125–8.475	90	6	10.4–12.5	120 (TM) & 60 (ETM+)	10 (TIR-1)	10.60–11.19	100			
	11	8.475–8.825										
	12	8.925–9.275					11 (TIR-2)	11.50–12.51	100			
	13	10.25–10.95										
	14	10.95–11.65										

(NIR) (750–1100 nm) and short-wave infrared (SWIR) (1550–2400 nm) ranges. The number of bands and the band spectral width in the VNIR (visible nearer infrared) and SWIR spectral ranges are dictated by atmospheric windows and sensor design. For instance, the spectral width of the SWIR bands needs to be much wider than the visible bands if the same spatial resolution is to be achieved. This is the case of TM/ETM+ band 5 and 7, because the solar radiation in the SWIR is significantly weaker than that in the visible spectral range.

In general, the term 'broadband' means that the spectral range is significantly wider than a few nanometres, as in the case of the hyperspectral sensor system described later. Broadband reflective multi-spectral sensor systems are a successful compromise between spatial resolution and spectral resolution. With relatively broad spectral bands, a sensor system offers reasonable spatial resolution with high signal-to-noise ratio (SNR) and, while operating in a wide spectral range from VNIR to SWIR, can provide images of multi-spectral bands enabling the identification of major ground objects and the discrimination of various land cover types. With dramatic improvements in sensor technology, from mechanical scanners to push-broom scanners, and to digital cameras, the spatial resolution of broadband multi-spectral imagery is improving all the time. For sun-synchronous near-polar orbiting satellites, spatial resolution has been improved from 80 m (Landsat MSS) in the 1970s to a few metres and then sub-metres, on current systems, as shown by the few examples shown in Table A.2.

The VNIR spectral range is used by nearly all the broadband reflective multi-spectral sensor systems. This spectral range is within the solar radiation peak and thus allows generation of high resolution and high SNR images. It also covers diagnostic features of major ground objects, for instance:

- Vegetation: minor reflection peak in green, absorption in red and then significant reflection peak in NIR. The phenomenon is often called 'red edge'.
- Water: strong diffusion and penetration in blue and green and nearly complete absorption in NIR.
- Iron oxide (red soils, gossans etc.): absorption in blue and high reflectance in red.

Many satellite sensor systems choose not to use the blue band to avoid the very strong Rayleigh scattering effects of the atmosphere that make an image 'hazy'. A popular configuration is to offer three broad spectral bands in green, red and NIR, such as the case of SPOT and many recent high spatial resolution spaceborne sensors (Tables A.1 and A.2).

The SWIR spectral range is regarded as the most effective for lithological and mineral mapping because most rock types have high reflectance in 1.55–1.75 μm and because hydrous (clay) minerals (often products of hydrothermal alteration) have diagnostic absorption features in the spectral range 2.0–2.4 μm. These two SWIR spectral ranges correspond to Landsat TM/ETM+ band 5 and 7, and to ASTER (Advanced Spaceborne Thermal Emission and Reflection Radiometer) band 4 and bands 5–9 (Table A.1). SWIR sensor systems are technically more difficult and complicated because the SWIR detectors have to operate at very low temperatures, which therefore require a cooling system.

With six broad reflective spectral bands, Landsat TM has provided the best spectral resolution of the broadband sensor systems for quite some time. The six broad reflective spectral bands are very effective for the discrimination of various ground objects, but they are not adequate for specific identification of rock types and mineral assemblies (of the hydrated type mentioned above, which are pathfinders for economic mineral deposits). Here lies the demand of a sensor system with a much higher spectral resolution, at a few nanometres bandwidth, to detect the very subtle spectral signatures of materials of land surface. This demand has led to the development of hyperspectral sensor systems.

ASTER, a push-broom scanner for VNIR and SWIR bands, represents a 'transitional' sensor system, somewhere between broadband multi-spectral and hyperspectral narrow band sensing. It is an integrated system of three scanners: a VNIR push-broom scanner with three broad spectral bands; a SWIR push-broom scanner with six narrow spectral bands; and a thermal infrared (TIR) across-track mechanical scanner with five thermal bands (Table A.1). The system combines good spatial resolution in the VNIR bands with high spectral resolution in SWIR and multi-spectral thermal bands, which are very useful in wide applications of earth observation. The three 15 m resolution VNIR bands are adequate for distinguishing broad categories of land surface such as vegetation, water, soils, urban areas, superficial deposits and general rock outcrops, whilst the six narrow SWIR bands of 30 m resolution and five TIR bands of 90 m resolution have potential for mapping major mineral assemblages (rock forming and alteration)

Table A.2 Some satellite-borne very high resolution broadband sensor systems.

Satellite	Launch Time and Status	Spatial Resolution (m)		Spectral Range (μm)	
		Pan	MS	Pan	MS
GeoEye-2 (WorldView-4) [S]	2016	0.34	1.36	0.45–0.80	0.45–0.51 0.51–0.58 0.655–0.69 0.78–0.92
WorldView-3 [S]	2014	0.31	1.24 VNIR 3.70 SWIR 30.0 CAVIS	0.45–0.8	0.405–2.04 (8 bands) 0.1.195–2.365 (8 bands) 0.405–2.245 (11 CAVIS bands)
SkySat-1 & -2 [V]	2013 & 2014	0.9	2.0	0.4–0.9	0..45–0.515 0.515–0.595 0.605–0.695 0.74–0.90
SPOT-6 & -7 [S]	2012 & 2014	1.5	6.0	0.48–0.7	0..455–0.525 0.53–0.59 0.625–0.695 0.76–0.89
Pleiades 1A & 1B [S]	2012	0.5	2.0	0.48–0.83	0..43–0.55 0.49–0.61 0.6–0.72 0.75–0.95
WorldView-2 [S]	2009	0.46	1.8	0.45–0.8	0.4–0.45 0.45–51 0.51–0.58 0.585–0.625 0.63–0.69 0.705–0.745 0.77–0.895
GeoEye-1 [S]	2008	0.41	1.64	0.45 – 0.80	0.45–0.51 0.51–0.58 0.655–0.69 0.78–0.92
Worldview-1	2007 In operation	0.46	–	0.45–0.90	–
Orbview-3	2003 In operation	1	4	0.45–0.90	0.45–0.52 0.52–0.60 0.625–0.695 0.76–0.90
QuickBird	2001 In operation	0.62	4	0.45–0.90	0.45–0.52 0.52–0.60 0.63–0.69 0.76–0.89
Ikonos 2 [S]	1999 In operation	0.82	4	0.45–0.90	0.45–0.53 0.52–0.61 0.64–0.72 0.77–0.88

[S], stereo capability; [V], onboard HD video.

and lithologies. Another unique advantage of ASTER is that it has along-track stereo capability. The VNIR scanner has a backward viewing telescope to take NIR images in addition to its nadir telescope for the three VNIR bands. Thus nadir and backward viewing NIR images are taken simultaneously, forming along-track stereo image pairs. The along-track stereo image pairs enable generation of digital elevation model (DEM) data. The along-track stereo imaging has become a common sensor configuration for stereo imaging for many recent very high resolution sensing satellites.

Thus far, we have not mentioned the panchromatic band image specifically. We can regard panchromatic imagery as a special case of broadband reflective multi-spectral imagery with a wide spectral range, covering a large part of the VNIR spectral range, which can achieve a high spatial resolution.

A.2 Broadband multi-spectral sensors

Aerial photography, using a large-format camera, is the earliest operational remote sensing technology for topographic surveying. Spaceborne remote sensing, in earth observation, began on Landsat 1 with its MSS (Multi-spectral Scanner) and RBV (Return Beam Vidicon) camera, which was launched on 23 July 1972. These instruments captured and transmitted images of the earth electronically; these images were then distributed to users in a digital format as digital image data for the first time. The concept and technology of digital imagery gave birth to satellite remote sensing. An earth observation satellite images the earth surface continuously from its orbit and sends the images back to the receiving stations on the ground electronically.

The development of sensor technology is now mainly focused on improving spatial resolution and spectral resolution. For a given sensor system, its spatial resolution is dictated by the minimal energy level of EMR that can make a signal distinguishable from the electronic background noise of the instrument, that is, the dark current. This minimum energy of EMR is proportional to the product of radiation intensity over a spectral range, IFOV (instant field of view) and the dwell time (equivalent to exposure time of a camera).

The IFOV is decided by the spatial sample density of an optical sensor system and it determines the pixel size of the image. For a given dwell time and spectral range, the larger the IFOV, the more energy will be received by the sensor, but the spatial resolution will be lower. To improve spatial resolution, the IFOV must be reduced, but to maintain the same energy level, either the dwell time or spectral range, or both, must be increased. When a sensor, which has a dwell time fixed by the sensor design and platform orbit parameters, receives reflected solar radiation from the earth, it may record the energy in a broad spectral range as a single image, that is, a panchromatic image, at a relatively high resolution. It may also split the light into several spectral bands and record them separately into several images of narrower spectral range, that is, multi-spectral images. In this case, the energy that reaches the detectors of each narrow spectral range is significantly weaker than in the panchromatic mode. To achieve the same energy level, one solution is to increase the size of IFOV, that is, to reduce spatial resolution. This is the reason that nearly all optical imaging systems achieve higher resolution in their panchromatic band than in the multi-spectral bands (Tables A.1 and A.2). For instance, SPOT 1-3 HRV panchromatic band has 10 m resolution while XS (multi-spectral) bands 20 m resolution.

The other way to improve the spatial resolution in both panchromatic and multi-spectral imagery is to increase the dwell or exposure time. This has been an important consideration in sensor design although the capacity for increasing the dwell time is very limited, for both airborne and spaceborne remote sensing, since the image is taken from a moving platform; long exposure time will blur the image.

A.2.1 Digital camera

With few exceptions, passive sensor systems are essentially optical camera systems. As shown in Fig. A.1, the incoming energy to the sensor goes through an optical lens and is focused onto the rear focal plane of the lens, where the energy is recorded by radiation-sensitive media or a sensor device, such as a CCD (charge-coupled device).

A digital camera is built on a full 2D (two-dimensional) CCD panel; it records an image through a 2D CCD panel linking to a memory chip. With great advances in multi-spectral 2D CCD technology, the new generation of passive sensors will be largely based on the digital camera mechanism that takes an image in an instantaneous frame rather than scanning line by line. The consequence is that the constraints on platform flight parameters can be relaxed, image resolution (spatial and spectral) can be improved and image geometric correction processing can be streamlined.

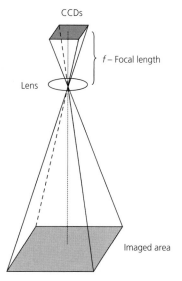

Figure A.1 Basic structure of an optical camera system.

A.2.2 Across-track mechanical scanner

The early design of spaceborne passive sensor systems was constrained by the technology of the primary sensor unit, the CCD. The mechanical scanner has been the ideal solution to achieve multi-spectral imaging at relatively high resolution, since it is based on a simple mechanical device that uses only a few CCDs. Figure A.2 is a schematic diagram showing the principle of an across-track mechanical multi-spectral scanner. The optical part of a scanner is essentially a camera but the image is formed pixel by pixel, scanned by a rotating mirror, and line by line, as the platform (aircraft or satellite) passes over an area. The range of a scan line is called the *swath*. The light reflected from the land surface reaches the rotating mirror that rotates at a designated speed and thus views different positions on the ground along a swath during its rotation scan cycle. The rotating mirror diverts the incident light through the scanner optics and then the light is dispersed into several spectral beams by a spectral splitting device (a prism or interference filters). The multi-spectral spectral beams are then received by a group of CCDs that sample the light at a regular time interval. By the time the scanner finishes scanning one swath, the satellite or aircraft has moved forward along its track to the

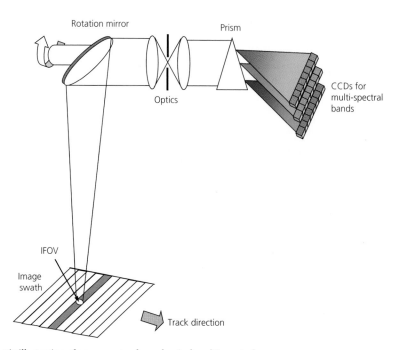

Figure A.2 A schematic illustration of an across-track mechanical multi-spectral scanner.

position for the next scan. One scan swath can be a single image line or several image lines depending on the sensor design and the synchronisation between flying speed, swath width, altitude of the satellite or aircraft and required image resolution. In this way, a scanner can achieve a very large image using limited CCD; although mechanically rather complex, a scanner relies less on the CCD technology.

The MSS, with four spectral bands on board Landsat 1–3, is a classic example of mechanical scanner. It is a one-way scanner that scans in one direction of mirror rotation only, and with an empty return run. Such a design makes compensation for earth rotation easier, since the earth rotates a fixed distance along the swath direction in each scanning cycle for a satellite imaging from a circular near-polar sun-synchronous orbit. The inactive return runs waste valuable time for imaging and cause a shorter dwell time in the active runs, thus reducing image spatial resolution. The TM, onboard Landsat 4–7, is a significantly improved scanner with six reflective spectral bands and one thermal band (Table A.1). It is a two-way scanner that scans in both directions. So for the same width of swath, the two-way scan allows the mirror to rotate more slowly, thus increasing the dwell time of the CCDs at each sample position. This configuration improves both spatial and spectral resolution. To compensate for the earth's rotation effects, the geometric correction for TM is more complicated than that for MSS because one scan direction is *for*, and the other *against*, the earth's rotation.

A.2.3 Along-track push-broom scanner

With rapid developments in CCD technology, a more advanced push-broom scanner has become dominant in broadband multi-spectral sensor design since the successful launch of SPOT-1 on 22 February 1986. As shown in Fig. A.3, the key difference between a push-broom scanner and an across-track mechanical scanner is that it does not have a mechanical part for pixel-by-pixel scanning along the swath direction. Instead of a rotating mirror, a push-broom scanner has a steerable fixed mirror to enable the sensor to image its swath either at nadir or off nadir. A line array panel of CCDs covering the whole imaging swath is mounted at the rear of the spectral dispersion device. The push-broom scanner images an area, line by line, along the track when the sensor platform (a satellite or an aircraft) passes over, just like pushing a broom forward to sweep the floor.

Since one swath of image is generated simultaneously, the dwelling time for each CCD representing an image pixel can be as long as a whole swath scanning time for a mechanical scanner. With significantly increased dwell

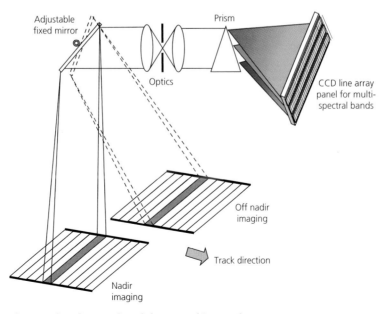

Figure A.3 A schematic diagram of an along-track push-broom multi-spectral scanner.

time, a push-broom scanner achieves much higher resolution. Based on advanced CCD technology, the push-broom scanner is also much simpler than the mechanical scanner in structure, and the data geometric correction is less complicated. Without a mechanical part, the system is robust and reliable. The number of CCDs in the line array, in swath direction, decides the size and the resolution of the image data generated. For instance, the SPOT HRV has 6000 CCDs per line for its panchromatic band and 3000 CCDs per line for its multi-spectral bands. The system therefore produces panchromatic images of 6000 pixels per line at 10 m resolution and multi-spectral images of 3000 pixels per line at 20 m resolution.

A.3 Thermal sensing and TIR sensors

A thermal sensor is also passive but the radiation energy that the sensor receives is emitted from the earth's surface rather than reflected from it. Thermal sensing does not therefore need an illumination source since the target itself is the illumination source. The earth's surface can be approximated as a blackbody of 300 K and, using Wien's law, we can calculate that the radiation peak of the earth is at about 10 μm. In this spectral range, radiation can be sensed and measured by temperature rather than visible brightness; it is thus called thermal sensing. Different natural materials on the land surface have different thermal radiation properties, and thermal sensors are therefore a useful tool for geological and environmental studies.

There are quite a few airborne TIR sensor systems; for example, the Thermal Infrared Multispectral Scanner (TIMS) with six bands in the 8.2–12.2 μm spectral region, which was developed in 1982. Landsat TM and ETM+ have a broad thermal band at wavelength of 10.4–12.5 μm. The ASTER on board of the Terra-1 satellite comprises a multi-spectral thermal system with five narrow thermal bands as shown in Table A.1.

In general, a broadband TIR sensor operating in the 8–14 μm spectral range images the radiation temperature of the land surface while the narrower band multi-spectral thermal image data present the thermal spectral signatures of materials of land surface. It is important to know that daytime thermal images are fundamentally different from those acquired at night. Daytime thermal images are dominated by topography, as governed by

the geometry between slopes and solar radiation, in the same way as in reflective multi-spectral images, whereas the pre-dawn night thermal images, which are nearly solely determined by emission from the earth surface, show better the thermal properties of ground materials.

In both systems, TM/ETM+ and ASTER, the spatial resolution of thermal bands is significantly lower than that of the reflective multi-spectral bands, as shown in Table A.1. One reason is that the interaction between thermal energy (or heat) and atmosphere is more complex than in the case of VNIR and SWIR energy. Heat can be transmitted in the air not only by radiation but also by air circulation. Second, the solar radiation impinging on the earth in the TIR spectral range and the direct thermal emission from the earth are both very weak compared to the energy intensity of the earth-reflected solar radiation in VNIR and SWIR spectral ranges.

So far most thermal sensors are of across-track mechanical scanner type as shown in Fig. A.4. The major difference of a thermal scanner from a reflective multi-spectral scanner is that it needs a cooling system to maintain the TIR detector at very low temperature for maximum sensitivity. For instance, the thermal sensor of the Landsat TM is surrounded by liquid nitrogen at 77 K and stored in an insulated vessel. In the ASTER system, a cryo-cooler is used to maintain the TIR detectors at 80 K. A blackbody plate is used as an onboard calibration reference that is viewed before and after each scan cycle, thus providing estimation of instrument drift. This is essential to maintain the accuracy and consistency of a TIR instrument. The temperature sensitivity of a modern TIR sensor system can be as high as 0.1 K. To fully represent the sensitivity, many thermal IR sensors use 10–12 bits quantization to record data, such as ASTER multi-spectral thermal band images, which are 12-bit integer data.

A.4 Hyperspectral sensors (imaging spectrometers)

Passive sensor technological development is continually aiming at higher spatial and spectral resolutions. Hyperspectral sensor systems represent a revolutionary development in the progress of optical sensor spectral resolution, which may be as high as a few nanometres, and can generate nearly continuous spectral profiles of

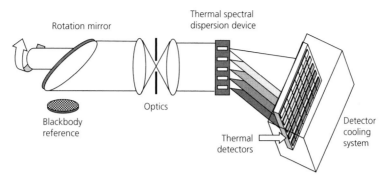

Figure A.4 A schematic diagram of a thermal scanner.

Table A.3 Some hyperspectral sensors (airborne and satellite-borne*).

Instrument	Spectral Range (nm)	Bandwidth (nm)	No. Bands	Spatial Resolution (m)
AisaEAGLE	400–970	3.3	488	Variable
AisaFENIX	380–2500	3.4	620	Variable
Aisa-HAWK	970–2500	6.3	256	Variable
AisaOWL	7500–12500	53	96	Variable
AVIRIS	400–2400	9.6	224	20
CASI-2	405–950	1.8	288	Variable
CHRIS	415–1050	1.3–12	19–63	Variable
DAIS 7915	400–12600	1–45	79	Variable
HIRIS	400–2500	10–20	128	100
HJ-1A	400–950	5	128	Variable
HyMap	450–2480	15–17	126	Variable
Hyperion*	400–2500	10	242	30
Hyperspec VNIR-SWIR	400–2500	5 & 10	384–1600	Variable
HySI*	400–950	15	80	Variable
MARS CRISM	363–3920	6.55	Variable	Variable
MIVIS	430–1270	20–54	102	Variable
PROBE-1	400–2500	12–16	128	Variable
SEBASS	3000–5500 & 7500–13500	Variable	128	Variable
VNIR-640	400–1000	5	128	Variable

land surface materials. A hyperspectral sensor system is a combination of the spatial imaging capacity of an imaging system with the spectral analytical capabilities of a spectrometer. Such a sensor system may have several hundred narrow spectral bands with a spectral resolution of the order of 10 nm or narrower. Imaging spectrometers produce a complete spectrum for every pixel in the image; the dataset is truly a 3D data cube that allows identification of materials, rather than mere discrimination as with broadband sensor systems. The data processing methodology and strategy are therefore different from broadband images in many aspects. It is

more important to analyse the spectral signature for each pixel rather than to enhance the image to improve visualisation, although the latter is still essential in examining the imagery data and presenting the results.

One of the earliest and the most representative hyperspectral systems is JPL's Advanced Visible Infrared Image Spectrometer (AVIRIS) (Table A.3). Figure A.5 shows the general principle of hyperspectral systems. The incoming EMR from the land surface goes through the sensor optics and is then split into hundreds (e.g. 224 for AVIRIS) of very narrow spectral beams by a spectral dispersion device (e.g. interference filters), and finally

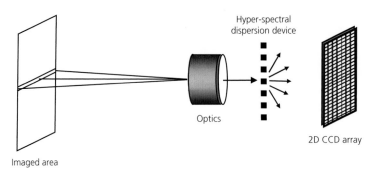

Figure A.5 Principle of an imaging spectrometer (hyperspectral system).

the spectral beams are detected by arrays of CCDs corresponding to, for instance, 224 spectral bands. A hyperspectral system can be either an across-track mechanical scanner, with a small number of detectors for each band, or an along-track push-broom scanner, with a panel of hundreds of line arrays of CCDs.

A.5 Passive microwave sensors

The land surface is also an effective radiator at microwave range, although microwave emission is significantly weaker than thermal emission. As another type of passive sensor, microwave radiometers are designed to image the emitted radiation from the earth surface at this spectral range.

Thermal radiation from natural surfaces, such as the land surface, extends from its peak in the thermal infrared region into the microwave region. An earth observation microwave imagining radiometer operates in this spectral region to receive microwave radiation from the earth. As a passive sensor system, it is important to understand that a microwave radiometer is fundamentally different from a radar sensor, which is a *ranging system*. The only similarity between the two is that they both operate in the microwave spectral range. A passive microwave sensor system, a microwave imaging radiometer, works more like a thermal sensor system. It collects emitted energy radiated from the earth in the microwave spectral range and provides useful information relating to surface temperature, roughness and material dielectric properties. This type of sensor has been used for global temperature mapping, polar ice mapping and regional soil moisture monitoring.

A spaceborne microwave imaging radiometer is often a multi-channel scanner such as the SMMR (Scanning

Multichannel Microwave Radiometer), onboard Seasat and Nimbus (1978), and the TRMM Microwave Imager, onboard the Tropical Rainfall Measuring Mission (TRMM) satellite (1997). It consists of an antenna together with its scanning mechanism, a receiver and a data handling system. The received emitted microwave signals are closely related to the observation angle and the path length in atmosphere. Ensuring that these scanning parameters are constant can significantly increase the accuracy of the derivation of the surface parameters from microwave brightness temperature. A conical scan configuration is popular for passive microwave scanners. As shown in Fig. A.6, the antenna observation direction is offset at a fixed angle from nadir, rotating its scan around the vertical (nadir) axis and thus sweeping the surface of a cone. If the scan is configured for

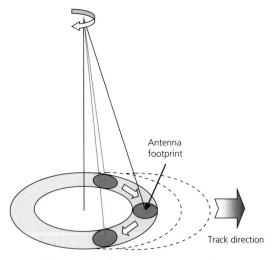

Figure A.6 The passive microwave sensor conical scanning mechanism.

full 360°, double coverage fore and aft of the spacecraft is obtained. With the forward motion of the satellite along its orbit, a belt of land surface is imaged. Obviously, in a conical scanning geometry, the observation angle and distance to any scanned position are constants. Spaceborne passive microwave scanners are usually of low spatial resolution, from several kilometres to several tens of kilometres, because of the weak signal in microwave spectral range.

A.6 Active sensing: SAR imaging systems

Radar is essentially a ranging or distance-measuring device. Nearly all the imaging radar systems are configured as *side-looking*; referred to as *side-looking radar* (SLR). The reason is that as a ranging system, radar forms an image by recording the position of return signals based on time. If a radar system is configured to view the both sides of the platform (aircraft and satellite) symmetrically, the return signals from both sides in an equal distance will be received at the same time, causing ambiguity.

A radar system transmits microwave pulses in a carrier wavelength/frequency and then receives the echoes of these pulses scattered back by ground surface objects. Commonly used microwave pulse carriers' wavelength and frequency (radar bands) are listed in Table A.4. The code letters for radar bands in the table were given during World War II and remain to this day.

Radar image data is configured in relation to two coordinates: slant range and azimuth. Slant range corresponds to the two-way signal delay time. By measuring the time delay between the transmission of a pulse and the reception of the backscattered 'echo' from different

Table A.4 Radar bands, wavelengths and frequencies.

Band	Wavelength λ (cm)	Frequency MH_z (10^6 cycles sec^{-1})
K_a	0.75–1.1	40,000–26,500
K	1.1–1.67	26,500–18,000
K_u	1.67–2.4	18,000–12,500
X	2.4–3.75	12,500–8,000
C	3.75–7.5	8,000–4,000
S	7.5–15	4,000–2,000
L	15–30	2,000–1,000
P	30–100	1,000–300

targets, their distance to the radar and thus their location can be determined, and in this way, a radar image is built in the slant range. In the azimuth direction, the image is built according to the pulse number sequence. As the radar platform moves forward, it transmits microwave pulse beams to scan on one side of its flight path, strip by strip, and simultaneously records the backscattered signals. As such, a 2D radar image is built up.

The azimuth resolution R_a of a *real aperture* SLR is a function of radar wavelength λ, the slant range S and the radar antenna length, D_r:

$$R_a = \frac{S\lambda}{D_r} \tag{A.2}$$

According to the formula above, the azimuth resolution R_a is inversely proportional to the length of radar antenna D_r. For a give radar wavelength and slant range, the longer the antenna the higher the azimuth resolution will be. There is, however, a physical limit to the length of a radar antenna onboard an aircraft or satellite, and that constrains the spatial resolution.

Synthetic aperture radar is a technology that solves this problem. Compared with conventional, real aperture radar, SAR achieves high along-track (azimuth) resolution by synthesising a virtual long antenna by the motion of a very short antenna, by intensive data processing of the coherent backscattered radar signals based on the information of the Doppler frequency shift.

As illustrated in Fig. A.7, while the SAR platform is moving along its path at an altitude H, it transmits microwave pulses into the antenna's footprint (the illuminated area by radar beam pulse), at the rate of the pulse repetition frequency (PRF), and receives the echoes of each pulse backscattered from the target surface. The short antenna of a SAR produces a wide footprint on the ground. If the platform motion is small in comparison with the width of footprint at the rate of PRF, a pile of consecutive footprints overlap. The typical PRFs for SAR systems are in the range of 1–10 kHz, which is relatively high in relation to the travel speed of a platform, and thus each point on the ground is illuminated many times by a sequence of consecutive footprints while a SAR platform passes over. The echoes of these repeat illuminations of the same point will be received by the antenna at a slightly higher frequency than the carrier frequency as $(f_c + \delta f)$, when the SAR is approaching the point, and slightly lower frequency than the carrier frequency as $(f_c - \delta f)$, when it is moving away from the point,

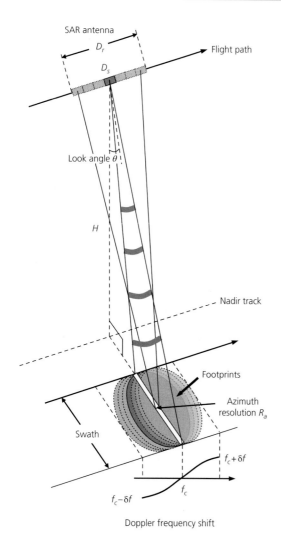

Figure A.7 Principle of SAR. The motion of the SAR antenna with a small length of D_s along its flight path simulates a virtual long antenna D_r that enables a high azimuth resolution R_a much smaller than the SAR footprint width, via matched filtering processing on the overlapped footprints, based on Doppler frequency shift δf from the SAR carrier frequency f_c.

according to Doppler frequency shift effects. Thus depending on its position in the overlapped stack of footprints, a point is uniquely coded by its Doppler frequency shift signature. Consequently, the signal processing of matched filtering from the Doppler effects can achieve very high azimuth resolution on the scale of a fraction of the SAR footprint width. The effect is equivalent to connecting a sequence of positions of a short antenna

while a SAR is travelling along its path, corresponding to the overlapped footprints over a ground point, to formulate a very long virtual antenna as though it were a very long real aperture antenna to focus on the point. Obviously, a short SAR antenna means a wide footprint and thus allows more footprints to overlap, forming a long virtual antenna to achieve high azimuth resolution. It can be proved that the azimuth resolution of SAR is half of the length of its antenna (Curlander and McDonough 1991), or the shorter antenna diameter D_s of SAR achieves higher azimuth resolution:

$$R_a = \frac{D_s}{2} \tag{A.3}$$

For a high slant range resolution, SAR emits a chirp pulse with a bandwidth B_v of 10s of *MHz* modulating a carrier wave frequency f_c (the nominal frequency of the SAR). Depending on the increase or decrease in chirp frequencies, there are ascending and descending chirps. For the case of an ascending chirp, the exact frequency of a radar pulse reaching the ground is higher in the far ground range than in near ground range, and so are the echoes from different ground ranges. The same applies to a descending chirp. The returned signal is then demodulated with the chirp form and sampled based on the chirp frequency shift from the nominal frequency of the SAR, via matched filtering, to achieve high range resolution.

When discussing microwave energy, the *polarization* of the radiation is also important. Polarization refers to the orientation of the electronic field. Most radar systems are designed to transmit microwave radiation either horizontally polarized (H) or vertically polarized (V). Similarly, the antenna receives either the horizontally or vertically polarized backscattered energy, and some radar systems can receive both. These two polarization states are designated by the letters H for horizontal and V for vertical. Thus, there can be four combinations of both transmit and receive polarizations as follows:

- HH – for horizontal transmission and horizontal receipt;
- VV – for vertical transmission and vertical receipt;
- HV – for horizontal transmission and vertical receipt;
- VH – for vertical transmission and horizontal receipt.

The first two polarization combinations are referred to as *like-polarized* because the transmitted and received polarizations are the same. The last two combinations are referred to as *cross-polarized* because the transmitted and received polarizations are orthogonal.

Some past, present and future spaceborne SAR systems are listed in Table A.5. SAR image data are supplied in several different formats. Typically, single look complex (SLC) data are in 8-byte complex numbers, while multi-look (intensity) images are in 16-bit unsigned integers.

Table A.5 Some past, present and future spaceborne SAR systems.

Sensor	Country	Mission Period	Band; Polarisation	Look Angle (°)	Antenna Size (m)	Altitude (km)	Swath (km)
SEASAT-A	USA	6/1978, 105 days	L; HH	20	10.8 × 2.2	795	100
SIR-A	USA	11/1981, 2.5 days	L; HH	47	9.4 × 2.2	260	50
SIR-B	USA	10/1984, 8.3 days	L; HH	15–60	10.8 × 2.2	224, 257, 360	20–40
SIR-C/ X-SAR	USA, Germany, Italy	4/1994, 11 days	L, C, X; Multipol	20–55	12 × 2.9/ 12 × 0.7/ 12 × 0.4	225	15–90, 225
ALMAZ-1	Russia	3/1991, 2.5 yrs	S; HH	20–65	12 × 1.5	300–70	30–45
ERS-1	EU	17/7/1991–3/2000	C; VV	23	10 × 1	780	100
ERS-2	EU	21/4/1995–5/9/2011	C; VV	23	10 × 1	780	100
JERS-1	Japan	2/1992–10/1998	L; HH	38	12 × 2.4	570	75
RADARSAT-1	Canada	4/11/1995–3/2013	C; HH	20–60	15 × 1.5	790–820	50–500
SRTM	USA Germany	11/2/2000, 11 days	C, X; HH, VV	20–60		233	56–225
ENVISAT	EU	1/3/2002–8/4/2012	C; HH, VV, VH/HV	15–45	10 × 1	800	57–400
CosmoSkymed	Italy	8/6/2007–present	X, Multipol		1.4 × 5.7		10–100
RADARSAT-2	Canada	14/11/2007–present	C;Multipol	variable	15 × 1.5	790–820	20–500
ALOS PALSAR	Japan	1/2006–5/2011	L;Multipol	8–60	8.9 × 2.9	692	30–350
TerraSAR-X	Germany/UK	15/6/2007–present	X;Multipol	20–55	4.8 m × 0.7 m	514	5 ×10 to 100 × 150
TerraSAR-2 TerraSAR-3	Germany/UK	2010–2015	X, Multipol	20–55	4.8	512–530	10–100
TanDEM-X	Germany/ UK	21/6/2010–present	X, Multipol	20–55	4.8 × 0.7	514	10–100
Sentinel-1	EU	3/4/2014–present	C, Multipol	20–46	12.3 × 0.821	693	20, 80, 250 & 400

APPENDIX B

Online resources for information, software and data

Here we have compiled a list of what we hope are useful resources available on or via the internet. These include links to the sites of proprietary software suites and to those providing programs that are *shareware*, or low cost, or entirely free of license. These sites are in themselves often rich sources of background information and technical help. Second, we include links to online information and technical resources that are largely independent of any allegiance to particular software or are provided by independent (often charitable) organisations. Third, a list of online data sources is provided, some of which allow download of data (either free of charge or with some payment system in place), and others that merely enable browsing of available data as quick looks or listings.

B.1 Software – proprietary, low cost and free (shareware)

Autodesk	www.autodesk.com
ERDAS/ERMapper	www.hexagongeospatial.com
ESRI	www.esri.com
FME Safe Software	www.safe.com
Geotools	geotools.codehaus.org
GlobalMapper	www.globalmapper.com
GRASS	grass.itc.it
Idrisi	www.clarklabs.org
ILWISS	www.itc.nl/ilwis
Exelis ENVI	www.exelisvis.com/envi
JUMP GIS	www.openjump.org
Landserf	www.landserf.org
MapInfo	www.mapinfo.com
Map Window	www.mapwindow.org
PCI Geomatics	www.pcigeomatics.com
Quantum opensource	www.qgis.org
SAGA GIS	www.saga-gis.org
Surfer	www.goldensoftware.com/products/surfer
Various independent	www.rockware.com

B.2 Information and technical information on standards, best practice, formats, techniques and various publications

Association for Geographic Information (AGI)	www.agi.org.uk
British Geological Survey (BGS)	www.bgs.ac.uk
Committee on Earth Observation Satellites (CEOS)	ceos.org
Digital Earth	www.dgeo.org
Digital National Framework	www.dnf.org
ESRI ArcUser online	www.esri.com/news/arcuser
GDAL	www.gdal.org
Geospatial Analysis online	www.spatialanalysisonline.com
Geospatial Information and Technology Association (GITA)	www.gita.org
GIS Day	www.gisday.com
GIS Research UK (GISRUK)	www.geo.ed.ac.uk/gisruk
International Association of Photogrammetry & Remote Sensing	www.isprs.org
Isovist Analyst	www.casa.ucl.ac.uk/software/isovist.asp
MAF/TIGER background documents	www.census.gov/geo/www/tiger/index.html
Open Geospatial Consortium	www.opengeospatial.org
Ordnance Survey (OS)	www.ordnancesurvey.co.uk

Image Processing and GIS for Remote Sensing: Techniques and Applications, Second Edition. Jian Guo Liu and Philippa J. Mason.
© 2016 John Wiley & Sons, Ltd. Published 2016 by John Wiley & Sons, Ltd.

Remote Sensing and Photogrammetry Society (RSPSoc)	www.rspsoc.org
UKGeoForum (umbrella organisation)	www.ukgeoforum.org.uk
Web 3D Consortium	www.geovrml.org
World Wide Web Consortium	www.w3.org

B.3 Data sources including online satellite imagery from major suppliers, DEM data plus GIS maps and data of all kinds

ALOS data search	cross.restec.or.jp/en
ArcGIS Online	www.arcgis.com/home
Asia Pacific Natural Hazards Network	www.pdc.org/mde/explorer.jsp

ASTER/PALSAR data	gds.ersdac.jspacesystems.or.jp
ASTER GDEM	gdem.ersdac.jspacesystems.or.jp/
Digital Globe (Quickbird & Worldview)	browse.digitalglobe.com
DigitalGlobe (GeoFUSE)	geofuse.geoeye.com/landing
EarthExplorer	earthexplorer.usgs.gov
ESA EOLI catalogues	earth.esa.int
GeoCommunity GIS free data depot	data.geocomm.com
GIS data depot	www.gisdepot.com
GIS Lounge	gislounge.com
Glovis	glovis.usgs.gov
NASA EO system	eospso.nasa.gov
NASA EOSDIS	earthdata.nasa.gov/data
SPOT catalogue	www.geo-airbusds.com/geostore/
SRTM Public Data	www2.jpl.nasa.gov/srtm/cbanddataproducts.html

References

In this book, we have organised the references to ensure optimum readership:

1 The references are listed in three parts: general references, Part I references and further reading and Part II references and further reading. The references relating to Part III are given at the end of each of the case studies.

2 The general references are divided into three sections, listing some of the most useful referenced books on image processing, GIS and remote sensing. These books cover much of the common ground with each other and with this book. Many of the widely used techniques and algorithms described in this book are derived from these books in conjunction with the authors' personal experience, judgement and modifications. For this reason, the general references are not generally directly cited in this book, since they can be relevant to any part of the book.

3 The publications listed in the references and further reading sections of Parts I and II are referred to in the text if they are directly relevant to particular algorithms, techniques or applications. Publications that are of general relevance to one or several parts of the book are not directly referred to in the text but are cross-linked to the related contents via the index. All publications are directly referenced, or are indexed, or both.

General References

Image processing

Castleman, K.R. (1996) *Digital Image Processing*, second edn. Prentice-Hall, Upper Saddle River, NJ.

Gonzalez, R.C. & Woods, R.E. (2002) *Digital Image Processing*, second edn. Pearson Education, Singapore.

Gonzalez, R.C., Woods, R.E. & Eddins, S.L. (2004) *Digital Image Processing Using MATLAB*. Pearson Prentice-Hall, Upper Saddle River, NJ.

Jensen, J.R. (2005) *Introductory Digital Image Processing – A Remote Sensing Perspective*, third edn. Prentice-Hall, Upper Saddle River, NJ.

Mather, P.M. (2004) *Computer Processing of Remotely-Sensed Images: An Introduction*, third edn. John Wiley & Sons, Chichester, UK.

Niblack, W. (1986) *An Introduction to Digital Image Processing*. Prentice-Hall, Englewood Cliffs, NJ.

Richards, J.A. (2006) *Remote Sensing Digital Image Analysis: An Introduction*, fourth edn. Springer, Berlin.

Schowengerdt, R.A. (1997) *Remote Sensing: Models and Methods for Image Processing*, second edn. Academic Press, London.

GIS

Bonham-Carter, G.F. (2002) *Geographic Information Systems for Geoscientists: Modelling with GIS*. Pergamon, New York.

Burrough, P.A. (1998) *Principles of Geographic Information Systems*, second edn. Clarendon Press, Oxford.

Burrough, P.A. & McDonell, R.A. (1998) *Principles of Geographical Information Systems*. Oxford University Press, New York.

Demers, M.N. (2009) *Fundamentals of Geographic Information Systems*, fourth edn. John Wiley & Sons, New York.

Dikau, R. & Saurer, H. (1999) *GIS for Earth Surface Systems*. Gebrűder Borntraeger, Berlin.

Foody, G.M. & Atkinson, P.M. (eds). (2002) *Uncertainty in Remote Sensing and GIS*. Wiley, Chichester, UK.

Iliffe, J.C. (2000) *Datums and Map Projections for Remote Sensing, GIS and Surveying*. Whittles Publishing, Caithness, Scotland.

Kelly, R.E., Drake, N. & Barr, S.L. (eds). (2004) *Spatial Modeling of the Terrestrial Environment*. John Wiley & Sons, New York.

Longley, P.A., Goodchild, M.F., Maguire, D.J. & Rhind, D. W. (2005) *Geographic Information Systems and Science*, second edn. John Wiley & Sons, New York.

Malcewski, J. (1999) *GIS and Multicriteria Decision Analysis*. John Wiley & Sons, New York.

Morain, S. & Lopez Baros, S. (eds). (1996) *Raster imagery in Geographical Information Systems*. Onword Press, Santa Fe, NM.

Petrie, G. & Kennie, T.J.M. (eds). (1990) *Terrain Modelling in Surveying and Civil Engineering*. Whittles Publishing, Caithness, Scotland.

Tomlin, C.D. (1990) *Geographic Information Systems and Cartographic Modeling*. Prentice-Hall, Englewood Cliffs, NJ.

Wilson, J.P. & Gallant, J.C. (eds). (2000) *Terrain Analysis: Principles and Applications*. John Wiley & Sons, New York.

Image Processing and GIS for Remote Sensing: Techniques and Applications, Second Edition. Jian Guo Liu and Philippa J. Mason.
© 2016 John Wiley & Sons, Ltd. Published 2016 by John Wiley & Sons, Ltd.

Remote sensing

Colwell, R.N. (ed). (1983) *Manual of Remote Sensing – Theory, Instruments and Techniques*, Vol. 1, second edn. American Society of Photogrammetry, Falls Church, Va..

Drury, S.A. (2001) *Image Interpretation in Geology*, third edn. Blackwell Science, Cheltenham, UK.

Elachi, C. (1987) *Introduction to the Physics and Techniques of Remote Sensing*. John Wiley & Sons, New York.

Lilesand, T.M. & Kiefer, R.W. (2000) *Remote Sensing and Image Processing*, sixth edn. John Wiley & Sons, New York.

Sabins, F.F. (1996) *Remote Sensing: Principles and Interpretation*, third edn. W.H. Freeman, Basingstoke, UK.

Part I References and Further Reading

Anandan, P. (1989) A computational framework and an algorithm for the measurement of visual motion. *International Journal of Computer Vision*, 2, 283–310.

Anderberg, M.R. (1973) *Cluster Analysis for Applications*. Academic Press, New York.

Balci, M. & Foroosh, H. (2005a) Estimating sub-pixel shifts directly from the phase difference. In: *IEEE International Conference on Image Processing*, Italy.

Balci, M. & Foroosh, H. (2005b) Inferring motion from the rank constraint of the phase matrix. In: *Proceedings of the IEEE International Conference on Acoustics, Speech, and Signal Processing (ICASSP 2005)*, vol. 2, pp. 925–8. 18–23 March, 2005, Philadelphia, PA.

Ball, G.H. (1965) Data analysis in social sciences: What about details? In: *Proceedings of the Fall Joint Computer Conference*, pp. 533–9.

Ball, G.H. & Hall, D.J. (1967) A clustering technique for summarizing multivariable data. *Behavioural Science*, 12, 153–5.

Barron, J.L., Fleet, D.J. & Beauchemin, S.S. (1994) Performance of optical flow techniques. *International Journal of Computer Vision*, 12 (no. 1), 43–77.

Bay, H., Ess, A., Tuytelaars, T. & Van Gool, L. (2008) SURF: Speeded up robust features. *Computer Vision and Image Understanding*, 110 (no. 3), 346–59.

Black, M.J. & Anandan, P. (1996) The robust estimation of multiple motions: Affine and piecewise-smooth flow fields. *Computer Vision and Image Understanding* 63 (no. 1), 75–104.

Black, M.J. & Jepson, A.D. (1996) Estimating optical flow in segmented images using variable-order parametric models with local deformations. *IEEE Transactions on Pattern Analysis and Machine Intelligence*, 18 (no. 10), 972–86.

Brown, M.Z., Burschka, D. & Hager, G.D. (2003) Advances in computational stereo. *IEEE Transactions on Pattern Analysis\ and. Machine Intelligence*, 25, 993–1008.

Brox, T., Bruhn, A., Papenberg, N. & Weickert, J. (2004) High accuracy optical flow estimation based on a theory for warping. In: *European Conference on Computer Vision* (ECCV), pp. 25–36.

Chavez, P.S. Jr, (1989) Extracting spectral contrast in Landsat Thematic Mapper image data using Selective Principle Component Analysis. *Photogrammetric Engineering and Remote Sensing*, 55 (no. 3), 339–48.

Chavez, P.S. Jr, Berlin, G.L. and Sowers, L.B. (1982) Statistical method for selecting Landsat MSS ratios. *Journal of Applied Photographic Engineering*, 8, 23–30.

Chavez, P.S. Jr, Guptill, S.C. & Bowell, J. (1984) Image processing techniques for Thematic Mapper Data. In: *Proceedings of the 50th Annual ASP-ACSM Symposium*, pp. 728–43. American Society of Photogrammetry, Washington, DC.

Cooper, J., Venkatesh, S. & Kitchen, L. (1993) Early jump-out corner detectors. *IEEE Transactions on Pattern Analysis and Machine Intelligence*, 15, 823–8.

Corr, D.G. & Whitehouse, S.W. (1996) Automatic change detection in spaceborne SAR imagery. In: *Proceedings of the AGARD Conference – Remote Sensing: A Valuable Source of Information*, AGARD-CP-582, Paper No. 39. October, NATO, Brussels.

Crippen, R.E. (1989) Selection of Landsat TM band and band ratiocombinations to maximize lithological information in color composite displays. In: *Proceedings of the 7th Thematic Conference on Remote Sensing for Exploration Geology*, vol. 2, pp. 917–9. Calgary, Alberta, Canada, 2–6 October.

Crist, E.P. & Cicone, R.C. (1984) A physically-based transformation of thematic mapper data – the TM tasselled cap. *IEEE Transactions on Geoscience and Remote Sensing*, 22 (no. 3), 256–63.

Crosta, A.P. & Moore, J.M. (1989) Enhancement of Landsat Thematic Mapper imagery for residual soil mapping in SW Minas Gerais State, Brazil: A prospecting case history in Greenstone Belt terrain. In: *Proceedings of the 7th Thematic Conference on Remote Sensing for Exploration Geology*, vol. 2, pp. 1173–87. Calgary, Alberta, Canada, 2–6 October.

Crosta, A.P., Sabine, C. & Taranik, J.V. (1998) Hydrothermal alteration mapping at Bodie, California using AVIRIS hyperspectral data. *Remote Sensing of Environment*, 65 (no. 3), 309–19.

Crosta, A.P., De Souza Filho, C.R., Azevedo, F. & Brodie, C. (2003) Targeting key alteration minerals in epithermal deposits in Patagonia, Argentina using ASTER imagery and principal component analysis. *International Journal of Remote Sensing*, 24 (no. 21), 4233–40.

Curlander, J.C. & McDonough, R.N. (1991) *Synthetic Aperture Radar, System and Signal Processing*. John Wiley & Sons, New York.

Delon, J. & Rougé, B. (2007) Small baseline stereovision. *Journal of Mathematical Imaging and Vision*, 28, 209–23.

Devijver, P.A. & Kittler, J. (1982) *Pattern Recognition – A Statistical Approach*. Prentice-Hall International, London.

Diday, E. & Simon, J.C. (1976) Cluster analysis. In: K.S. Fu (ed), *Digital Pattern Recognition*, pp. 47–74. Springer-Verlang, Berlin.

Dougherty, L., Asmuth, J.C., Blom, A.S., Axel, L. & Kumar, R. (1999) Validation of an optical flow method for tag displacement estimation. *IEEE Transactions on Medical Imaging*, 18 (no. 4), 359–63.

Foroosh, H., Zerubia, J.B. & Berthod, M. (2002) Extension of phase correlation to subpixel registration. *IEEE Transactions on Image Processing*, 11 (no. 3), 188–200.

Förstner, W. (1994) A framework for low level feature extraction. In: *Proceedings of the 3rd European Conference on Computer Vision*, pp. 383–94. Stockholm, Sweden.

Fraster, R.S. (1975) Interaction mechanisms within the atmosphere. In: R.G. Reeves (ed), *Manual of Remote Sensing*, pp. 181–233. American Society of Photogrammetry, Falls Church, VA.

Gabriel, A.K., Goldstein, R.M. & Zebker, H.A. (1989) Mapping small elevation changes over large areas: Differential radar interferometry. *Journal of Geophysical Research*, 94 (no. B7), 9183–91.

Gens, R. & van Genderen, J.L. (1996) SAR interferometry – issues, techniques, applications. *International Journal of Remote Sensing*, 17, 1803–35.

Gillespie, A.R., Kahle, A.B. & Walker, R.E. (1986) Color enhancement of highly correlated images – I. Decorrelation and HSI contrast stretches. *Remote Sensing of Environment*, 20, 209–35.

Goetz, A.F.H., Billingsley, F.C., Gillespie, A.R., Abrams, M.J., Squires, R.L., Shoemaker, E.M., Lucchita, I. & Elston, D.P. (1975) *Applications of ERTS Images and Image Processing to Regional Geologic Problems and Geologic Mapping in Northern Arizona*. JPL Technical Report 32-1597. Jet Propulsion Laboratory, Pasadena, California.

Goldstein, R.M., Engelhardt, H., Kamb, B. & Frohlich, R.M. (1993) Satellite radar interferometry for monitoring ice sheet motion: Application to an Antarctic ice stream. *Science*, 262, 1525–634.

Green, A.A. & Craig, M.D. (1985) Analysis of aircraft spectrometer data with logarithmic residuals. In: *Proceedings of the Airborne Imaging Spectrometer Data Analysis Workshop*, JPL Publication, pp. 85–41. Jet Propulsion Laboratory, Pasadena, CA.

Hagberg, J.O., Ulander, L.M.H. & Askne, J. (1995) Repeat-pass SAR-interferometry over forested terrain. *IEEE Transactions on Geoscience & Remote Sensing*, 33, 331–40.

Harris, C. & Stephens, M. (1988) A combined corner and edge detector. In: *Alvey Vision Conference*, pp. 147–51. Manchester.

Heo, Y.S., Lee, K.M. & Lee, S.U. (2011) Robust stereo matching using adaptive normalized cross correlation. *IEEE Transactions on Pattern Analysis and Machine Intelligence*, 33 (no. 4), 807–22.

Hoge, W.S. (2003) Subspace identification extension to the phase correlation method. *IEEE Transactions on Medical Imaging*, 22 (no. 2), 277–80.

Horn, B.K.P. & Schunck, B.G. (1981) Determining optical flow. *Artificial Intelligence*, 17, 185–203.

Huang, C., Wylie, B., Yang, L., Horner, C. & Zylstra, G. (2002) Derivation of a tasselled cap transformation based on Landsat 7 at-satellite reflectance. *International Journal of Remote Sensing*, 23 (no. 8), 1741–8.

Ichoku, C., Karnieli, A., Arkin, Y., Chorowicz, J., Fleury, T. & Rudant, J.P. (1998) Exploring the utility potential of SAR interferometric coherence images. *International Journal of Remote Sensing*, 19, 1147–60.

Igual, J.P.L, Garrido, L., Almansa, A., Caselles, V. & Rougé, B. (2007) Automatic low baseline stereo in urban areas. *Inverse Problems and Imaging*, 1, 319–48.

Kauth, R.J. & Thomas, G. (1976) The tasselled cap – a graphic description of the spectral-temporal development of agriculture crops as seen by Landsat. In: *Proceedings of the Symposium on Machine Processing of Remotely-Sensed Data*, 4B, pp. 41–51. Purdue University, West Lafayette, Ind.

Kittler, J. & Pairman, D. (1988) Optimality of reassignment rules in dynamic clustering. *Pattern Recognition*, 21, 169–74.

Klette, R., Schlüns, K. & Koschan, A. (1998) *Computer Vision: Three-Dimensional Data from Images*. Springer, Singapore, pp. 29–33.

Kruger, S. & Calway, A. (1998) Image registration using multiresolution frequency domain correlation. In: *Proceedings of the British Machine Vision Conference*, pp. 316–25. 14–17 September 1998, Southampton, UK.

Kruse, F.A., Lefkoff, A.B. & Dietz, J.B. (1993) The Spectral Image Processing System (SIPS) – interaction visualisation and analysis of imaging spectrometer data. *Remote Sensing of Environment*, special issue on AVIRIS, 44, 145–63.

Kuglin, C.D. & Hines, D.C. (1975) The phase correlation image alignment method. In: *Proceedings of IEEE International Conference on Cybernetics and Society*, pp. 163–5. New York.

Lai, S.H. & Vemuri, B.C. (1998) Reliable and efficient computation of optical flow. *International Journal of Computer Vision*, 29 (no. 2), 87–105.

Lee, H. & Liu, J.G. (2001) Analysis of topographic decorrelation in SAR interferometry using ratio coherence imagery. *IEEE Transactions on Geoscience and Remote Sensing*, 39 (no. 2), 223–32.

Liu, C., Yuen, J. & Torralba, A. (2011) SIFT flow: Dense correspondence across scenes and its applications, *IEEE Transactions on Pattern Analysis and Machine Intelligence*, 33 (no. 5), 978–94.

Liu, J.G. (1991) Balance contrast enhancement technique and its application in image colour composition. *International Journal of Remote Sensing*, 12, 2133–51.

Liu, J.G. (2000) Smoothing filter based intensity modulation: A spectral preserve image fusion technique for improving spatial details. *International Journal of Remote Sensing*, 21 (no. 18), 3461–72.

Liu, J.G. & Moore, J. M. (1990) Hue image colour composition – a simple technique for shadow suppression and spectral enhancement. *International Journal of Remote Sensing*, 11, 1521–30.

Liu, J.G. & Haigh, J.D. (1994) A three-dimensional feature space iterative clustering method for multi-spectral image classification. *International Journal of Remote Sensing*, 15 (no. 3), 633–44.

Liu, J.G. & Moore, J.M. (1996) Direct decorrelation stretch technique for RGB colour composition. *International Journal of Remote Sensing*, 17, 1005–18.

Liu, J.G. & Morgan, G. (2006) FFT selective and adaptive filtering for removal of systematic noise in ETM+ imageodesy images. *IEEE Transactions on Geoscience and Remote Sensing*, 44 (no. 12), 3716–24.

Liu, J.G. & Yan, H. (2006) Robust phase correlation methods for sub-pixel feature matching. In: *Proceedings of the 1st Annual Conference of Systems Engineering for Autonomous Systems*, p. A13. July, Defence Technology Centre, Edinburgh.

Liu, J.G. & Yan, H. (2008) Phase correlation pixel-to-pixel image co-registration based on optical flow and median shift propagation. *International Journal of Remote Sensing*, 29 no. (20), 5943–56.

Liu, J.G., Capes, R., Haynes, M. & Moore, J.M. (1997a) ERS SAR multi-temporal coherence image as a tool for sand desert study (dune movement, sand encroachment and erosion). In: *Proceedings of the 12th International Conference and Workshop on Applied Geologic Remote Sensing*, pp. 478–85. 17–19 November, Denver, CO.

Liu, J.G., Moore, J.M. & Haigh, J.D. (1997b) Simulated reflectance technique for ATM image enhancement. *International Journal of Remote Sensing*, 18, 243–55.

Liu, J.G., Black, A., Lee, H., Hanaizumi, H. & Moore, J.M. (2001) Land surface change detection in a desert area in Algeria using multi-temporal ERS SAR coherence images. *International Journal of Remote Sensing*, 22 (no. 13), 2463–77.

Liu, J.G., Mason, P.J., Hilton, F. & Lee, H. (2004) Detection of rapid erosion in SE Spain: A GIS approach based on ERS SAR coherence imagery. *Photogrammetric Engineering and Remote Sensing*, 70 (no. 10), 1179–85.

Liu, J.G., Mason, P.J. & Ma, J. (2006) Measurement of the left-lateral displacement of Ms 8.1 Kunlun earthquake on 14th November 2001 using Landsat-7 ETM+ imagery. *International Journal of Remote Sensing*, 27 (no. 10), 1875–91.

Liu, J.G., Yan, H. & Morgan, G.L.K. (2012) PCIAS subpixel technology. *Measurement + Control*, 45 (no. 7), 207–11.

Lowe, D. (2004) Distinctive image features from scale-invariant keypoints, *International Journal of Computer Vision*, 60, 91–110.

Lucas, B. & Kanade, T. (1981) An iterative image registration technique with application to stereo vision. In: *Proceedings of the 7th International Joint Conference on Artificial Intelligence*, pp. 674–9. 24–28 August 1981, University of British Columbia, Vancouver, B.C., Canada,

MacQueen, J. (1967) Some method for classification and analysis of multi-variate observations. In: *Proceedings of the 5th Berkeley Symposium on Mathematics, Statistics and Probability*, pp. 281–99. University of California Press, Berkeley.

Massonnet, D. & Adragna, F. (1993) A full scale validation of radar interferometry with ERS-1: The Landers earthquake. *Earth Observation Quarterly*, no. 41, 1–5.

Massonnet, D., Rossi, M., Carmona, C., Adragna, F., Peltzer, G., Feigl, K. & Rabaut, T. (1993) The displacement of the Landaus earthquake mapped by radar interferometry. *Nature*, 364, 138–42.

Massonnet, D., Feigl, K., Rossi, M. & Adragna, F. (1994) Radar interferometric mapping of deformation in the year after the Landers earthquake. *Nature*, 369, 227–30.

Massonnet, D., Briole, P. & Arnaud, A. (1995) Deflation of Mount Etna monitored by spaceborne radar interferometry. *Nature*, 375, 567–70.

Mikhail, E.M., Bethel, J.S. & McGlone, J.C. (2001) *Introduction to Modern Photogrammetry*. John Wiley & Sons, New York, p. 180.

Morgan, G., Liu, J.G. & Yan, H. (2010) Precise sub-pixel disparity measurement from very narrow baseline stereo.

IEEE Transactions on Geoscience and Remote Sensing, 48 (no. 9), 3424–33.

Olmsted, C. (1993) Alaska SAR Facility Scientific User's Guide. http://www.asf.alaska.edu/content/reference/SciSARuser Guide.pdf , 57p.

Reddy, B.S. & Chatterji, B.N. (1996) An FFT-based technique for translation, rotation, and scale-invariant image registration. *IEEE Transactions on Image Processing*, 5 (no. 8), 1266–71.

Ren, J., Jiang, J. & Vlachos, T. (2010) High-accuracy sub-pixel motion estimation from noisy images in Fourier domain. *IEEE Transactions on Pattern Analysis and Machine Intelligence*, 19 (no. 5), 1379–84.

Robinson, N. (1966) *Solar Radiation*. Elsevier, Amsterdam.

Rosen, P.A., Hensley, S., Zebker, H.A., Webb, F.H. & Fielding, E.J. (1996) Surface deformation and coherence measurements of Kilauea volcano, Hawaii, from SIR-C radar interferometry. *Journal of Geophysical Research – Planets*, 101 (no. E10), 23109–25.

Schmid, C., Mohr, R. & Bauckhage, C. (2000) Evaluation of interest point detectors. *International Journal of Computer Vision*, 27 (no. 2), 151–72.

Sclove, S.L. (1983) Application of conditional population-mixture model to image segmentation. *IEEE Transactions on Pattern Analysis and Machine Intelligence*, PAMI-5 (no. 4), 428–33.

Sheffield, C. (1985) Selecting band combinations from multi-spectral data. *Photogrammetric Engineering and Remote Sensing*, 51 (no. 6), 681–7.

Shi, J. & Tomasi, C. (1994) Good features to track. In: *IEEE Conference on Computer Vision and Pattern Recognition*, pp. 593–600. June, Seattle, WA.

Shimizu, M. & Okutomi, M. (2005) Sub-pixel estimation error cancellation on area-based matching. *International Journal of Computer Vision*, 63 (no. 3), 207–24.

Smith, A.R. (1978) Colour gamut transform pairs. In: *Proceedings of SIGGRAPH 1978 Conference Held in Atlanta, Georgia, on 23–25 August 1978*, 3, pp. 12–19. Association of Machinery, New York.

Soha, J.M. & Schwartz, A.A. (1978) Multispectral histogram normalisation contrast enhancement. In: *Proceedings of the 5th Canadian Symposium on Remote Sensing Held in Victoria, BC, Canada, on 28–31 August 1978*, pp. 86–93.

Stiller, C. & Konrad, J. (1999) Estimating motion in image sequences. *IEEE Signal Processing Magazine*, 16 (no. 4), 70–91.

Stone, H.S., Orchard, M.T., Chang, E.C. & Martucci, S.A. (2001) A fast direct Fourier-based algorithm for subpixel registration of image. *IEEE Transactions on Geoscience and Remote Sensing*, 39 (no. 10), 2235–43.

Szeliski, R. & Coughlan, J. (1997) Spline-based image registration. *International Journal of Computer Vision*, 22 (no. 3), 199–218.

Taylor, M.M. (1973) Principal component colour display of ERTS imagery. In: *The Third Earth Resources Technology Satellite-1 Symposium, Held on 10–14 December 1973*. NASA SP-351, pp. 1877–97. NASA, Washington, DC. .

Turner, S., Liu, J.G., Cosgrove, J.W. & Mason, P.J. (2006) Envisat ASAR interferometry measurement of earthquake deformation in the Siberian Altai. In: *AGU Western Pacific Geophysics Meeting Conference*. 24–27 July, Beijing G21A-03.

Wang, F., Prinet, V. & Ma, S. (2001) A vector filtering technique for SAR interferometric phase images. In: *Proceedings of Applied Informatics (AI2001)*, Innsbruck, Austria, 19–22 February 2001, pp. 566–70.

Zebker, H. & Goldstein, R.M. (1986) Topography mapping from interferometric synthetic aperture radar observations. *Journal of Geophysics Research*, 91, 4993–9.

Zebker, H. & Villasenor, J. (1992) Decorrelation in interferometric radar echoes. *IEEE Transactions on Geosciences and Remote Sensing*, 30 (no. 5), 950–9.

Zebker, H., Rosen, P., Goldstein, R., Gabriel, A.K. & Werner, C. (1994a) On the derivation of coseismic displacement fields using differential radar interferometry: The Landers earthquake. *Journal of Geophysics Research*, 99, 19617–34.

Zebker, H., Werner, C., Rosen, P. & Hensley, S. (1994b) Accuracy of topographic maps derived from ERS-1 interferometric radar. *IEEE Transactions on Geosciences and Remote Sensing*, 32, 823–36.

Part II References and Further Reading

Agterberg, F.P. (1974) *Geomathematics, Mathematical Background and Geo-Science Applications*, Elsevier, Amsterdam.

Agumya, A. & Hunter, G.J. (2002) Responding to the consequences of uncertainty in geographical data. *International Journal of Geographical Information Science*, 16 (no. 5), 405–17.

Ambroi, T. & Turk, G. (2003) Prediction of subsidence due to underground mining by artificial neural networks. *Computers & Geosciences*, 29 (no. 5), 627–37.

An, P., Moon, W.M. & Rencz, A. (1991) Application of fuzzy set theory for integration of geological, geophysical and remote sensing data. *Canadian Journal of Exploration Geophysics*, 27, 1–11.

Band, L.E. (1986) Topographic partition of watersheds with digital elevation models. *Water Resources Research*, 22 (no. 1), 15–24.

Batson, R.M., Edwards, K. & Eliason, E.M. (1975) Computer-generated shaded-relief images. *Journal of Research of the US Geological Survey*, 3 (no. 4), 401–408.

Burrough, P.A. (1981) Fractal dimensions of landscapes and other environmental data. *Nature*, 294, 240–42.

Burrough, P.A. & Frank, A.U. (eds). (1996) *Geographic Objects with Indeterminate Boundaries*, Taylor & Francis, London.

Carranza, E.J.M. (2004) Weights of evidence modeling of mineral potential: A case study using small number of prospects, Abra Philippines. *Natural Resources Research*, 13, 173–87.

Carranza, E.J.M. (2009) Controls on mineral deposit occurrence inferred from analysis of their spatial pattern and spatial association with geological features. *Ore Geology Reviews*, 35, 383–400

Carranza, E.J.M. (2010) Improved wildcat modelling of mineral prospectivity. *Resource Geology*, 60, 129–49.

Carranza, E.J.M. (2014) Data-driven evidential belief modeling of mineral potential using few prospects and evidence with missing values. *Natural Resources Research*, 24 (no. 3), 291–304. doi.org/10.1007/s11053-014-9250-z.

Carter, J. (1992) The effect of data precision on the calculation of slope and aspect using gridded DEMs. *Cartographica*, 29 (no. 1), 22–34.

Carver, S. (2007) Integrating multi-criteria evaluation with geographical information systems. *International Journal of Geographical Information Systems*, 5 (no. 3), 321–39. doi:10.1080/02693799108927858.

Chen, H., Wood, M.D., Linstead, C. & Maltby, E. (2011) Uncertainty analysis in a GIS-based multi-criteria analysis tool for river catchment management. *Environmental Modelling & Software*, 26, (no. 4), 395–405.

Chen, Y., Yu, J. & Khan, S. (2010) Spatial sensitivity analysis of multi-criteria weights in GIS-based land suitability evaluation. *Environmental Modelling & Software*, 25 (no. 12), 1582–91.

Cheng, Q. (2012) Singularity theory and methods for mapping geochemical anomalies caused by buried sources and for predicting undiscovered mineral deposits in covered areas. *Journal of Geochemical Exploration*, 122, 55–70.

Chiles, J.P. & Delfiner, P. (1999) *Geostatistics: Modeling Spatial Uncertainty*, John Wiley & Sons, New York.

Chung, C.F. & Fabbri, A.G. (2003) Validation of spatial prediction models for landslide hazard mapping. *Natural Hazards*, 30, 451–72.

Chung, C.F., Fabbri, A.G. & Chi, K.H. (2002) A strategy for sustainable development of nonrenewable resources using spatial prediction models. In: A.G. Fabbri, G. Gáal, & R.B. McCammnon (eds), *Geoenvironmental Deposit Models for Resource Exploitation and Environmental Security*. Kluwer Academic, Dordrecht, the Netherlands.

Chung, C.J. (2006) Using likelihood ratio functions for modeling the conditional probability of occurrence of future landslides for risk assessment. *Computers & Geosciences*, 32 (no. 8), 1052–68.

Chung, C.J. & Keating, P.B. (2002) Mineral potential evaluation based on airborne geophysical data. *Exploration Geophysics*, 33, 28–34.

Chung, C.J. & Fabbri, A. G. (2007) Predicting future landslides for risk analysis – models and cross-validation of their results. *Geomorphology*, 94 (nos. 3–4), 438–52.

Ciampalini, A., Garfagnoli, F., Devl Ventisette, C. & Moretti, S. (2013) Potential use of remote sensing techniques for exploration of iron deposits in western Sahara and southwest of Algeria. *Natural Resources Research*, 22, 179–90.

Cicerone, S. & Clementini, E. (2003) Efficient estimation of qualitative topological relations based on the weighted walk-throughs model. *GeoInformatica*, 7 (no. 2), 211–27.

Clark, I. (1979) *Practical Geostatistics*. Applied Sciences Publishers, London.

Corripio, J.G. (2003) Vectorial algebra algorithms for calculating terrain parameters from DEMs and solar radiation modelling in mountainous terrain. *International Journal of Geographical Information Science*, 17 (no. 1), 1–23.

Cressie, N.A.C. (1993) *Statistics for Spatial Data*. John Wiley & Sons, New York.

Danneels, G., Havenith, H.B., Caceres, F., Outal, S. & Pirard, E. (2010) Filtering of ASTER DEMs using mathematical morphology. *Geological Society London Special Publications*, 345 (no. 1), 33–42.

Dempster, A.P. (1967) Upper and lower probabilities induced by a multi-valued mapping. *Annals of Mathematics Statistics*, 38, 325–39.

Dempster, A.P. (1968) A generalization of Bayesian inference. *Journal of the Royal Statistical Society, Series B*, 30, 205–47.

Dodgson, J., Spackman, M., Pearman, A.D. & Phillips, L.D. (2000) *Multi-Criteria Analysis: A Manual*. Department of the Environment, Transport and the Regions, London (2000).

Douglas, D.H. & Peucker, T.K. (1973) Algorithms for the reduction of the number of points required to represent a digitized line or its caricature. *Canadian Cartographer*, 10 (no. 2), 112–22.

Fabbri, A.G. (1984) *Image Processing of Geological Data*, Van Nostrand Reinhold, New York.

Feizizadeh, B. & Blaschke, T. (2014) An uncertainty and sensitivity analysis approach for GIS-based multicriteria landslide susceptibility mapping. *International Journal of Geographical Information Science*, 28 (no. 3), 610–38.

Flacke, W. & Kraus, B. (2005) *Working with Projections and Datum Transformations in ArcGIS*. Points Verlag Norden, Halmstad, Sweden.

Florent, J., Theriault, M. & Musy, A. (2010) Using GIS and outranking multicriteria analysis for land-use suitability assessment. *International Journal of Geographical Information Science*, 15 (no. 2), 153–74.

Gaddy, D.E. (2003) *Introduction to GIS for the Petroleum Industry*, Penwell Books, Tulsa, OK.

Goodchild, M.F. (1980) Fractals and the accuracy of geographical measures. *Mathematical Geology*, 12 (no. 2), 85–98.

Goodchild, M.F. & Gopal, S. (1989) *The Accuracy of Spatial Databases*. Taylor & Francis, New York.

Gordon, J. & Shortliffe, E.H. (1985) A method for managing evidential reasoning in a hierarchical hypothesis space. *Artificial Intelligence*, 26, 323–57.

Hart, B.S. & Sagan, J.A. (2007) Curvature for visualization of seismic geomorphology. *Geological Society Special Publications*, 277, 139–49.

Hart, J.F. (1954) Central tendency in aerial distributions. *Economic Geography*, 30 (no. 1), 48–59.

He, C., Jiang, J., Han, G. & Chen, J. (2004) An algorithm for building full topology. *Proceedings of the XXth ISPRS Congress*. 12–23 July, Istanbul.

Hunter, G.J. & Goodchild, M.F. (1997) Modeling the uncertainty of slope and aspect estimates derived from spatial databases. *Geographical Analysis*, 29 (no. 1), 35–49.

Isaaks, E.H. & Srivastava, R.M. (1989) *An Introduction to Applied Geostatistics*. Oxford University Press, Oxford.

Jenson, S. & Domingue, J. (1988) Extracting topographic structure from digital elevation data for geographic information system analysis. *Photogrammetric Engineering and Remote Sensing*, 54 (no. 11), 1593–1600.

Jiang, H. & Eastman, J.R. (2000) Application of fuzzy measures in multi-criteria evaluation in GIS. *International Journal of Geographical Information Science*, 14 (no. 2), 173–84.

Jones, K.H. (1998) A comparison of algorithms used to compute hill slope as a property of the DEM. *Computer and Geosciences*, 24 (no. 4), 315–23.

Keeney, R.L. & Raiffa, H. (1976) *Decisions with Multiple Objectives: Preferences and Value Tradeoffs*. John Wiley & Sons, New York.

Knox-Robinson, C.M. (2000) Vectorial fuzzy logic: A novel technique for enhanced mineral prospectivity mapping, with reference to the orogenic gold mineralisation potential of the Kalgoorlie Terrane, Western Australia. *Australian Journal of Earth Sciences*, 47 (no. 5), 929–41.

Knox-Robinson, C.M. & Wyborn, L.A.I. (1997) Towards a holistic exploration strategy: Using Geographic Information System as a tool to enhance exploration. *Australian Journal of Earth Sciences*, 44, 453–63.

Krige, Daniel G. (1951) A statistical approach to some basic mine valuation problems on the Witwatersrand. *Journal of the Chemistry, Metallurgy and Mining Society of South Africa*, 52 (no. 6), 119–39.

Lam, N. & De Cola, L. (1993) *Fractals in Geography*. Prentice-Hall, Englewood Cliffs, NJ.

Liu, J.G., Mason, P.J., Clerici, N., Chen, S., Davis, A., Miao, F., Deng, H. & Liang, L. (2004a). Landslide hazard assessment in the Three Gorges Area of the Yangtze River using ASTER imagery: Zigui-Badong. *Geomorphology*, 61, 171–87.

Liu, J.G., Mason, P.J., Hilton, F. & Lee, H. (2004b) Detection of rapid erosion in SE Spain: A GIS approach based on ERS SAR coherence imagery. *Photogrammetric Engineering and Remote Sensing*, 70 (no. 10), 1179–85.

Lodwick, W.A. (1990) Analysis of structure in fuzzy linear programming. *Fuzzy Sets and Systems*, 38 (no. 1), 15–26.

Malczewski J. (2000) On the use of weighted linear combination method in GIS: Common and best practice approaches. *Transactions in GIS*, 4, 5–22.

Malczewski, J. (2007) GIS based multicriteria decision analysis: A survey of the literature. *International Journal of Geographical Information Science*, 20 (no. 7), 703–26, doi:10.1080/13658810600661508.

Maling, D.H. (1992) *Coordinate Systems and Map Projections*, second edn. Pergamon Press, Oxford.

Mason, P.J. (1998) Landslide hazard assessment using remote sensing and GIS techniques. PhD thesis, University of London.

Mason, P.J. & Rosenbaum, M.S. (2002) Geohazard mapping for predicting landslides: The Langhe Hills in Piemonte, NW Italy. *Quarterly Journal of Engineering Geology & Hydrology*, 35, 317–26.

Matheron, G. (1963) Principles of geostatistics. *Economic Geology*, 58, 1246–66.

Matheron, G. (1975) *Random Sets and Integral Geometry*. John Wiley & Sons, New York.

Meisels, A., Raizman, S. & Karnieli, A. (1995) Skeletonizing a DEM into a drainage network. *Computers and Geosciences*, 21 (no. 1), 187–96.

Mineter, M.J. (2003) A software framework to create vector-topology in parallel GIS operations. *International Journal of GIS*, 17 (no. 3), 203–22.

Minkowski, H. (1911) *Gesammelte Abhandlungen*. HRSG, von D Hilbert, Leipzig.

Parker, J.R. (1997) *Algorithms for Image Processing and Computer Vision*. John Wiley & Sons, New York.

Openshaw, S., Charlton, M. & Carver, S. (1991) Error propagation: A Monte Carlo simulation. In: I. Masser and M. Blakemore (eds), *Handling Geography Information: Methodology and Potential Applications*, pp. 102–14. Wiley, New York.

Opricovic, S. & Tzeng, G.H. (2003) Fuzzy multicriteria model for post-earthquake land-use planning. *Natural Hazards Review*, 4 (no. 2), 59–64.

Ross, T.J. (1995) *Fuzzy Logic with Engineering Applications*. John Wiley & Sons, New York.

Saaty, T.L. (1990) *The Analytic Hierarchy Process: Planning, Priority Setting, Resource Allocation*, second edn. RWS Publications, Pittsburgh, PA.

Sentz, K. & Ferson, S. (2002) *Combination of Evidence in Dempster-Shafer Theory*. SANDIA Tech. Report.

Serra, J. (1982) *Image Analysis and Mathematical Morphology*. Academic Press, London.

Shafer, G. (1976) *A Mathematical Theory of Evidence*. Princeton University Press, Princeton, NJ.

Skidmore, A.K. (1989) A comparison of techniques for calculating gradient and aspect from a gridded digital elevation model. *International Journal of Geographical Information Systems*, 3, 323–34.

Snyder, J.P. (1997) *Map Projections – a Working Manual*. U.S. Geological Survey Professional Paper 1395. USGS, Washington, DC.

Snyder, J.P. & Voxland, P.M. (1989) *An Album of Map Projections*. U.S. Geological Survey Professional Paper 1453. USGS, Washington DC.

Soille, P. (ed). (2003) *Morphological Image Analysis: Principles and Applications*, second edn. Springer-Verlag, Berlin.

Stein, A., van der Meer, F.D. & Gorte, B.G.H. (eds). (1999) Spatial statistics for remote sensing. In: *Remote Sensing and Digital Image Processing*. Kluwer Academic, Dordrecht.

Sui, D. & Goodchild, M. (2011) The convergence of GIS and social media: Challenges for GIScience. *International Journal of Geographical Information Science*, 25 (no. 11), 1737–48.

Tangestani, M.H. & Moore, F. (2001) Porphyry copper potential mapping using the weights-of evidence model in a GIS, northern Shahr-e-Babak, Iran. *Australian Journal of Earth Sciences*, 48, 695–701.

Tobler, W. (1970) A computer movie simulating urban growth in Detroit region. *Economic Geography*, 46, 234–40.

Unwin, D.J. (1989) Fractals in the geosciences. *Computers and Geosciences*, 15 (no. 2), 163–5.

Varnes, D. (1984) *Landslide Hazard Zonation: A Review of Principles and Practice*. Commission on Landslides of the International Association of Engineering Geology, United Nations Educational Social and Cultural Organisation, Natural Hazards, No. 3.

Vincent, L. (1993) Morphological greyscale reconstruction in image analysis: Applications and efficient algorithms. *IEEE Transactions in Geosciences and Remote Sensing*, 2 (no. 2), 176–201.

Wadge, G. (1998) The potential of GIS modeling of gravity flows and slope instabilities. *International Journal of Geographical Information Science*, 2 (no. 2), 143–52.

Wang, J.J., Robinson, G.J. & White, K. (1996) A fast solution to local viewshed computation using grid-based digital elevation models. *Photogrammetric Engineering and Remote Sensing*, 62, 1157–64.

Watson, D.F. (1992) *Contouring: A Guide to the Analysis and Display of Spatial Data*. Pergamon Press, Oxford.

Welch, R., Jordan, T., Lang, H. & Murakami, H. (1998) ASTER as a source for topographic data in the late 1990s. *IEEE Transactions on Geoscience and Remote Sensing*, 36, 1282–89.

Whitney, H. (1932) Congruent graphs and the connectivity of graphs. *American Journal of Mathematics*, 54, 150–68.

Wiechel, H. (1878) Theorie und Darstellung der Beleuchtung von nicht gesetzmassig gebildeten Flachen mit Rucksicht auf die Bergzeichnung, *Civilingenieur*, 24.

Wood, J. (1996) The geomorphological characterisation of digital elevation models. PhD thesis, University of Leicester, UK.

Wolf, P.R. & Dewitt, B.A. (2000) *Elements of Photogrammetry (with Applications in GIS)*, third edn. McGraw Hill, New York.

Wright, D.F. & Bonham-Carter, G.F. (1996) VHMS favourability mapping with GIS-based integration models, Chisel Lake–Anderson Lake area. In: G. Bonham-Carter, A.G. Galley, G.E.M. Hall (eds), *EXTECHI: A Multidisciplinary Approach to Massive Sulfide Research in the Rusty Lake–Snow Lake Greenstone Belts, Manitoba*. Geological Survey of Canada Bulletin, 426, 339–76.

Wu, Y. (2000) R2V Conversion: Why and how? *GeoInformatics*, 3 (no. 6), 28–31.

Xia, Y. & Ho, A.T.S. (2000) 3D vector topology model in the visualization system. *Proceedings of the Symposium on Geoscience and Remote Sensing* (IGARSS 2000, IEEE 2000 International), 7, 3000–2.

Yang, Q., Snyder, J. & Tobler, W. (2000) *Map Projection Transformation: Principles and Applications*, Taylor & Francis, London.

Zadeh, L.A. (1965) Fuzzy sets. *IEEE Information and Control*, 8, 338–53.

Zevenbergen, L.W. & Thorne, C.R. (1987) Quantitative analysis of land surface topography. *Earth Surface Processes and Landforms*, 12, 47–56.

Ziadat, F.M. (2007) Effect of contour intervals and grid cell size on the accuracy of DEMs and slope derivatives. *Transactions in GIS*, 11 (no. 1), 67–81.

Index

Image Processing and GIS for Remote Sensing: Techniques and Applications, Second Edition. Jian Guo Liu and Philippa J. Mason.
© 2016 John Wiley & Sons, Ltd. Published 2016 by John Wiley & Sons, Ltd.